Library
Liberty Baptist College
Lynchburg, Virginia 24506

HISTORY OF
EAST AFRICA

TO THE READER

This book represents the wealth of knowledge i... today. Liberty Baptist ... ludes this knowledge ... er to standardize the work an ... the credits of the colle... this volume by Liberty Ba... lege is not an endorsement of its contents. The position of the College on the fundam... the faith and the separa... ...an life is well known

LIBERTY BAPTIST COLLEGE
Lynchburg, Virginia

HISTORY OF
EAST AFRICA

Edited by

ROLAND OLIVER

READER IN THE HISTORY OF AFRICA
IN THE UNIVERSITY OF LONDON

and

GERVASE MATHEW

FELLOW OF BALLIOL COLLEGE
AND LECTURER IN BYZANTINE STUDIES
IN THE UNIVERSITY OF OXFORD

VOLUME I

OXFORD
AT THE CLARENDON PRESS

Oxford University Press, Ely House, London W.1

GLASGOW NEW YORK TORONTO MELBOURNE WELLINGTON
CAPE TOWN SALISBURY IBADAN NAIROBI LUSAKA ADDIS ABABA
BOMBAY CALCUTTA MADRAS KARACHI LAHORE DACCA
KUALA LUMPUR HONG KONG

© *The Colonial Office, London, 1963*

FIRST PUBLISHED 1963
REPRINTED LITHOGRAPHICALLY AT THE
UNIVERSITY PRESS, OXFORD
FROM CORRECTED SHEETS OF THE FIRST EDITION
1966

PRINTED IN GREAT BRITAIN

PREFATORY NOTE

THE three-volume history of East Africa of which this is the first instalment is the outcome of a parallel initiative on the part of the Governments of Tanganyika and Uganda and of the Colonial Social Science Research Council. The publication of a comprehensive history of the region is one of the essential contributions which United Kingdom funds could make to the future of the new East African states. Special thanks are due to the Colonial Office and more recently to the Department of Technical Co-operation for allowing Colonial Development and Welfare funds to be used for this purpose.

Any work on African history on this scale must be in some measure a collaborative study; and in this series not only historians, but archaeologists, anthropologists, and geographers have made their contribution. For the most part the present volume directly conveys results of original research.

This venture owes much to the generous enthusiasm and unflagging efforts of Mrs. E. M. Chilver and Miss Alison Smith, who have done detailed work beyond the usual limits of subeditorship. At various stages it has been indebted to many of the staff of the Oxford Institute of Commonwealth Studies, in particular to Mrs. Mary Holdsworth. Among those who have given their specialist advice are Dr. J. H. M. Beattie, Dr. L. W. Hollingsworth, Mrs. Kerslake, Dr. L. S. B. Leakey, Dr. Geoffrey Lewis, Dr. Peter Lienhardt, Mr. A. T. Matson, Mr. J. K. Mbazira, Dr. Merrick Posnansky, and Professor Paul Wheatley.

In the spelling of Arabic words, the 'ain, hamza, diacriticals, and indications of vowel-length have been disregarded. The assimilation of the definite article has not been shown; thus al-Shaikh not ash-Shaikh. Where the spelling departs from the normal transliteration of Arabic, e.g. in the use of bin for ibn, the purpose has been to reproduce the Swahili as distinct from the Arabic pronunciation.

Accepted English forms of Arabic place names (e.g. Mecca not Makka) are used where they exist.

In Bantu and other African tribal names the spelling follows the practice of the International African Institute, except in one or two cases where these are so unfamiliar as to be likely to perplex the general reader. The Bantu prefixes have in general been omitted.

The place names are generally in accordance with the usage of the Oxford Economic Atlas.

R. O.
G. M.

CONTENTS

List of Maps	ix
Abbreviations	x
Introduction	xi
I. The East African Environment S. J. K. BAKER	1
II. The Stone Age of East Africa SONIA COLE	23
III. The Peopling of the Interior of East Africa by its Modern Inhabitants G. W. B. HUNTINGFORD	58
IV. The East African Coast until the Coming of the Portuguese GERVASE MATHEW	94
V. The Coast, 1498–1840 G. S. P. FREEMAN-GRENVILLE	129
VI. Discernible Developments in the Interior *c.* 1500–1840 ROLAND OLIVER	169
VII. Zanzibar and the Coastal Belt, 1840–84 SIR JOHN GRAY	212
VIII. The Southern Section of the Interior, 1840–84 ALISON SMITH	253
IX. The Northern Interior, 1840–84 D. A. LOW	297
X. The Wider Background to Partition and Colonial Occupation JOHN FLINT	352

XI. The British Sphere, 1884–94 391
MARIE DE KIEWIET HEMPHILL

XII. The German Sphere, 1884–98 433
G. S. P. FREEMAN-GRENVILLE

Epilogue 454
ROLAND OLIVER

Bibliography 457

Index 481

LIST OF MAPS

1. East Africa: relief	3
2. East Africa: mean annual rainfall	11
3. Main Stone Age sites in Kenya, Uganda, and Tanganyika	43
4. Sketch map of the distribution of Hamites, Negroes, Nilotes, and Nilo-Hamites and of the T/K languages, in East Africa	59
5. Tentative reconstruction of Nilotic, Hamitic, and Bantu movements in East Africa	85
6. The Indian Ocean and its trade, c. 10th–15th centuries	109
7. The East African coast, showing the principal places mentioned in the text	128
8. Approximate movements of Lwo-speaking Nilotic peoples and central Nilo-Hamites, c. 15th–18th centuries	174
9. Extension of Bito and Nyiginya dynasties in the Interlacustrine region, 15th–18th centuries	183
10. The growth of Buganda, 16th–19th centuries	188
11. Probable spread of Ntemi chieftainships from Nyamwezi–Sukuma area	198
12. Approximate lines of the main trade routes to the interior based on Zanzibar, 1840–1884	252
13. The north-eastern interior, 1840–1884	298
14. The north-western interior, 1840–1884	299
15. Partition of East Africa, 1884–1891	375
16. German and British advance into the interior, 1884–1893	392

ABBREVIATIONS OF PERIODICAL TITLES

J. Af. Soc.	*Journal of the African Society.*
J. E. Afr. (Ug.) Nat. Hist. Soc.	*Journal of the East Africa (and Uganda) Natural History Society.*
J. E. Afr. Swahili Committee	*Journal of the East African Swahili Committee.*
J. Roy. Anthrop. Inst.	*Journal of the Royal Anthropological Institute.*
J.R.G.S.	*Journal of the Royal Geographical Society.*
Num. Chron.	*Numismatic Chronicle.*
Proc. R.G.S.	*Proceedings of the Royal Geographical Society.*
S-W. J. Anthrop.	*Southwestern Journal of Anthropology.*
Sudan Notes	*Sudan Notes and Records.*
Tang. Notes	*Tanganyika Notes and Records.*
Uganda J.	*Uganda Journal.*

INTRODUCTION

This volume is an experiment. Only a few years ago it would have been regarded as axiomatic that a History of East Africa would be almost wholly a history of the colonial period, with an introductory chapter or two on the exploration of the area by Europeans and the build-up of European interests prior to political occupation. The history of the colonial period itself would have been almost solely a record of European initiatives in policy and administration, economic development, technical and social services, education, and religious proselytism. If pressed, the planners of such a history would probably have defended themselves on the technical grounds that these were the only questions which the available evidence permitted them to examine. In fact, however, they would not have been seriously pressed, for, technical grounds apart, their views would have corresponded with a general habit of contemporary thought.

Today, the habit of thought has changed. It has become realistic to visualize a time when the majority of the readers of such a history will be East Africans, who will be looking at the colonial period in retrospect as an episode, albeit a most important episode, in the history of their own countries. Even of the colonial period, such readers will want to know primarily what it has done to *them*. The detailed evolution of colonial policy, the personalities of successive Governors and Secretaries of State, all the various aspects of external challenge, will rightly yield pride of place to the history of internal response. And in the measurement of internal response the antecedents of the colonial period assume a new and vital importance. To assess what the colonial period has done to the people of East Africa, it is more than ever necessary to know where they stood when it began.

It is with considerations such as these in mind that the present volume, and indeed the whole of this History, has been planned. It is experimental, because clearly so radical a change of approach towards African history must, as it gains acceptance, set in motion a new tide of research in hitherto almost neglected fields. The archaeologist, preoccupied till now primarily with the greater problem of human origins, will turn increasingly to the multifarious problems of the last two thousand years. The historian, overcoming the barriers of language and making use of new techniques of recording and analysis, will sound out at deeper levels the still scarcely charted ocean of traditional history, tell-tale custom and ceremony, tribal law, family nomenclature and genealogy, place names, and all other substitutes for literary evidence. Even within the modern period, the political historian will venture forth from the Secretariat archives to examine in, say, the records of school attendance, the origins of African nationalism; and the economic historian will leave the official reports on trade and commerce for the family papers of the shopkeeper or the coffee-farmer.

All this will take time. Meanwhile what is here offered must be mainly an exercise in looking at old evidence in a new way. This will perhaps be seen most obviously in the second half of the volume, which deals with the half-century preceding the European occupation. The authors of these chapters have relied almost entirely on written evidence, much of it the familiar published record of European exploration. But such materials have been used not for the purpose of reconstructing individual journeys or for singing the fame of those who made them, but mainly for what they have to say about the East African peoples and their way of life. Where historical tradition handed down by Africans or Arabs has later been recorded in writing, this evidence also has been used alongside the contemporary written record. The resulting outline of the second half of the nineteenth century may seem surprisingly full. But on the traditional side there is much evidence still to come, which in two or three decades will greatly modify the picture here presented. To take one example only, M. Jan Vansina and the Abbé Alexis Kagame, who are engaged on a joint study of the traditional history of Rwanda, have forecast that for the whole of the nineteenth century it will be possible to reconstruct a year-by-year account of local events. The example of Rwanda is doubtless an unusually favourable one, but it affords at least an indication of what may be possible elsewhere.

If much remains to be revealed about the nineteenth century, the same is *a fortiori* true of earlier periods. The two chapters

on the earlier history of the East Coast, while using such documentary evidence as is available, draw also upon archaeological evidence which is as yet scarcely more than the result of preliminary survey: The chapter on the traditional history of the interior is largely an analysis of the by-products of ethnographic inquiry. In the earlier chapters still, a geographer, a prehistorian, and an ethnographer have tried to set out those facts from their own fields of study which seem most relevant by way of introduction. The result is, as the authors and editors realize, a most imperfect and ill-balanced attempt at East African history, but at least it attempts the history of Africa and not only that of its invaders.

R. O.
G. M.

I

THE EAST AFRICAN ENVIRONMENT

S. J. K. BAKER

THE East African plateaulands lie within the division of the African continent which is known to geographers as High Africa and which comprises a series of plateaux, mostly over 4,000 feet above sea-level, extending over the southern and eastern part of the continent. On the north the low plateaux of northern Kenya, which are continued into the neighbouring plateaux of Somalia, form a zone of negative relief between the highlands of East Africa and those of Ethiopia; but on the south it is more difficult to set a limit to the physical environment of East Africa. It is perhaps best to consider the plateaux as extending southwards to the distinct break formed by the valley of the middle and lower Zambezi, although one is thereby led into an area in which the geographical conditions are in some ways transitional to those of southern Africa.

From its frontage on the Indian Ocean the East African region stretches inland to the western rift valley, to the west of which the well-marked basin of the Congo, with its floor at 1,500 feet or less above sea-level, belongs to the different terrain of Low Africa, as does the basin of the Bahr el Ghazal to the north-west of East Africa. Frontage on the Indian Ocean comprises an important element in the space relations of East Africa. For many centuries there have been maritime contacts between the various coastlands on the periphery of the Indian Ocean; and the coastal strip of East Africa could be properly described as the western shore of the Indian Ocean, as well as the eastern coastland of its own continent.

Structure and Relief

The fundamental rock material of the East African plateaux, and the one most widely exposed, is that of the Archaean Basement. Over parts of the interior plateau of East Africa it has been affected by the processes of granitization and by granitic intrusion, and in such areas tumbled masses of granite boulders may form a striking feature of the landscape. Elsewhere on the plateau the Basement is covered with sedimentary rocks of pre-Cambrian age, of which the schists and quartzites of the Karagwe-Ankolean system are the best known. Over the low foreland plateaux of the eastern lowlands there are extensive exposures of the Karroo system, and in south-western Tanganyika rocks of the same Permian-Triassic system are found faulted into the Basement. Nearer to the coast successive marine transgressions have resulted in the deposition of sedimentary rocks of Jurassic, Cretaceous, and Tertiary age.

Intensive folding occurred in East Africa in pre-Cambrian times, but for the remainder of its history the structural character of the region has been resistant, and earth movements have been confined to faulting and to regional uplift and warping. The history of the East African plateaux has been one of long-continued denudation interrupted by shorter phases of tectonic activity. There is evidence that the beginnings of rifting go back into pre-Cambrian times, and that later faults of post-Karroo, post-Jurassic, and Tertiary or later times have in many instances rejuvenated earlier faults. There can be no doubt, though, that the rift valley landscape in East Africa owes its freshness to a quite late expression of the movement, and that the main faulting of the Kenya rift valley occurred in Middle to Upper Pleistocene times, well after the first appearance of man on the East African scene. It is still far from clear whether the rift valleys themselves resulted from absolute depression, or only from failure to rise to the same extent as the plateaux on either side, but there is no room for uncertainty about the elevation of the flanking highlands as a part of the process of rifting. Similarly, the Tertiary to Recent volcanics to be found especially over vast areas in Kenya comprise an evident accompaniment of the tectonic movements; and the fact that equilibrium has not yet been reached is demonstrated by contemporary vulcanism on the northern shores of Lake Kivu in the western rift valley.

The rocks of East Africa provided early man with a wide variety of materials, including especially obsidian, for the manufacture of his implements; and widely scattered small-scale deposits of iron later provided the raw material of tools and weapons. Salt played its part in the lives of the people and served the needs of their cattle.

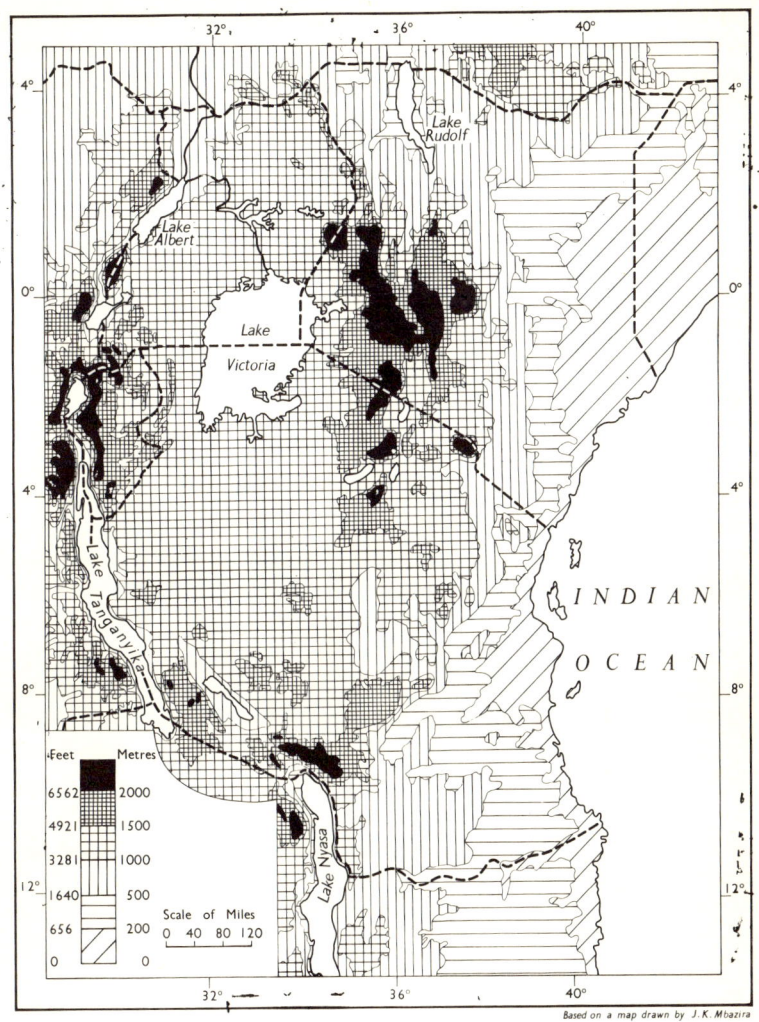

Map 1. East Africa: relief

Against the geological background which has been briefly sketched, the main relief regions of East Africa may now be considered. (See Map 1.)

The narrow fringe of *coastal lowland* below the 500-feet contour is for the most part less than twenty miles in width, but its conditions tend to extend inland along the lower courses of the rivers, more especially along the Ruvu and the Rufiji, in the central section of the Tanganyika coastal belt. The exposures of the marine sedimentary rocks are thus interrupted in places by the alluvium of the lower reaches of the rivers; but the flood plain which plays so large a part in the physiography of eastern and southern Asia has a very limited place in the East African environment. The islands of Mafia, Zanzibar, and Pemba may be included in this region of the coastal lowlands, the first two being situated on the continental shelf, whereas Pemba is separated from the mainland by a deeper, presumably fractured, trough.

Low foreland plateaux in Tanganyika occupy the area between the coastal lowlands and the north-east to south-west trending rim mountains of the interior plateau. The region of low plateaux is thus widest in the south, along the Ruvuma river, beyond which it continues into northern Mozambique; and it is narrowest in the north, where the Usambara Mountains come close to the coast. Young sedimentary rocks are found in the east, but towards the west the Archaean Basement emerges from beneath this cover. The height rises gradually to over 3,000 feet in the west, but the characteristic level is nearer 2,000 feet, in a landscape of minor plateaux in which the *inselberg* is sometimes in evidence. Some of the changes of level occur along fault-lines. Two important alluvium-filled lowlands lie at the foot of the rim mountains escarpment, the Kilombero valley, bounded on the east by the Mahenge plateau, and the Mkata plains, situated to the west of the Uluguru mountains.

In Kenya a comparable area occurs between the coastlands and the highlands, in this case a narrow belt in the south, widening northwards to cover the whole of northern Kenya. The slope up to the highlands is gradual, and not confined in the west by any relief corresponding to the rim mountains of Tanganyika, but in the south it is broken by the residual mass of the Teita hills. In the north the landscape of low plateau continues right up to the edge of the escarpment overlooking Lake Rudolf. The Basement outcrops widely in the west, but eastwards it is overlain by sandstones of Karroo age in the south and by Tertiary sediments in the north. In the area on either side of Lake Rudolf these sedimentary rocks are in turn covered by the wide spread of Tertiary volcanics. Elsewhere the lavas rest directly on the Basement, as in the long line of the Yatta plateau and of the Chyulu range, in the area to the southeast of the Kenya highlands.

The Kenya highlands, the northern highlands of Tanganyika, and their associated rift valleys may all be included in the *eastern rift highlands*. This region is largely covered with Tertiary volcanics, and most of its fault-lines occur within these lavas. Farther west the detached exposures of western Nyanza and of Elgon, Kadam, and Moroto belong to the same field. The altitude of the highlands and of the higher parts of the rift valley floor is to be explained in terms of the elevational movements involved in the process of rifting, the presence of great thicknesses of plateau lava, and the superimposition of volcanic cones upon the general level of the landscape. The region may have a certain unity of structure and rock material, but, with its composite character, it is best described in relation to its main sub-regions.

In the East Kenya highlands the Aberdare range, which rises in several places to more than 12,000 feet above sea-level, presents a steep and complex face to the central portion of the Kenya rift valley, and shows on the east a gentler slope, furrowed in its southern section by the valleys of numerous parallel headwater streams. The Nyeri–Nanyuki corridor separates the Aberdares from the volcanic cone of Mount Kenya, which reaches in the higher of its rugged twin peaks an altitude of 17,058 feet. To the north of the East Kenya highlands the Laikipia plateau flanks the Lake Baringo section of the rift valley; whilst to the south Nairobi, at a height of about 5,500 feet, is situated on the south-eastern edge of the highlands.

In the West Kenya highlands the Mau escarpment is continued northwards in the Elgeyo escarpment; whilst the Uasin–Gishu plateau and the Kitale corridor lie between Elgeyo and Mount Elgon (14,178 feet). The West Kenya highlands contain considerably more land above 6,500 feet than do the highlands to the east of the rift valley.

The highest part of the floor of the Kenya rift valley is in the central portion, where Lakes Naivasha, Elmenteita, and Nakuru are situated at about 6,000 feet above sea-level. Here the rift valley is about 40 miles wide, and its boundary scarps rise 2,000 feet or more above the floor. Northwards from this central area the level of the floor decreases and the valley sides become lower and less impressive, though the magnificent cliff of Elgeyo, dominating the Arorr valley and the Kamasia horst, constitutes an outstanding exception to this generalization. The surface of Lake Baringo is about 3,200 feet above sea-level, whilst Lake Rudolf lies at an altitude of only 1,230 feet, on a floor of Quaternary alluvium with the wide broken plain of Turkana on its western side. Similar decreases in altitude occur southwards, and Lake Magadi rests at a level of 1,900 feet, with no pronounced escarpment on the eastern side. All these lakes lie within areas of inland drainage, and with the exception of Lakes Baringo and Naivasha they have a high alkaline content; and this is true of their neighbours in northern Tanganyika.

Lakes Natron and Manyara lie in the southern continuation of the rift; Lake Eyasi lies in a south-western branch; whilst the Pangani valley comprises an essentially structural feature along north-west to south-east trend-lines. To the north-east of the Pangani valley rise the Pare and Usambara mountains and, to the south-west of Lake Manyara, the Mbulu highlands. Vulcanism was responsible for the tangled highlands which occupy the area between Lakes Natron, Manyara, and Eyasi and, farther east, for Meru (14,980 feet) and Kilimanjaro (19,340 feet).

The 'rise to the rift' is as evident on both flanks of the *western rift highlands* as in the case of their eastern counterpart. In Ruanda-Urundi, together with the fringing areas in south-western Uganda and north-western Tanganyika, the presence of resistant quartzites within the Karagwe-Ankolean system reinforces the tectonic factors favourable to altitude; and the characteristic relief is thus that of a much-dissected highland, with altitudes above 5,000 feet in the east and above 6,500 feet nearer to the rift valley. Farther north the highland character of the landscape is maintained in the plateaux of western Toro, between 4,000 and 5,000 feet above sea-level, and in the Ruwenzori Mountains, the highest peak of which is 16,763 feet. The Ruwenzori massif is unique among the major mountains of East Africa, in that it is not composed of young volcanics, but of rock material of pre-Cambrian age. Vulcanism is on a less massive scale than in the eastern rift highlands, but it has thrown a line of craters across the valley to the north of Lake Kivu, has produced the plateau craters of north-west Ankole, and is responsible for the numerous explosion craters in the rift valley on the northern side of the Kazinga channel.

The western rift valley is divided into well-defined basins, of which the most northerly is that of Lake Albert, where the lake lies at a height of about 2,000 feet above sea-level on a floor of sedimentary Kaiso beds of Quaternary age. The boundary scarps are here well defined, but they peter out in the country to the north-east of Lake Albert. The Ruwenzori massif separates the Lake Albert basin from that of Lake Edward and Lake George, although the River Semliki provides a somewhat tenuous link between them. The present-day floor of the Lake Edward–Lake George rift is about 3,000 feet above sea-level, and it is strewn with shallow craters, of which one provides the natural basis of the Lake Katwe salt industry. South of the Mufumbiro volcanoes Lake Kivu lies at 4,788 feet in the highest part of the western rift valley, set in an amphitheatre of hills. The Ruzizi carries the overflow of Lake Kivu southwards to the 2,534-feet surface level of Lake Tanganyika, the longest and the deepest of the East African rift valley lakes; whilst the Lukuga channel provides an outlet from Lake Tanganyika westwards into the Congo drainage system. The cliff sides of the rift come close to the shores of Tanganyika, allowing little development of coastal plain. A feature of the western rift valley as a whole is the high proportion of its surface occupied by lake water, in contrast to the very small area of water surface at the present day in the eastern rift valley.

The *southern rift highlands*, a complex of highlands, plateaux, and valleys, comprise a composite physiographic unit which extends well beyond the boundaries of Tanganyika southwards to the Zambezi valley. It is a Y-shaped area of which the two arms are in Tanganyika; on the north-west the Ufipa highlands comprise a horst between the southern section of Lake Tanganyika and the offset rift of Lake Rukwa, whilst on the north-east the Iringa highlands lie between the valley of the great Ruaha and that of the Kilombero. At the meeting-place of the two arms with the stem of the Y an extrusion of young volcanics occurs in the area to the north-west of Lake Nyasa, and culminates in Rungwe mountain (9,713 feet). Highlands flank Lake Nyasa on both sides, including the Kipengere range and the Livingstone mountains on the eastern or Tanganyika side of the lake. The lake itself is at a level of approximately 1,550 feet above sea-level.

The *interior plateau* is a horseshoe-shaped region set amid the three groups of rift highlands; but in eastern Tanganyika it is bordered by its own rim mountains, to which, in a sense, the Iringa highlands belong. The dissected rim, designated the African Ghats by the nineteenth-century travellers, includes the Usagara mountains, between the Great Ruaha and the Mkondoa breaks, and farther north the Nguru mountains. After a wide gap the Usambara and Pare highlands provide a north-eastern rim to the Masai plain.

Developed on rocks of the Basement system at a height of 3,000 to 4,000 feet, the Masai plain is situated between the rim mountains and the higher ground associated with the southern prolongation of the eastern rift. In the south the plateau faces upon the Ruaha and Rukwa rifts, whilst in the west plateau conditions extend right up to the eastern edge of the western rift valley, in the break between the Ufipa highlands and those of Ruanda-Urundi. There is a wide exposure of Basement rocks over the southern part of the plateau; but farther north granites and granitoid rocks occupy much of the surface, diversifying the landscape with their characteristic piles of boulders and acting as water bearers.

In the depression formed by the gentle down-warping of the interior plateau, the shallow waters of Lake Victoria have collected at a surface level of 3,720 feet over an area roughly comparable in size with Scotland. On each side of the lake a corridor leads from the interior plateau of Tanganyika to that of Uganda. The western corridor through Karagwe, Bukoba, and Masaka provides a topographically easier route than that through the broad dissected plateaux in the area to the west of the Kenya highlands and on either side of the fault-bordered Kavirondo gulf and Kano plains. Around the northern and western shores of the lake the Buganda peneplain, at a height of 4,300 to 4,500 feet, provides the typical even sky-lines of a landscape which is now well dissected by the steep-sided valleys of a later cycle of erosion. Levels rise gradually towards the western rift highlands, although granitic outcrops add variety to the relief, especially in Mubende; but northwards there is a slight fall of level and a reduction in the scale of relief to the Lake Kyoga system, at 3,400 feet above sea-level. The ungraded course of the Victoria Nile between Lakes Victoria and Kyoga thus negotiates a drop of over 300 feet. Beyond the drowned river system of Lake Kyoga the landscapes of northern Uganda are very uniform. In the north-east, though, the approach to the Kenya highlands brings to Karamoja a diversification of relief dependent partly upon the 'rise to the rift' and partly upon the presence of eruptive cones. The lowest land in Uganda is in the flat swampy valley which the Albert Nile occupies between Lake Albert and its exit from the country at Nimule.

Climate

Situated on the eastern side of the continent of Africa and between latitudes 4° 30′ N. and 11° 30′ S., the East African countries are mainly affected by air streams coming from the Indian Ocean. From May to September the south-east trades blow as a moderate to strong wind along the coast, changing their direction to south-west as, north of the equator, they are caught up in the oceanic monsoon of the Indian Ocean and southern Asia. From November to March the south-east trades are replaced by the steady light wind of the north-east trades, in character and direction fitting into the system of the continental monsoon of Asia. The transition months of April and October present light variable winds, mainly from an easterly direction. The winds of the Indian Ocean and its margins were of great importance in the development of contacts between the peripheral coastlands, for full advantage was taken of them in the operation of the traditional dhow traffic. The vessels reached the East African coast during the north-east monsoon, and departed with the setting-in of the south-west monsoon, as happens to the present day. The ocean currents of the shores of East Africa fit into this wind system but do not coincide with it.

The main source of East African rainfall is the air stream of the south-east trades. The north-east trades are dry in their northern section, but south of the equator they become moister after their long passage over the equatorial latitudes of the Indian Ocean. The inter-tropical convergence zone where these two air masses meet is held to be an important source of rainfall; and the seasonal distribution of East Africa's rainfall is linked with the oscillating movement of this junction between its January position to the south of the territories and its July position to the north. Another source of moisture for East Africa is westerly or Congo air, which causes occasional general rain over the western areas, especially in its zone of convergence with the easterly air streams. A drier air stream is that from the Sudan, which appears in Uganda at irregular intervals during the December to February period.

The most cursory glance at any map of rainfall distribution over East Africa shows the close connexion of rainfall and relief; and on such a map the areas of high relief stand out as well-watered patches amid the vast surrounding zones of moderate or insufficient rainfall. (See Map 2.) This inadequacy of the rainfall in total amount, seasonal distribution, and general reliability is a salient feature of the climate of much of East Africa, especially when the low latitudes, with their intense solar radiation and high surface temperatures, are taken into account. In the moister parts of East Africa potential evaporation varies, according to locality, from 45 to 50 inches per annum, and still higher values obtain over the arid plains of northern Kenya. In general, then, potential evaporation exceeds rainfall; and for a large part of the year the upper layers of the soil may be without moisture, to the consequent detriment of plant growth. In the report of the East Africa Royal Commission, 1953–1955, considerable attention is devoted to the reliability factor in East African rainfall; and two maps attached to the report show the probability of receiving each year respective totals of not less than 20 inches and not less than 30 inches. These maps, and the discussion of them in the text of the report, deserve the close study of those who are interested in the East African environment.

Three different régimes of seasonal rainfall occur in East Africa. That of the equatorial zone extends from the extreme north to 3° S. on the plateau and to 7° S. in the coastal belt.

The rains fall from March to May and from mid-October to December, the first rains being in most places and in most years

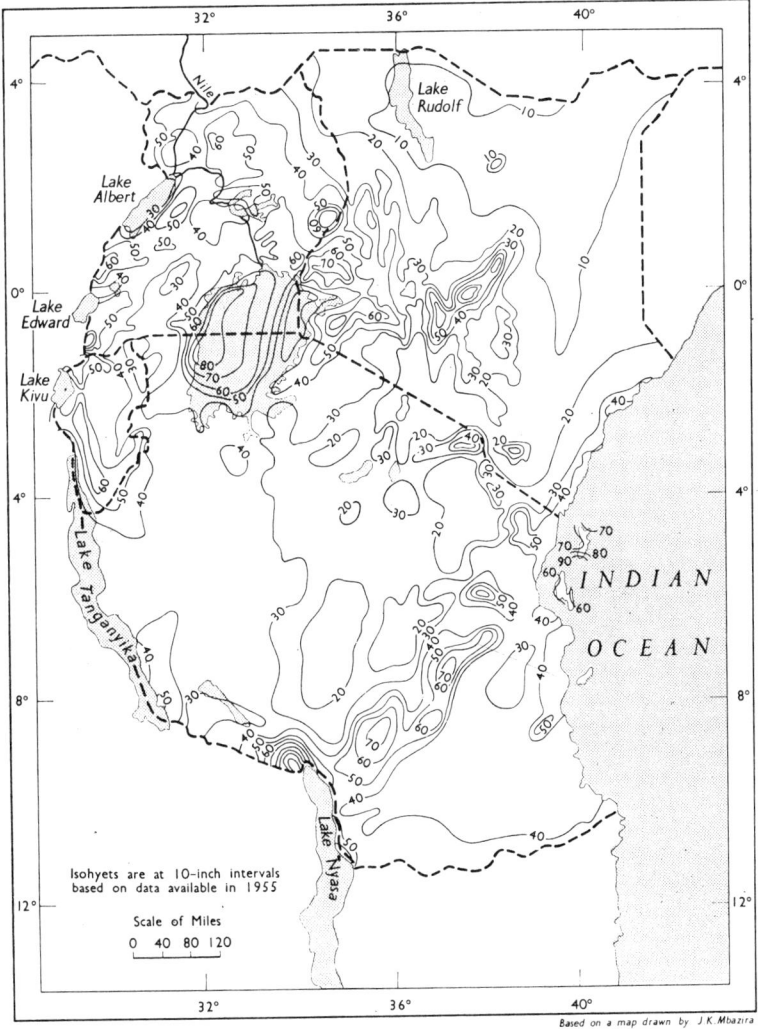

MAP 2. East Africa: mean annual rainfall

longer and heavier than the second rains. The drier seasons, which are relative rather than absolute in the more humid areas, occur during the remaining months of the year. In the extreme north of East Africa the beginnings of the régime of the inner tropics of the northern hemisphere are marked by the coalescence of the rainy seasons and by the consequent development of a single dry season. The southern hemisphere version of the same régime occurs over most of Tanganyika, including the coastal belt south of Dar es Salaam. The rains begin in November or December and continue until April; and they are followed by the strongly contrasted conditions of the dry season, which occupies the rest of the year. The régime continues southwards, well beyond the boundaries of Tanganyika, to extend over the basin of the Zambezi.

At the coast the mean annual temperature is in the neighbourhood of 80°, with a mean annual range of 7° and an average diurnal range of from 10° to 15°. These temperatures are equatorial in their characteristics, and they are accompanied by high relative humidities; but the rainfall totals are only moderate. In most parts the average annual fall is over 40 inches, but north of Lamu the rainfall diminishes towards the Somalia boundary, and smaller totals are received south of Kilwa Kivinje. Mafia, Zanzibar, and Pemba have an appreciably higher rainfall than the coast, totals of 60 inches or more being a usual feature. North of Dar es Salaam, over half the annual rainfall occurs in March, April, and May and there is a secondary maximum between October and November, although rain falls throughout the year. A hot season occurs before the main rains, but after them the cooler months from July to September form the pleasantest season of the year. The nights are often still and sultry, but fresh ocean breezes temper the heat of the day. South of Dar es Salaam a single rainy season from November to April is associated with the tropical régime of the southern hemisphere. Sunshine is everywhere abundant at the coast, although there is appreciable cloud during the period of the south-east trades.

Semi-arid to arid conditions prevail over the low foreland plateaux, especially over the wide tracts covered by this landscape in Kenya. Rainfall is least in northern Kenya, where the unreliable average is under 10 inches; but in eastern Kenya a wide tongue with less than 20 inches protrudes southwards almost to the Tanganyika border. The occasional *inselberge* and volcanic cones stand out as islands of higher rainfall. These general conditions extend westwards to include a narrow strip in eastern Karamoja, and southwards for some distance along the course of the rift valley; and there is a similar prolongation of the type around the southern side of the Kenya highlands into the Magadi–Natron section of the rift valley. The low foreland plateaux of south-eastern Tanganyika have somewhat better rainfall conditions, for their average is from 30 to 40 inches per annum.

The rim mountains of the interior plateau of Tanganyika catch the rain-bearing winds on their eastward-facing scarps, standing out as high rainfall areas; and a number of areas receive over 70 inches of rainfall a year. The western half of the plateau receives between 30 and 40 inches per annum, but there are considerable areas in the eastern part with less than 20 inches. There is a little cloud during the rainy season, but the dry season is marked by a fresh wind, cloudless skies, and powerful sunshine. At Tabora the mean annual temperature is 73°, the mean annual range is 8°, and the diurnal range varies according to season from 18° to 24°.

Of the lands marginal to Lake Victoria it is those on the west, north, and north-east of the lake which are most affected by the 'maritime' conditions which it creates. This lacustrine area comprises a narrow belt stretching round from south of Bukoba, expanding to slightly greater width north of the lake, and then extending south-eastwards to include the plateaux of north-western Kenya. The mean annual rainfall of this zone is everywhere over 40 inches and in places, such as the islands of Buganda and the Kakamega and Kisii areas of Kenya it is over 60 inches. Mean monthly temperatures range from 69° to 72°, and the diurnal range varies between 12° and 16°. Day temperatures seldom exceed 80°, and night temperatures rarely fall below 60°.

Most of Uganda has an average of between 40 and 50 inches of rainfall per annum, seasonally well distributed; and the country is more fortunate than her neighbours Kenya and Tanganyika in that the rainfall is less variable as well as more abundant. Over the north of Uganda there is an accentuation of the hot, dry season and an increase in diurnal range of temperature; but reasonably favourable rainfall conditions prevail over much of plateau Uganda. There is, however, a dry zone stretching discontinuously across the country from south-west to north-east in which the variable annual fall averages less than 40 inches. This zone includes the dry belt of eastern Ankole and western Masaka, continued southwards into western Bukoba and Karagwe; a dry area over northern Buganda and south-eastern Bunyoro; and, thirdly, the Karamoja area adjacent to the dry zone of northern Kenya.

The climatic conditions of the highlands, above 5,500 feet, may be best described in relation to the eastern rift highlands, in which altitude and aspect affect each of the elements of climate. At Nairobi, on the edge of the highlands, the rainfall averages 33 inches a year, but, with an annual variability which has included 19 inches and 61 inches in its range, the prospect of securing at least 30 inches in any given year is less than good. The amount and frequency of rainfall increases with increasing altitude up to 8,000 feet. Considerable areas over the Kenya highlands receive from 40 to 50 inches a year, and the Kericho highlands, the Aberdares, Mount Kenya, the Nyambeni range, and Kilimanjaro have averages in excess of 60 inches. In the eastern rift highlands the eastern slopes of mountains are generally wetter and cloudier than their western flanks. Altitude affects the character of the rain as well as its amount, for light or moderate continuous rain takes its place with the heavy showery type in the highlands; and above 7,500 feet cloud tends to lie on the surface of the land. The mean annual temperature of Kabete, just under 6,000 feet above sea-level, is 64°, the mean annual range is 7°, and the mean diurnal range varies according to season from 16° to 24°. Above 7,000 feet frost occurs occasionally, and above 8,000 feet it is a frequent occurrence; whilst glaciers descend to 15,000 feet on Mount Kenya and to 16,000 feet on Kilimanjaro.

The western rift highlands have good rains with annual totals of 50 inches or more, well distributed over the year, and with a significant contribution from Congo sources of moisture. In the Uganda section of the highlands mean annual temperature varies according to altitude from 66° at Fort Portal to 62° at Kabale; whilst over the snow-capped peaks of the Ruwenzori mountains frigid conditions prevail. The western rift valley is on the whole better watered than the eastern, although in the northern sections the shores of Lake George and Lake

Albert receive on the average barely 30 inches a year. The Albertine rift and the valley of the Albert Nile comprise the hottest part of Uganda. The southern rift highlands are well watered, with average rainfalls of over 60 inches per annum; and the northern end of the Nyasan rift valley shares with the adjacent highlands some of the highest totals of East Africa. Temperatures are reduced by altitude, and with their somewhat higher latitude the southern rift highlands show a slightly higher mean annual range than areas farther north.

At the end of this account of the climates of East Africa as they are at the present day, it is wise to remember that important changes of climate have occurred in Pleistocene and Recent times, in that wetter and cooler periods have alternated with drier and warmer periods. Dr. L. S. B. Leakey has recognized four pluvial periods in East Africa: Kageran, Kamasian, Kanjeran, and Gamblian, followed by two wet phases known as Makalian and Nakuran. The first of the pluvial periods is ascribed to the Lower Pleistocene, the second and third, which may be the two maxima of a single pluvial period, to the Middle Pleistocene, and the fourth to the Upper Pleistocene; while the Makalian wet phase is regarded as epi-Pleistocene and the Nakuran as Recent. In the inter-pluvial periods the climate was drier and warmer, and an arid phase separates the Makalian and Nakuran wet phases. The degree of climatic change involved in the fluctuations cannot be easily assessed, but wherever conditions were in any sense marginal, the effects of the changes may well have been considerable. On Kilimanjaro and Kenya, for instance, the glaciers reached down to 10,500 feet and 11,500 feet respectively during Middle Pleistocene times, and the Aberdare range and Elgon, now without ice, were glaciated; and much of the highlands would then become too cold for habitation. On the other hand, a relatively small increase in precipitation would result in a large expansion in the size of the shallow lakes on the floor of the eastern rift valley and would enhance the habitability of the remaining areas of the floor. The converse effects would occur in the inter-pluvial periods, when the lower and drier terrains would become less attractive, and man would tend to move up to the higher altitudes in search of permanent water-supplies. It must be added that H. B. S. Cooke[1] calls for a critical re-examination of the evidence for this interpretation of the climatic sequence of Pleistocene and Recent times.

Biogeography

The coastal lowlands contain a limited area of tropical rain forest along the lower courses of the rivers, and mangrove-swamp forest occurs in tidal estuaries and on lagoons. Elsewhere there are stands of high bush, wooded grassland, or extensive open glades, and the occasional baobab may form a striking feature of the scene. In certain areas the landscape has been considerably affected by centuries of Arab cultivation, and the coconut palm and the mango tree provide evidence of human occupation, past and present. During the course of the last 130 years the clove tree has become an established feature of the cultural landscape of the western parts of Zanzibar and Pemba.

Semi-desert, with acacia shrubs and bunch grass, occupies the whole of eastern and northern Kenya in the hinterland of the coastal strip, and extends north-eastwards through the adjoining territory of Somalia. In its southern portion this landscape is known as the Nyika, a Swahili word with the connotation of 'wilderness'. It laps around the Kenya highlands and extends into the northern and southern sections of the rift valley, and in a broadly comparable type it crosses the international boundary into the Masai plain of north-eastern Tanganyika. In the Lake Rudolf basin this type degenerates into desert scrub, from which the occasional taller tree and the bunch grass are missing, and in which the bare soil gives its colour to the landscape throughout the year. On the more humid side of the formation slightly more favourable conditions may cause the development of a thorn thicket. This Nyika zone, lacking water and other supplies for man and beast, and lying between the coastal strip and the Kenya highlands, constituted one of the main barriers to the penetration of the interior from the northern part of the East African coast.

The *Acacia–Themeda* formation consists of flat-topped acacia thorns, 6 to 8 feet high, scattered through an even cover of short grass, 3 to 4 feet in height. There are occasional trees,

[1] H. B. S. Cooke, 'Observations Relating to Quaternary Environments in East and Southern Africa', *Geological Society of South Africa*, Annexure to vol. 60, 1957.

and in some localities the candelabra euphorbia and other succulents comprise a prominent element in the landscape. The grassland is characterized by the species *Themeda triandra* (red oat grass), and along with the scattered-tree grassland there are wide tracts of open treeless sward. The *Acacia–Themeda* formation occurs in dry areas at altitudes from 4,000 to 6,500 feet on the margins of the highlands; and it is well developed over the Laikipia and in the plateaux on either side of the Magadi section of the rift valley, including the Machakos area.

The next formation provides a richer manifestation of the scattered-tree grassland. It consists of a high-grass cover 5 to 8 feet high, with *Hyparrhenia* as an important constituent, densely scattered with low trees, 10 to 15 feet in height, in which *Combretum splendens* is the dominant. In addition, there are isolated areas in which forest trees, 50 to 90 feet in height, occur individually or in groups. The *Combretum–Hyparrhenia* formation is considered to be a fire sub-climax, in which the maintenance of the grassland depends on fairly frequent burning. The association is found at levels between 3,000 and 6,000 feet, and it is well represented in a zone to the south and east of Mount Kenya stretching as far south-eastwards as the upper courses of the Tana and Athi rivers.

The highland grassland and forest unit provides a patchy distribution of comparatively small areas of forest in wide expanses of undulating open grassland, with a tendency for the forest to occupy the eastern slopes. The altitudinal limits are roughly 6,500 to 9,000 feet, and above the evergreen forest a bamboo zone extends up to about 10,000 feet before the appearance of the mountain moorland. On the lower edge of the forest, under rather drier conditions, the cedar (*Juniperus procera*) and the olive (*Olea* spp.) are found. The grassland is mainly dominated by *Themeda triandra* and *Pennisetum clandestinum* (Kikuyu grass). The reduction in the amount of forest is the result of forest clearance, and the maintenance of the *Themeda* grassland depends upon periodic burning. The vegetation unit occurs on the slopes of Elgon, in the West Kenya highlands and in the Aberdares, with a narrow connecting link across the central part of the rift valley, on the slopes of Mount Kenya, and on the various components of the northern highlands of Tanganyika. In the last-mentioned area the mountains rise like islands directly out of the sea of semi-arid bush which surrounds them, and there is less scope for the appearance of the wooded grassland formations which are developed in areas below 6,500 feet on the margins of the Kenya highlands.

The highland grassland and forest unit of the West Kenya highlands gives way to the *Combretum–Hyparrhenia*, high-grass low-tree formation over the western plateaux, and to the drier *Acacia–Themeda* type over the Kano plains. Drier conditions tend to accompany the coastal strip of Nyanza region and to continue across the international boundary into the extreme south-east of Uganda. In the Lake Victoria zone of Buganda and its continuation into north-eastern Bukoba the high-grass low-tree formation is at its richest, and *Pennisetum purpureum* (elephant grass) replaces *Hyparrhenia*. The evidence that this association represents a degeneration from woodland, under the influence of cultivation, is strong, and within the formation there are considerable areas of closed forest and scattered isolated trees such as mvule (*Chlorophora excelsa*). In Buganda there is a well-developed catenary complex comprising shorter grass on the plateau tops, richer grass and woodland on the valley sides, and swamp forest and swamp in the valley bottoms; but the present appearance of the landscape depends largely upon the activities of man as a cultivator.

In the south-west of the interior dry belt western Masaka and eastern Ankole present a vegetation of short grass, scattered acacias, and euphorbias, with ubiquitous thicket-covered termite mounds, which is closely comparable with the *Acacia–Themeda* formation of Kenya. In the eastern part of the Lake Kyoga basin, there is an area of wooded grassland which, although its grasses are long as compared with those of Ankole, deserves none the less to be classed with the short-grass country. Scattered-tree short-grass formations prevail over northern Uganda, developed under dry conditions in Karamoja and in the valley of the Albert Nile, and under somewhat moister circumstances in Acholi and in the West Nile highlands.

In the Uganda section of the western rift highlands elephant grass is found, along with considerable areas of closed forest. South of the Uganda boundary highland grassland and forest occurs in Ruanda and Urundi, where there is a notable development of rich open grassland on the plateau tops. The rift valley vegetation is luxuriant by comparison with that found in the Uganda sections of the rift. Highland grassland and forest is found in the attractive landscapes of the southern rift highlands, and it is present in patches in the eastern rim mountains.

A scattered-tree short-grass formation occupies a good deal of the central and eastern parts of the interior plateau of Tanganyika, including the Serengeti plains. Sometimes, as in the Itigi thicket, there is a local intensification of the woodland element in the association. Westwards and southwards the scattered-tree short-grass formation gives way to the open woodland of the 'miombo', in which the woody species are more in evidence and the grasses less so. This *Brachystegia–* Other Species woodland covers nearly half the land surface of Tanganyika, for apart from the western distribution, with a substantial outlier in the Central region, there is a fairly continuous cover over the low foreland plateaux of the southeast. This last-mentioned region, although not rich, is thus better endowed than the corresponding parts of Kenya. Valley grassland occurs in the areas where there is seasonal flooding, such as the floor of the Rukwa rift and the Kilombero and Wami flats, as well as along numerous smaller river valleys.

The more open types of vegetation as, for example, the *Acacia–Themeda*, have for ages been the grazing grounds of numerous herds of herbivores, such as antelopes, buffaloes, zebras, and elephants, and the hunting grounds of such carnivores as lions and leopards. Hunting for food and as a sport was thus a common activity, until the recent expansion of population and settlement caused a severe reduction in the availability of game. From the hunting of the elephant there resulted the important East African export of ivory to the lands bordering the Indian Ocean and beyond, from at least the ninth century A.D. to the present day.

The members of the East Africa Royal Commission, 1953–1955, whilst recognizing the variety of conditions obtaining over East Africa, come to the conclusion that on the whole the environment is a harsh one. Using rainfall reliability maps as a guide, they estimate that about half of East Africa has potentialities only for an unintensive type of pastoralism, and that two-thirds of Kenya, one-third of Tanganyika, and small areas of Uganda are occupied by this type of country. They place a quarter of East Africa in a marginal category between the pastoral and arable zones, noting that the widest expanse of such country is in western and southern Tanganyika. Rather less than a quarter of East Africa has a good chance of obtaining more than 30 inches of rainfall each year and is therefore, as far as the rainfall factor is concerned, suited to intensive arable cultivation. Kenya has little of this type of country, Uganda much, and Tanganyika is in an intermediate position between the other two countries.

In terms of the vegetation types discussed in this chapter the areas of high agricultural potential comprise, apart from limited areas in the coastal belt, the highland grassland and forest and the high-grass low-tree formation. Some of the moister parts of the various short-grass associations are also potential areas of cultivation, but under more uncertain conditions. The *Acacia–Themeda* type of Kenya and Uganda, and the corresponding areas of dry savannah of Tanganyika, are well adapted to light pastoral utilization; whilst the semi-desert with acacia shrub and bunch grass allows of occupation by pastoral tribes only on a nomadic basis. It should be added that the absence of adequate water-supplies for use by man and for the watering of stock constitutes a major problem over vast areas of East Africa. It is only in modern times that policies of water development have somewhat alleviated this shortcoming of the primitive environment.

The soils of East Africa pose their problems to the potential user. Those soils which are derived from the old hard rocks are often thin and poor, although under generous conditions of climate and vegetation, for example, under an elephant-grass coverage, soils may be deepened and enriched. The sedimentary rocks of the eastern lowlands yield patches of fertile soil, as do some of the recent alluvia, once the problems of drainage have been mastered; and the soils derived from the young volcanic rocks show a natural fertility and some degree of durability. The climatic conditions of soil formation help to account for deficiencies of minerals or of humus, and the soils are as a whole more friable and less durable than those of cooler latitudes. In the high-rainfall areas a protective cover of vegetation soon returns to the soil from which a crop has been harvested; but over much of East Africa the thin covering of vegetation is not easily replaced after grazing or cultivation, and the soil is open to the ravages of erosion. The study of the soil factor in the East African environment illustrates the element of fragility in that environment; but this was a quality which the unintensive utilization of the land in the period prior to the twentieth century did not seriously test.

A significant aspect of the biological environment is the presence of numerous diseases of man and animals, some of the effects of which may be exemplified by a brief reference to trypanosomiasis. The trypanosome diseases are carried by various species of tsetse fly (*Glossina* spp.), whose habitats have at least in part been created by processes of human intervention which have operated through the centuries. At the present day about 40 per cent. of the land area of East Africa is tsetse-infested, that is about two-thirds of Tanganyika, one-third of Uganda, and one-tenth of Kenya, there being few occurrences of the tsetse fly in desert and semi-desert areas. An effect of human trypanosomiasis, or sleeping sickness, is to be seen in the one devastating epidemic of which we have knowledge, that striking Uganda at the beginning of the twentieth century, when out of a lakeside population of 300,000 in the areas principally affected, 200,000 succumbed. The effect of the animal disease, nagana, on human development is perhaps greater than that of sleeping sickness; for where nagana occurs there can be few cattle and horses, or none at all, and before the advent of mechanical power, cultivation could only be by the hoe and transport could only be by human porters. Yet another serious effect is the denial of the animal proteins of meat and milk to populations suffering from malnutrition.

In the early phases of agricultural development, East Africa appears to have had a range of basic food crops which included sorghum (*Sorghum vulgare*), millet (*Eleusine coracana*), the Bambara ground-nut (*Voandzeia subterranea*), yams (*Dioscorea* spp.), and other root crops. Although wild forms of the banana occur in East Africa, the varieties of plantain principally used for cooking belong to the species *Musa paradisiaca*, which appears to be of Indo-Malayan provenance and a very early immigrant into East Africa. Like other regions, then, East Africa benefited by the introduction of food crops from outside. The Arabs introduced the citrus fruits and the sugar-cane (*Saccharum officinarum*) into East Africa from Asia, and by way of return gave India sorghum. Asian rices (*Oryza sativa*) were introduced by the Arabs to the coast between the eighth and tenth centuries A.D., but they seem not to have spread inland. The debt to the tropical Americas is heavier, and among the items included in the list are maize (*Zea mays*), cassava (*Manihot utilissima*), ground-nuts (*Arachis hypogaea*), sweet potatoes (*Ipomaea batatas*), and the papaw (*Carica papaya*). All these food crops were brought into tropical Africa after the Columbian discovery of the Americas; and their spread over East Africa was incomplete when the modern administrations were set up at the end of the nineteenth century. With the foundation of experimental stations and research institutions, and with the development of commercial agriculture in the tropical and the temperate landscapes of East Africa, the diversification and enrichment of the resources of the East African territories has continued with accelerated tempo during the twentieth century.

II

THE STONE AGE OF EAST AFRICA

SONIA COLE

The Earlier Stone Age

THE unique importance of Africa in the story of man's origins has only recently been recognized. Between the 1920's and the early 1940's Asia had generally been accepted as the 'cradle of humanity'; but from that time on the many discoveries made in Africa shifted attention to that continent.

The story begins 20 million years ago, long before man himself had appeared. At that time the great Lake Victoria had not been formed, but part of its basin was occupied by other lakes, bordered by active volcanoes. The ancient lake deposits, which alternate with beds of volcanic ash or tuff, have preserved the fossil remains of many kinds of animals and plants of the Lower Miocene period. The first ape fragments, consisting mostly of isolated teeth, were found as long ago as 1923; about ten years later palaeontologists had enough material to describe three new genera of apes. An almost complete mandible of one of these, *Proconsul africanus*, was found on Rusinga island in 1942 and a whole skull in 1948.[1] This is the earliest ape skull known and is therefore of great interest and importance.

Proconsul, of which there were three species, was essentially unspecialized. Exaggerated features such as pronounced browridges were evidently late specializations of the ape family; the hominid—or human—family, on the other hand, retained many of the characteristics of its generalized ancestors. *Proconsul's* limb bones, too, were of a form that could have given rise either to the extreme kind adapted for brachiation—as seen in the modern apes—or to those suitable for an erect posture, assumed eventually by the hominids.[2] It seems probable, then,

[1] L. S. B. Leakey, 'Skull of *Proconsul* from Rusinga Island', *Nature*, 162, 1948, p. 688.
[2] W. E. Le Gros Clark, and L. S. B. Leakey, *The Miocene Hominoidea of East Africa*, 1951.

that both the anthropoid apes and man shared the same kind of remote ancestor, which was something like *Proconsul*.

The Australopithecines, whose remains have been found so abundantly in the Transvaal, lived over half a million years ago and it now seems certain that they represent one of the stages in man's family tree. Two main groups of these 'near-men' are known: the unspecialized *Australopithecus* and the more robust *Paranthropus*, which had powerful jaws and a sagittal crest running along the top of its skull, like a gorilla.

Because of certain features such as the human appearance of the teeth, and the evidence from limb bones and pelvis which shows that the Australopithecines walked upright, anatomists grouped them with the hominid family. Man has been defined as the tool-maker and, until recently, there was little evidence that the Australopithecines were capable of *making* tools for specific purposes, though it seemed fairly certain that they *used* tools, particularly bones and horns. Then some crude pebble-tools were found in the Australopithecine breccias of the Transvaal, which suggested that these 'near-men' had deliberately shaped pebbles.[1] The question was still controversial, however, since there was always the possibility that the tools had been made by true men living contemporaneously with the Australopithecines—men who might, in fact, have preyed on the Australopithecines.

The argument that it was the Australopithecines themselves who made pebble-tools has been strengthened considerably by a remarkable find from East Africa. Until 1959 the only possible remains of Australopithecines known outside South Africa were some fragmentary jaws from Sangiran in Java, a piece of upper jaw containing two teeth from the Laetolil beds in northern Tanganyika, and two large milk teeth from Olduvai gorge, also in northern Tanganyika.[2] Then, in July 1959, Dr. and Mrs. Leakey discovered a skull[3] in the lowest bed of this famous gorge, Bed I, which was nicknamed 'Nutcracker man' because of the enormous size of his molars. He has been called *Zinjanthropus boisei*, Zinj being the ancient name for East

---

[1] J. T. Robinson, 'Occurrence of Stone Artefacts with *Australopithecus* at Sterkfontein', *Nature*, 180, 1957, p. 521.

[2] L. S. B. Leakey, 'Recent Discoveries at Olduvai Gorge, Tanganyika', *Nature*, 181, 1958, pp. 1099–1103.

[3] L. S. B. Leakey, 'A New Fossil Skull from Olduvai', *Nature*, 184, 1959, p. 491.

Africa; but in the strongly developed sagittal crest and other features the resemblance to *Paranthropus* seems fairly marked.

The skull was associated with a living floor which contained numerous waste flakes knocked off in making Oldowan choppers, a broken hammer stone, and fragments of bones of small animals and birds, which formed part of 'Nutcracker' man's diet.

Some eighteen months later, Leakey reported[1] the discovery of remains of a juvenile creature even older than *Zinjanthropus*, as well as fragments of an adult; both were associated with a living floor in a lower level of Bed I. The child, who was not more than twelve years old when he died, is represented by a clavicle, bones of the foot and hand, a lower jaw, and two parietal bones. The parietals are larger than those of the adult *Zinjanthropus*, suggesting that this hominid had a bigger brain.

A remarkable discovery at this living floor was a bone tool, polished on one side through repeated use, which would have been suitable for working leather, perhaps thongs. There were also stone tools of the Oldowan pebble culture and the remains of the creatures' meals, mostly tortoises and catfish (both easy to catch).

The age of *Zinjanthropus* was determined by Drs. J. F. Evernden and G. H. Curtis of the University of California in July 1961.[2] They used the new potassium-argon method of absolute dating, which depends on the disintegration of the radioactive isotope potassium-40 into argon-40. The samples used were crystals from the volcanic tuffs which make up Bed I. The mean of a number of dates obtained for the *Zinjanthropus* level was 1,750,000 years. This is about three times as old as previous estimates and radically alters ideas on the rate of evolution of the hominids and of the duration of the Pleistocene period.

Pebble-tools have been recognized in many different parts of Africa. The first were noticed in Uganda in 1919, in terraces hundreds of feet above the present level of rivers such as the Kafu and the Kagera. They were found in abundance and

---

[1] L. S. B. Leakey, 'Recent Discoveries at Olduvai Gorge', *Nature*, 188, 1960, pp. 1050–1; L. S. B. Leakey, 'New Finds at Olduvai Gorge', *Nature*, 189, 1961, pp. 649–50.

[2] L. S. B. Leakey, J. F. Evernden, and G. H. Curtis, 'The Age of Bed I, Olduvai Gorge, Tanganyika', *Nature*, 191, 1961, p. 478.

were believed to date from the Lower Pleistocene, about half a million years ago. This so-called Kafuan culture, consisting of pebbles chipped at one end and on one face only, was thought to be the oldest evidence of man's tool-making capacities in the world. Now, however, there is some doubt as to whether the majority of Kafuan 'pebble-tools' were, in fact, shaped by human agency, or whether they were produced by natural causes—for instance as the result of boulders being chipped as they fell over waterfalls or down the sides of steep-sided valleys and gorges.[1]

Although the authenticity of the earlier Kafuan pebble-tools is open to question, the more advanced Oldowan pebble culture can be attributed to man (or at least to hominids) without hesitation. The characteristic chopping tools consist of pebbles chipped on two surfaces to form an effective cutting edge at the intersection of the flaked surfaces. They were first recognized in Bed I of Olduvai gorge, one of the most important early prehistoric sites in the world, where there is an amazing abundance of stone tools in well-stratified lake deposits, accompanied by fossil fauna which make it possible to date the tools.

In Bed II the majority of tools are still Oldowan choppers, but they are accompanied by a few hand-axes of the Chellean stage of the Chelles–Acheul culture, a sequence which clearly originates in the Oldowan pebble culture and which becomes more evolved in higher beds in the gorge. The main difference between Chellean and Acheulean hand-axes is that the latter were shaped by the 'cylinder hammer technique', that is, by using the side of a cylindrical piece of bone or wood rather than the point of a hammer-stone; this makes it possible to remove flatter flakes and hence to produce finer shaped implements. The Chelles–Acheul culture was first named from sites in northern France in the middle of the nineteenth century; but the typology of hand-axes from western Europe, from parts of southern Asia, and from all over Africa is basically the same. Central Africa lies more or less in the middle of this area of distribution. Hand-axes, moreover, are more abundant in Africa than elsewhere. These two considerations make it seem

---

[1] J. D. Clark, 'The Natural Fracture of Pebbles from the Batoka Gorge, Northern Rhodesia, and its bearing on the Kafuan Industries of Africa', *Proceedings of the Prehistoric Society*, N.S. 24, 1958, pp. 64–77.

likely that the culture first arose in Africa and spread later to the peripheral areas of Europe and Asia.

The men of hand-axe times may have been mainly scavengers, seizing carcases from hyaenas and finishing off dying animals; but evidence from Olduvai shows that they were no mean hunters. Most of their prey was of gigantic proportions: horn-cores of extinct sheep indicate that these animals were as large as modern buffaloes; there were also pigs the size of a rhinoceros and baboons as big as chimpanzees. Although there is no evidence as yet that *Zinjanthropus* killed these huge animals in Oldowan times, it is certain that the Chellean people hunted them. A concentration of fossil bones of these huge herbivorous animals in one part of Olduvai gorge suggests that the beasts had been driven into a swamp: limb-bones were found sticking in the clay in a vertical position, having lain undisturbed since the animals were killed. Other bones, split open to extract the marrow, and accompanied by stone tools on the edge of the clay, seem to indicate a butchering spot.[1]

The human milk molar found associated with the first stage of the Chellean culture at Olduvai is about two and a half times as big as a modern infant's molar. The molars of *Zinjanthropus* are the largest of any known hominid. This need not necessarily mean that the men of those times were of gigantic proportions, like so many of the contemporary herbivorous animals; although it seems likely that they were heavily built, very powerful jaws need not go with an enormous body. The huge grinding teeth were necessary for a diet that was mainly vegetarian. Pieces of red ochre were found associated with the Chellean culture; these must have been brought from not less than fifty miles away, the nearest known source, suggesting that people at this time already took a pride in their appearance —the ochre was presumably used for decorating the body or hair.

Apart from *Zinjanthropus*, another possible maker of pebble-tools in East Africa is the controversial individual whose jaw was discovered at Kanam on the southern shores of the Kavirondo Gulf of Lake Victoria. This mandible at first seemed to contradict all other evidence by apparently having a chin. The few other early Pleistocene hominids known, such as *Pithecanthropus* from Java and *Atlanthropus* from Algeria, do not have

[1] L. S. B. Leakey in *Nature*, 1958.

prominent chins. Some of the Australopithecines from South Africa, however, are now known to have fairly pronounced chins, but the Kanam individual suffered from a bone tumour, a pathological condition which undoubtedly exaggerated the depth of the chin, pushing it forward until it appeared to be vertical. It now seems more likely that the Kanam jaw belonged to Rhodesian Man or one of his near relations (see p. 38).

There now seems considerable doubt as to whether the Kanam jaw was really contemporary with pebble-tools and associated early Pleistocene mammals; but its degree of mineralization is similar to the latter and, even if it was not contemporary, it must certainly be of high antiquity. Recent tests[1] have shown that the radioactivity (uranium content) of the Kanam mandible is very low compared with that of all the undoubtedly Lower Pleistocene fossils from the site, which makes it seem unlikely that the human jaw is Lower Pleistocene.

At Kanjera, quite close to Kanam, fragments of four human skulls were found associated with Acheulean hand-axes. Most of the pieces were collected from the surface, but a few were *in situ* in the ancient lake beds, which contained stone tools and fossil fauna. The radioactivity of the Kanjera skull fragments was similar to that of associated extinct fauna,[2] presumed to be of Middle Pleistocene age, which strongly suggests that the skulls were contemporaneous. Amazingly, they are quite modern in appearance. They are ultra-dolichocephalic—extremely long and narrow—and a small piece from the frontal region shows that the brow was quite smooth, as in modern man, with no traces of brow-ridges. This character and certain other features of the skulls show certain similarities with the Bushmen.

The remains of hand-axe makers are extremely rare, the reason being that they did not bury their dead, nor did they live in caves—or only very exceptionally before the discovery of fire—so that their bones were not sealed under cave floors, as in the case of the Neanderthal people. Part of the skull of one of the hand-axe makers was found at Swanscombe in Kent, but unfortunately the important frontal region is

[1] K. P. Oakley, 'Dating the Stages of Hominoid Evolution', *The Leech*, 28, 1958, pp. 112–15. [2] Ibid.

missing. Because the occipital and parietal bones are similar to those of modern man, it has been assumed that the Swanscombe skull is *Homo sapiens*. The Kanjera finds also seem to support the theory that the hand-axe makers were little different from the people of today. What, then, was to be made of the jaw fragments of *Atlanthropus* from Algeria, which were associated with hand-axes, and which are very similar to the jaws of *Pithecanthropus* of Java and Pekin, hominids which had a low vault to the cranium and strongly developed browridges? It seemed that hand-axes were not necessarily made by men of *Homo sapiens* type, though they could, of course, have been made by different types of men in different areas.

A most important find bearing on this problem was announced by Dr. Leakey early in 1961.[1] It consists of the skullcap of an individual from Bed II at Olduvai, in the same horizon as hand-axes of the third Chellean stage. It has enormous brow-ridges and is very like *Pithecanthropus* of Java and Pekin. The vault of the skull, however, is higher, implying that the brain was considerably larger. Here, then, is good evidence of what the makers of Chellean hand-axes were like. But what of the much later makers of the Acheulean handaxes? Were the modern-looking Kanjera people really contemporary with hand-axes and a Middle Pleistocene fauna? Or were the bones washed into a fissure in the lake beds at some later period? If the first view is correct, as the recent radioactivity tests seem to confirm, then the smooth browridges are unique for this period as far as present knowledge goes—and admittedly the evidence is scanty. It does indeed seem a cruel stroke of fate that the Kanjera skulls should have no lower jaws and that the *Atlanthropus* mandibles should be accompanied by no crania; until the *complete* remains of an Acheulean hand-axe maker are found, the intriguing question of his appearance must be guesswork.

Apart from Olduvai, there are several sites in East Africa where the hand-axe culture is represented in abundance. One of the most important, because, like Olduvai, it contains many fossil mammal remains as well as thousands of implements, is Olorgesailie in the Kenya Rift Valley, some forty miles southwest of Nairobi. The gigantic Middle Pleistocene fauna is well

[1] L. S. B. Leakey, *Nature*, 189, 1961, pp. 649–50.

represented here; many of the bones had been split to extract the marrow and skulls were smashed to get at the brains. But no human remains have ever been found at Olorgesailie. As well as the ubiquitous hand-axes, straight-ended axe-shaped tools known as cleavers are fairly common both at Olduvai and Olorgesailie; they were probably used mainly for skinning and for chopping meat off carcases. Stone balls are also found

TABLE I. Approximate correlation of Early and Middle Stone Age cultures of East Africa with Europe.

Thousands of years	Europe	E. Africa
0–		
100–	MICOQUIAN / MOUSTERIAN	STILLBAY
200–	CHELLEAN–ACHEULEAN (ABBEVILLEAN) / CLACTONIAN TAYACIAN LEVALLOISIAN	ACHEULEAN / FAURESMITH SANGOAN
300–		CHELLEAN
400–		
1,000–		OLDOWAN

at both sites; at Olorgesailie they were discovered on several occasions in groups of three, from which it has been suggested that they were used as bolas. This weapon, made by the Patagonians in recent times, consists of three balls of different weights enclosed in leather bags attached together by thongs of different lengths; when thrown, the bolas became entangled in the legs of running animals, bringing them down. There are, of course, many other more obvious uses of stone balls—as pounders, as anvils for shaping stone tools, or simply as missiles —and the apparent distribution in groups of three may have been fortuitous.[1]

An industry known as the Hope Fountain has also been found stratified at Olorgesailie.[2] It has been recognized also in Uganda and Tanganyika and is particularly common in the Rhodesias. Since it occurs associated with several different cultures at various periods, it probably represents a particular occupational phase;[3] the implements consist mainly of small, crude flake-tools which would be suitable for scraping, cutting, and boring and which might also have been used for woodworking. Although wooden implements dating from so many thousands of years ago are preserved only under most exceptional circumstances—such as in the water-logged conditions at Kalambo Falls, Northern Rhodesia, for instance—perishable equipment made of bone, wood, and other vegetable material must have played a very important part in the lives of Stone Age people, probably just as important as the stone tools which are generally all that survive for us to study.

Other hand-axe sites in East Africa include Kariandusi near Gilgil in the Kenya Rift Valley; Nsongezi on the Kagera river in Uganda;[4] and Isimila south of Iringa in Tanganyika.[5] Abundant though the stone tools are at Olduvai and Olorgesailie, they exist in even greater profusion at Nsongezi and Isimila. The abundance of implements at certain sites suggests that these were camps near drinking-places where game would congregate at certain times of the year. Stone Age hunters had to travel light and, provided that the raw material was available, they probably discarded their weapons when they moved to another area and made new ones as necessary. The hand-axe people were great craftsmen who evidently took a pride in their work, rejecting tools which did not come up to the required standard. At Isimila, some of the finest hand-axes are over

[1] S. Cole, *The Prehistory of East Africa*, 1954, pp. 142–4.
[2] M. Posnansky, 'A Hope Fountain Site at Olorgesailie, Kenya Colony', *South African Archaeological Bulletin*, 14, 55, 1959, pp. 83–89.
[3] J. D. Clark, *Prehistory of Southern Africa*, 1959, p. 40.
[4] C. H. van Riet Lowe, *The Pleistocene Geology and Prehistory of Uganda* (Pt. II, *Geological Survey of Uganda*), Memo. No. 6, 1952.
[5] F. C. Howell, 'A Preliminary Note on a Prehistoric Donga (Maclennan's Donga) in Central Tanganyika', *South African Archaeological Bulletin*, 10, 38, 1955, pp. 43–52.

15 inches long and weigh over 9 lb.; it is difficult to see how they could have been put to any practical use unless the men who wielded them had enormous strength.

The unique importance of East Africa's early Stone Age sites is due to the presence of abundant fossil fauna, which not only gives an insight into the habits of early man—his diet and his skill as a hunter, for instance—but also is of the greatest assistance in dating. Although mammals tended to survive longer in Africa than in other continents—partly because most of Africa was cut off by the Sahara desert and therefore became something of an evolutionary cul-de-sac—broadly speaking, the fauna may be taken to be roughly contemporaneous with similar forms in other parts of the world which have been dated by various means.

Stone Age people, like all hunters, must have been largely nomadic, following the game to the best grazing according to where rain had fallen. That they were able to live for long periods at places like Olduvai and Olorgesailie, which at present have no permanent water-supply, was due to the fact that there were lakes at both these localities during Middle Pleistocene times. This suggests that the rainfall must have been greater than it is today. But the vegetation must even then have been grassland, rather than bush or savannah, as beasts with the immense horn-spans indicated by the fossil remains could not have lived except in open country. Thus the rainfall, though it must have been greater than today, could not have been heavy enough to change the vegetation pattern to any considerable extent.

There is, in fact, considerable evidence to show that there were long periods during the Pleistocene when the rainfall in many parts of Africa was greater than it is today—or at least that it was more regular and more evenly distributed throughout the year. These so-called pluvial phases were separated by interpluvials, when the rainfall was less than during the pluvials though not necessarily always less than it is today. Evidence for such past climatic changes is seen in high terraces signifying former greater extension of lakes, in the presence of wind-blown sands which suggests aridity, and in various soil formations such as ferricrete which indicate strongly seasonal rainfall.[1]

[1] R. F. Flint, 'Pleistocene Climates in Eastern and Southern Africa', *Bulletin of the Geological Society of America*, 70, 1959, pp. 343–74.

The climatic succession was first worked out in detail in the Kenya Rift Valley and it has generally been accepted that there were four main pluvial periods during the Pleistocene—the Kageran, Kamasian, Kanjeran, and Gamblian—followed by two post-pluvial wet phases—the Makalian and Nakuran—each separated by interpluvials or drier phases.[1] Some geologists[2] believe that the evidence for the earlier pluvials is inconclusive in the type areas, but these climatic terms have been widely used over most of Africa south of the Sahara. It has, however, been recommended that they should only be applied outside East Africa where correlation is firmly attested by at least two of the three lines of evidence: palaeontological, archaeological, or the geological setting.[3]

It has also been assumed by most workers that pluvials in equatorial Africa were contemporary with glacials in high latitudes and interpluvials with interglacials. This has been disputed on meteorological grounds[4] and in reality the situation is probably far more complicated. Long-range correlations are by no means satisfactory, but the sequence established for East Africa has proved very useful in establishing the relative ages of Stone Age cultures. The tentative 'absolute' dates suggested seem, on the whole, to have been supported by the few radiocarbon dates now available (see Table II).

In southern Africa the Stone Age is divided into Earlier, Middle, and Later. These terms are not synonymous with Palaeolithic, Mesolithic, and Neolithic—the Later Stone Age, for instance, very seldom implies a Neolithic way of life in Africa south of the Sahara. The Earlier Stone Age ends with the final Acheulean hand-axe industries, at the end of the third or Kanjeran pluvial. A carbon-14 date for a late hand-axe

[1] L. S. B. Leakey, 'The Climatic Sequence of the Pleistocene in East Africa', *IIe Congrès Panafricain de Préhistoire, 1952*, Paris, 1955, p. 293.

[2] H. B. S. Cooke, 'Observations Relating to Quaternary Environments in East and South Africa', *Geological Society of South Africa*, Annexure to vol. 60, 1957; R. F. Flint, 'On the Basis of Pleistocene Correlation in East Africa', *Geological Magazine*, 96, 1959, pp. 265–84.

[3] J. D. Clark, *Proceedings of the 3rd Pan-African Congress on Prehistory, Livingstone, 1955*, London, 1957.

[4] G. C. Simpson, 'Further Studies in World Climate', *Quarterly Journal of the Royal Meteorological Society*, 83, 1957, p. 468; E. A. Bernard, 'Les Climats d'insolation des latitudes tropicales au Quaternaire', *Bulletin de l'Académie royale des sciences coloniales*, 1959, pp. 344–64.

TABLE II. *Later Stone Age cultures of East Africa, with carbon-14 dates from other parts of Africa.*

Age (Thousands of years)	N. Africa, Egypt, and Sudan carbon-14 dates	E., Central, and S. Africa carbon-14 dates	Presumed duration of other E. African cultures	Climatic phases
1		Njoro River Cave 960 B.C.	WILTON — HYRAX HILL — GUMBAN A, B	Nakuran: 2nd wet phase
2				Drier
3	Neolithic of Capsian tradition 3,050 B.C.			
4	Shaheinab Neolithic 3,500 B.C.			Makalian: 1st post-pluvial wet phase
4	Fayum A Neolithic 4,300 B.C.	Nachikufan I 4,350 B.C.	ELMENTEITAN	
5	Upper Capsian 5,050 B.C.			
6	Capsian 'typique' 6,450 B.C.	Ishangian 6,500 B.C.	MAGOSIAN — Upper Kenya Capsian?	
7				Drier
8				
9				
10		Lupembo/Tshitolian 10,230 B.C.		
11			STILLBAY	
12				Gamblian pluvial
13		Late Lupemban 12,600 B.C.		

industry at Kalambo Falls, Northern Rhodesia, is more than 57,000 years before the present; evidence from fossil pollens indicates that the climate was then colder and wetter than today.[1]

Incomplete though our knowledge of the first inhabitants of East Africa is, it is nevertheless more detailed perhaps than in any other part of the world at a comparable period. Sites such as Olduvai and Olorgesailie are of unique importance in gaining an understanding of the way of life of the early hunters. There are many outstanding problems which can only be answered when further discoveries are made: were the later Acheulian hand-axe makers, for instance, still of the *Pithecanthropus-Atlanthropus* type, were they 'Rhodesioids' (see p. 38), or were they already fully *Homo sapiens* like the Kanjera skulls? Or were the Kanjera skulls exceptions, or do they date from after Acheulian times? Until very recently the other intriguing question was: did the Australopithecines make pebble-tools? East Africa has supplied a positive answer on this point. There is every reason to hope that the riddle of the hand-axe makers' appearance will also be answered soon, for there is still a vast field of exploration awaiting the prehistorian in Africa.

The First Intermediate period and the Middle Stone Age

After the decline of the third, or Kanjeran, pluvial the Earlier Stone Age came to an end. There were also great changes in man's environment in East Africa. Not only did most of the gigantic fauna become extinct, to be replaced largely by species still living today, but there were extensive earth movements and volcanic eruptions which altered the topography of the country considerably and gave the Rift Valley its present form.

About 50,000 years ago the ubiquitous hand-axe industries were replaced by several regional variants, which differed according to ecological conditions and the raw materials available. Implements became smaller and were adapted for special purposes. Although hand-axes had been made on both cores and flakes, and a number of small flake-tools for cutting and scraping have been found with hand-axe industries, the increase in 'flake tools' is most marked in Middle Stone Age

[1] Clark, *Prehistory of Southern Africa*, p. 130.

times. Both the 'unprepared core' and the 'faceted platform' technique were used; briefly, the latter consists in removing two series of flakes from a core—the first to block out the desired shape of the tool and the second to prepare a platform at one end of the core—and then, by striking a blow on the prepared platform, one large, fine flake or blade was removed and shaped into the required tool.

To a large extent the choice between tools made on cores or flakes depended on the quality of the raw material; small river boulders are particularly suitable for core tools, while fine flakes can be obtained readily from lumps of homogeneous rocks such as chert or obsidian. These flakes could then be shaped into a variety of blades, points, or other delicate tools which could not easily be made from the core itself. Light equipment is obviously essential to hunters depending on mobility and speed and the invention of composite tools such as spears, lances, and eventually arrows was a tremendous step forward. These weapons were made by hafting small stone points or barbs on to wooden shafts; they were gummed into slots with mastic and then bound with animal sinews or vegetable fibres.

Scrapers of various kinds were an essential part of the equipment of all prehistoric people and probably most of these, too, were hafted. Hollow scrapers would have been used for shaping wooden shafts, end-scrapers and side-scrapers mainly for leather dressing. From the evidence of carpenters' tools and tools designed for leather working, it is clear that equipment of wood, skins, and other perishable materials must have played an ever more important part during the Middle Stone Age. The age of specialization, in fact, had dawned.

During the First Intermediate period, starting during the interpluvial which succeeded the Kanjeran pluvial, two cultures succeeded the hand-axe culture in different parts of East Africa. The Fauresmith occurs in the drier, grassland country of the east; the Sangoan is found in the wetter, forested country farther west. The Fauresmith died out after this interpluvial, though the Sangoan developed in various forms throughout the Gamblian pluvial. The Stillbay, which started only in Gamblian times, was characteristic of the drier parts of East Africa during the Middle Stone Age.

Implements of the Fauresmith culture tended to be conservative, continuing the Acheulean tradition of hand-axes, cleavers, and stone balls, though on a smaller scale, with the addition of rather crude points and scrapers made on flakes. Living as they did in open country, these people must have been mainly hunters and their equipment includes knives for cutting up meat and scrapers for preparing skins. The Fauresmith is fairly common in South Africa, where it was first recognized, is not known at all in the Rhodesias, and then turns up again in the highlands of Kenya and Ethiopia. In Kenya the culture has been found mostly at an altitude of above 7,000 feet, on the slopes of Mount Kenya and the Aberdares; probably these camps were situated beside permanent streams which would not have dried up during the interpluvial.

The Sangoan culture is found in the forested parts of central and western Africa, areas which have over 40 inches mean annual rainfall at the present time; it is particularly common in the Congo and in Angola and also occurs in Uganda and western Kenya. Although the Fauresmith seems a logical development from the Acheulean culture, a certain degeneration in the stone-working technique is noticeable; this is even more marked in the early stages of the Sangoan, where characteristic tools include rather crude hand-axes, points, and massive picks. Possibly this may be accounted for by the fact that many of the stone tools must have been used to make finer equipment in other materials, such as wood, bone, and leather. The forest-living Sangoans must surely have made great use of wood, and many of their stone tools would have been most suitable for shaping and scraping wooden implements, as well as for cutting, chopping, and gouging. Animals may have been caught in traps, which, again, needed wood-working tools for their construction.[1]

In the later stages of the 'Sangoan' from the 'O horizon' and the surface at Nsongezi,[2] now known to be of the Later Stone Age and referred to in the Congo as the Lupemban and Tshitolian stages, some of the stone tools are very fine indeed. They include long, delicately flaked lance-heads and tranchets; the latter are roughly the same shape as cleavers and were made

[1] Clark, *Prehistory of Southern Africa*, pp. 132–4.
[2] Lowe, *Pleistocene Geology of Uganda*, pp. 80–88, 90–91.

in various sizes—the smaller ones were presumably arrowheads, the transverse ends of which would wound animals and cause them to leave a blood spoor which could easily be followed in wooded country,[1] and the chisel-ends of the larger ones would have been useful for wood-working.

The Sangoan culture was first discovered at Sango Bay, on the western shore of Lake Victoria. It is also common at Nsongezi on the Kagera river and at various other sites in Uganda where there are extensive outcrops of quartzite; this coarse-grained rock is unsuitable for making very fine implements and must partly account for the clumsy appearance of some of the heavier picks and other early Sangoan tools. The Sangoan is found also at many sites in the Nyanza Province of western Kenya,[2] which was linked more with Uganda than with the rest of Kenya in prehistoric times—as, indeed, it still is.

Finds from South Africa, Northern Rhodesia, and Tanganyika have shown that men with heavy brow-ridges and receding foreheads were widespread in early Middle Stone Age times. At Elandsfontein, near Hopefield, Saldanha Bay, Cape Province, a skull of this type was associated with a Fauresmith industry. It resembles the well-known *Homo rhodesiensis* from Broken Hill, who was accompanied by a Proto-Stillbay industry. Skull fragments from Eyasi, northern Tanganyika, were found with an industry which has been called 'Developed Levalloisian'[3] and which consists of small flake tools struck from prepared cores. A very different individual from Singa on the Blue Nile, who may have been aberrant, was associated with an industry which may be Proto-Stillbay or possibly Sangoan. It is difficult to be more precise about these industries, which have been called by various names but which were very much alike in early Middle Stone Age times.[4]

The human remains from Eyasi consist of parts of three crania. The outline of one of these was reconstructed from a

---

[1] Clark, *Prehistory of Southern Africa*, p. 224.
[2] Formerly known as the 'Tumbian' culture. L. S. B. Leakey and W. E. Owen, 'A Contribution to the Study of the Tumbian Culture in East Africa', *Coryndon Museum Occasional Papers*, 1, Nairobi, 1945.
[3] L. S. B. Leakey and W. H. Reeve, 'Report on a Visit to the Site of the Eyasi Skull', *J. E. Afr. Nat. Hist. Soc.*, 19, 1 and 2, Nairobi, 1943, pp. 40–50.
[4] Cole, *Prehistory of East Africa*, p. 157.

large number of pieces, which include the important frontal region. This showed that the brow-ridges were extremely massive. The occipital bone of one of the fragmentary skulls is thicker than that of the more complete one and is nearer to Rhodesian man; possibly this individual was a male and the more complete skull a female. The foramen magnum of the more complete skull is inclined backwards at an angle which implies that the head was hung rather forward, whereas Rhodesian man held his head erect. No mandibles were found, but a piece of maxilla containing the broken sockets of several teeth was recovered.

The Eyasi individuals were named *Africanthropus njarensis* when first described, but this generic name had already been given to a quite different type of skull from Florisbad in South Africa, so they are now generally referred to simply as 'Eyasi man'. Although there are many differences between the Eyasi and Broken Hill skulls, obviously they are of the same general type and 'Eyasi man' should also be described as *Homo rhodesiensis*, though possibly he may represent a distinctive race. These African equivalents—and contemporaries—of Neanderthal man have been called 'Rhodesioids'; they are distinguished by exaggerated brow-ridges and by thick walls and a rather flat vault to the cranium. Since their geographical range was wide—from the Cape to Tanganyika—it seems reasonable to suppose that this type was responsible for the industries of early Middle Stone Age times, whether Fauresmith, Sangoan, or Proto-Stillbay, which started slightly later than the other two.

The Proto-Stillbay and Stillbay industries, characteristic of the fourth or Gamblian pluvial, extend right down the eastern side of the continent from the Horn to the Cape. Typical implements are triangular or leaf-shaped points, which presumably were hafted as spear-heads; during the Proto-Stillbay stage these points were worked only on one surface, but during the Stillbay proper they were often beautifully flaked over both surfaces. The Stillbay people must have been nomadic hunters; probably they hunted in fairly large bands, driving and surrounding the animals as well as trapping them.

Proto-Stillbay and Stillbay industries are particularly common in the Kenya Rift Valley, where they are apparently succeeded by the Kenya Capsian. Presumably the two cultures were made by different groups; the makers of the Kenya Capsian are well known and were modern *Homo sapiens*, and probably by the time of the Stillbay culture the heavy-browed Rhodesioids, associated with the Proto-Stillbay, had developed also into the *sapiens* form. The Kenya Capsians seem to have been more sedentary than the Stillbay people and relied more on fishing than hunting.

The Lower Kenya Capsian[1] first appears in deposits said to date from the early part of the Gamblian pluvial at sites near Naivasha in the Kenya Rift Valley. The Upper Kenya Capsian is now generally believed to have followed the Stillbay in the same area and is a very much more advanced culture; in European terminology it would be regarded as typically Mesolithic.

The presence of bone harpoons with the Kenya Capsian, very similar to those associated with the Ishangian Mesolithic culture from the western shore of Lake Edward, suggests that the two cultures may be contemporary. The Ishangian has been dated by carbon 14 to about 6,500 B.C. by extrapolation,[2] a much later date than has been supposed for the Upper Kenya Capsian by its discoverer.[3] It is controversial whether the Kenya Capsian arose independently in East Africa or whether it was intrusive from Palestine and Arabia via a land bridge at the southern end of the Red Sea. Its connexions, if any, with the very similar Capsian of Tunisia and Algeria are also problematical; radio carbon dates of about 6,450 B.C. for the Capsian 'typique' of Tunisia[4] are again very much later than those supposed for the Kenya Capsian by Dr. Leakey.

The Upper Kenya Capsian is particularly well known from excavations at Gamble's Cave, Elmenteita, in the Kenya Rift Valley.[5] The 'cave' is really a rock-shelter, cut by the waters of a lake which at the time of the Gamblian pluvial stood 510 feet above the present level of Lake Nakuru.

The first prehistoric level contained an Elmenteitan industry (p. 47), the second was said to be Stillbay. The third

---

[1] Formerly described as the Lower Kenya Aurignacian. L. S. B. Leakey, *The Stone Age Cultures of Kenya Colony*, 1931, p. 90.

[2] J. de Heinzelin, 'Les Fouilles d'Ishango', *Institut des Parcs Nationaux du Congo Belge*, 1957, p. 19.

[3] L. S. B. Leakey, *Adam's Ancestors*, 4th edn., 1953, p. 130.

[4] M. Rubin and H. E. Suess, 'U.S. Geological Survey Radiocarbon Dates III', *Science*, 123, 1956, pp. 442–8.

[5] Leakey, *Stone Age Cultures of Kenya Colony*, pp. 91–109.

and fourth occupation levels contained Upper Kenya Capsian industries. Stone artefacts of the latter include backed-blades, gravers, microlithic crescents or lunates, and end-scrapers. The blades were made on long, narrow flakes with one edge left sharp and the other 'backed' by chipping off tiny flakes so as to produce a blunted edge on which the fingers could press, on the same principle as modern penknives. Burins or gravers are small tools with a chisel-end made by removing a flake from the side of a blade; in western Europe they are thought to have been used mainly for engraving on rock surfaces and bone, but in East Africa, where engravings of this period are unknown, they were probably used mainly for working grooves in wooden shafts. The tiny half-moon-shaped lunates must have been barbs for spears, harpoons, or perhaps arrows: on several occasions a number of lunates were found lying in groups, arranged in such a way as to leave little doubt that they had been fixed into wooden shafts which have, of course, perished with time.

Nearly all the tools of the Upper Kenya Capsian, as well as those of other cultures in this part of Kenya from Middle Stone Age times onwards, were made of obsidian, a black volcanic glass common in the Rift Valley which was apparently traded for hundreds of miles during prehistoric times. Owing to its homogeneous composition obsidian is easy to flake and gives an extremely sharp edge. Evidently it was as much prized by Stone Age man in East Africa as the best-quality flint was in Europe.

The presence of end-scrapers and bone awls in Gamble's Cave makes it certain that skins were dressed to make clothing or leather containers. Obsidian tools with jagged edges may have been used for fraying sinews for thread. Bone was used extensively, both for awls and for harpoons. Beads of ostrich eggshell were also common; in some cases both sides of the bead are polished, suggesting that these were threaded and worn next to the skin, but generally only one side is polished, which perhaps means that the beads were sewn on to clothing as ornaments. Pendants were made from lake shells, some decorated with perforations. Numerous lumps of red ochre were found in the deposits beneath the floor of the shelter, as well as stained stones used for grinding it. Ochre was used extensively in prehistoric times in connexion with the burial of the dead and was presumably also used for decorating the body and hair, as the Masai and other warriors use it today.

A few fragments of pottery were found, with a pattern which shows that clay had been smeared on the inside of a basket, which was then fired. At the time when Gamble's Cave was excavated, between 1926 and 1929, pottery was regarded as characteristic of the Neolithic and its discovery at Gamble's Cave occasioned considerable surprise; pottery has since been found occasionally with Mesolithic cultures, but the fragments associated with the Upper Kenya Capsian may still be among the earliest known.

Five skeletons were buried 14 feet beneath the floor of Gamble's Cave. Four of the bodies lay on their right side in the contracted position, with the knees drawn up; the faces were turned towards the opening of the rock shelter. The backs had been pushed up against the rock face and large stones had been laid over and around them. There were numerous traces of ochre in the surrounding earth.

The individuals are of the Caucasoid race and have no Negroid characters. The skulls are dolichocephalic and both chins and noses are prominent. The limb bones indicate that these people were over 5 feet 10 inches tall. Very similar skeletons, associated with the same kind of industries, were found near Naivasha and also at Olduvai. The latter skeleton was discovered as long ago as 1913 and was the first prehistoric remains to be found in Africa south of the Sahara. It was buried in the second lowest bed of the gorge and for many years was believed to be contemporary with the Middle Pleistocene fauna of Bed II. Later work, however, made it clear that the body had been buried long after the two layers which once covered Bed II had been removed by erosion.[1]

From the finds at Gamble's Cave it is possible to deduce a good deal about the way of life of these Upper Kenya Capsian people, who were by no means typical of others of the Mesolithic or Later Stone Age in East Africa. They worked bone, leather, and wood and for their stone tools they used obsidian, which they mined close by and probably traded over great distances. Their main source of food was probably fish, though they

[1] L. S. B. Leakey, *Olduvai Gorge*, pp. 159–60.

also hunted and may possibly have been among the first people to use bows and arrows. Living a fairly sedentary life by the lake shore, they were able to make pottery and had enough leisure to take a pride in their appearance, decorating themselves with

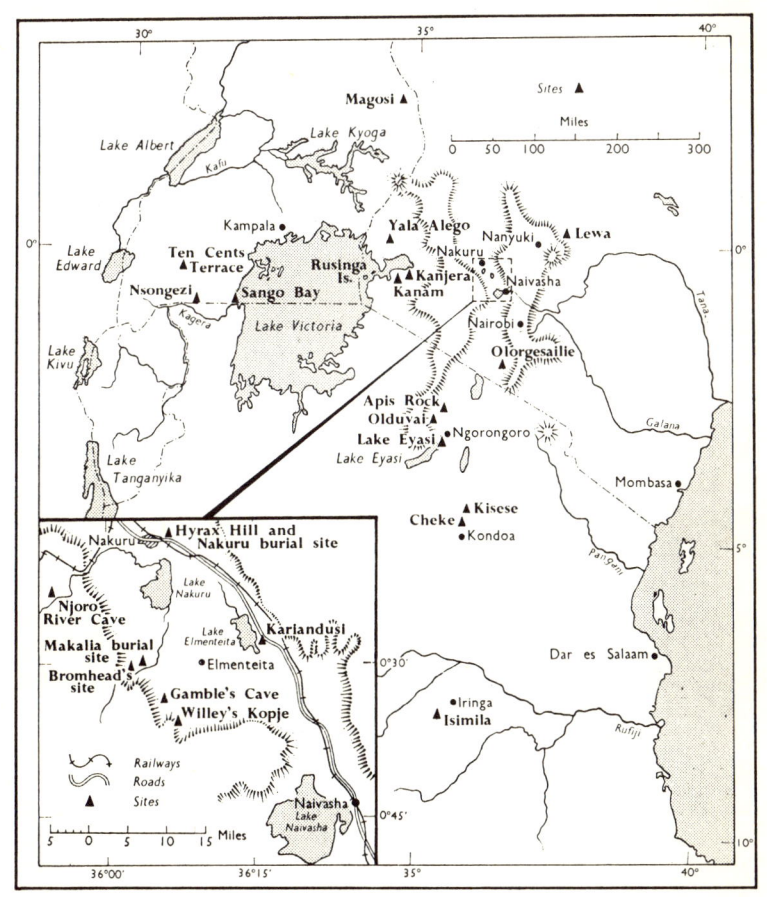

MAP 3. Main Stone Age sites in Kenya, Uganda, and Tanganyika

ochre, beads, and pendants. They also took pains over the burial of their dead and covered them with ochre. In character the culture is typically 'Mesolithic'. If it is proved to be contemporary with the Stillbay, of the time of the fourth pluvial in the Upper Pleistocene, as its discoverer believes, it is unique for this period.

The Second Intermediate period and the Later Stone Age

Just as the Middle Stone Age heralded new techniques in producing flake tools for specialized equipment, so the Later Stone Age is characterized by revolutionary ideas. The most important was the emphasis on microliths—tiny implements such as blades, scrapers, points, and arrow-barbs; these were hafted and many of them formed part of composite tools and weapons. Lunates had already appeared in the Upper Kenya Capsian and the microlithic element was important in the Magosian culture of the Second Intermediate period, but essentially it is characteristic of the Later Stone Age. Arrows barbed with lunates, probably poisoned, must have been very effective weapons and made the hunter's life far less precarious. This allowed him more time to perfect other skills, such as the manufacture of pottery, stone bowls, beads, and ornaments. Microlithic scrapers and adze-blades, hafted in wood or bone, must have made fine wood-working possible. For the manufacture of microliths, homogeneous rock-like chert or obsidian was essential and must have been collected over long distances; trading meant contacts with other groups and the exchange of new ideas and discoveries.

The regional specialization which started in Middle Stone Age times became even more marked during the Later Stone Age. The Sangoan of the forested regions of central and western Africa was succeeded by the Lupemban and then the Tshitolian; these cultures are particularly characteristic of the Congo, but appear just within our area, in Uganda. The Stillbay of the drier country of eastern Africa was followed by the Magosian and later by the Wilton in most parts of East Africa, extending right down to the Cape and continuing from about 5,000 B.C. into recent times. In Northern Rhodesia and in parts of Tanganyika, the Nachikufan is found in place of the Wilton; both are associated with rock paintings. During Later Stone Age times in South Africa two cultures existed side by side: the microlithic Wilton and the more macrolithic Smithfield; the latter has been recognized also in western Kenya. Finally, the Upper Kenya Capsian of the Rift Valley was followed by the Elmenteitan, equally restricted in distribution and succeeded by variants of the Stone Bowl culture, of which the earliest is the only evidence of a Neolithic way of life in East Africa before the coming of iron.

The cultures of the Second Intermediate period—between the Middle and Later Stone Age—started during the dry period which followed the Gamblian pluvial, perhaps about 10,000 years ago, and continued into the first post-pluvial wet phase. They were the Tshitolian in the west, the Magosian in the east.

A Lupembo/Tshitolian industry found in the 'O horizon' at Nsongezi, on the southern border of Uganda, was originally described as 'Upper Sangoan' (see pp. 37–38); it shows a gradual development from the Sangoan and the Lupemban at this site. In Angola, a Lupembo/Tshitolian industry was dated by carbon 14 to be about 11,200 years. In superficial loams overlying the 'O horizon' at Nsongezi, and on the surface, is a Tshitolian industry which includes pottery, grindstones, slender bifaced lance-heads, tranchets, and a few tanged arrow-heads very similar to those of the Aterian culture of North Africa.

The Magosian was first discovered at a water-hole called Magosi in Karamoja, eastern Uganda;[1] since then it has been recognized at many sites in East Africa, the Rhodesias, and as far south as the Cape. The Magosians seem to have lived mostly on open sites rather than in shelters, so that the position of their culture in the sequence was not known until it was found stratified between the Stillbay and the Wilton at Apis Rock in northern Tanganyika.[2]

Magosian tools are similar to those of the preceding Stillbay, but are generally more microlithic. They include backed-blades, points, lunates, burins, and end-scrapers. Bored stones and mullers were also made, anticipating cultures of the Later Stone Age. Beads of ostrich eggshell were found at Apis Rock, as well as pottery.

Perhaps about 7,000 years ago the Magosian of eastern and southern Africa was succeeded by regional variants of the Wilton and Nachikufan cultures. The Wilton is, on the whole, more characteristic of the drier, open country, while the Nachikufan is found in the savannah and woodlands of Northern Rhodesia and in parts of Tanganyika. Both cultures are

[1] E. J. Wayland and M. C. Burkitt, 'The Magosian Culture of Uganda', *J. Roy. Anthrop. Inst.*, 62, 1932, pp. 369–90.
[2] L. S. B. Leakey, *Stone Age Africa*, pp. 64–67.

essentially microlithic; the Nachikufan is characterized by trapezes and U-shaped forms, while the most typical microliths of the Wilton are the tiny thumb-nail scrapers which were probably used as adze-blades for wood-working or for dressing leather.

A Nachikufan I industry at the Chifubwa Stream shelter on the Zambezi/Congo watershed was dated by carbon 14 to about 4,350 B.C.[1] At the type site, Nachikufu Caves in Northern Rhodesia, three Nachikufan industries known as I, II, and III were found stratified in a large painted rock shelter. Excavations made in Kisese rock shelter in Tanganyika (see p. 54) have revealed a Nachikufan industry associated with rock paintings, overlying a Magosian industry.

The Wilton culture in East Africa has been divided into three phases, 'A', 'B', and 'C';[2] these phases were partly contemporary and are distinguished mainly on account of the different ways of life of the people concerned. Wilton 'A' people lived in the open, Wilton 'B' people mostly in rock shelters, and Wilton 'C' people on the shores of lakes. Their equipment differed little from that of the Magosians, though the microlithic element was more pronounced.

Wilton 'A' is known from many scattered sites in Kenya and northern Tanganyika and has not been found in deposits later than the first post-pluvial wet phase. Wilton 'B' is a direct derivative of the Magosian, as seen for instance at Apis Rock; it has been found in several rock shelters near Nairobi and also in western Uganda. The tools associated with both these so-called 'phases' are very similar, with thumb-nail scrapers and lunates predominating; possibly the open sites and rock shelters were seasonal camps of the same people.

Wilton 'C' is found with huge middens of shell-fish beside the lakes of East and Central Africa, as well as on the sea coast of South Africa. Many skeletons were associated with shell mounds near Kanam, on the Kavirondo Gulf of Lake Victoria. They were heavily built individuals with large, dolichocephalic skulls and small faces. The skulls are similar to those of the Bushmen who may, in fact, be descended from these people. The teeth, even of the very old, are hardly worn at all, presumably owing to the diet of shell fish which needed little chewing.

[1] Clark, *Prehistory of Southern Africa*, p. 188.   [2] Leakey, *Stone Age Africa*, p. 68.

The makers of the Wilton culture are renowned throughout southern Africa as artists who decorated the walls of rock shelters. They overlap in places with the Smithfield people, who also engraved and painted rocks; some of their industries, which are centred in the basins of the Vaal and Upper Orange rivers, are connected with the Bushmen. North of the Transvaal the only known industries of the Smithfield type come from rock shelters near Beit Bridge and far away at sites in western Kenya. This 'Kavirondo Smithfield' was discovered by the late Archdeacon Owen, who found so many Sangoan and Lupemban (as they would now be called) industries in this most interesting area between the characteristic cultures of the Kenya Rift Valley to the east and the very different ones of Uganda to the west.

The very localized Upper Kenya Capsian of the Rift Valley was followed by the Elmenteitan which was apparently even more restricted in its distribution. It started early in the first post-pluvial wet phase, a time when there were many volcanic eruptions in the Rift Valley which covered the old lake beds in ash. In time, the ash solidified and the Makalia river, which flows into Lake Nakuru, cut a gorge through this rock. The Elmenteitans buried their dead along the sides of the gorge, placing the bodies in holes in the rock and piling up a few large stones in front of them. Twenty-eight skeletons were found at Bromhead's site in the Makalia valley. Two racial types were present: one tall, with dolichocephalic skulls and long faces, like the Upper Kenya Capsians; the other shorter, with rounder heads and broader faces. These two types continued in this part of East Africa throughout the Later Stone Age.

The equipment of the Elmenteitans included long, two-edged blades, which replaced the backed-blades of the Kenya Capsians; in some cases the thick butt had been trimmed away, presumably to facilitate hafting. Other tools are similar to those of the Upper Kenya Capsians, though burins are rare and the lunates are more symmetrical. In the top occupation level of Gamble's Cave many potsherds were found with the Elmenteitan industry. Some of the jars must have been very large, with a diameter of over one foot at the neck; other pots have holes drilled on either side, through which cord or sinew must have been threaded for carrying purposes.

The Neolithic

One of the most astonishing features of the Stone Age in Africa south of the Sahara was the almost total absence of a Neolithic way of life before the coming of iron. This was no doubt largely because much of Africa was cut off from outside influences by desert and by the sudd barrage on the Nile; because there were few rivers suitable for irrigating crops; and because tsetse fly and other pests made large areas impossible for keeping stock. It is quite possible, though, that vegetables and crops other than grain may have been cultivated long before there is any certain evidence that this was done.

In south-west Asia, the domestication of animals and cereal cultivation started before 7,000 B.C., for instance at Jericho. The new way of life had spread to the Fayum in Egypt by 4,500 B.C. and to Shaheinab in the Sudan by about 3,500 B.C.[1] At this site on the banks of the Nile the people lived mainly by hunting and fishing, but they had domesticated a dwarf goat similar to one known from Algeria.[2] This Neolithic culture followed the Khartoum Mesolithic[3] which may possibly be connected with the Upper Kenya Capsian: both have pottery, lunates for arrow-barbs, and bone harpoons for fishing. Whereas the Upper Kenya Capsians were of the Caucasoid race, however, the people of Mesolithic Khartoum were Negroids. Together with a roughly contemporary skull from Asselar, north-east of Timbuktu, these are the earliest recognizable Negroids in Africa. The great spread of the Negroids over Africa does not seem to have begun before the Iron Age.

The early 'Neolithic' of the Sahara, which occurred during a period of increased rainfall, has pottery and grindstones, which suggest a fairly settled way of life, but there is no certain evidence of food production until the time of the cattle paintings associated with the later Saharan Neolithic, towards the end of the 3rd millennium B.C.

Connexions between the Shaheinab Neolithic and Upper Egypt—Neolithic Fayum and predynastic Badari—are seen in the stone gouges and other carpenters' tools and in the pottery and stone mace-heads. Beads of imported green amazon stone both at Shaheinab and the Fayum suggest trade with

[1] W. F. Libby, 'Chicago Radiocarbon Dates III', *Science*, 116, 1952, pp. 673–81.
[2] A. J. Arkell, *Shaheinab*.   [3] A. J. Arkell, *Early Khartoum*.

Tibesti, where this stone is found. Gradually it seems that influences from Egypt and the Sudan—and through them the Sahara—reached East Africa some time after 3,500 B.C., the date established for Shaheinab by carbon 14.

We have seen that the Upper Kenya Capsians, and later the Elmenteitans, were settled beside the Rift Valley lakes making pottery and burying their dead with some ceremony, although there is no evidence of food production. The people of Mesolithic Khartoum are comparable; they too were probably dependent more on fishing than hunting and so could remain in one place for longer periods. It is not surprising that the Upper Kenya Capsians and Elmenteitans as well as the Khartoum Mesolithic people should have been followed by still more settled communities which apparently had a true Neolithic way of life.

The earliest makers of the Stone Bowl culture, known mainly from Hyrax Hill in the Rift Valley of Kenya seem to have been exceptional not only in East Africa but in Africa south of the Sahara as a whole. In other areas, the Nachikufans and Wiltons never changed their way of life until the coming of iron. Although the Nachikufans and Wiltons were primarily hunters, their bored stones were presumably weights for digging sticks, which indicates that they collected wild roots of some kind even if they did not actually cultivate vegetables. Strangely enough these bored stones, which are so common farther south, are very rare in East Africa, though one was found at Magosi. Polished stone axes, which suggest forest clearance for agriculture but may have been used mainly to obtain timber for huts and other purposes, have been found on the surface at widely scattered sites in Kenya and Tanganyika but were not associated with any particular culture. Possibly they were made by the Stone Bowl people, who were called 'Neolithic' mainly because they lived in villages, which presupposes some form of food production. Bones of animals which may have been domesticated have been found and the characteristic stone bowls made by these people may have been used for grinding grain or vegetable food of some kind. But in the early stages the grain may have been wild rather than cultivated and the bowls may also have been used for cooking purposes or for grinding ochre.

The Stone Bowl culture has been divided into four variants: the Hyrax Hill, Gumban 'A', Gumban 'B', and Njoro River Cave variants. The earliest of these, the Hyrax Hill variant, was found at a living site and cemetery on the south-eastern side of a rocky hill of that name a few miles east of Nakuru.[1] The living site lies directly above a terrace left by Lake Nakuru at a time when it stood 335 feet above its present level. During the first post-pluvial wet phase, the highest level of the lake was 375 feet. At the time of the Hyrax Hill variant the level of the lake was falling; the succeeding dry phase started about 2,500 B.C. and the living site was presumed by its excavators to have been occupied before this date.

Parts of eighteen skeletons were found in the cemetery. The female burials were all accompanied by stone platters, which are shallow and about 14 inches in diameter. In two cases there were traces of burning on the inside, suggesting that the platters might have been used for food offerings at the burial ceremonies. Both the long-headed and the broad-headed types of skull, already present amongst the Elmenteitans, are represented at Hyrax Hill and indeed continued throughout the period of the Stone Bowl culture.

In the centre of the living site pebbles had been levelled to form a roughly cobbled floor, on which lay quantities of stone tools and potsherds. The stone equipment is very like that of the Upper Kenya Capsians, though more microlithic. The commonest form of pottery is the ovoid beaker, usually with a conical base and often decorated inside a wide band beneath the rim.

The Gumban 'A' seems to be a direct derivative from the earlier variant, though it is probably associated with iron. Burials accompany the industry at two sites in the Rift Valley: the Makalia burial site is situated on a flat-topped hill overlooking the valley of the Makalia river; Willey's kopje is on the northern slopes of Eburru mountain. The low, stone-covered burial mounds at the Makalia site are almost identical with those of the Hyrax Hill variant. Most of the pottery, too, is very similar, though a unique feature is an overall decoration

[1] M. D. Leakey, 'Report on the Excavations at Hyrax Hill, Nakuru, Kenya Colony, 1937–1938' (with contributions by L. S. B. Leakey and P. M. Game), *Transactions of the Royal Society of South Africa*, 30, 1945, pp. 271–409.

looking like basket-work on the outside of the pot and deep, irregular lines scratched on the inside.

At Willey's kopje there were three skeletons, all old males. They had lost most of their teeth before death but, from the way in which the bone had healed, it is apparent that the two central incisors in the lower jaw had been removed in early youth—a custom which still prevails amongst some African tribes today. The skulls are considerably longer and narrower than those of any of the present races of Africa and the nose must have been prominent.

The Gumban 'A' variant was dated by its discoverers to the second post-pluvial wet phase, believed to have started about 850 B.C., but it is probably much later. Traces of iron and beads of apparently Egyptian origin are reported to have been associated with Gumban 'A' industries.[1]

The Gumban 'B' variant, certainly Iron Age, was found at the Nakuru burial site; on the north-eastern slopes of Hyrax Hill; near Nanyuki on the slopes of Mount Kenya; and in the huge crater of Ngorongoro in northern Tanganyika. The Gumban 'B' people had a most peculiar method of burial. A rather similar industry from Lanet near Hyrax Hill has been dated by carbon 14 to about A.D. 1600. At the Nakuru burial site eight skeletons had been thrown amongst a pile of rocks lying against the face of a cliff; but at the bottom was one complete skeleton which had been buried carefully and covered with red ochre. Presumably this was a chief or some important person and the dismembered skeletons may have been wives —or slaves—killed at the time of his burial. Strangely enough, though, another burial excavated near by was that of a youth of about fifteen years of age and it is difficult to believe that he was a person of importance; his knees had been tied up close to the forehead so that the whole body, from top of the head to the feet, was compressed into a length of under two feet.[2]

By the time of the north-east village at Hyrax Hill, trade with the coast had been established: beads found at this site are similar to ones from Zanzibar and Pemba. Thirteen pit dwellings in this village are particularly interesting because

[1] L. S. B. Leakey, *Stone Age Africa*, p. 70.
[2] Cole, *Prehistory of East Africa*, p. 108.

they are paralleled by those made by certain East African people in modern times, for instance the tribe traditionally known as the Gumba who lived in the parts of Kenya occupied by the Kikuyu when they arrived about 500 years ago. The Iraqw of northern Tanganyika still live in pit dwellings today.

Stone bowls were not found with the Gumban 'B' variant at the north-east village of Hyrax Hill, though they are common at the Nakuru burial site. These bowls are shaped like a pudding basin and are much deeper than the platters of the early Hyrax Hill variant. An unfinished bowl shows that the method of manufacture was first to peck round the edge, leaving a central core, and then, when the surrounding trench was deep enough, the core must have been knocked off with a sharp blow.

Apart from the stone bowls, the presence of pestles and mortars and saddle-querns makes it fairly certain that the Gumban 'B' people ground grain of some kind. Bones of domestic cattle and sheep were found in rubbish heaps outside the pit dwellings at Hyrax Hill.

The fourth variant of the Stone Bowl culture was named after Njoro River Cave[1] and has been dated by carbon 14 to about 2,920 years ago, or 960 B.C.[2] Such an early date is rather surprising for the following reasons. The large rock shelter in which the industry was found, about 12 miles west of Lake Nakuru, was cut into volcanic ash laid down before the first post-pluvial wet phase; the river, therefore, must have hollowed out the cave during a succeeding wet phase. The floor of the shelter lies only 5 feet above the present level of the river and, it is supposed, would have been flooded when the second wet phase was at its maximum; this would have removed any archaeological content, which suggests that the shelter was occupied after the second wet phase reached its height in about 850 B.C. The margin of error quoted for the carbon-14 dating is 80 years plus or minus; the youngest date, therefore, would be 2,920 minus 80 years before the present, or about 880 B.C. There are three possible ways of explaining this anomaly: the interpretation of events may be wrong and the cave may have been occupied before the maximum of the Nakuran wet phase

[1] M. D. Leakey and L. S. B. Leakey, *Excavations at the Njoro River Cave*, 1950.
[2] G. W. Barendsen, E. S. Deevey, and J. L. Gralenski, 'Yale Natural Radiocarbon Measurements III', *Science*, 126, 1957, p. 908.

when it may, in fact, not have been flooded; the carbon-14 date may be wrong through contamination of the samples or other errors; or the second post-pluvial wet phase may have started earlier than has been supposed.

The Njoro River Cave people practised cremation, a custom unknown to any of the present tribes of East Africa, though the Gisu of Mount Elgon occasionally burnt the bones of their dead after ritual cannibalism. The process of closed combustion turned many perishable objects into charcoal and thus preserved them. These soft materials include skin clothing around the bodies, string bags woven from vegetable fibres, plaited cords, gourds, and a carved wooden vessel decorated with an all-over honeycomb pattern.

About eighty individuals were cremated and each adult—whether male or female—was accompanied by a stone bowl, pestle, and mortar. The bodies were tied in the contracted position with cord, covered with red ochre, and then placed in shallow graves. The graves were covered over with soil and a fire was lit on top of them. In some cases necklaces and other ornaments had been left on the bodies. Over 800 beads were collected in the rock shelter, some in groups forming complete necklaces. The beads were made of materials mostly found locally, such as agate and chalcedony. In some necklaces sedge-seed beads were threaded between the stone beads. Although many modern African people use seeds of various kinds for necklaces, the only tribe known to use sedge-seeds are the Gisu. Bone pendants were also found, some decorated with lines, notches, or perforations, and there were a few cylindrical bone beads and bone awls.

Both the obsidian tools and the pottery of the Njoro River Cave variant are similar to those of the Elmenteitan culture. Some of the pots are shallow, with flat bases, while others are rounded with pointed bases—suitable for standing between three stones. One large pot has three holes drilled after baking: presumably cords were threaded through the holes to make the pot easier to carry.

The skulls of these people range from ultra-dolichocephalic to brachycephalic and fall into two distinct groups, as with the Elmenteitans. The custom of removing the two lower central incisors was practised. These individuals have no Negroid characteristics; in fact, both the occipital and nasal indices are quite unlike those of the present inhabitants of Kenya. They were of the Caucasoid race and may have resembled the Hamitic people of the Horn of Africa.

Rock Paintings

The rock paintings of Tanganyika, though less well known than those of South Africa, the Rhodesias, and the Sahara, are often of a high quality. The earliest probably date back to Later Stone Age times and may have been the work of the Nachikufans, while the 'late white' paintings are quite recent.

Most of the paintings are centred on Kondoa, about half-way between Dodoma and Arusha, where several hundred sites are known to exist. At least sixteen styles of Stone Age times are said to have been distinguished,[1] though this need not necessarily imply great intervals of time between each, and allowance should be made for the individuality of different artists. The earliest paintings of all are invisible unless the silica film which covers them is sprayed with water, when it becomes temporarily transparent.

The subjects are very commonly herbivorous—edible—animals, such as antelopes of all kinds, elephants, giraffes, rhinoceroses, and also ostriches; this is only to be expected, since the hunters who painted them were absorbed in the problem of filling the larder. In some styles the animals are naturalistic and extremely well drawn, though there are no true polychromes as in South Africa and Southern Rhodesia, while in other styles they are crude and stylized. The human figure is sometimes portrayed but is always stylized. Generally the animals are single figures, superimposed one on top of the other, though occasionally they are arranged in groups or friezes.

The two most famous painted shelters are Cheke and Kisese, about twenty and thirty miles north of Kondoa respectively. At Cheke there is a frieze of elands and giraffes and an elephant caught in a trap surrounded by dancing human figures, one of whom seems to be doing a handstand for joy at the outcome of the chase. At Kisese a remarkably well-drawn female

[1] H. A. Fosbrooke, *et al.*, 'Tanganyika Rock Paintings', *Tang. Notes*, 29, 1950, p. 17.

rhinoceros pursues a male. A delightful scene at Kolo, south of Cheke, has been called 'the abduction': a woman is held by two men by one arm, while two masked figures try to drag her away by the other.

Excavations at Kisese[1] revealed a microlithic industry believed to be Nachikufan overlying a Magosian industry. A slab of rock had fallen from the side of the shelter after the paintings were done and was found buried 2 feet beneath the surface; it was associated with late Nachikufan tools immediately below the layer containing the first pieces of iron.

Engravings are far less common in East Africa than they are farther south. Some have been found in Tanganyika and Uganda, but are probably not more than a few hundred years old; they consist of schematic lines and patterns which in some cases are said to have been associated with initiation ceremonies and there are also cup marks used in the game of *bau*.

Although links between rock art and the modern East African tribes are tenuous, some of it may be attributed to the ancestors of the Tindiga, or Hadzapi, and Sandawe people who wander between the regions of Lake Eyasi and Kondoa. Their languages have clicks, like the Bushmen and Hottentots. The Hadza culture is based on hunting and food-collecting like that of the Bushmen, while the Sandawe are mainly pastoral, like the Hottentots. Skeletons found buried beneath large cairns at Naberera in northern Tanganyika have been compared with the 'proto-Hottentot' Iron Age burials at Mapungubwe in the Transvaal.[2] It seems probable that people of Bushman and Hottentot stock once occupied much of Africa, until they were surrounded by the Bantu.

Summary and Conclusions

The Earlier Stone Age probably lasted from over a million to 50,000 years ago. The first tools were pebbles chipped to form a cutting and chopping edge and were made by the Australopithecines. The skull of one of these early hominids was found at Olduvai Gorge, Tanganyika, in 1959. The pebble culture

[1] R. Inskeep, 'The Age of the Kondoa Rock Paintings', *Actes du IVe Congrès Panafricain de Préhistoire (Leopoldville, 1959)* ed. G. Mortelmans, 1962, pp. 249–56.
[2] L. Fouché (ed.), *Mapungubwe. Ancient Bantu Civilisations on the Limpopo*, Cambridge, 1937.

gradually evolved into the Chelles–Acheul hand-axe culture, whose whole sequence can be studied from the series of deposits at Olduvai.

The first skull of a hand-axe maker was found at Olduvai in 1960. Jaws of the later *Atlanthropus* associated with hand-axes in Algeria are very robust and suggest that he was also of the *Pithecanthropus* type. This stage in the development of the hominids follows logically from the Australopithecines and is, in turn, obviously ancestral to Rhodesian Man. The crania of *Homo sapiens* type associated with hand-axes at Kanjera are an apparent anomaly not yet explained.

Whereas in Earlier Stone Age times the hand-axe culture was ubiquitous, from Middle Stone Age times onwards regional specialization prevailed. Smaller implements made on flakes—many of which must have been hafted—were differentiated into hunting weapons, wood-working and leather-working tools, according to different ecological conditions. The Sangoan–Lupemban–Tshitolian sequence of the forested parts of central Africa, with Uganda coming in to the eastern boundary of this region, shows an emphasis on wood-working. The Stillbay–Magosian–Wilton and Nachikufan people of the drier parts of eastern Africa were nomadic hunters who did not change their way of life until the coming of iron; the Wiltons and Nachikufans were responsible for the earliest known rock paintings, in Later Stone Age times.

In the early part of the Middle Stone Age, the specialized 'Rhodesioids' known from Broken Hill, Northern Rhodesia, Saldanha, Cape Province, and from Eyasi, northern Tanganyika, were the African equivalents of Neanderthal man. A 'proto-Bushmanoid' stock may also have appeared early, represented perhaps by the Kanjera skulls, and certainly by numerous Middle Stone Age remains in South Africa. The Caucasoid stock is first represented in East Africa by the Upper Kenya Capsian people of Gamble's Cave and it continued to be important—and perhaps unique—in the Rift Valley until the coming of iron. Characteristic Negroids first appear with the Khartoum Mesolithic but are unknown in East Africa before the Iron Age.

The localized Upper Kenya Capsian of the Rift Valley, the date of which is still uncertain, was succeeded by the

Elementeitan and then by four variants of the Stone Bowl culture. The earliest of these variants is evidence of a Neolithic way of life in East Africa and seems to have been exceptional in Africa south of the Sahara as a whole. The Stone Bowl culture continued throughout the Iron Age until at least the sixteenth century A.D.

III

THE PEOPLING OF THE INTERIOR OF EAST AFRICA BY ITS MODERN INHABITANTS

G. W. B. HUNTINGFORD

THE movements of ancient peoples cannot be contained within the boundaries of modern political territories. The study of such movements in East Africa cannot be confined to the British sphere of administration, and we shall be constrained to include the whole of north-east Africa in our survey. The evidence on which our conclusions and inferences are based is slight indeed, and the origins of the East African ethnic groups go back far beyond the range of tradition. Even the classification of the modern peoples is complicated by the existence of an unusually clear-cut linguistic grouping which, though it does happen to correspond substantially with the ethnic grouping, has proved more of a hindrance than a help in dealing with the so-called Bantu peoples, who, viewed as a single group, are in the main connected by speech alone.

Blood-grouping and the incidence of the sickle-cell gene should prove helpful at a later stage, when research has gone much farther, since on the whole the material at present available neither covers a wide enough area nor provides sufficiently detailed information, for the samples are largely random, and in the literature the distinctions of 'tribe' are somewhat uncertain. The value of the evidence which blood-groups and sickle-cells may be able to provide when the material is much more abundant and methodical will lie in its application as a test of traditional history. The sickle-shaped red blood-cell is normal in Negroes. African ethnic groups which have mixed with Negroes, however, show traces of sickling which vary according to the amount of Negro blood they have. Enough preliminary work has been done in Uganda to enable the broad conclusion to be drawn that there the Nilotic peoples have the

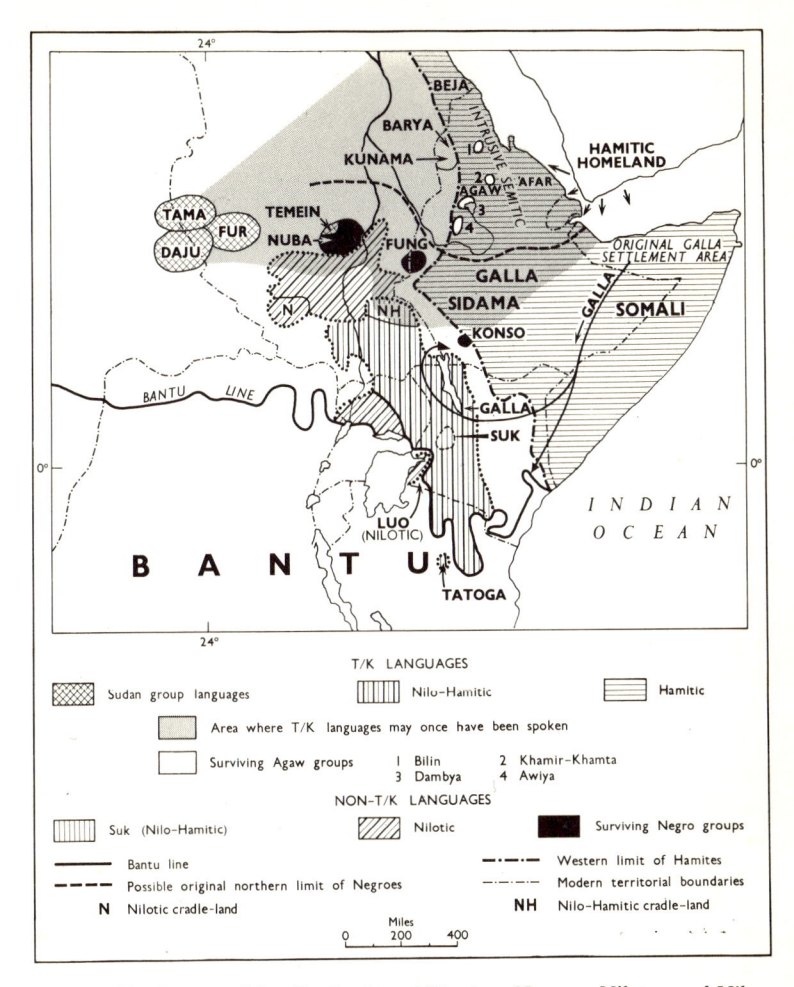

MAP 4. Sketch map of the distribution of Hamites, Negroes, Nilotes, and Nilo-Hamites and of the T/K languages, in East Africa

highest incidence of sickling, the 'Bantu' the next highest, and the Nilo-Hamites the least,[1] a result which suggests that the 'Bantu' and Nilo-Hamites have a larger non-Negro element in them than the Nilotes, and confirms the grouping made on the basis of other evidence.

On the anthropometric side, again, the material is quite inadequate, but we may perhaps generalize to the extent of saying that the Hamites are mesocephalic and mesorrhine, the Nilotes dolichocephalic and platyrrhine, and the 'Bantu' both dolichocephalic and mesocephalic as regards head-shape and largely platyrrhine as regards the nose.[2]

The first attempt at classifying the peoples of north-east Africa seems to have been made by the grammarian Agatharchides of Cnidus,[3] who wrote: 'To the south of Egypt are four great races of mankind. First are the riverain peoples who sow millet and sesamum. Second are the marsh-dwellers, who live on water plants and tender shoots. Third, the people who wander here and there without plan, feeding on meat and milk. Fourth are those who live by fishing on the sea-shore.' The third group is later defined by the name Trogodyte. This is a remarkably acute classification for so early a period. 'Riverain peoples' clearly suggests people like those of the Nile Valley; 'marsh-dwellers' calls to mind the peoples of Watta type;[4] the Trogodytes are, I believe, Hamitic pastoralists, while the fishers on the sea-shore suggest people like the Strandloopers of South Africa. According to some modern ethnographers the present inhabitants of East Africa may be divided into five main ethnic groups, Bushmen, Symbiotic Hunters, Negroes, Hamites, and the Nile Valley Peoples. In addition to these the Bantu peoples, in whom are probably Bushman, Negro, Hamitic, and other elements, form a broad linguistic grouping with many cultural and indeed ethnic divisions. In fact, the word

---

[1] H. Lehmann, 'Distribution of the Sickle Cell Gene', *Eugenics Review*, 46, 1954, pp. 3 sqq.

[2] These terms refer to the cephalic (head) and nasal indices, obtained by multiplying the breadth by 100 and dividing the result by the length. Heads are classed as dolichocephalic or long when the index is below 75, as mesocephalic or medium when it is between 75 and 80, and as brachycephalic or short when it is over 80. Similarly, leptorrhine or narrow noses have an index of 55 to 70, mesorrhine or medium an index of 71 to 85, and platyrrhine or broad of 86 to 100.

[3] *Fl.* 130 B.C.; in C. Müller, *Geographi Graeci Minores*, 1855, i. 111–95.

[4] For these see p. 62.

'Bantu' really refers to language alone, and its use in connexion with physical and cultural features is incorrect, though common because it is convenient.

It is possible that the oldest of the surviving groups is that of the Bushmen. The most recent conclusions[1] are that while their physical features set them 'apart from all other peoples, except some Hottentots', their ancestors are to be sought in south-central Africa, the more northerly extensions of Bushman types being later rather than earlier. Of actual Bushmen in East Africa, the only survivors seem to be the Hadzapi of Lake Eyasi in Tanganyika (called Tindega, Kindiga, and Kangeju by their neighbours), now numbering only a few hundred souls. They live entirely by hunting and collecting, grow no crops, and have no animals but dogs. They represent a cultural level even simpler than that of Agatharchides' marsh-dwellers. It is to be noted that certain physical features usually associated with Bushmen do not seem to occur among them: there is only a slight tendency to steatopygia, and the skin is much darker, though these differences may be due to admixture of Negroid blood.[2] Of organization, social or political, they seem to have practically nothing, and their religion so far as our information goes is very rudimentary. The language of the Hadzapi[3] may be related to that of the Central Bushmen of South Africa,[4] and contains clicks.

About seventy miles south of Lake Eyasi and the area over which the Hadzapi wander are the Sandawe, who speak a click language related to Hottentot. Though much in advance of the Hadzapi, with an acephalous[5] organization of clans, they do not seem to have reached the tribal stage till the coming of the European, and then only under his guidance; and until quite recently were far behind the other peoples of East Africa, the Hadzapi excepted, in general culture. Formerly hunters, fishers, and collectors, they took to agriculture, using a long-

---

[1] P. V. Tobias, 'Bushmen of the Kalahari', *Man*, 1957, p. 36.
[2] C. G. Seligman, *Races of Africa*, 1957, p. 16. He actually wrote 'Bantu'.
[3] D. F. Bleek, 'The Hadzapi or Watindega of Tanganyika Territory', *Africa*, 4, 1931, pp. 273-86.
[4] 'The validity of the classification of Hadza [singular of Hadzapi] with Bushman is by no means established' (E. O. J. Westphal in A. N. Tucker and M. A. Bryan, *The Non-Bantu Languages of North-eastern Africa*, London, 1956, p. 170).
[5] i.e. without a chief or head.

handled hoe, and cattle-keeping in comparatively recent times.[1] There is no doubt that both physically and linguistically they are related to the Hottentots, and are not merely analogous to them. The establishment of this relationship suggests at first sight that they might be a later northern extension of the Hottentots, just as the Hadzapi seem to be a later northern extension of the Bushmen of South Africa. But we do not yet know enough about them to feel certain about this aspect of the problem, and the process may in fact have been the reverse, as suggested by Seligman.[2] It should be added that the pastoralism of the Hottentots *seems* to be older and more deep-seated than that of the Sandawe, which suggests a fairly recent borrowing from possibly Nilo-Hamitic sources, for there are cultural differences that may be significant[3] and tend to support such a view. The Sandawe, unlike the Hadzapi, are receptive of outside culture elements, and they have changed a good deal from the rather primitive people they were forty or fifty years ago.

Somewhat different from the Bushmen are the Symbiotic Hunters, scattered survivors of hunting peoples who occupied much of East Africa before the present inhabitants arrived. There are at least four main groups of them: the Watta or Wayto of Ethiopia, the Sanye or Ariangulo of the East Coast hinterland, the Midgan of Somaliland, and the Dorobo of Kenya and Tanganyika. Though hunters and collectors, they (especially the Dorobo) differ from the Bushmen firstly in that they are forest people, while the Bushmen are people of the open or bush country; and secondly in that they are able and willing to acquire elements from higher cultures with which they come in contact—so much so that in many cases they have lost their own languages and adopted those of the people with whom they are in closest contact, as well as much of their culture. Thus, the Watta speak Amharic, Galla, and Kafa, though the group in Kafa called Manjo still possess a language of their own; the Sanye speak Galla, while the Midgan besides speaking Somali have also their own language. The Dorobo all speak forms of Nandi.

[1] J. C. Trevor, 'The Physical Characters of the Sandawe', *J. Roy. Anthrop. Inst.*, 77, 1947, pp. 61–78.
[2] Seligman, *Races of Africa*, p. 23.
[3] e.g. among the Sandawe, men milked, but among the Hottentots—if we may generalize from the Nama—only women.

A significant characteristic of these people is that they have adopted so much of the cultures of their neighbours that except for the hunting aspect of their life they are often outwardly indistinguishable from them. They are also largely dependent upon these neighbours for certain things (like ironwork, cereals, and so forth) which they do not produce themselves; and to this extent they are quite properly termed 'symbiotic.'[1] The Bushmen, on the other hand, have shown themselves unable or unwilling to modify their own very primitive culture in any respect, and they are not symbiotic.

In the traditions of their neighbours these hunters are usually regarded as the original inhabitants of the country: 'they were the original lords and free peoples' of Ethiopia;[2] while to the Nandi and Masai the Dorobo were the original human inhabitants of the world who were dispossessed by the incoming Nandi and Masai.[3] Though on the whole shorter than their neighbours, they are in no way pygmy;[4] and while they have without doubt a foreign strain in their make-up, they are not to be classed simply as 'modified Hamites' or 'modified Nilo-Hamites' as some have been inclined to think. This is strongly suggested by both their measurable and non-measurable characters. Nor are there any grounds for supposing[5] that they belong to the same stock as the Hadzapi. Possibly we may regard them as survivals of one of the early stocks which have contributed to the formation of the rest of the modern peoples of East Africa.

The Negro group, though it has almost everywhere left its mark on the peoples of East Africa to a greater or lesser extent, does not exist in any sort of 'purity' in East Africa, and the surviving groups which are regarded as being of Negro stock—the so-called Nuba of Kordofan and certain of the inhabitants of Dar Fung and others opposite them across the Nile on the west of Ethiopia who have been called Shangalla by the

---

[1] Seligman, *Races of Africa*, p. 111.
[2] G. Massaja, *I miei trentacinque anni di missione nell'alta Etiopia*, 1885, viii. 9.
[3] G. W. B. Huntingford, 'The Social Institutions of the Dorobo', *Anthropos*, 46, i–ii, 1951, p. 4.
[4] The average stature of the Nandi is 1,700 mm., and of the Dorobo 1,660 mm. G. W. B. Huntingford, *The Northern Nilo-Hamites*, 1953, p. 17.
[5] As does K. H. Honea, *A Contribution to the History of the Hamitic Peoples of Africa*, 1958, p. 9.

Ethiopians from at least as early as the fifteenth century—are certainly not 'pure' types, though they show some of the main features of what may be called Negro culture.[1] Typically, East African Negro culture is marked, as among the Nuba, by a concentration on agriculture,[2] the use of irrigation, skill in pottery, scanty clothing but much cicatrization, spirit-possession of prophets, and the absence of both cattle and circumcision. Physically the Nuba are mesocephalic, though with a fair amount of dolichocephaly and even some brachycephaly, and they are hyperplatyrrhine (very broad-nosed). In Dar Fung a form of kingship occurs, with ceremonial king-killing in some places. Like Dar Nuba, Dar Fung contains a number of unrelated languages.

The origin of the Negro is far from certain. Some have supposed that he originated in Asia, and it might be said that the occurrence of sickle-cells in India supports this view.[3] But the distinctive Negro type, with its prognathism, bridgeless nose, and thick and everted lips, separates it physically from other races, and it is likely (though perhaps unprovable) that the Negro originated somewhere in tropical Africa, possibly in the Lake Region of East Africa, if the sickle-cell trait is any guide, for it is here that the highest recorded incidence occurs.[4] The earliest historical evidence for the existence of Negroes in north-east Africa comes from Egyptian slate palettes dating from about 3000 B.C., while burials of people of Negro type from Khartoum take them back to about 10,000 B.C. But the Negro may have been there long before this, if the skull from Asselar, some 220 miles north-east of Timbuktu, which has marked Negro or Negroid characters, can be considered to be of Upper Pleistocene (i.e. 'Palaeolithic') date.

The Hamites, also known as Cushites,[5] form a group, broadly

[1] The rest of the dark-skinned, woolly-haired peoples of East Africa, who owe these features to the Negro element in their ancestry, are 'Negroid', which is not the same thing as 'Negro'.
[2] Though without the plough, which in pre-European Africa was confined to the Hamites.
[3] Lehmann, *Eugenics Review*, 1954, p. 12; R. N. Shukla and B. R. Solanki, 'Sickle-cell Trait in Central India', *Lancet*, 8 Feb. 1958, pp. 297-8.
[4] Lehmann, *Eugenics Review*, 1954, fig. 11.
[5] Though the term Hamitic has been in use for many years, some linguists, notably Greenberg (J. H. Greenberg, *Studies in African Linguistic Classification*, reprinted from *S.W. Anthrop.*, New Haven, 1955, pp. 51–55), now object to its use.

homogeneous in language and basic culture, spread over a large part of north-east Africa in Ethiopia and Somaliland. It has been suggested that their ancestral stock came from Arabia a long time ago—if Honea is right, they began to appear in Africa in Upper Palaeolithic times. In any case they are of considerable antiquity in East Africa, and quite distinct from Bushmen, Hunters, and Negroes, being Europoids without any Negroid physical characters except those acquired since their arrival in Africa through contact with Negroes or Negroids.

In his recent study Honea has included as Hamites not only those now living in north-east Africa—the Beja, Agaw, Galla-Somali, and Sidama groups—but the Stone Age peoples as well.[1] This admission of Stone Age peoples to the ranks of the Hamites may be doubted, and another scheme has been suggested[2] in which three main waves of Eastern Hamitic immigrants are distinguished. The first wave consists of Proto-Hamites (A-Hamites), perhaps represented today by the Barya of northern Ethiopia, whose language has many Hamitic characteristics, and possibly also by the near-by Kunama, whose language, however, shows few if any Hamitic traces.[3] The second wave (B-Hamites) is represented by the Beja, Agaw, and Sidama. The Beja seem to be free from Negro physical influence, and it is probable that the Negroes never penetrated their country. The same may apply also to the Agaw. But with the Sidama of south-western Ethiopia the case is different, for they have certain Negroid characters which suggest that there were Negroes in the area when they arrived with whom they mixed

One of the difficulties raised by the objectors is that the Semitic and Hamitic languages belong to the same family, and a single term is therefore needed to cover both; and some, e.g. Greenberg, now deny that the non-Semitic languages ('Hamitic') of this great family form a 'linguistic unity' as against Semitic. In recent years 'Hamito-Semitic' has been used as a convenient, if awkward, term, and it has been suggested lately that this should be replaced by 'Hamitic', though Greenberg would prefer 'Afro-asiatic', with the complete elimination of 'Hamitic'. It is odd that 'Cushitic' should have been chosen as a substitute, for historically 'Cush' belongs to the ancient Egyptian province of Nubia which eventually became the kingdom of Meroe, and the word is used in Ethiopic (Geʻez) inscriptions with reference to people of this region in the form *Kasu*.

[1] Honea, *Hamitic Peoples of Africa*, pl. i.
[2] C. F. Beckingham and G. W. B. Huntingford, *Some Records of Ethiopia, 1593–1646*, 1954, pp. 1–11.
[3] But they could still be Hamites speaking a non-Hamitic language.

to a limited extent. The third wave (C-Hamites) comprises the Galla, who, though they are less Negroid than the Sidama, do show traces of Negroid influence. They were followed by the Somali and Afar. The Somali in their traditions consistently refer to the Galla as their immediate predecessors, but, like the Beja, they and the Afar lack Negro physical influence. The conclusions which may be drawn are that the Sidama, coming first, met Negroes and eventually drove them out or absorbed them. The Galla coming next may have encountered the last remnants of the Negroes still left in the area; though on the other hand it is possible, in spite of what has been suggested above, that Negroid characters in the Galla may have been acquired quite late, even after the sixteenth century. The last survivals of the Negro population are to be found in Dar Fung, Dar Nuba, and some other small groups scattered along the western edge of Ethiopia. The Somali, coming last, found the Horn of Africa occupied by the Galla, whom they gradually expelled,[1] preserving their Europoid physical type.[2] Most of the present-day peoples of East Africa have been formed through mixture of Bushman, Hamite, and Negro in varying proportions—how varied may be seen by contrasting such peoples as the Bantu Kavirondo, who might be described loosely if not accurately as 'typical Negroes', with the Hima of Uganda, whose hair and skin-colour indicate unmistakable Negro influence, while their facial characteristics are Hamitic.

It is to be noted that among the Bushman and Symbiotic Hunter groups the tribal stage, as we know it in East Africa, has not been reached, for the 'tribe' as a coherent social and political entity is the product of a somewhat higher and more settled stage, in which cattle and cultivation play a part. Moreover, the tribe as we find it in East Africa can be defined with precision: it is a group united by a common name in which

[1] Beckingham and Huntingford, *Records of Ethiopia*, pp. lxxi, lxxii.
[2] There is, however, another possibility. Some of the Stone Age peoples whose skulls are of non-Negroid type may conceivably have formed part of the ancestry of some of the earlier Hamitic peoples. This possibility might be admitted in the case of the Sidama group of the second wave, whom we *might* visualize as descended on one side from a Stone Age Hamitoid group—using this word without any implication of *relationship* with Hamites—which produced a Hamitoid people who became the Sidama. This would account for the divergence in some of the Sidama groups from the Hamitic norm. But that *all* those whom we call Hamites came from Stone Age Hamitoid stocks I do not believe.

the members take a pride, by a common language, by a common territory, and by a feeling that all who do not share this name are outsiders, 'enemies' in fact.[1] The Hamites perhaps introduced political as well as social organization on a wider scale. Not only does the tribe exist, but it appears as a member of a group which we may fairly call a nation. Thus the Galla, Somali, and Afar are not just three tribes, but three major groupings each recognizing its common or national name and consisting of many tribes, the number of Galla and Somali tribes running into three figures. Within the tribe, the political structure may reach a relatively high level, as among the Kafa; a lesser level (though complex) among the Galla; while it may be even rudimentary, as among the Somali, whose way of life has been largely determined by semi-desert conditions.

Certain broad generalizations may be made which show that there is a correlation between ethnic groups, language, occupation, political organization, and religion. I stress the fact that these are 'broad', because there are within each ethnic group exceptions to or modifications of the group pattern; but this pattern does seem to exist. Starting from the assumption that the general language pattern—the glossotype—of a group is broadly homogeneous (except in the Negro group, where linguistic allogeneity rather than homogeneity occurs), we can briefly summarize and contrast the culture types of the Hamites, Negroes, and Nile Valley Peoples as follows. In occupation the Hamites are basically pastoral, with a fairly well developed form of agriculture including the use of the plough. The Nile Valley Peoples are strongly pastoral, with hoe culture as a subsidiary. The Negroes are basically agricultural, with hoe culture, and few or no cattle. In the sphere of political organization, the Hamites are patrilineal, and acephalous units tend to be replaced by chiefdoms; some develop into kingdoms. Two-class social systems occur, and there are submerged classes.[2]

[1] Hence the absence of a word for 'tribe' from the vocabularies of many East African tribes; for while the tribal *idea* is strong, the *word* to define it is unnecessary: 'we' are the 'tribe', all other people are 'not of us' (cf. Huntingford, *The Northern Nilo-Hamites*, p. 14).

[2] Submerged classes are the sometimes servile but usually despised members of a tribe who practise menial occupations—ironworking, hunting, tanning, fishing—and are not regarded as being either of the same stock or of the same status as the rest of the tribe.

Among the Negroes, both patriliny and matriliny occur, but, as among the Hamites, acephalous units exist side by side with kingship, and the rain-maker is politically important; there are no submerged classes. The Nile Valley Peoples are patrilineal, with mainly acephalous units, though in some there is a divine king, and in others the rain-maker becomes the equivalent of a chief; there are submerged classes. The religion of the Hamites centres on a sky-god, with non-ancestral spirit cults, and sporadic spirit-possession. The Negroes, too, have spirit-cults, with a type of spirit-possession sometimes described as 'shamanism'. The Nile Valley Peoples have strong ancestor-cults associated with belief in a supreme being, but normally they lack the non-ancestral spirit cults and spirit-possession found in other groups.

Though we do not know how long ago Hamitic peoples began to enter north-east Africa, they were certainly established in the highlands of Ethiopia when the first Semitic immigrants from Yemen (who eventually founded the kingdom of Aksum) began to settle there about the seventh century B.C., if not earlier. Their culture-pattern was different from that of the Semitic peoples, and in those early times was probably of the Beduin type, involving nomadic pastoralism like that of the Beja and many Somali tribes. A second cultural stage was reached in which agriculture combined with livestock led to a more settled life, with an eventual development of trade, as among the Agaw of central Ethiopia and the Kafa of south-west Ethiopia.

The languages of these Hamites may be divided into four zones, Beja, Agaw, Galla-Somali, and Sidama, corresponding roughly to northern, central, eastern, and south-western Ethiopia respectively. Briefly, their words tend to consist of the pattern consonant-vowel-consonant, with often a formative prefix; there is also a common triliteral pattern (with three radical consonants) of the Semitic type. Plurals may be made by a suffix, or by change of vowel in the word itself. There may be an article, which sometimes but not always indicates gender as well. There are many significant formative elements, usually suffixes, which denote such ideas as 'man-*kind*', 'bright-*ness*', 'witch-*craft*', 'man-*hood*', the italicized elements representing suffixes in Hamitic. Gender, the grammatical distinction between masculine and feminine, is shown in varying degrees, and may be indicated by affixes, e.g. masc. K, fem. T; masc. O, fem. E; masc. A, fem. I; and may occur in nouns, pronouns, and verbs, the latter showing gender usually in the third person singular only. The tense is formed normally by a suffix to the stem which combines the notion of time and person. Negation is expressed, among other ways, by particles which are combined with the verb to form a negative conjugation.[1] The sentence order is normally subject—object—verb. An analysis of Hamitic vocabularies shows that there is a small number of words common to all four zones, though not found in every language of each zone: such words are those for eye, fowl, heart, I (ego), knee, tooth, two, twenty. It shows also that in vocabulary Agaw, Galla-Somali, and Sidama have more in common with each other than with Beja.

Greenberg[2] has denied that these languages form a linguistic unity as against Semitic. In actual fact, they form four groups each sufficiently homogeneous internally, yet contrasting externally with each other. But it is plain that while they present this inter-zone diversity, they

> yet possess a sufficiently strong over-all resemblance due to relationship which contrasts strongly with the Semitic languages of the Ethiopic group.... Of course all are members of the great Hamito-Semitic family; but to deny linguistic unity to such a group as the Ethiopian (north-eastern) Hamitic would lead to as many misconceptions as would arise if the derivatives of Latin among European languages were denied a linguistic unity as against the Slavonic or Teutonic languages. This is not a denial of *relationship*, which is not the same thing as linguistic unity.[3]

Greenberg objects also to a correlation between 'pastoral' life and the speaking of Hamitic languages.[4] His reasons, which have a 'racialist' flavour, are not very convincing, for he has apparently failed to realize the extent to which the Hamites are pastoral, and points out that it is the Nilotes and Nilo-Hamites who place the greatest cultural emphasis on cattle.

[1] The negative particle *ma*, which is found in Somali and Saho, occurs also in Nilo-Hamitic languages (Nandi, Masai, Teso).
[2] Greenberg, *S-W. J. Anthrop.* 1955, p. 54.
[3] G. W. B. Huntingford, 'The "Nilo-Hamitic" Languages', *S-W. J. Anthrop.* 12, 1956, p. 202.   [4] Greenberg, *S-W. J. Anthrop.* 1955, p. 54.

This today is quite true: but the Hamites have the ritual machinery for a 'cattle cult', which may have in the past provided a foundation for the intensity of the Nile Valley cultures, even if they have been agriculturalists for a long time and on a bigger scale than the Nile Valley Peoples.

Evidence for devotion of an East African pastoral people to cattle comes from as far back as the second century B.C., for Agatharchides wrote of the Trogodytes that they lived on blood mixed with milk, that they called those who slaughtered cattle for food 'unclean', and that they regarded the bull as their 'father' and the cow as their 'mother'; to these he adds the other relevant details that they practised circumcision, that they buried the dead in a contracted position to the accompaniment of laughter, and that the body was covered by a mound of stones on top of which was placed an antelope horn on a post.[1] These features occur today among Hamites and Nilo-Hamites. Moreover, observances to preserve the high ritual status of cattle and milk are as much fundamental to the Hamites as to the Nile Valley Peoples. Thus, among such Hamitic usages are the reluctance to count cattle; the use of cattle in sacrifices; the taboos on boiling milk, on putting it in any container not made of vegetable material (wood, fibre baskets, gourds), and on drinking milk immediately after milking; the high ritual value of cow-dung and grass; and the wearing of insignia to denote the ownership of large herds.[2]

These customs all stress the high pastoral status of the Hamites, even if the extreme features that occur among some of the Nile Valley Peoples are absent. Not all Hamites have remained 'Beduin'; most of them have adopted the plough, and among them some highly organized political systems came into

[1] Müller, *Geographi Graeci Minores*, p. 54.
[2] As regards cattle, it may be noted that the four linguistic zones show a linkage formed by the words for *cow*, *grass*, and *milk*. There are at least four Hamitic words for 'cow': *sa* and **la* in Zones 2 and 3, **la*, *kem*, and *miy-an* in Zone 2, and what seems at first sight to be an entirely different word, *mi-za*, in the western part of Zone 4, but is probably no more than a form of the word found in Agaw (Qemant) *miy-an* meaning 'young cow'. This suggests that the two earliest words may be *sa* and *kem*, and that **la* may be a later word, borrowed by the Agaw of Zone 2. These cattle-words are linked with the words for 'grass' and 'milk': in Zones 1 and 4 (eastern), 'milk' is *ado* and 'grass' **ays*: in Zone 2, 'milk' is *shab* and 'grass' *shanka*; in Zone 3, 'milk' is **an* and 'grass' **ays*; in Zone 4 (western), 'milk' is **matta*, though the words for 'grass' in this zone show no lexical correspondence with those in other zones.

being. For although, among the eastern Hamites,[1] the Somali and Afar have no very highly developed political organization, it was not from them that ideas of political growth spread westwards. The Galla had a complex system of age-sets and age-grades which regulated in each tribe or tribal group the periodical choice of a ruler who had legislative powers and was assisted by a number of officials. And among the Sidama[2] the Kafa kingdom is pre-eminent as a well-developed monarchy with a hierarchy of sub-kings, court officials, and provincial governors all revolving round the king. Other Sidama states like Janjero, Damot, Enarya, Hadya, Kambatta, and Wolamo—to name a few—were also ruled by kings, and had comparable if smaller and less complex political structures.

When the Semitic peoples arrived from the Yemen, they found the Ethiopian highlands occupied by the ancestors of the Agaw, the hot lowlands being inhabited, as now, by nomadic pastoralists of the same type as the modern Afar. The Semites probably brought with them the plough, which spread throughout the Hamitic area. Among other cultural features ascribed to the Hamites are the small round shield and the use of trousers, both of which are shown on the ancient drawing of an Aksumite monument. The contempt of blacksmiths may be an early feature, and the practice of circumcision is certainly older than Islam, for it is mentioned by both Agatharchides and Strabo. Certain features connected with cattle, like the use of wooden milk vessels, taboos on milk, and the drinking of blood, are also ancient. The use of horses is widespread; it is not known when the Somali began to use them (though both the Ethiopians and the Somali had them in the fourteenth century);[3] the Galla did not take to riding horses till about 1554–62.[4]

Though the Galla are today the only unequivocally Hamitic people who live within East Africa, there is evidence which suggests that people of a Hamitic type occupied, till a few centuries ago, a large part of the East African Highlands. All over these highlands there exist the remains of structural and other works made by peoples who preceded the modern inhabitants. Broadly, these consist of dwellings, walled enclosures,

---

[1] C-Hamites, see p. 66. [2] B-Hamites, see p. 65.
[3] According to the *Chronicle of 'Amda Tseyon*, referring to the year 1329.
[4] Beckingham and Huntingford, *Records of Ethiopia*, p. 117.

agricultural and irrigation works, and roads.[1] The dwellings consist in the main of circular hollows excavated on slopes, with nearly level floors and an entrance on the lower side protected by banks; the sloping sides and backs are often revetted with stone, and the hearth is usually near the door. The large numbers of these 'hut-circles' in some areas, notably the Nandi and Uasin Gishu Districts of Kenya, suggest that the population at the time when they were inhabited was much larger than it is today. The people who lived in these dwellings may have been agricultural, for what seems to be a hoe made of a piece of grey-green micaceous stone $8\frac{1}{2}$ inches long by $3\frac{1}{2}$ inches broad and shaped to fit the hand was found in a hut-circle in Uasin Gishu.[2] Moreover, the size of the population when these sites were occupied suggests that the people were not purely pastoral. A possible estimate for the inhabited areas in Nandi Uasin Gishu is a density of about 70 to 80 to the square mile, in contrast to which it may be noted that the population density in the Nandi Native Land Unit in 1932 was estimated at 62 to the square mile, though early counts suggest that a hundred years ago it was probably nearer 40 to the square mile. While we know something of the types and distribution of these structures, we know nothing definite about their origin or date except that they are earlier than the present inhabitants, who not only do not make such things, but have no authenticated traditions of having come in contact with the people who made them.

Little excavation has been done in these sites, a notable exception being Hyrax Hill near Nakuru (Kenya), which was excavated by Mrs. Leakey in 1937–8. Here was an Iron Age settlement consisting of dry-stone walled enclosures and hut-circles of the same type as other sites in the highlands, which yielded among other things material from the coast which can be dated to any time between the sixth and sixteenth centuries A.D.[3] Though this gives an upper terminus of a sort, it does

---

[1] G. W. B. Huntingford, 'Local Archaeology in Kenya Colony', *J.E. Afr. Ug. Nat. Hist. Soc.*, 24, 1926, pp. 3–25; 'The Azanian Civilization of Kenya', *Antiquity*, 7, 1933, pp. 153–65. C. Gillman, 'An Annotated List of Ancient and Modern Indigenous Stone Structures in Eastern Africa', *Tang. Notes*, 17, 1944, pp. 44–55.

[2] Huntingford, *J.E. Afr. Ug. Nat. Hist. Soc.* 1926, p. 23.

[3] M. D. Leakey, 'Report on the Excavations at Hyrax Hill, Kenya Colony, 1937–1938', *Transactions of the Royal Society of South Africa*, 30, 1945, pp. 366 and 372.

not enable us to fix any limit to the possible time of occupation before the sixth century.

For the culture or cultures which these remains represent the name Azanian[1] has been suggested,[2] and their origin ascribed to Hamitic peoples who came from the north. Tradition and the use of stone for building suggest a northern origin, and from the traditional evidence current in the first quarter of this century we gain a picture of these 'Azanians' as tall, bearded, long-haired, and 'red' in colour. Moreover, they are usually said to have come from the north and, in some traditions, to have had long-horned cattle.[3] Taken as a whole, these accounts suggest a people of Hamitic rather than Negro affinities.[4] It is uncertain how extensive the use of stone-built dwellings was in the Hamitic area, since our knowledge of the surface archaeology of southern Ethiopia is very limited; but there

[1] From Azania, the name given to the interior of East Africa south-west of Cape Guardafui in the *Periplus of the Erythraean Sea* and Ptolemy's *Geography*.

[2] Huntingford, *Antiquity*, 1933, p. 153.

[3] For Uasin Gishu and Nandi see Huntingford, *J.E. Afr. Ug. Nat. Hist. Soc.* 1926; *Antiquity*, 1933; and *Nandi Work and Culture*, 1950, pp. 6–10; for long-haired people see C. W. Hobley, 'Notes on the Dorobo People', *Man*, 1906, 78; and for northern Tanganyika see G. E. H. Wilson, 'The Ancient Civilization of the Rift Valley', *Man*, 1932, 298. Long-horned cattle occur in Ethiopia, especially among the Galla; they are also represented on ancient Egyptian monuments. Such cattle, with horns often of enormous size, are found in Ankole, and it may be inferred that the ancestors of the Hima group brought them down from Ethiopia. But it does not appear that any deeper cultural significance can be attached to this than to any other element in the Hima-Bito culture. That the long-horned type is to be found elsewhere in east and south-east Africa does not necessarily prove direct contact, though in the case of the Kenya traditions it gives support to a Hamitic origin for the 'Azanian' culture.

[4] It should be noted that in some traditions, recorded from the Dorobo by Hobley (1906) and Kikuyu by Dundas ('Notes on the Origin and History of the Kikuyu and Dorobo Tribes', *Man*, 1908, 76; see also L. S. B. Leakey, *The Stone Age Cultures of Kenya Colony*, 1931, p. 198) there are references to clever dwarfs, called Gumba by the Kikuyu, who say that they lived in holes in the ground (i.e. hut-circles) and made pottery. Leakey (quoted in H. E. Lambert, *The Systems of Land Tenure in the Kikuyu Land Unit, Pt. 1: History of the Tribal Occupation of the Land*, 1950, p. 46) thought it possible that they might be the remains of a Stone Age people. Their name has been adopted by prehistorians as a label for two East African neolithic cultures, Gumban 'A' and 'B' (L. S. B. Leakey, *Stone Age Africa*, 1936, pp. 70, 174); see pp. 50 sqq. above. In both these cultures pottery was made, that of Gumban 'A' being characterized by the scoring of the inside of many pots with deep lines before baking, a practice so far without parallel. The Gumban 'A' people buried the dead in shallow holes under small cairns, while those of Gumban 'B' buried the body against a cliff or slope and erected a large cairn over it. The Gumbans were of Europoid physical type; unlike the Gumba of tradition, however, they were tall (Leakey, *Stone Age Africa*, pp. 70, 174).

are circular stone sites in eastern Galla land, and terraces are common.

However, such use of stone clearly belongs to relatively recent times, viz. the Iron Age; and if the introduction of iron-working into Africa outside Egypt cannot be placed earlier than the first century A.D., this must be the extreme backward limit.[1] On the other hand, the extensive ruined town at Engaruka north-east of Lake Eyasi, built on terraces on the steep slope of the Rift Valley and containing more than 6,000 stone houses, has been tentatively given an age of not more than 300 to 150 years by Leakey.[2] This date is possible if the Masai did not arrive till about 1600; and his suggestion that the town should be attributed to the 'Mbulu' (Iraqw) is also a possibility. The Iraqw are a people of whom comparatively little is known, and their ethnic and linguistic affinities are uncertain, though their language has some resemblances to Hamitic. It is possible that they may be the last survivors of the 'Azanian' peoples, and they construct mainly the sunken, flat-roofed dwelling known as *tembe*, though it is said that they claim this to be a fairly recent adoption as a form of habitation more suitable for defence against their enemies.[3] The *tembe* is a pit, covered with a flat roof of mud and dung, not thatch, which is rarely above ground level. It is confined to an area to the south and west of Lake Eyasi in northern and central Tanganyika.

In the Nile Valley there came into being a mixed group, here referred to as the Nile Valley Peoples, which stretched on both sides of the Nile from Sennar to Lake Victoria, and whose influence certainly penetrated deeper into East Africa than the distribution of their languages would by itself suggest. There are two major divisions of these, the Nilotes and the Nilo-Hamites, distinguished from each other on both cultural and linguistic grounds, the Nilotes including the Dinka and Nuer, and the Shilluk and Luo, and the Nilo-Hamites the Bari, Nandi, and Masai. It is among these Nile Valley Peoples that the cattle-complex occurs in its most pronounced form,

---

[1] See p. 81 below.
[2] L. S. B. Leakey, 'Preliminary Report on an Examination of the Engaruka Ruins', *Tang. Notes*, 1, 1936, pp. 57–60.
[3] M. D. Leakey, 'Report on Hyrax Hill,' pp. 366, 372. Some of the other sites, like Hyrax Hill, are likely to be much older than this.

especially among the Dinka and Nuer, though they in common with most Nilotes and Nilo-Hamites practise some form of agriculture, the Pastoral Masai being the only exception. In this group the normal unit is the tribe, usually acephalous, though the Dinka, Nuer, and Masai consist of aggregations of tribes sharing a common group-name but with individual tribe-names as well.

The chief characteristic of the Nilotes is of course their cattle-complex, one might almost say cattle-cult, with its peculiar and even exaggerated attitude to cattle. Material features include the use of very large cattle byres and parrying-shields. In some of their members, like the Shilluk and Dinka, there was a divine king or a rain-maker who was put to death when his bodily powers failed. The Acholi of Uganda have broken away from the northern Nilotic culture-type, and though their rain-maker was politically prominent, he was not a divine king. Shield and hut shapes are different, wrist-knives occur, and they have age-sets.[1] The southern Nilotes—the Luo of Kenya—are more strongly agricultural than other Nilotes; they used to live in walled villages, and today have hedged farmsteads.

In the five cultural sub-groups found among the Nilo-Hamites, one may note first that age-sets, or the grouping of the males of the tribe into age-categories, occur throughout. We shall have something to say about these later on, since they are important for dating. In the Bari sub-group there is a great development of the rain-maker as a public authority; and the bow is the chief weapon, there being no shield. There are age-sets and submerged classes, and blacksmiths are despised. In the Lotuko sub-group the culture is in some ways like that of the Bari; but wrist- and finger-knives are used, the long-handled hoe occurs, and stone circles are set up as sitting-places. The Lotuko have three cultural features of their own. The first is the *nametere*, a roughly cylindrical bundle of bamboo and dry grass, representing the body of a dead person, used at funerals. The second is an unusual form of triple initiation. The third is the use of tall wooden towers connected in some way with initiation, though their exact purpose is not clear. The Didinga sub-group by its general character suggests that it should be

[1] See p. 76 below.

placed among that of the Nilo-Hamites, though the language is not regarded as Nilo-Hamitic. A distinctive feature was a way of felting the hair so that it looked like an inverted pudding-bowl on the wearer's head. Round Lake Rudolf is the fourth sub-group, distinguished by its use of lip-plugs (which are an ornament of some antiquity, and occur in great quantities at Jebel Moya in the Sudan), wrist- and finger-knives, and sharp-cornered shields. Circumcision does not occur among the Nilotes, nor among any of the Nilo-Hamites whom we have mentioned so far. The southernmost Nilo-Hamitic sub-group is that of the Rift Valley, where the Nandi and Masai have very distinctive spears, shields, pig-tails, and horn arm-ornaments. They practise circumcision, and have a well-developed system of age-sets, of which something must now be said.

Age-sets are named groups of people whose ages all fall within certain limits; every male in a tribe belongs to a set, which he enters either automatically at birth or later when a set is formed, and in which he remains for the rest of his life. This system is of some antiquity among the Galla, for it is described in an Ethiopic document written about 1593.[1] It is here associated with a system of five named grades through which each set passes as a whole, remaining in each grade for eight years. In East Africa, especially among the Nilo-Hamites, the sets usually occur without the grades, though occasionally, among the Nyika, for example, the grades seem to be more prominent than the sets. The Galla entered Ethiopia from the Lake Rudolf region[2] about A.D. 1522–30, an event which can be dated from the evidence of the Ethiopic document of 1593, confirmed by the Ethiopian chronicles. It was undoubtedly from the Galla that the age-set system spread to the Nilo-Hamites and the Nyika, the Nandi and Masai being those among whom it has developed most strongly and in whose political structure it has become integrated as a central feature. The known presence of the Galla in the Lake Rudolf region in the late fifteenth and early sixteenth century means that

---

[1] This, known as 'The History of the Galla', was written in Ge'ez by a priest named Bahrey; see Beckingham and Huntingford, *Records of Ethiopia*, pp. 111–29; Huntingford, *The Galla of Ethiopia*, pp. 41–53.

[2] 'The Galla came from the west and crossed the river of their country which is called Galana, to the frontier of Bali' (Bahrey, ch. 1, in Beckingham and Huntingford, *Records of Ethiopia*).

there could have been contact between them and the ancestors of the Nandi and Masai, for this was in or near their cradle-land. There are indications that the Nandi had the system when they first settled in their present country about the beginning of the seventeenth century; and the regularity of the periods among the Nandi (each set being the warrior set for fifteen years) makes it possible to use them for dating historical traditions in which the names of age-sets are mentioned. On the coast, where some of the Nyika tribes have acquired the system from the Galla, who came to the Tana Valley between 1200 and 1500, the Galla word *luba* is used to describe it.

Turning now to language, we find that excluding the Bushman and Bantu languages there are three language groupings or 'Larger Units', as Tucker and Bryan[1] call them, which dominate East Africa north of a zigzag line drawn from east to west along the equatorial region; this is the 'Bantu line', and with a few minor exceptions, no Bantu languages are spoken to the north of it, and only Bantu to the south. The first of these Larger Units is the Hamitic, of which we have already spoken. The second Larger Unit is the Nilotic, spoken in the Nile Valley from Uganda northwards. Nilotic languages show little resemblance to Hamitic. The common word-pattern is consonant–vowel–consonant, but formative elements are few, and there are no noun-prefixes. Plurals are formed by vowel-change, by change of tone, and sometimes by suffixes. There is no gender. Tense-formation is marked by tense and person prefixes, and sometimes by the absence of inflexion for person. The sentence order is subject—object—verb.

The third Larger Unit is the Nilo-Hamitic, which on the other hand shows considerable signs of Hamitic influence, though lexically these languages are closer to Nilotic. Word patterns are consonant–vowel–consonant, consonant–vowel, and vowel–consonant, with many tri-literal and polysyllabic words. There is often a formative suffix, and plurals are normally indicated by suffixes, though vowel-change may occur simultaneously. In some languages there is an article, either prefixed (as in Karamojong and Masai, both showing also gender) or suffixed (as in Nandi, which has T in the singular and K in the plural). Significant formative elements are

[1] Tucker and Bryan, *Non-Bantu Languages*, p. viii.

common, as in Hamitic, of the type 'white-*ness*', 'trad-*ing*', 'man-*hood*', 'senior-*ity*'. Grammatical gender is shown by the elements O, fem. E, A, as in Hamitic, and by L, fem. N. The distinction of sex gender appears in some of the languages in the form of an article; and it may occur in possessive and demonstrative pronouns, but not in the verb (as it does in Hamitic and Semitic). The sentence order is verb—subject—object. A feature which differentiates Nandi from the other Nilo-Hamitic languages is that the article, sing. T, plur. K, is used as a secondary suffix, and is separable, so that each noun has normally two forms to each number, *punyo*, 'an enemy', *punyoT*, 'the enemy', plural *pun*, 'enemies', *puniK*, 'the enemies'. This secondary suffix occurs also in Suk and Tatog.

It used to be thought that the Nilo-Hamites represented a mixture of Nilotic and Hamitic, and from this it was deduced that the Nilo-Hamitic languages had simply a Nilotic base with certain additions from Hamitic sources. Further work on these languages, however, has shown that the problem of their relationship is not so simple, but that 'it may be that the Nilo-Hamites are not the hybrid Nilotic people with Hamitic elements that some suppose them to be, nor even the Nilotes that others consider them, but a separate ethnic group which arose from a mixture of some form of Sidama type with a people yet to be discovered, and developed its own language and culture. From this group it is possible that the ancestors of the Nilotes came, and from this group therefore the Nilotic language type also. Such a view would account for the large proportion of words which are common within the Nilo-Hamitic group but not found in either Hamitic or Nilotic; it would also account for the large measure of agreement both morphological and lexical between Nilo-Hamitic and Nilotic.'[1] This last fact must be stressed: there is a large lexical agreement between Nilo-Hamitic and Nilotic, especially in basic words. Yet there is one apparently minor element which proves to be common to Hamitic, Nilo-Hamitic, *and several other languages hitherto considered as unrelated to Nilo-Hamitic and Hamitic*,[2] but *not to Nilotic*. This element is T and K used as singular and plural determinants,

[1] Huntingford, *S-W. J. Anthrop.* 1956, p. 208.
[2] i.e. certain languages of Dar Nuba, Dar Fur, Kordofan, and the Lake Chad region.

with T also as a significant determinative ('particularizing') affix to nouns. A preliminary statement of the problems involved has been made by Miss Bryan,[1] who concludes from the evidence that there is a 'common substratum underlying a number of African languages, Language Groups, and Larger Units'; that the 'peoples yet to be discovered' who may have produced the Nilo-Hamites by mixing with some form of Sidama ethnotype[2] spoke languages in which this opposition of T and K was a characteristic feature; and that the Nilo-Hamitic languages were formed by 'the impact of Nilotic upon these T/K languages', plus cross-fertilization from the Cushitic (Hamitic) languages, which are themselves based upon the same T/K sub-stratum.[3] The problem is further complicated by the existence in Hamitic of a T which is the Hamito-Semitic feminine determinative. Miss Bryan summarizes her findings by saying that the Hamitic languages are formed on a T/K substratum, a mainly Hamito-Semitic corpus, morphological and lexical, and other, unknown, elements. The Nilo-Hamitic languages are likewise formed on a T/K substratum, with many Nilotic elements (mainly lexical), some Hamitic elements, and other elements known and unknown. The Nilotic languages, on the other hand, are not based on the T/K substratum, but they have lexical affinities with Nilo-Hamitic.[4] The position of the Nilotic languages is most obscure, but they have some morphological affinities with Nilo-Hamitic, e.g. in Luo, tense-formation, the use of suffixed object-pronouns (me, thee, &c.), the use of the relative to connect an adjective with the noun it qualifies, and the existence of a verb meaning 'to be in a place' as opposed to mere being.

The speakers of the T/K languages have still to be identified, and it is possible that we may never be able to do this with any certainty. The present situation appears therefore to be that certain Hamitic and Negro ethnotypes produced the Nilotic peoples; another Hamitic ethnotype (perhaps a Sidama type?) added to a people who spoke a T/K language produced the Nilo-Hamites.

---

[1] M. A. Bryan, 'The T/K Languages: a New Substratum', *Africa*, 29, 1959, pp. 1–21.   [2] Minor ethnic division within a larger ethnic unit.
[3] Bryan, *Africa*, 1959, pp. 1, 8.   [4] Ibid., p. 20.

One of the many difficult problems in the earlier history of Africa is that of the origin and spread of the peoples who speak Bantu languages. It is complicated by the fact that the Bantu-speakers, who cover nearly three-quarters of the continent south of the Sahara, do not constitute genetically an ethnic group, for there is no homogeneity of physical type among them, and they range from the Nande of Ruwenzori to the Europoid types found among the Interlacustrines of western Uganda and among the Zulu. We have to study them broadly as a linguistic group, and must note first that the Bantu languages are united by two outstanding features not found in other African languages. The first is the grouping together of words in classes distinguished by pairs of singular and plural prefixes (so that the beginning of the word changes to form the plural),[1] a typical word, common to all Bantu languages, being *UMUntu*, 'person', plural *ABAntu*,[2] 'people' (from which Dr. Bleek a hundred years ago coined the term 'Bantu'). The second feature is the agreement of adjectives, pronouns, and verbs with the noun to which they refer by means of a prefix derived from the class-prefix of the governing noun, a method known as 'concord'. This problem of the spread of the Bantu-speakers was first attacked by Sir Harry Johnston.[3] Starting with the assumption that the Bantu languages have their nearest affinities with the 'Semi-Bantu' languages of the Benue, Lower Niger, and Calabar–Cameroons areas, he held that from an original nucleus round the upper watershed of the Welle river the Bantu spread over southern Africa in ten migrations, beginning with an east–west movement from the Welle–Ubangi basin to the Cameroons, a movement which produced the Semi-Bantu languages. A second, which resulted in a great concentration of Bantu somewhere about Lake Albert,[4] went southwards, round Lake Victoria and down Lake Tanganyika, while the remainder spread eastwards and southwards to cover the

---

[1] It is as if in Old English we had *Aox*, plural *ANox*, 'ox', instead of *oxA*, plural *oxAN*, or in modern English, *ox*, plural *ENox*.

[2] Other derivatives from the radical *ntu* are *EKIntu*, 'non-person, thing', plural *EVIntu*; *UBUntu*, 'being, existence'.

[3] H. H. Johnston, 'A Survey of the Ethnography of Africa: and the former racial and tribal migrations in that continent', *J. Roy. Anthrop. Inst.* 43, 1913, pp. 375–421, especially pp. 391–5.

[4] H. H. Johnston, *The Opening Up of Africa*, 1911, pp. 134–5.

whole of Bantu Africa. These Bantu speakers brought with them, he thought, the ox, perhaps the domestic fowl, and a knowledge of metallurgy. There was, he believed, a distinct Bantu language family, with a group in the west associated with it by grammatical affinities—the so-called Semi-Bantu languages.

Greenberg sees[1] the Bantu as merely a branch of a great 'Niger-Congo' family which covers Africa south of the Sahara from Cape Verde to the Cape of Good Hope. The Semi-Bantu languages disappear from his classification, since they do not, in his opinion, 'form a genetic unity as against the Bantu languages'. His original Bantu nucleus seems to be 'the general area of Nigeria and the Cameroons'. Guthrie, more recently, on purely linguistic grounds, sees the original Bantu nucleus much farther south, on the upper reaches of the Congo, which may have been the home of a people who acquired from the north a knowledge of iron-working which spread along with the Bantu language. Though there appear to be at least two distinct words each for 'spear', 'axe', and 'hoe'—one in the east and one in the west—the word for 'forge iron' is universal except in parts of the north-east where it has been replaced by other words.[2] It has been suggested that the knowledge of iron-working spread southwards from Meroe; and if, as the linguistic evidence suggests, the Bantu were carriers of this knowledge, their dispersal from the Congo may be dated to the early centuries of the Christian era.

It is possible to supplement the linguistic evidence by certain cultural features. First, the distribution of the usually tanged and heart-shaped iron hoe, which is found in one form or another over a large part of the Bantu area, suggests that this, in spite of the existence of two radicals for 'hoe', was the pattern which the Bantu took with them; it also occurs in the area south-west of Lake Chad, together with a certain type of bellows. The distribution of bellows, secondly, shows that there are two basic types used in Africa, the drum type, in which a solid chamber is covered with a skin diaphragm; and the bag type, consisting of a skin bag with a solid mouthpiece.

[1] Greenberg, *S-W. J. Anthrop.* 1955.
[2] Professor Guthrie's theory was put forward to a Seminar at the School of Oriental and African Studies in 1959; it has not yet been published.

According to Cline[1] the drum type seems to have been the older pattern in much of the Congo area, where in some parts it was replaced by the bag type. This latter type occurs west and south of Lake Tanganyika, and up the east coast and inland to Lake Victoria, with three enclaves where both bag and drum types occur together. It appears again in north-east Africa, and among the Shilluk on the Nile. The drum type is found throughout the Lake Region, among the Interlacustrine Bantu, Nilotes, and Nilo-Hamites. This distribution suggests that the occurrence of bag bellows may be due to borrowing from two separate sources: in the north-east the Arabian bag-bellows have possibly been the prototype; elsewhere their presence may be due to Arab sources by way of the Zambezi, from which this type spread both northwards and southwards, bag bellows being the prevailing type south of the Zambezi. Drum bellows seem to have been the prevailing type in the centre of Africa. Their use is recorded in the Lake Chad–Northern Nigeria region, and it is possible that they reached the Bantu from here together with the knowledge of iron-working, which, as has been noted, spread from Meroe. Dr. Arkell has suggested[2] that when the kingdom of Meroe fell, the Meroitic royal family may have gone westward into Dar Fur, and they may have taken with them, as a royal prerogative, the knowledge of iron-working; it then spread to the adjoining Lake Chad region and then farther west. Carbon-14 dating suggests that this knowledge may have reached the iron-working centre at Nok in Nigeria by the first century A.D., if not a little earlier. It may therefore be suggested that when the Bantu moved eastwards from the Congo they had the drum type of bellows together with a matrilineal social system,[3] and that the drum type was replaced by the bag type after their occupation of most of what is now Tanganyika and southern Kenya. The use of drum bellows may thus be correlated with the practice of matriliny, though the latter, like the drum bellows, was later displaced in some areas by patriliny. Thus the apparent correlation between bag bellows and matriliny is deceptive and results from a later development.

[1] W. Cline, *Mining and Metallurgy in Negro Africa*, 1937.
[2] In R. A. Hamilton (ed.), *History and Archaeology in Africa: report of a conference held . . . at the School of Oriental and African Studies*, 1955, p. 71.
[3] See p. 92 below.

THE PEOPLES 83

We have seen[1] that we have to deal with three 'foundation stocks' in East Africa: Bushman, Negro, and Hamite,[2] with probably a leavening of Stone Age blood, though of course we cannot give a name to any possible Stone Age ancestral groups. These stocks have in turn given rise to the other ethnic groups of East Africa, groups which may be called hybrid in that they are formed from the mixture of two or more dissimilar stocks. We do not know exactly where the mixing took place, nor do we know either the precise nature of the stocks involved, or when the mixing process began. But since 'race formation' (as opposed to the physical alteration of an existing unit) and the spreading of ethnic groups are necessarily slow processes, the time factor must be allowed to run into hundreds of years. And the movements which resulted in these stocks, as well as the later movements of the stocks, must have taken place during a long period, and are in no way to be compared with the relatively quick invasions by conquering tribes such as took place in historical times. We are not justified in assuming that events like the penetration by the Ngoni of south-east Africa after their revolt from Chaka about 1823 were typical of the way in which new ethnic groups were formed, acquired land, and developed into the peoples we know today.

Vast stretches of East Africa even now are thinly populated, the total African population of Kenya, Uganda, and Tanganyika being about 16 million, which in an area of approximately 800,000 square miles gives an average of twenty to the square mile. In early times, however, the population must have been very much smaller.[3] New stocks could form, languages could develop and fade out, and culture elements could evolve and be disseminated more easily than after a certain degree of fixity had been reached.

[1] See pp. 61–68 above.
[2] It should perhaps be made clear that the position which it appears proper to assign to the Hamites is not due to a mere 'belief in the inherent superiority of the Hamite element', as Greenberg puts it (*S-W. J. Anthrop.* 1955, pp. 52, 54), but to their actual and potential abilities as demonstrated by factual evidence. If we took the Afar as representative of the Hamites in general, any claim made for them as civilizers would be absurd. But the Afar are not the only Hamites; and just as there are many European ethnotypes, so there are several Hamitic ethnotypes linked by a broad homogeneity of language and culture (cf. p. 69).
[3] The Dorobo population in the East African highlands a thousand years ago was perhaps about 1 person to 12 square miles.

Some attempt must be made to deal with the problems referred to above—the places where the secondary ethnic stocks were formed, and what elements entered into their composition. As to when this took place, the only answer that we can give at present is that, in the broadest terms, it must have been after the Hamitic penetration of north-east Africa.

The Hamites seem to have formed part of the ancestry of both the Nile Valley Peoples and the 'Bantu'. In view of the vast extent of the latter, I would suggest that Hamitic penetration took place in at least two major series of movements, series A going westwards far beyond the Nile and producing the early Bantu, series B being responsible for the Nile Valley Peoples; in both cases the stock upon which the Hamites impinged being Negro, plus other minor elements—Bushmen, Stone Age peoples, Hunters.

In the southern Sudan between latitude 12° and 4° north and longitude 36° and 27° east, there is an area of some 360,000 square miles, possibly extending even farther west than longitude 27°, which such evidence as we possess suggests was the cradle-land of the Nile Valley Peoples. Perhaps somewhere here, successive waves of B-Hamites from Ethiopia produced by mixture with Negroes the Nile Valley Peoples, one Hamitic ethnotype giving rise to the Nilo-Hamites, and another to the Nilotes; we do not yet know which came first. There are in the modern B-Hamites at least three ethnotypes, and there is enough difference between Nilo-Hamites and Nilotes to suggest that the Hamitic side of their ancestry came from different ethnotypes, rather than that (as used to be thought) the Nilo-Hamites were the result of a mixture of Nilotic and Hamitic, the Nilotes themselves being supposed to be an earlier Hamitic–Negro cross.

From the Nile Valley the two ethnic groups may have spread gradually from dispersal centres, that of the Nilotes having been probably in the western part of the area, in the south-east Bahr al Ghazal region, that of the Nilo-Hamites somewhere to the north or north-west of Lake Rudolf, centred perhaps on the modern district of Kapoeta. The Nilotes, moving north-east from their original homeland, reached the present Shilluk country, where part of them broke away and turned southwards, a movement which ended in the eventual occupation

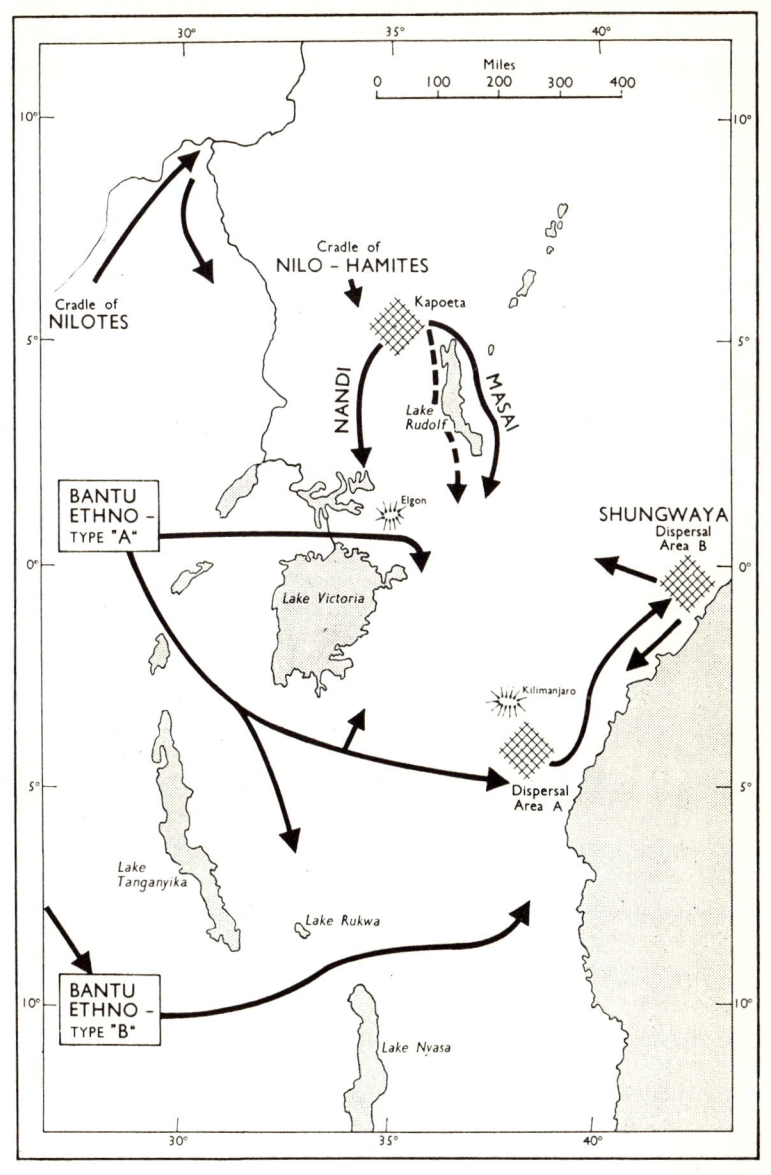

MAP 5. Tentative reconstruction of Nilotic, Hamitic, and Bantu movements in East Africa

of the rest of the Nilotic area.[1] The Nilo-Hamites, whose cradle was probably near the present Ethiopian frontier, dispersed in different directions, some going west (the northern division), some going south (the southern division), and the rest (the central division) staying in the Kapoeta–Lake Rudolf region.[2] They entered what is now Kenya and Uganda in two distinct movements, one (the Nandi branch) reaching Mount Elgon and thence western Kenya by way of a secondary dispersal centre,[3] the other (the Masai branch) from a more easterly point, possibly down Lake Rudolf.[4]

The Bantu peoples have somewhat fuller traditions than some of the Nile Valley Peoples, though it is clear that even the earliest of these refer only to secondary movements. There are three zones of Bantu in Kenya and Uganda: (1) the Lake region of western Uganda, where comparatively recent immigrant Hamitic–Nilotic ruling groups dominate an older layer of Bantu; (2) the Lake Region of western Kenya, where the Bantu Kavirondo or Luhya lie between the Nilo-Hamites and Lake Victoria; and (3) the eastern Kenya region where there are two rather widely separated Bantu groups.

Among the western Uganda Bantu the traditional political situation is that an ethnically distinct aristocracy known as Hima (Huma) in the north and Tutsi in the south dominated the original Bantu inhabitants, who in the Hima–Tutsi states were known as Iru and Hutu. We have to distinguish two layers superimposed on the earlier Bantu inhabitants. The first, including those known to Interlacustrine tradition as Chwezi, seems to have been of Hamitic stock; their ethnotype is uncertain, but it is now clear that it was not Galla, as used to be thought,[5] and it is much more likely that they were akin to the Sidama peoples of south-western Ethiopia, and to have come from that region. The distinctive elements in their culture, however, are not specifically Sidama, but can be

---

[1] J. P. Crazzolara, *The Lwoo, Pt. 1: Lwoo migrations; Pt. 2: Lwoo traditions*, 1950–1.
[2] Huntingford, *The Northern Nilo-Hamites*, p. 10.
[3] According to tradition Mount Kamalinga (Nepak), forty-five miles northwest of Mount Elgon.
[4] Huntingford, *The Northern Nilo-Hamites*, p. 13; *The Southern Nilo-Hamites*, 1953, p. 14.
[5] R. A. Oliver, 'Oral Tradition: East Africa', in Hamilton, *History and Archaeology in Africa*, 1955, p. 18.

paralleled from other parts of the Hamitic area; these elements include a two-class social system (Hamitic overlords and Bantu serfs), spears with blood-courses, small shields with central bosses, wooden milk vessels, and a much fuller style of dress than that worn by most of the Negroids. Two-class systems are found among the Afar, who are divided into the 'Red' Afar or nobles, and the 'White' Afar or commoners, who are tributary clients of the nobles; and to a lesser degree among the Kafa. Spears with blood-courses, now a distinctive Interlacustrine feature, occur less frequently, but significantly, among the Somali and Afar, and certain Hamitic-like tribes in south-western Ethiopia, though not apparently among the Galla or Kafa; and wooden milk vessels are used by the Somali. But the Interlacustrine peoples do not circumcise, they have no horses, and they did not bring the plough with them—or if they did, it failed to survive. About the turn of the fifteenth century the Chwezi were displaced by Nilotic invaders, members of the Lwo branch, called in tradition Bito, who, after establishing themselves firmly, completed the formation of a group whose cultures were an integrated mixture of Hamitic and Nilotic elements. We have not even a faint linguistic clue to their origins. One of the features of this group is the presence of a divine king, who among the Ankole was formerly put to death when his bodily powers failed. It is uncertain whether this was brought in by the Hamitic or by the Nilotic immigrants, for although a divine king used to exist in western Ethiopia, in Kafa, he was not put to death and his body was considered to be the habitation of the ruling spirit of the Kafa spirit-world. On the other hand, the king of the Nilotic Shilluk, though believed to be the habitation of the spirit of Nyikang, the Shilluk 'culture-hero', was nevertheless killed when his bodily powers failed; and it is perhaps to Bito influence that the divine king in western Uganda is to be attributed.

Superficially, the organization of the Hima-Bito kingdoms has a closer resemblance to those of the Sidama—especially Kafa and Janjero, on which alone we have detailed information —than to any others. The status and powers of the king, the large and well organized hierarchy of court and state officials (this in turn possibly due to the influence of Semitic Ethiopia), the system of provincial administration, the position of the king's mother, the method of choosing his successor, and the royal burial rites are all closely paralleled in the Sidama kingdoms. Indeed, the similarity of certain of the burial rites is too striking to be merely coincidental. The Jesuit Antonio Fernandes, writing of Janjero in 1613,[1] says 'when the king dies they wrap his body in rich clothes, kill a cow, and put the body in its hide ... on the seventh day after the old king's death they bring a maggot which they say comes from his nostrils, wrap it in a piece of silk, and make the new king kill it, squeezing it between his teeth'. In Ankole the dead king's body was laid in a cow-skin and left to decompose. Milk was mixed with the fluids which came from it, and the resulting mixture produced maggots, one of which was said to be the king reborn.[2] In Bunyoro, too, the king's body was put in a bull's hide.[3] Other features, like the king's sacred fire, the custom of the king marrying his half-sister, and, of course, the ritual nature of his death, do not occur in the Sidama kingdoms, but they are found among the Shilluk, members of the ancestral stock of the Bito, and politically the most highly organized of the Nilotes. Taking all the rather slender evidence into account, it would seem that the Interlacustrine kingdoms represent the Hima-Bito development of a Hamito-Semitic type of kingdom, with a form of kingship introduced by the Nilotes (Shilluk), though we still have to look for the ultimate source of the Shilluk kingship. The latter problem is particularly formidable, since we have virtually no clue. The obvious source, one might suggest, is Meroe; but there seems to be no evidence that Meroitic kings were killed like those of the Shilluk or of the Interlacustrine peoples,[4] and it was certainly not the custom in ancient Egypt. That certain ritual practices and beliefs found in Equatorial Africa are of Egyptian origin need not reasonably be doubted. The annual ceremony called 'shooting the nations', for example, in which the king of Bunyoro shot an arrow towards each of the four quarters of the globe saying 'I shoot the nations to overcome them',[5] is so closely paralleled

[1] Beckingham and Huntingford, *Records of Ethiopia*, pp. 158–60.
[2] J. Roscoe, *The Banyankole*, 1923, p. 53.
[3] J. Roscoe, *The Bakitara*, 1933, p. 121.
[4] Unless a statement by Strabo can be construed as such; the priests at Meroe, he says, in old times could order the king to die, till a later king broke the custom by killing the priests instead (Strabo xvii, ii). [5] Roscoe, *The Bakitara*, p. 134.

by the ancient Egyptian *sed* festival in which the king shot arrows towards the four cardinal points that it is permissible to conclude, as Seligman does,[1] that the Nyoro rite is derived from the Egyptian. But we are no nearer to finding the source of the custom of king-killing in East Africa.

In the Lake region of Kenya, the peoples of the Luhya group, including the Gisu and other tribes round Mount Elgon, mostly claim to have come in from the west, that is, from Uganda along the northern side of the lake. One tribe, the Hanga (Wanga), formerly had a divine king of the Ankole type who was put to death, as well as other elements of the Hima-Bito type of organization; and the origin of the Hanga, confirmed by tradition, is clearly the Interlacustrine peoples of western Uganda.[2] Some of the Luhya, however, have more affinity with the Gusii (Kisii) of South Kavirondo and plainly represent a northerly extension from the area south-east of Lake Victoria which had no direct connexions with western Uganda. The Luhya (or Bantu Kavirondo) are agricultural with cattle, and have a general homogeneity of speech and culture. They used to wear very scanty clothing, and lived in walled villages; they occasionally wore lip-plugs, and were noted as bridge-builders and quail-catchers. Some of them have age-sets borrowed from the Nandi, but of small political significance.

According to oral tradition the ancestors of the North-eastern Bantu came from a dispersal area in the Taita hills near Mount Kilimanjaro. Having migrated to the coast, they went northwards as far as a region called in their traditions Shungwaya, which was somewhere between the Tana and Juba rivers.[3] This became a second dispersal area, and they began to move away from it on the arrival of the Galla, which may be put tentatively at about A.D. 1200–1300,[4] though some, like Dr. Gervase Mathew, prefer a date some 300 years later. From this centre they broke up into the groups which became known as

[1] Ibid., p. 18.
[2] K. R. Dundas, 'The Wawanga and other Tribes of the Elgon District, British East Africa', *J. Roy. Anthrop. Inst.* 43, 1913, pp. 20 sqq.
[3] V. L. Grottanelli ('A Lost African Metropolis', *Afrikanistische Studien*, Berlin, 1955, p. 236) identifies it with Birikau (Bur Gao) about 260 miles north-east of Mombasa, and roughly half-way between the Tana and the Juba.
[4] G. W. B. Huntingford, 'Bantu Peoples of Eastern Kenya and North-eastern Tanganyika', in Hamilton, *History and Archaeology in Africa*, pp. 48–49.

Pokomo along the Tana, and Nyika down the coast southwards, and much farther inland they became the Kamba and Kikuyu in the region of Mount Kenya. The Nyika are agricultural with cattle, though the Pokomo of the Tana have no cattle. The Digo and Duruma are matrilineal; the Kauma were formerly. Age-sets of Galla origin occur, with male circumcision, and there has been much Galla influence among the Pokomo. A feature which links in a rather loose way the coast Bantu with those round Mount Kenya is the existence in both groups of a form of tribal council with a name based on the stem -ama (ki-ama, nj-ama, nz-ama), and an associated rank-class system. The Kikuyu-Kamba group are agricultural with cattle, the Kikuyu keeping large numbers of goats and sheep. The Kamba use the bow but no shield or spear, and formerly for ritual reasons used only wooden digging-sticks for cultivation. The Kikuyu use both bow and spear, and have imitated extensively various Masai fashions in weapons and ornament. Male and female circumcision are practised.

The movements of the Mount Kenya peoples, as deduced by Lambert,[1] may have been as follows. The Kamba and Chuka were perhaps the first to leave the coast, the former up the Athi, the latter along the Tana. The forerunners of the Kikuyu also went up the Tana and first settled in the Mbere country south-east of Mount Kenya and near the Tana, whence they gradually spread southwards. Tentative dates for the first arrival in their present homes, based on age-sets and traditional generations, have been worked out by Lambert, who suggests that the Chuka arrived about 1300; the Embu, about 1425; and that the Kikuyu, reaching the Fort Hall area about 1545, arrived in the Kiambu district about 1800.

The tribes of north-eastern Tanganyika (Chaga, Gweno, Shambaa, and others) seem also to have come immediately from the Taita dispersal area. Those of western and central Tanganyika may belong to the same ethnotype as the rest of the Bantu peoples of Uganda and Kenya; in fact, they perhaps correspond to Johnston's 'third migration'. The tribes of southern and eastern Tanganyika, among some of which matriliny occurs, seem to belong to a different Bantu ethnotype which may have entered the country from the Congo south of or even

[1] Lambert, *Kikuyu Land Unit*, chs. i–iii.

across Lake Tanganyika. What we know of the movements and cultural features of these groups suggests that the Bantu of Uganda, Kenya, and Tanganyika belong to at least two ethnotypes, one (*A*) arising somewhere west or north-west of Lake Victoria, the other (*B*) coming into our area from the Congo west of Lake Tanganyika. To *A* belong the Luhya, some of the peoples of northern Tanganyika and the Taita dispersal area, and the western Tanganyika tribes down to about Lake Rukwa. To *B* belong the matrilineal peoples of southern Tanganyika and the coast. The peoples of *A* were either patrilineal or became so through contact with non-Bantu patrilineal societies; while those of *B* passed on their matriliny to the Nyika peoples on the coast to the north of them.

It is difficult to generalize on Bantu cultures. As regards occupation, while it is true that all Bantu groups are basically agricultural, some have taken to cattle-keeping. Excluding the pastoral Interlacustrine—Hima—peoples of western Uganda, it may be said safely that none of the Bantu of East Africa are pastoral in the same sense that the Hamites and Nile Valley Peoples are. To these Bantu, cattle are little more than a profitable and useful adjunct to agriculture and a source of increased prestige and wealth. Some of the Bantu cattle-keeping tribes, while lacking the pastoral outlook of the true pastoralist, do, however, own large herds of cattle, and among them the number of cattle paid as bride-wealth may be considerably larger than that paid by pastoral tribes. In some tribes, cattle-keeping is a recent development, as among the Ganda, deliberately brought about for reasons of wealth and prestige. Many of the tribes of the south-eastern part of Tanganyika, on the other hand, are purely agricultural with no livestock.

In political organization there is no consistent type, though it is noticeable that, of the tribes which in pre-European days had no chiefs, the majority are north of the territorial frontier between Kenya and Tanganyika, i.e. nearer to the Nilo-Hamitic peoples, who as we have noted tend to tribal acephaly, the acephalous tribes of Tanganyika being along the eastern side and towards the sea. Out of 101 tribes in Kenya, Uganda, and Tanganyika, 45 had no chiefs, 47 had district chiefs, and 9 had central chiefs. It would appear that acephaly is an earlier form than rule by district or central chiefs; one can conceive of the adoption of rule by a chief through culture contact or conquest, but the reverse process is not so likely. If this is true, then it may be inferred that in the north-eastern Bantu ethnotype rule by a chief was not an ethnic characteristic; that the kings of the Uganda peoples were due (as in fact is reasonably certain) to Hamitic and Nilotic influence; and that the chiefs in the Tanganyika area developed through influence from western Bantu ethnotypes.

In religion, however, there is some degree of homogeneity, for most of the East African Bantu have ancestor-cults associated with belief in a supreme being, the latter somewhat vague among some peoples, but more clearly defined where there is strong Hamitic or Nilotic influence, as among the Ankole of Uganda.

A further cultural distinction that should be noted is the occurrence in East Africa of both patrilineal and matrilineal systems of descent. The Hamites, Nilotes, and Nilo-Hamites are patrilineal, as are most of the Bantu of Kenya and Uganda. But the area with which this history is concerned is bordered on the west and south by Bantu matrilineal societies, as can be seen from Baumann's map.[1] What does seem possible is that Hamitic patriliny has influenced a large part of East Africa, though the Negroes show both patriliny and matriliny.[2] The bulk of the Bantu peoples outside East Africa (except those classed by Torday as 'Central Bantu—Forest') are matrilineal, and this system was presumably a characteristic of the core from which the various Bantu groups were derived, the East African Bantu being affected by Hamitic patriliny.

Summing up all the evidence and suggestions here examined it may perhaps be concluded that at the end of the Stone Age most of the area of what is now called East Africa was shared by Bushmen and Proto-Hamites, both hunters and collectors, though the latter were beginning to be pastoralists and cultivators. Negroes may have been in the area from the tenth

[1] H. Baumann, and D. Westermann, *Les Peuples et les civilisations de l'Afrique*, traduction française par L. Homburger, 1948, p. 56, where the word is miswritten 'Matriarcat'.

[2] Cf. C. G. Seligman and B. Z. Seligman, *Pagan Tribes of the Nilotic Sudan*, 1932, pp. 383–4, 431; E. Torday, *Descriptive Sociology: African Races*, 1930, pp. 75–77 and tables v, vi.

millennium B.C., though possibly Negroid peoples may not have been much in evidence before the Iron Age. The true Hamites, beginning with those referred to as B-Hamites, seem to have come in after the Negroes had appeared, and before the Bantu. Those responsible for the Azanian culture probably occupied the comparatively dry central highlands of the modern Kenya and Tanganyika at the time when the Bantu peoples were expanding into the areas of higher rainfall to the east and to the west of them. The Nilotes and Nilo-Hamites are probably the most recent arrivals in the area of modern East Africa, absorbing all but a few of the older Hamitic stocks to the north of the Bantu Line outside Ethiopia, while the Bantu absorbed those to the south.

IV

THE EAST AFRICAN COAST UNTIL THE COMING OF THE PORTUGUESE

GERVASE MATHEW

THE *Periplus of the Erythrean Sea*[1] is the earliest surviving description of the coast of East Africa. This is a Greek commercial handbook of the late first or early second century. Its most likely date is approximately A.D. 110.[2] It seems clear that it is based on the author's own experience. Possibly it is the report of a rather illiterate supercargo who had travelled the Indian Ocean as a representative for some firm at Berenice on the Red Sea.

It describes in some detail the ports along the shore of what was to become Somaliland.[3] There were emporia at Avalites, on the north side of the bay of Zeila, at Malao, which is certainly the modern Berbera, at Mundos, which is probably Bandar Hais, and at Mosyllon, which has not yet been identified. Then, as the ships rounded Cape Guardafui and passed south into the Indian Ocean, they came to the emporium of Aromata, the Market of Spices, probably at Olok. Farther south there were harbours at Tabae and at Pano, and at last, 400 stades beyond Pano, there was the emporium of Opone, now Ras Hafun. In the *Periplus*, 'emporium' seems a technical term implying a fixed commercial organization and regular customs dues.[4] The list of imports suggests the existence of a fairly evolved culture. The ships sailed direct to these markets from western India, bringing cotton cloth and girdles and grain, oil and sugar and ghee. Graeco-Roman ships came down the Red

---

[1] Ed. H. Frisk, Goteborg, 1927. The *Periplus* of W. H. Schoff, New York, 1912 should be used with caution; it is a free translation of an imaginatively emended text.

[2] The date of the *Periplus* and the identity of its harbours will be discussed in detail in a forthcoming edition from the Hakluyt Society.

[3] *Periplus*, § 7–14.

[4] For the meaning of 'emporia' cf. Sir Mortimer Wheeler, *Rome beyond the Imperial Frontiers*, 1954, p. 124.

Sea, bringing dyed cloaks and tunics, copper and tin and worked silver and wine and drinking-cups. In return there was an export or re-export of cinnamon, frankincense, and fragrant gums, a great quantity of tortoise-shell, and a little ivory. From Opone an increasing number of slaves were sent north to Egypt. Each market-town had its own ruler; there was no single suzerain.[1]

From Opone it was necessary to sail south for some weeks before reaching the next African emporium, that of Rhapta, somewhere on the Tanganyika coast. A number of place-names are mentioned on the journey: Serapion, probably a few miles north of Merca, Nikon, probably Port Durnford, the Pyralean Islands, which seem to be the Lamu group off Kenya, the place called The Channel, which, if taken disjunctively from the Pyraleans, might be a watering place on the west side of Mombasa, and Menouthias Island, which is very likely Zanzibar. But none of these are referred to explicitly as trading centres, and trade at Rhapta is organized quite differently from that at Opone; the primary import consists of iron weapons and implements, the primary export of a great quantity of ivory.[2] It would seem to have been long organized as an entrepôt of the ivory trade by Arab merchants. Charibael, the King of Mapharitis in South Arabia, rules as its suzerain[3] by ancient right, and farms its trade to the men of Muza,[4] somewhere near the present Mocha; Arab agents settle among the Rhapta savages, intermarry with them and speak their language. Since Charibael had become a subordinate ally of Rome, he is given the title of 'friend of the Emperors'.[5] Rhapta belongs in some fashion to the Roman sphere, but the inhabitants are clearly far more primitive than those of Opone. Small gifts of wine and wheat win their goodwill; the list of imports suggests that they may have not yet learnt to smelt iron. Their own chiefs rule them under Charibael. They are clearly contrasted with men of the north, whom it is customary to suppose were 'Hamites'. They are of great height, but there is nothing to suggest that they are negroid. They are seafarers and pirates.[6] It is possible that they represented a transient racial amalgam, in which there was an Indonesian element

[1] *Periplus*, § 14.
[2] Ibid., § 17.
[3] Ibid., § 31.
[4] Ibid., § 16.
[5] Ibid., § 23.
[6] Ibid., § 16.

derived from a wave of emigration that seems to have reached Madagascar from Indonesia.

The *Geography* of Ptolemy[1] contains the next description of the coast to have survived to us. This was long assumed to be a second-century document from Alexandria.[2] Later it was suggested that it was a Byzantine encyclopaedia. Most probably much of it is basically second-century and by Claudius Ptolemaus of Alexandria, notably the theoretics and the section on India, but probably the text was edited by geographers in fifth-century Alexandria and new sections were added to it, in especial those dealing with Malaya and with East Africa. Certainly it seems safest to treat the East African section as representing the sum of knowledge acquired in the Mediterranean world by the close of the fourth century A.D.

There are marked changes in the area described by the *Periplus* and by Ptolemy. The economically developed northern zone now stretched continuously south towards the Kenya border. There was a new emporium in Somalia called Essina, which cannot yet be identified; there are several references to the harbour of Serapion. Nikon or the Niki had become an emporium.[3] From there ships sailed direct to Rhapta. Rhapta had increased in importance; it is now called a 'metropolis'.[4] The town lay some way up a river. There is no reference to any Arab suzerainty and, to judge by 'Ptolemaic' usage, the term metropolis suggests that it was conceived as the capital of a state. It had clearly become a centre of trade contacts that stretched far south and far inland. To the author of the *Periplus*, Rhapta is the end of the known world: 'beyond it the unexplored ocean curves round towards the west'.[5] But Ptolemy knew that to the south there lived another people, 'man-eating Barbarians', or 'man-eating Ethiopians',[6] who lived round a wide shallow bay, and that by sailing south-east it was possible to reach Prason,[7] which, with its promontory, is most likely Cape Delgado. A great snow mountain is described as lying

---

[1] *C. Ptolemaei geographia*, ed. C. Müller, Book I, 1883, ch. 14; Book IV, 1901, ch. 7.
[2] Cf. L. Bagrov, 'The Origin of Ptolemy's Geographia', *Geografiska Annaler*, 27, Stockholm, 1945, pp. 318–87.
[3] Ptolemy i, § 17. 5.
[4] Ibid.
[5] *Periplus*, § 18.
[6] Ptolemy iv, § 8, 2; cf. i, § 17. 5.
[7] Id. § 8. 1.

inland from Rhapta,[1] and this is surely Kilimanjaro. Beyond it and far inland are great lakes that form the source of the Nile.[2]

There is nothing in the little archaeological evidence that we possess to contradict the history derived from the geographers. Too much has been built on the coin of Ptolemy VIII (116–80 B.C.) found in association with a dagger not far from Dar es Salaam. It may be as insignificant a 'stray' as the Kenya coin of Victorinus.[3] The group of coins now in the Beit al-Amani Museum at Zanzibar is more important, for though, characteristically, there is no exact record of their provenance, earth and sand still adhered to them when they were examined in the museum by Dr. Freeman-Grenville, and this at least suggests that they were a local find. Some of these may be 'strays', but surely not all. They include coins of the emperors Diocletian, Licinius, Justin, and Justinian, five Parthian and Sassanian pieces from the mint of Ctesiphon, of which the latest is one of Ardashir I (A.D. 212–41), and two Hellenistic coins of about the second century B.C.[4] Taken in conjunction with the glass beads in the museum that have been identified as of Roman manufacture, they suggest some possible association with Menouthias. Perhaps Menouthias may yet be discovered on the west coast of Zanzibar.

The site of Rhapta is still quite unknown; possibly it was up the old course of the Pangani, which, as an aerial survey shows, was some distance to the south of the present river bed. More probably it lies lost in the Rufiji delta.

By far the most important find so far reported was made at Port Durnford, the most likely site of the southernmost emporium of the northern zone. This included at least forty-six Roman coins of the first half of the fourth century A.D., struck at the mints of Alexandria, Rome, Thessalonica, Antioch, Cyzicus, Nicomedia, and Constantinople. But there are also six coins of Roman emperors from Nero to Antoninus Pius struck at the Alexandrian mint, and sixteen Ptolemaic coins

---

[1] Ibid. iv, § 8. 3.   [2] Ibid. i, § 17. 6.

[3] M. P. Charlesworth, 'A Roman Imperial Coin from Nairobi', *Num. Chron.* vi. 9, 1949, pp. 107–10.

[4] For a full account of these coins cf. G. S. P. Freeman-Grenville, *The Medieval History of the Tanganyika Coast*, 1962, ch. i.

from Egypt, six of which date from Ptolemy III.[1] All this is stated to have been found below surface in 'a walled fortress enclosing five acres of ground', possibly the walled-in Muslim burial place on the summit of the rising ground above the estuary.

It is perhaps significant that an analysis of this find largely corresponds with reports on coins from India. There again, fourth-century coins of the house of Constantine are very numerous; there is a well-established sequence from Nero to Antoninus Pius, and a curious scarcity from the late second to the late third century, which perhaps coincided with the development of overland routes from the far east. Only the Ptolemaic section is unique.

It may be hazarded that Greek ships first pressed round Cape Guardafui to the south in the reign of Ptolemy III,[2] as part of the search for ivory. This and the spice trade explain the development of the ports along the Horn of Africa. The Graeco-Roman exploitation of the coast, like that of the Indian markets, both of which followed in the first century, was a result of the discovery of the monsoon winds attributed to Hippalus. Until the coming of steam the monsoons conditioned the economic development of East African ports, and from the first to the early nineteenth century the East African coast shared in each changing rhythm in Indian ocean trade.

The period from the late fourth until the late seventh century is the most obscure in the history of the coast, and perhaps one of the most crucial. It is clear that some time in the late fourth and early fifth centuries direct contacts between the Mediterranean and the Indian Ocean first faded and then vanished. This can be illustrated by the sixth-century *Christian Topography of Cosmas Indicopleustes*.[3] Graeco-Roman fleets had once sailed annually to India, but now a sailing journey to India had become so rare that it could provide an author with a surname. Cosmas gives a detailed description of the great commercial port at Adulis, in what was to be later Eritrea. But though he passed on to Ceylon, he knew nothing of what lay below Cape Guardafui except that a land named Zingion might lie thereabouts.

[1] H. Mattingly, 'Coins from a Site Find in British East Africa', *Num. Chron.* v. 12, 1932, p. 175.   [2] 247–221 B.C.   [3] Ed. E. O. Winstedt, 1909.

It is clear also that the eclipse of direct Graeco-Roman trade with East Africa was due to two political factors: the dominance of Persian sea power in the west Indian Ocean, and the economic development of the empire of Axum. These were interconnected.

The Sassanian kings who ruled Mesopotamia and the Persian area had built up a considerable sea power in the fourth century. Perhaps this had been stimulated by the rivalry between the pearl fisheries at Bahrein and those in southern India. At the death of King Shapur II in 379 Persia seems to be already a dominant sea power. By the reign of Bahram V (420–39) Persia had control of the sea-ways in the western half of the Indian Ocean. This was maintained until the early seventh century; it only vanished completely when the Sassanian empire was destroyed by the Muslim in 643.

When the Roman ships were being driven from the Indian Ocean the farthest outpost of the Roman sphere was the allied kingdom of Axum in the uplands of modern Ethiopia. In 336 the Emperor Constantius declared that its subjects merited to be treated as if they were Romans. Shortly before this the alliance had been sealed by the official acceptance by the Axumite rulers of Christianity, the new state religion of the empire. Axum possessed the old Ptolemaic port of Adulis on the Red Sea. Inevitably Adulis became the entrepôt where Indian Ocean goods were transhipped for the ports of Roman Egypt and so reached the Alexandrian market. Axum had been increasing in power since the late third century; in the fourth its wealth was symbolized by the common use of a gold coinage; in the fifth it was at the height of its prosperity; in the seventh it began to disintegrate; by the eighth its trade routes were broken and its coastal lands lost to the Muslim.

It is still occasionally asserted that at one time its empire included the Eastern African coast, but this was only a suggestion of Dr. Glaser based on quite inadequate epigraphic evidence. It is certain from Cosmas that by the sixth century the trade route Adulis–Colloe–Axum had extended far into the interior of Africa. It may soon be proved that it reached the gold-bearing area in Kenya. But these lines of trade contacts tended to run roughly parallel to the coast, from which they were often divided by great areas of impenetrable bush.

Echoes of Axumite influence, transmitted through subordinate kingdoms in Kaffa and Sidamo, may be found as far south as southern Rhodesia. The fact that they are devoid of Christian association suggests they may have their origin in this period of commercial expansion—for Ethiopia did not become integrally Christian until the late sixth century. But they certainly passed overland. Much has been written about the 'Azanian culture'. This will probably be found to consist of a series of separate cultures beginning in the early Iron Age and ending in the mid-eighteenth century, and associated with quite different racial groupings. Some of these were almost certainly stimulated by influences that came ultimately from Ethiopia. Sometimes they may well have been affected by influences that came ultimately from Meroe in the Sudan, passing through the kingdoms in the south Sudan and then by way of the great lakes. Traces of them are common enough in Kenya and Tanganyika. Nothing has so far been found to link them with the Kenya or Tanganyika coast.

There is one exception. The medieval custom of erecting pillars, frequently phallic in shape, seems to have spread southward along the coast from the Port Durnford area, the Shungwaya of African tradition. They are most likely derived from the phallic pillars so common in Sidamo in the south Ethiopian zone. And it has been noted that there are two pillars close to Port Durnford far higher than is common on the coast and at least vaguely reminiscent of the obelisks at Axum. It is possible therefore that about this period there was a line of trade contacts that came south from Axum to what had been the emporium at Nikon.

If there is no evidence for any Axumite control of the coast, there is likewise none of any Persian. The suggestion that Shirazi traditions reflect a period of Persian rule probably implies a too great faith in racial memory; the Persian customs that have been noted among the Swahili may well have filtered in during the late medieval and post-medieval period. Trade presumably continued, since the demand for soft ivory, tortoiseshell, and slaves was constant. Indian influences are apparent at Adulis and at Axum. It is possible, therefore, that Indian traders took an increasing share in this traffic. But there is so far no archaeological or documentary evidence for this.[1]

[1] It is sometimes stated that the Purana contain detailed accounts of East

It is likely that there were fresh exports to the Persian Gulf. In Mesopotamia there is evidence for a considerable population of black warrior slaves from East Africa. They first revolted in 696. It is probable enough that the late Sassanian kings bought slaves from East Africa to be their Mamelukes.

This raises the problem as to the change in the population of the coast during this period. There is nothing in any Graeco-Roman source to suggest that the inhabitants of Rhapta or Zingion or Aromata were negroid. But early medieval geographers describe the natives of these areas as fully negro, and the few words they have preserved from their speech are Bantu in root.

According to one theory, the upper Congo–Zambezi watershed was a diffusion centre for Bantu-speaking negroid stocks. If this should be correct, it is not impossible that Ptolemy's 'man-eating Ethiopians', who lived to the south of the people of Rhapta,[1] were early Bantu migrants. They could have been described as Ethiopian because of their colour, and 'man-eating' because of a ritual. They might have pressed north up the coast during the succeeding centuries. But this is only the most likely theory among many. It seems certain, however, that by the beginning of the Middle Ages a negroid belt had come into existence along all the coastal area from Sofala[2] in Portuguese East Africa to north of Mogadishu in Somalia.[3] This will roughly correspond with the land of Zanj of the Arab geographers.

It seems clear that a Bushmanoid people survived in pockets within this belt, perhaps often as isolated hunters like some of the modern Boni. Probably they had been there before ever there was a Rhapta or an Aromata, and their continued presence on portions of the Kenya coast may have given rise to Ptolemy's name for that unfrequented shore, 'Troglodytice'. As so often in Africa, the newcomers may have tended to absorb rather than to displace, and quite different cultural levels

---

Africa. This strange myth originated with Lieutenant Wilford in 1801. See the first edition of J. H. Speke, *Journal of the Discovery of the Source of the Nile*, London, 1863. The passage was omitted in subsequent editions.

[1] Ptolemy iv, § 8. 3.
[2] Cf. E. Axelson, *South East Africa (1488–1530)*, 1940, p. 3.
[3] Cf. E. Cerulli, 'Gruppi etnici negri nella Somalia', *Archivio di Antropologia ed Etnologia*, 54, 1934.

could survive in proximity. A primary need of the archaeology of the East African coast is a close study of middens.[1]

According to one hypothesis, it would be possible to construct an exactly dated documentary history of the coast from the late seventh century. It is true that we now possess the documents. The problem is the extent to which they are historically reliable.

According to the *Kitab al-Zanuj*, there were a number of pre-Islamic Arab settlements along the coast, beginning at Mogadishu and ending at Kilwa. They accepted Islam, and the coast became Muslim in the seventh century under Khalifa Omar ibn al-Khattab. The Khalifa Abd al-Malik ibn Marwan (A.D. 685–705) sent Syrians to settle along the coast under the leadership of the Amir Musa. This is corroborated by the *Chronicle of Lamu*, which claims Abd al-Malik ibn Marwan as the city's founder, and by the *Chronicle of Pate*, which adds a list of thirty-five coastal towns founded by his Syrians in A.D. 696. The *Annals of Oman* add the story of two brothers, Sulaiman and Said, who fled with their followers from Oman to East Africa when their country was conquered for Abd al-Malik in A.D. 692. João de Barros, in his *Asia*, has preserved the story of an heretical sect that he names the 'Emozaidij', who fled to the East African coast and then into the interior. They have been identified with the followers of the Shiite pretender, Said ibn Ali, who was killed in A.D. 739. The *Kitab al-Zanuj* writes of the doubtful loyalty of the coastal towns to the Khalifa al-Mansur and of the expedition that he sent against them in 766–7. It also gives a list of cities in East Africa in which governors were placed by the Khalifa Harun al-Rashid (786–809). De Barros ascribes the expedition of the Seven Brothers of Laçah to the ninth century; it took place, perhaps, in about the year 887. He states that they came from the Persian Gulf, clearly from near Al Hasa, settled with their followers at Mogadishu and at Brava, and then opened the gold trade with Sofala. Both he and the *Chronicle of Kilwa* tell, with some variants, the story of the emigration of the Seven Princes of Shiraz, which he places seventy years later. According to the Kilwa chronicler, al-Hasan ibn Ali Sultan of Shiraz sailed from

[1] Cf. J. D. Clark, *The Prehistory of Southern Africa*, 1959, p. 311.

Persia with his six sons and with followers some time in the tenth century, and founded settlements on the East African coast and islands. His son Ali is stated to have become the first ruler of Kilwa island in 956.

This is a consistent story, but it is not corroborated by archaeology, and it seems incompatible with the descriptions of early Arab geographers. It is worth analysing the historical reliability of its sources.

The most recently discovered of these is the *Kitab al-Zanuj*.[1] It is a late-nineteenth-century redaction of an earlier compendium. There is much to be said for the suggestion that the compendium was constructed in the Shungwaya area in Somalia. It would seem to be based on three documents, a chronicle in Shungwaya, the chronicle of Mombasa, and an Arabic treatise on the conversion of the coast to Islam; this last would have been the source for the statements that have been quoted. It is clearly the work of a learned man, but there is nothing to suggest that it was written as early as the medieval period.[2]

The *Chronicle of Lamu*[3] is an early-twentieth-century collection of nineteenth-century oral tradition on Lamu island.

The story of Sulaiman and Said and of the eighth-century immigration from Oman became current in literature on East Africa after 1871, when the Rev. G. P. Badger published his translation of the *History of the Imams and Seyyids of Oman* by Salil ibn Raziq. Salil ibn Raziq died in 1873. He had derived the story from the *Annals of Oman*, written by Sirhan ibn Said ibn Sirhan in 1720. It cannot be traced farther.[4]

João de Barros became factor of the India House at Lisbon in 1532. His account of the history of the African coast[5] is based primarily on the *Chronicle of Kilwa*, but it seems clear that he had also another source, possibly sixteenth-century oral tradition gathered at the Portuguese factory at Malindi. The stories of the Seven Brothers from Laçah and of the Emozaidij come from this other source. 'The Seven Brothers' seems

[1] First published by E. Cerulli in *Somalia, scritti vari editi ed inediti*, 1957.
[2] For the *Kitab al-Zanuj* cf. A. H. J. Prins, 'On Swahili Historiography', *J.E. Afr. Swahili Committee* 28, 1958, pp. 26–40.
[3] Translated and edited by W. Hichens, *Bantu Studies*, 12, 1938, pp. 1–33.
[4] Cf. J. Schacht, *Ars Orientalis*, ii, 1957, p. 167.
[5] J. de Barros, *Decadas da Asia*, i, ed. A. Baião, Coimbra, 1930.

a common folk-tale. But the tale of the Emozaidij must have been recounted by an Arab who used the odd phrase *Umma zaydiyya*; possibly it was told in sixteenth-century Malindi as the explanation of a name of a tribe in the interior.[1]

The *Chronicles* of Pate and of Kilwa are of greater importance. It seems certain that there was a court chronicle in Pate Island[2] that dealt with the exploits of its Nabhani kings. It was destroyed during the bombardment of Witu in 1890. A member of the royal house, 'Bwaṇa Kitini', dictated part of its contents from memory to Captain Stigand and wrote a Swahili abstract in 1903, commonly known as the Hollis MS.[3] An increasing number of such versions are being recorded, and a copy of the original manuscript may yet be found. This may be treated, with caution, as an historic source from the time of the founding of the Nabhani dynasty in the thirteenth century; archaeology seems to be providing its fourteenth-century section with some curious corroborations. Its earlier details on coastal history are surely historically valueless.

The *Chronicle of Kilwa* is an incomparably more valuable source. The author states that he was born on the 2nd Shawwal 904 A.H. (13 May 1499) and that he was commissioned to write his history by Sultan Muhammad ibn Sultan al-Husain, who succeeded to Kilwa some time after 1512. Perhaps it was written in about 1530. Long excerpts from it are preserved in an Arab manuscript,[4] and a Portuguese summary of the *Chronicle* was first printed in 1552, in the *Asia* of de Barros.[5] The author came from a family of hereditary learning; he states that he was related to the Faqih Muflah al-Malindi and he writes of 'one on whom I rely as an historian', very possibly an earlier court chronicler. Coin finds along the coast closely corroborate the *Chronicle* from about the end of the thirteenth century. But when all this has been admitted, the passage dealing with the tenth-century Sultan of Shiraz can

---

[1] I am indebted for this suggestion to Dr. Schacht.
[2] A. J. H. Prins, *J.E. Afr. Swahili Committee*, 1958.
[3] For the first version C. H. Stigand, *The Land of Zinj*, London, 1913; for the second, A. Werner, 'A History of Pate', *J. Af. Soc.* 14/15, 1914–15, pp. 148–61 and 278–97.
[4] S. A. Strong (ed.), 'The History of Kilwa', *Journal of the Royal Asiatic Society*, N.S. 27, 1895, pp. 385–430.
[5] For a detailed examination of the relation between the Arabic and Portuguese texts cf. Freeman-Grenville, *Medieval History of Tanganyika*.

only prove that the story was current at the sixteenth-century court of Kilwa.

It would seem to be equally mistaken either to accept the chronological context of this mass of late oral tradition or to ignore its factual content. In many Islamized societies there have been attempts to forge links with the Age of the Prophet, and to do so by artificially prolonging authentic traditions and by embroidering them with the great names of early Islamic history. This tendency may have been already apparent on the sixteenth-century coast. Few African societies have developed any time sense akin to the European. Often the authentic tradition of a single family group may be ascribed to a whole people or township. All this may be true, not only of the documents that have been analysed, but of the oral traditions of the coast and its islands, which are still slowly being gathered. It is possible that none of them give information as to what really happened in East Africa between the seventh and the twelfth centuries. It is probable that all contain facts of importance for the medieval and post-medieval history, even if these have been antedated by centuries and recorded in exaggerated form.[1]

An outline of the history of the coast from the tenth to the thirteenth century can be reconstructed from contemporary Arab geographers. All these suggest that it remained predominantly non-Islamic and negroid, although there were a number of Muslim trading posts and settlements of increasing importance. Ibn Hawqal writes, in the tenth century, 'The sea continues to the land of Zingbar, Aethiopia opposite Aden: thence it departs from the regions of Islam.'[2] He adds the Zanji to his list of non-Islamic peoples whom he does not wish to exalt by describing.[3] Al-Masudi,[4] who died in 956, writes that the Zanji worship many kinds of totem, as well as a god they name Maliknajlu, and apparently have a cult of

[1] The Swahili ballads on the hero Liongo Fumo may provide an example. They were first recorded in the nineteenth century. Their hero has since frequently been described as a Shirazi warrior of the tenth century. It seems more likely that his setting was the seventeenth-century resistance to the Galla.

[2] Ibn Hawqal (trans. William Ouseley), *The Oriental Geography of Ebn Haúkal*, London, 1880, p. 16.

[3] Ibid., p. 4.

[4] Trans. and ed. by A. C. Barbier de Meynard and P. de Courteille, 9 vols., Paris, 1861.

ancestral spirits. As late as the twelfth century, Idrisi gives the impression that the coast is still largely infidel; pillars are worshipped in the town of Brava, a one-sided drum is worshipped at El Banes. All geographers consistently describe the Zanjis as negroes. There is mention of their curious physical traits such as hanging lips and long ears, projecting eyebrows and crisp hair. They are so notoriously black that black rhubarb is called after them, and so is ebony. The few words of their language that have been preserved by Masudi seem akin to Bantu—'Maliknajlu' for their god, 'Waqlimi' or 'Flimi' for their rulers, perhaps 'Kalari' for a form of food. They wear iron or copper as ornaments. Somewhere in this area, apparently to the south, there is a paramount chief who, Masudi notes, is called The Son of the Great God and who is elected and occasionally killed by his people; this seems an echo of a divine kingship.[2]

Throughout this period trade seems to be increasing and taking new shapes. Masudi lays his emphasis on ivory, Idrisi on iron. The slave trade was probably a constant factor; there are many references to Zanji slaves. Besides these there was the export of rhinoceros horn and tortoise-shell, amber, and leopard skins.

Much of this trade must have been carried by Muslims, and it seems clear that there were Muslim trading posts. The oldest and most famous of these is Kanbalu. It has been suggested that this is in Zanzibar.[3] Masudi had visited it, and states that Muslims had taken the island and enslaved the Zanji who lived in it between the beginning of the Abbasid and the end of the Umayyad dynasty; that is, between A.D. 747 and 754. But though in the tenth century there was a dynasty of Muslim kings, and trade was in the hands of Arabs of the tribe of Azd, the island was still partly infidel and the Muslims must have become Africanized, since they now spoke the language of the Zanji.[4]

[1] *Textes relatifs à l'extrême Orient*, ed. G. Ferrand, Paris, 1913, pp. 172 sqq.
[2] For the statements of al-Masudi cf. ed. Barbier de Meynard and de Courteille, i. 164–5, 205, 231–5, 334; iii. 5–7, 26–31.
[3] Dr. J. Schacht regards this as proved (*Ars Orientalis*, ii. 1957, p. 169, citing L. M. Devic and de Goege). But according to Ahmad ibn Majid and Yakut it would appear to be in Pemba.
[4] *Al-Mas'udi*, ed. Barbier de Meynard and de Courteille, i. 205 and 232, iii. 31.

To the south of Kanbalu there were also apparently some Muslim trading posts in the Kilwa group of islands. The *Kilwa Chronicle* records some prosaic details about the settlement on the island by a Persian trader in cloth, and some humiliating facts about the wars in the eleventh century between Kilwa and its neighbour, 'Shagh'.[1] These may well be reliable. To the north of Kanbalu, Mogadishu had developed into a considerable Muslim entrepôt between the tenth and twelfth centuries. By the twelfth century the town of Malindi in Kenya was an important centre for the iron trade; it seems likely that its population was part infidel, part Muslim.

Much of this is corroborated by archaeology. The existence of Muslim rulers in Zanzibar is proved by an inscription in Kufic script in the mosque in Kizimkazi, on the east side of the island. It states that the Shaikh al-Sayyid Abi Amran al-Hasan ibn Muhammad ordered its building, on the first day of the month Dhul-Qada in the year 500 A.H. (27 July A.D. 1107).[2] Further, there are a number of Umayyad and Abbasid coins in the museum of Zanzibar[3] though their exact provenance is obscure. The development of Mogadishu can be illustrated by a quantity of inscriptions[4] unique on the coast.

Besides the trade contacts with Arabia and the shores of the Persian Gulf, there were others with the Far East. Already by the tenth century Chinese geographers had a detailed knowledge of Po Pa Li, a country on the western sea, and had heard of the customs of the black and ferocious inhabitants of Mo Lin. It has been suggested that Po Pa Li was somewhere on the Somali coast and that Mo Lin was the Malindi area. Such stories were probably hearsay, gathered in India or Indonesia. Besides, there was a steadily increasing slave traffic. In A.D. 614 envoys from Java brought two Seng-chi women as part of the Imperial tribute;[5] in the eighth

[1] Most probably the settlement on Songo Mnara island.
[2] Cf. S. Flury, 'The Kufic inscriptions of the Kisimkazi Mosque', *Journal of the Royal Asiatic Society*, 1922, pp. 257-69.
[3] G. S. P. Freeman-Grenville, 'Coinage in East Africa before Portuguese Times', *Num. Chron.* 17, 1957, pp. 171-2.
[4] E. Cerulli, 'Iscrizioni e documenti arabi per la storia della Somalia', *Rivista degli Studi Orientali*, ii. 1926, pp. 2 sqq.
[5] *Hsin T'ang Shu*, ch. 222 C, pp. 6a, 8a.

[For all my citations from Chinese geographers I am indebted to Professor Paul Wheatley.]

century two dwarf women and two Seng-chi women were presented to the Emperor by the ruler of Sri Vijaya. In 1119 most of the wealthy in Canton possessed negro slaves.[1] Ivory, rhinoceros horn, tortoise-shell, and ambergris were all known in China as products of the black Zanj country,[2] and one later source also mentions gold.[3]

Yet there is no good evidence or likelihood that Chinese ships had penetrated the western Indian Ocean. One possible intermediary is the kingdom of Sri Vijaya.

The southern capital of Sri Vijaya was at Palembang in Sumatra, the northern at Kedah on the west coast of Malaya. From the eighth to the twelfth centuries it controlled the sea routes from India to China. As late as the early thirteenth century some form of suzerainty was claimed over Ceylon. Muslim geographers knew its Maharajah as King of the Islands of Zabag. In the ninth and tenth centuries traders from Oman and from Siraf in Persia used Kedah as their entrepôt. But it is also possible that Sri Vijaya was already trading in the western half of the Indian Ocean. Ibn Hawqal had noted[4] that there was a 'race of white peoples in Zingbar who bring from other places articles of food and clothing'. This was some time in the tenth century. By the twelfth there had been a further development. Arab geographers are notoriously in need of critical editions, and none more than Idrisi. But Mr. Moorhead's statement that 'Idrisi tells us that there was an animated trade between the East African coast and Sri Vijaya'[5] is at least a plausible explanation of a difficult passage.[6]

Such trade seems to have been at times accompanied by settlement. Perhaps both passed west by the equatorial current. There were Malay settlers in Ceylon, 'Javankas', at the beginning of the Middle Ages. Some time about this period there was perhaps a second considerable Indonesian settlement in Madagascar. Indonesian influences have been found on the

---

[1] *P'ing-chou K'ot'an* (Shuo K'u ed., p. 78).
[2] *Chu-fan-chih* (Feng Ch'eng Chun ed., p. 55).
[3] *Tao-i chih-lüeh*, folios 17–18.
[4] Ibn Hawqal (trans. W. Ouseley), p. 4.
[5] F. J. Moorhead, *A History of Malaya and her Neighbours*, London, 1957, i. 85.
[6] For an interpretation of Idrisi which would imply the presence of twelfth-century Chinese traders as far west as the Comoro Islands cf. Axelson, *South East Africa*, pp. 4–5.

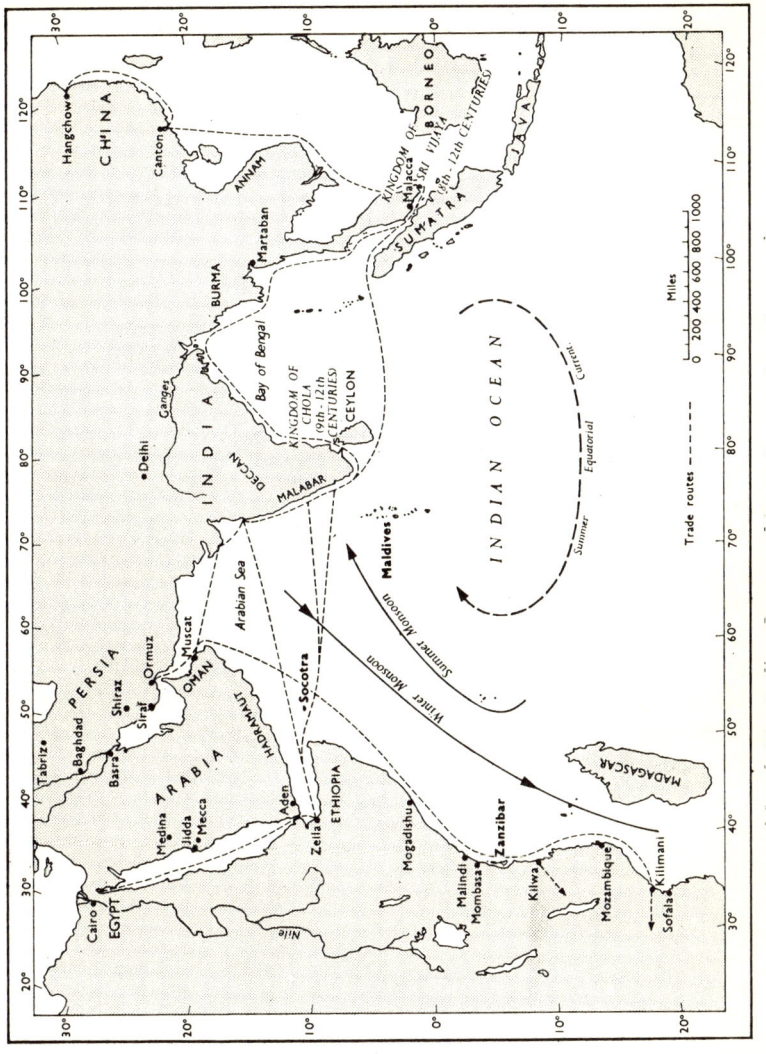

MAP 6. The Indian Ocean and its trade, c. 10th–15th centuries

East African coast, notably in South Somalia, and seem to have echoed far into the interior. Both the outrigger canoe and the banana[1] have been traced to Indonesia. It is at least not impossible that Indonesian traders first explored river trade routes that led towards the Rhodesian goldfield.

There was another Indian Ocean sea power which may have had contacts with the African coast. The Chola kingdom in south-east India had first become great in the ninth century. In origins it was essentially a coastal power, and in 1007 its ruler claimed that he had conquered 12,000 islands. It seems certain, at least, that he ruled the Maldives. In the fourteenth century Ibn Battuta found that the use of Maldive cowries as currency had spread across the African continent. The use of Indian systems of weights has been noted in distant parts of Africa. 'Trade wind' beads have been found in such quantity both on the coast and in Rhodesia that they must have been used as currency as well as ornament. Some at least of all this may have been due to Chola traders between the ninth and twelfth centuries.[2]

The Chola Kingdom had reached its height in the early eleventh century and vanished in the thirteenth; in the twelfth century Sri Vijaya was disintegrating and its successor state in East Java was reaching out towards the east,[3] not the west. By the thirteenth century the west Indian Ocean was becoming part of the Islamic sphere.

It now seems likely, on the evidence supplied by archaeology and by the Arab geographers, that the East African coast first became integrally a part of Islam in the thirteenth century. If

---

[1] Breadfruit, coconut, certain kinds of yam, taro, and banana have been listed as of Indonesian origin (G. P. Murdock, *Africa: its Peoples and their Culture History*, New York, 1959, p. 208). If this is so, they could well have been introduced by the first wave of Indonesian emigration and have been diffused in the first centuries A.D. through the expansion of the Bantu peoples.

[2] For this period the archaeological evidence is still slight. A number of Chinese coins from 713 to 1201 have been recorded from Somalia; a few from Zanzibar were recorded in 1939; two were found at Kisimani in the Mafia islands in 1954. The most considerable find, the hoard of 176 Chinese coins found at Kajengwa in 1945, cannot have been buried until after 1265. There is no porcelain that can be securely dated earlier than late Sung. Parallels have been found between the Sanje ya Kati group of ruins in Tanganyika and those at Kedah and Kotor Tinggi in Malaya. But this is an enigma best left for future excavation.

[3] Resemblances noted in Borneo and East Africa may be due to a common diffusion centre in the early Middle Ages.

so, once again it was following an Indian Ocean pattern. For the first time great Islamic states were developing as Indian Ocean powers.

On 26 March 1169 Saladin became the master of Egypt. On 5 February 1174 he sent his elder brother Turan Shah to conquer south-west Arabia. Aden, and therefore the outlet of the Red Sea traffic, was to remain under the control of the Ayyubid sultans of Egypt for the next fifty-five years. Egyptian influence was reasserted in the late fourteenth and early fifteenth centuries by the Burji Mamluk sultans, and in some sense Somalia and the north Kenya coast down to the Tana river can be considered as part of an Egyptian sphere until after the coming of the Portuguese.

In India in 1206 Qutb-ud-Din Aibak created an Islamic empire of the north-west with its capital at Delhi. Already in the thirteenth century the Delhi sultans were penetrating south-west to the coast, and between 1318 and 1320 they finally conquered the Deccan. In 1334 Daulatabad in the Deccan became for a short time the capital of their empire; it was taken by Ala-ud-din Bahmani in 1347. But the Bahmanid rulers of the Deccan were also to be champions of Islam. Hindu merchants could still flourish as their subjects, and it is possible that some may have traded in East African ports, but it is more probable that from the thirteenth century Indian influences on the east coast were officially and predominantly from Muslim India.

It is harder to calculate the effect of the developments at the head of the Persian Gulf. On 17 January 1258 the Mongols seized Baghdad and overthrew the Khalifate. But in Iran and Iraq they accepted Islam. Everywhere, by policy, they opened new trade routes and safeguarded trade, and late thirteenth- and early fourteenth-century coins of the Il Khans, the Mongol sultans of Tabriz, have been found in the Mafia Islands and in Zanzibar.[1]

As Muslim influence spread along the trade routes Islam began to be established in Indonesia. Marco Polo notes that the Kingdom of Perlak in Sumatra had become Saracen because of the many Saracen merchants who frequented it, and in 1281 the two envoys sent to Kublai Khan from Malayu, the last

[1] G. S. P. Freeman-Grenville, *Num. Chron.* 1957, pp. 151–71.

capital of the Sri Vijaya people, both bore Mohammedan names. In 1297 the King of Sumudra, on the Straits of Malacca, styles himself in an inscription Sultan Malik al-Salih. On both sides of the Indian Ocean Islam could spread by other ways than by emigration.

There was an additional reason for fresh Muslim trade interests in East Africa. All through the twelfth century the Crusades had been linking the west and east Mediterranean, and there was a rapidly expanding market in western Europe both for ivory and gold. Soft African ivory was best suited for medieval ivory carvings. In the thirteenth century the first gold coins to be struck at the London mint have been found by assay to be of African gold.[1]

Some time in the late twelfth century, Kilwa would seem to have gained control of the trade route that brought Rhodesian gold north from Sofala, in what is now Portuguese East Africa. This extension of influence is attributed by the *Kilwa Chronicle* to Sulaiman al-Hasan ibn Daud, who was held to have reigned from 1170 to 1188.[2] It was remembered of him in early sixteenth-century Kilwa[3] that with his father's help he had gained the mastery of the gold trade, that he had conquered as far north as Pemba Island, and that it was he who had first founded Kilwa citadel, building it with stone and lime and embellishing it with towers. It is likely enough that it was at this period that Kilwa developed from a small trading settlement into a fortified town.

De Barros also recorded a tradition that the men of Kilwa had captured the Sofala trade from men of 'Mogadoxo', possibly Mogadishu, but not impossibly Madagascar. It is described as trade with the Sofala 'Caffres'. No record or tradition suggests that Kilwa traders ever penetrated inland. A long trade route linked Sofala with the gold-bearing area in south Rhodesia[4] partly dominated by the Zimbabwe peoples.

[1] For this information I am indebted to Sir Maurice Powicke.
[2] Throughout, I quote the regnal years in the *Kilwa Chronicle* according to their computation by Dr. Freeman-Grenville.
[3] For this period there is a lacuna in the Arabic version of the *Kilwa Chronicle*. The statements cited are taken from the *Chronicle* in the sixteenth century by de Barros and appear in his *Asia* under the headings of 'Rei Daut' and 'Rei Soleiman Hacen'.
[4] Cf. the account of the journey of Antonio Fernandes from Sofala to the Monomotapa cited by Axelson, *South-East Africa*, pp. 137-45 .

Gold and ivory were carried to the coast, perhaps passing slowly through many intermediate tribes. Inevitably an increasing demand at Sofala stimulated an increasing supply. The traffic was at its height in the fifteenth century. Even in the early sixteenth century, when it was in decline, the Portuguese factor received annually an average of 51,000 pounds of ivory, 12,500 'miticals'[1] of gold, and 2,000 pounds of copper,[2] and this reckoning does not include the considerable contraband trade that went north on Muslim ships.[3] The Indian beads and Chinese porcelain that have been found at Zimbabwe[4] passed inland by this route. Cloth and beads were the primary trade goods brought by merchants.[5] Since both could be procured cheaply enough in the Indian Ocean ports, it is probable that no voyage during the Middle Ages could bring in such clear profit as the one to Sofala.

Certainly it helped to bring prosperity to the coast. By the fifteenth century there were thirty-seven towns along the coastal strip between the Kilwa group of islands and Mogadishu. Many were of considerable size and apparently wealthy. The majority, perhaps the great majority, of them had their origin about the twelfth or thirteenth century; often it would seem in a small fashion. Some of them may have grown rich through the possession of a local catchment area for ivory and slaves; this was very likely the case with Mombasa. The quite outstanding wealth of Kisimani in the Mafia Islands may have been due to a catchment area behind the mouth of the Rufiji. But directly or indirectly nearly all of them must have been affected by the growing importance of the Sofala trade route.

They were never to coalesce into a single state or confederation. 'The Zanj Empire' is a late-nineteenth- and early-twentieth-century fiction. Sometimes at feud, sometimes in league with their neighbours, they normally each had their own ruler and probably often their own oligarchy. But they had this in common—they were all Muslim. When Ibn Battuta describes his visit to the coast in 1331, he is very conscious

---
[1] It is impossible to determine the exact weight of a 'mitical' in the early sixteenth century.
[2] From the Accounts of Pedro Vaz Soares, factor at Sofala 1512–15, cited by Axelson, *South-East Africa*, pp. 126–7, 254.   [3] Ibid., pp. 121, 125, 142.
[4] G. Caton-Thompson, *The Zimbabwe Culture*, Oxford, 1931, in especial pp. 249–52.   [5] Axelson, *South East Africa*, p. 126. Cf. J. de Barros, *Asia*, loc. cit.

that he is in the land of Zanj, but even more conscious that he is still in Islam; the inhabitants are pious and devout, generous in almsgiving, orthodox Sunni, and obedient to the Sharia law.[1]

It is still difficult to decide how far the creation of this new Islamic culture along the coast caused a displacement of the previous population. Throughout Kenya and in parts of Tanganyika there are peoples[2] who claim to have come from the Shungwaya area. Shungwaya is now commonly agreed to have been in southern Somalia, the immediate hinterland of Port Durnford. On one hypothesis, the dispersal from there took place about the twelfth or thirteenth century. If this is correct the dispersal might have been the result of Islamic pressure, direct or more probably indirect. But on another theory the dispersal took place 500 or 600 years later. It is certain that in the early seventeenth century there was a great upsurge of the Galla and that they passed down the coast over the Tana river as far as Kilifi creek north of Mombasa.[3] It is certain that they destroyed the mainland Islamic settlements, likely enough that they drove whole peoples before them, and that these refugees in turn displaced other peoples, like the tribe that the Portuguese named 'Mosungalos', who appear to have had their centre in the Taita hills. It is possible that family groupings from such peoples took refuge among the less evolved tribes of the interior and brought with them oral traditions of their contact with the more elaborate culture of the coast.

On the whole it seems most likely that in the early medieval period the original coastal population was absorbed rather than displaced.

Again there are two hypotheses as to the origin of these settlements. It was long customary to refer to them as Persian or Arab colonies. This is not supported by any conclusive documentary or archaeological evidence. The buildings that survive from the late medieval period are not characteristically Persian or characteristically Arab. They seem to be a variant within Islamic architecture and the closest parallels

[1] For an analysis of the religious implication in Ibn Battuta's description cf. Freeman-Grenville, *Medieval History of Tanganyika*, ch. v.

[2] Segeju, Duruma, Giryama, Dabida, Digo, Pokomo, and in central Kenya, Meru and Kikuyu.

[3] Kirkman, *Gedi*, pp. 74–75.

to them will probably be found in Muslim India. Again it seems unlikely that colonies from an evolved Islamic society would forgo the use of money and the custom of funerary inscription. No medieval funerary inscription has yet been deciphered south of Merca in Somalia and coins do not seem to have been used either on the Kenya coast or in north Tanganyika. It seems excessively improbable that there was ever a Persian or Arabic colonization of the coast in the sense that there was a white settlement in Kenya, although there may have been numbers of Arab or Persian or Muslim Indian settlers.

A second hypothesis has been put forward more recently. According to this, trade had caused gradual development of a town life among the negroid population of the coast before the towns became really Islamized and their rulers began to build in masonry. The change would have come about through the presence of Arab, Persian, and Muslim Indian traders in such African entrepôts. African kingships which had become Mohammedan progressively acquired the technique and organization of Islamic states.

This is a possible development, for it occurred among the African peoples on the southern edge of the Sahara and can be paralleled on the east side of the Indian Ocean. But this development is not yet supported by archaeology. The documentary evidence in its support is slight and ambiguous. Idrisi seems to imply that there were many large towns along the coast that were peopled by pagans, but many of these may have been only seasonal town-sheltered anchorages to which the natives flocked each year to barter with the foreign traders. The conjunction between an African seasonal town and a permanent Muslim trading post would seem a likely origin for a medieval market town.

Both forms of the Africanist hypothesis could only apply to the coast and to the offshore islands. Such towns as those in the Mafia and the Comoro Islands must have developed on a quite different pattern. Possibly they were first temporary settlements where ships were beached between the monsoons, and then permanent trading ports with storage room for slaves and ivory.

Each of these hypotheses may yet be proved correct in a

particular case, for the medieval Islamic culture of the coast may well have grown haphazard. But in the fourteenth century there was probably a great deal that was characteristically African about these towns, whatever may have been their ultimate origin. Kilwa seems to have developed from a Persian trading settlement; in 1332 Ibn Battuta states that the inhabitants are black and have incised their faces (presumably for ornament). The Nabhani rulers of Pate had apparently come to the coast from Oman in the thirteenth century, but if the *Pate Chronicle* is to be trusted, they used what is apparently an African throne-name, 'Fumo Madi' or 'Fumo Mari', perhaps a reference to investiture with a spear. Mogadishu was the most highly evolved of the Islamic coastal cities. But Ibn Battuta's description of his visit there reads like an account of one of the negro Muslim states in western Africa. He notes that the king, who spoke a first language that was other than Arabic, had the four staves of his royal canopy surmounted by golden birds. A Chinese traveller who had apparently visited Mogadishu in 1417–19 writes of the townspeople there that 'the men wear their hair in rolls which hang down all round and wrap cotton cloths round their waists' and that the women 'apply a yellow varnish to their shaven crowns and hang several strings of disks from their ears and wear silver rings round their necks'.[1]

The official written language of these small Islamic states was presumably always Arabic. At the end of the period the *Kilwa Chronicle* was written in it. The King of Malindi used a rather corrupt Arabic for his correspondence with the King of Portugal.[2] But it was probably written rather seldom; the odd scarcity of inscriptions suggests that literacy was rare in the medieval coastal towns. It is impossible to tell how far it was spoken. Nicknames and the plays on words used by the *Kilwa Chronicle* suggest that the common speech of Kilwa was something akin to Swahili.[3] The same could be argued for the

---

[1] Fei-Hsin. The description of Mu-ku-tu-shu in the *Hsing-ch'a Sheng lan* (1436), ff. 5ʳ, ᵛ. Fei-Hsin had been a junior officer to Cheng Ho, who had been in charge of the voyages to East Africa in 1417–19 and 1421–2. For this information and the whole citation I am indebted to Professor Paul Wheatley.

[2] For this information I am indebted to Dr. Aubin.

[3] G. S. P. Freeman-Grenville, 'Medieval Evidences for Swahili', reprinted from the *J.E. Afr. Swahili Committee*, 29, 1959, pp. 4–8.

common speech at Pate[1]—so far as the *Pate Chronicle* is a reliable source. The Bantu language 'Chimbelazi' which survives in the town of Brava probably derives from the common speech of the coastal cities in Somalia.[2] It is therefore a possible hypothesis that the common speech along the coast consisted of variants of Bantu dialect charged perhaps in varying degrees with Arabic and, on occasion, Hindi loanwords.

The political history of the medieval coast can be reconstructed in rough outline. From the late twelfth century until the middle of the fourteenth Mogadishu and Kilwa were by far the most important towns in the area; they were the two that Ibn Battuta was to choose to visit during his world tours. Mogadishu was the larger and culturally the more evolved, to judge by the extent of the site and by surviving monuments. Kilwa was the more important as a commercial entrepôt. It not only controlled the sea route from Sofala, where it maintained an outpost until the late fifteenth century, but a network of trade contacts seems to have linked it overland with the remote interior. Mr. Roger Summers[3] has found traces of such a route joining Kilwa with the Ziwa people on the eastern borders of southern Rhodesia; it would seem to have crossed the Zambezi at Tete and to have led to the gold-producing area round Penhalonga. The Ziwa culture most probably belongs to the thirteenth and fourteenth centuries. It has been suggested that the route from Kilwa to Lake Nyasa was already in use in the medieval period, and even that there may have been a transcontinental route in the fourteenth century that linked Kilwa with West Africa and the Mali empire.

All this might be possible if by 'trade route' is meant 'trade contacts'. The penetration would still seem to have been from the coast to the interior. Ivory from many parts of inner Africa, Rhodesian gold, copper ingots from Katanga,[4] would have passed slowly to Kilwa to meet the demands of an unsated market.

[1] Ibid., pp. 9–11.
[2] Cf. I. M. Lewis, *Peoples of the Horn of Africa*, London, 1955, p. 43 (with references to E. Cerulli and N. Puccioni).
[3] Roger Summers (and others), *Inyanga, Prehistoric Settlements in Southern Rhodesia*, Cambridge, 1958, pp. 263, 151.
[4] For the significance of Katanga copper in early African trade cf. Clark, *Prehistory of Southern Africa*, pp. 307–9.

North of Kilwa there are as yet no indications of such traffic with the interior, except in Somaliland, where the Adal towns came into existence at the eastern end of the greatest of transcontinental African trade roads.

It was natural that Kilwa tradition should remember this period as one of prosperity. The Kilwa chronicler writes of the thirteenth-century ruler Khalid ibn Sulaiman that whatever he undertook he accomplished. It would also not be surprising if a new group of traders had attempted to seize the entrepôt.

Ali ibn Daud of Kilwa (c. 1263–77) was succeeded by al-Hasan ibn Talut (c. 1277–94). It is clear from the opening of the third chapter of the chronicle that this marks the beginning of a new dynasty: 'Then the throne departed from those that have been mentioned and was transferred to the house of Abul Mawahib (the giver of gifts).' 'With the help of his people he seized the kingdom by force.'

Al-Hasan ibn Talut is the earliest ruler in East Africa who is known to have struck coins. He inscribed a rhyme upon them:

> al Hasan ibn Talut
> yathiku bi Rabbi'l Malakut.

Such rhymed couplets were to be a distinctive mark of the coins of his dynasty.[1] He was associated with the introduction of a new type of architecture. He built the great mosque at Kilwa with its many domes; one dome in particular was called after him, the dome of Abu'l-Mawahib.

Similar rhymed couplets were inscribed on the coins of fourteenth-century Muslim kings in the Deccan, and the domes are reminiscent of Indian Muslim architecture. The geographer Ibn Said, who probably died about 1286, states that the inhabitants of Kilwa were brothers to those of India and Sind.[2] It is a possible hypothesis that al-Hasan ibn Talut was an adventurer who seized the throne with the aid of Indian Muslim traders. It is true that de Barros, in his *Asia*, writes of 'Hacen'

---

[1] Cf. Dr. J. Walker, 'The History and Coinage of the Sultans of Kilwa', *Num. Chron.* 16, 1936, pp. 43–81. Dr. Walker translates the full inscription on the coins of al-Hasan 'al Hasan son of Talut (may he be happy) trusts in the Lord of the Kingdom (He is glorious)'.

[2] For this information I am indebted to Dr. Freeman-Grenville, as also for the reference to G. Ferrand, *Textes relatifs à l'extrême Orient*, 1913, ii. 321 sqq.

as a grandson of 'Ale Daut', but this could well be a sixteenth-century dynastic fiction.

Little is known of al-Hasan's son Sulaiman al-Matun (*c.* 1294–1308), but one grandson, Hasan ibn Sulaiman (*c.* 1310–33), conquered the Mafia islands. The Kilwa chronicler wrote of him 'that he was the first to rule Mafia with absolute power' and another, Daud ibn Sulaiman (*c.* 1333–56), was known as 'The Master of the Princes'. Then some time in the second half of the fourteenth century the prosperity of Kilwa seems to have declined.[1]

One possible factor in that decline was the rise to power of the rulers of Pate island off the north coast of Kenya. According to the *Pate Chronicle* a Nabhani prince from Oman had established himself there in 1204. Until 1339 it would appear that Pate was only one small state among the many that lay between Mogadishu and Kilwa. But the *Pate Chronicle* states that Sultan Omar ibn Muhammad 'Fumo Mari', who died in 1392, first conquered the neighbouring island towns of Lamu and Manda, and then Malindi, that he made war against Kilwa and extended the Nabhani kingdom as far south as the Kerimba islands. It is possible that Songo Mnara, just to the south of Kilwa, was seized and fortified as a base from which to control the trade in Kilwa harbour. It may even have been temporarily the southern Nabhani capital; the great mosque there is still known as the Nabhani mosque and would appear to be late-fourteenth-century.[2]

Sultan Barquq, the first of the Burji Mamluks, ruled at Cairo from 1382 to 1399. His silver coins have been found in the Adal towns in Somaliland and his reign coincided with a fresh expansion of Egyptian influence towards the Indian Ocean and a new volume of Indian Ocean trade passing through Egypt towards Venice. Archaeologically, at least from the late fourteenth century, the Kenya towns north of Tana river seem oriented towards Egypt;[3] Pate was the chief among them. If the account in the *Pate Chronicle* is at all to be relied upon, it is not impossible that the sudden victories of

[1] The indications of such a decline are discussed fully by Dr. Freeman-Grenville, *Medieval History of Tanganyika*, ch. vi.
[2] Cf. A. G. Mathew, 'Songo Mnara'; essays presented to Sir John Gray, *Tang. Notes*, no. 53, 1959, pp. 155–60.
[3] I am indebted for this suggestion to Mr. J. S. Kirkman.

Pate were due, however indirectly, to new trade policies in Egypt. In any case they would seem to have been transient. There is no evidence that Pate was still a powerful state when the Portuguese reached East Africa, and all through the fifteenth century the coastal towns south of the Tana remained primarily oriented to the Persian Gulf and to India.

The climax of the prosperity of Indian Ocean trade was reached between the late fourteenth and the late fifteenth centuries, for this was the period between the eclipse of the overland routes from the Far East to the West and the discovery of the route round the Cape. The thirteenth-century Mongol empire had ensured the security of the great roads that went east from Caffa in the Crimea and from Tabriz. These became hazardous as the Mongol empire disintegrated. They must have been virtually impassable during the anarchy that accompanied the ravages of Timur, who died in 1405. They were never to regain their role in world trade.

In consequence a new route was developed which linked the Persian Gulf with the Black Sea. In the fifteenth century a Venetian fleet sailed annually to Trebizond to purchase Indian pepper. Another far older route was revived to join the Mediterranean and the Persian Gulf through the great market at Aleppo. The wealth of the Jalairid sultans of Baghdad (1358–1409) was largely based on this traffic. The Muzaffarid sultans of Shiraz and their Timurid successors who reigned there until 1467 became the greatest power along the Persian Gulf and their subjects traded as far east as Malacca. To the west, Egypt under the Burji sultans (1382–1517) entered into the closest trade contacts with Venice and had become by 1468 the richest of the Islamic states.

All this is explicable, for the European demand for pepper and for every form of spice had reached a new height with the fifteenth century. The need for ivory and gold was constant. The international court culture that came into existence in the west about 1380 provided a fresh market for luxury goods, while throughout the Middle East there was a demand not only for all these but for women slaves and for Malayan tin.

This trade expansion was joined to increased traffic from east to west in the Indian Ocean. The Ming dynasty had gained control of China in 1368. This led to an increase in

Chinese sea power, primarily in order to restore Imperial prestige. The third of the dynasty, the Yung-Lo emperor, sent fleets as far as East Africa in 1417–19 and in 1421–2. Mogadishu, Brava, and Juba were reckoned as Imperial tributaries. Under the Hsuan-Te emperor the Imperial fleet entered the Red Sea in 1431–2. The Mon dynasty in Burma had created a great entrepôt at Martaban and it was perhaps at its zenith from 1353 to 1472. Annam would seem to have been a centre for trade contacts under the Second Tran dynasty and the second Le (1407–59).

Ming porcelain has been found at every late medieval site on the East African coast, Martaban ware is not uncommon, and there are ceramic variants which are best explained as being Annamite.[1] But this should not lead to an over-emphasis on trade contacts between Africa and the Far East. The great portion of such wares was probably acquired by Muslim Indian merchants in western India and exchanged by them for African goods for the Indian market. There were new developments in Islamic India. In 1392 Zafar Khan founded the line of the kings of Gujerat who, to judge from Portuguese records and from the prevalence of 'Cambay' beads, were more closely linked with Africa than any other Indian dynasty. The Muslim Kingdom of Khandesh was founded in 1382 and that of Jaunpur in 1394. The Bahmanid kings of the Deccan would seem to have been at the height of their prosperity between 1378 and 1482. The traders of such kings would be the natural entrepreneurs to bring African goods far into India; soft ivory, gold, iron ore, probably copper, certainly slaves, especially as slave warriors.[2]

Inevitably there was a marked increase in the wealth of the African coastal towns. The 'stone cities' described by the Portuguese were a fifteenth-century creation. The huts of the townsmen were replaced by houses built of coral rag with cement of coral lime. The fifteenth-century towns were populous; even in Kenya, which was economically less developed than Tanganyika, the town walls of Pate, 'Ungwana', and

---

[1] Cf. A. G. Mathew, 'Chinese Porcelain in East Africa and on the Coast of South Arabia', *Oriental Art*, N.S. 2, ii. 1956, pp. 50–55.

[2] The Tanganyika markets were probably the centre of this slave trade. Along most of the Kenya coast it is more likely that slaves were imported than exported.

'Gedi' enclose areas of over 45 acres; Mombasa was probably much larger. It is possible through the excavations at Gedi to reconstruct much of their material culture[1]—the flat-roofed semi-detached houses with their elaborate sanitation and their narrow courts and their women's quarter, the taste for many beads, the common use of Chinese porcelain—celadon in the first period and then from about 1475 predominantly Ming blue and white.

A change in culture seems to have coincided with an increase of population. There is archaeological evidence for many smaller settlements between the towns—villages with only two buildings in coral rag; a mosque and a rich man's house, perhaps a trader's, perhaps a headman's. Fifteenth-century commercial prosperity may have led to a new influx of settlers; it almost certainly caused reorganization of trade.

On the other side of the Indian Ocean, Muslim traders, primarily Persian, had organized a great entrepôt at Malacca. Paramerswara, the ruler of this small Malay port, had accepted Islam in 1414 and had taken the name of Iskander, perhaps in compliment to the Iskander who reigned at Shiraz from 1410 until 1415. His dynasty retained the throne and were accorded the title of sultan. The real power rested with a group of great Muslim officials. The most important of these, known by the Indian title of Bendehara, virtually controlled the state and made and unmade sultans. But theoretically he was only the president of a council of four: there was the Temenggong who administered justice, the sultan's Treasurer, and the Shahbandar who supervised the harbour, the reception of foreign traders, and the collection of customs dues. Malacca became the wealthiest harbour town in the Indian Ocean and the offices of Bendehara and of Temenggong were retained within two family groupings until the coming of the Portuguese.

Possibly a similar process occurred at Kilwa. Muhammad ibn Sulaiman who reigned there from 1412 until 1421 was named 'The Just King' and 'The New Rain'. The *Kilwa Chronicle* records that he accepted an Amir and 'other notables of Government'. He is remembered in an oral tradition at Kilwa

---

[1] According to one hypothesis Gedi was the palace quarter of old Malindi in the fifteenth and sixteenth centuries. Even if this is not correct, at least it provides archaeological evidence for the Malindi area in this period.

because he gave the Amirate and the Office of Justice to men from Malindi; perhaps a confused memory of the fact that he had conferred them upon foreigners. He is the first ruler of Kilwa so far known to have used the title of sultan on his coinage. There is every evidence for a resurgence of prosperity in fifteenth-century Kilwa. But to judge from the *Chronicle* real power had passed to the Amirs who could depose and recall sultans and whose office was retained in one family grouping.

There was also another family grouping, the House of the Wazirs, who held an office perhaps equivalent to that of the Malacca Temenggong. It might be possible to discern the office of Shahbandar behind the Portuguese accounts of the activities of Muhammad 'Ankoni'.[1] The *Chronicle* states that during the reign of Ibrahim (1499-1506) the office of Treasurer was shared between two brothers from Dabhol in India.

In the sixteenth century, Portuguese accounts agree as to the prevalence of 'white Moors' at Kilwa. Perhaps the present oral traditions of the coast as to the 'Wa Dabuli' and the 'Wa Shirazi' are based in the main on fifteenth-century memories. Possibly the prestige of the fifteenth-century sultans of Shiraz led to the fabrication of pedigrees tracing the descent of African rulers to tenth-century Shiraz kings. But at its most international, Kilwa was still an Afro-Islamic state. The thirteenth-century founder of the dynasty of Abul Muwahib may possibly have come to Kilwa from Sind. It is probable that by the male line he was ultimately an Arab, since the chronicler gives him the Hadramauti clan name of al-Mahdali. By the sixteenth century his descendants were claiming a far longer royal descent from Persia. But in Kilwa there are still memories of their 'Siwa', the horn of elephant tusk that enshrined the strength and spirit of their people.[2]

Kilwa continued to prosper through most of the fifteenth century. Sultan Sulaiman ibn Muhammad rebuilt the Great Mosque there which was to be compared by the Portuguese with the Great Mosque at Cordoba. The output from the mint seems to have much increased. The last of the sultans known to have struck coins, al-Hasan ibn Sulaiman (1479-84 and

[1] 'Ankoni' is a Portuguese corruption of Rukn-ad-Din.
[2] For this information I am indebted to Dr. Peter Lienhardt, who also suggested to me the meaning of the term Mahdali.

1486–90), struck them in sixteen variants, and over 2,000 of them have been recovered. Although sultans rose and fell quickly—there were thirteen between 1442 and 1498—the amirate was stable; Muhammad Kiwab was amir from 1476 to 1495. Yet Kilwa never gained the same significance on the west side of the Indian Ocean as Malacca on the east.

This was primarily because Kilwa was never to gain control of the coastal cities to the north through which all her trade had to pass. From the fourteenth century Kilwa ruled the Mafia islands with their two capitals, Kisimani and Kua. It is possible that Tongoni on the mainland near Tanga was also a dependency.[1] But Zanzibar was an independent and at times hostile[2] sultanate which possessed its own mint in the fifteenth century,[3] and to the north there were Mombasa and Malindi. Either of these could control not only the coastal traffic towards the Red Sea but also the Kilwa trade with the Persian Gulf and with India, since they were the points of arrival or departure on the monsoon.

The rise of Mombasa has not yet been adequately explained. In the fourteenth century, when Ibn Battuta visited it, it was a small town of wooden houses. But the Portuguese describe it as a great city with high 'stone' houses, a palace, and a fortress. Its ruler was able to offer Almeida 50,000 'pardaos' of gold as a bribe in 1506 and it took fifteen days to load riches gained in the subsequent sack. Yet the state itself seems to have remained economically primitive; no medieval coin has yet been found there. Perhaps the explanation may lie in the many 'ships of Cambay' noted there by the Portuguese. Possibly it had become the headquaters of the Gujerat traders.

Malindi would seem to have remained in some fashion an ally of Kilwa united by a common hostility of Mombasa. But before Vasco da Gama rounded the Cape on the 22nd of November 1497 Mombasa was in the ascendant and Kilwa already in decline. Mombasa is described by the Portuguese as the wealthier and more populous, and no coins have been recorded from the mint of the last Kilwa sultans.

---

[1] The only evidence for this is from oral tradition at Tongoni.
[2] Cf. the account in the *Kilwa Chronicle* of the attempt of Sultan al-Hasan ibn Abu Bakr of Zanzibar to place his own candidate on the Kilwa throne in 1454.
[3] Cf. G. S. P. Freeman-Grenville, *Num. Chron.* 1957, pp. 155–71.

Medieval Mogadishu had been the successor state of the emporia of Aromata, Zanzibar had represented Menouthias, and Kilwa, Rhapta. But the significance of Mombasa lay in the future, not in the past. The economic pattern of the East African coast had been broken for the first time since the age of the *Periplus*.

NOTE TO CHAPTER IV

SINCE this chapter was completed five archaeological discoveries are being made which may come to affect some of its tentative conclusions. The most important of these is the excavation of the citadel of Kilwa, the Husuni, where a very elaborate palace complex is being uncovered by Mr. H. N. Chittick. There is no trace within it of any African influence. It is marked by a delight in geometrical precision by the use of the octagon, and by careful and sophisticated decoration. It belongs to an international type of early Islamic palace. The ruler for whom it was built must have been wealthy enough to procure craftsmen either from Mesopotamia or Egypt. There is already some evidence that it was only occupied for a brief period and that it was deserted by the fourteenth century. An inscription records the name of al-Mansur al-Hasan ibn Sulaiman. On my own tentative interpretation of the site, this may refer to al-Hasan ibn Sulaiman of Kilwa, 1191–1215, the brother of the Daud ibn Sulaiman whose building of a fortress is recorded in the Portuguese version of the Kilwa Chronicle. On this hypothesis the building of the palace was the result of the new wealth that came to Kilwa after 1170 with the control of the gold trade to the north, while its sophistication reflects the fresh influences from inner Islam which seem to mark the late-twelfth-century Indian Ocean world. But according to another interpretation the inscription refers to al-Hasan ibn Sulaiman, who ruled at Kilwa from 1005 to 1042. If this should prove to be correct Kilwa was the capital of a great power over a century before it controlled a trade route from Sofala.

Work has only begun on the small Husuni, which is divided from the palace complex by a gully. It seems already clear that both groups of building are of the same period. On my interpretation the Husuni Ndogo was the guard house of the palace garrison; this would be quite compatible with a late-twelfth-century dating. But it has been suggested that it was an open mosque. If this should prove to be correct, the early eleventh century is the latest date conceivable. Even in the eleventh century such a mosque form would be oddly archaic. The Byzantine influence on the palace architecture, so obvious in the use of the octagon, would be perfectly compatible with an early Abbasid date. A third possibility therefore cannot yet be completely discarded; the Husuni complex might be earlier than anything recorded in the Kilwa Chronicle. It could then be the remains of a fortified and perhaps uncompleted palace built by some forgotten Islamic conqueror in the late eighth or early ninth century.

The third archaeological discovery does not seem likely to provide many problems for the historian. As early as 1955 Dr. Freeman-Grenville in his survey of the Mafia islands suggested that the site there now known as Kisimani would prove to be of crucial importance for the mediaevalist. This has since been demonstrated by the excavations that are being directed by Mr. H. N. Chittick. It already seems almost certain that when these have been completed the Kisimani site will be found to represent one of the major commercial entrepôts in East Africa in the thirteenth and early fourteenth centuries; the very high proportion of imported wares to local pottery is already unparalleled on the coast. If my tentative interpretation of the Husuni site as late-twelfth-century should prove to be correct, the Kisimani discoveries could fit into the same historic sequence, On this hypothesis Kilwa had first grown wealthy in the twelfth century as the entrepôt on the routes that led south to Sofala. In the thirteenth century it is succeeded by the 'Kisimani' site on Mafia, which may also have been the terminus of a hitherto unsuspected route from the interior to the mouth of the Rufiji river. In the early fourteenth century a new dynasty of Kilwa rulers conquered Kisimani and perhaps established a mint there.

Some fresh but inconclusive evidence has been found which may yet help towards the solution of the two main problems discussed in the chapter; the possibility of a continuity between some of the Islamic trading cities and pre-Islamic settlements or seasonal markets, and the extent of Africanization in the town population of the fifteenth century.

Mr. Chittick has discovered the traces of an extensive late Stone Age settlement on the slopes of Nguruni below Kilwa citadel. On the other hand the excavations at the Kizimkazi site on Zanzibar suggest that, though the site had been inhabited for some time before a mosque was built there in 1107, it was probably only a small Islamic trading post.

Recent excavations close to the great mosque at Kilwa and by the 'Sultan's Mausoleum' have provided new evidence for the character of the Kilwa culture shortly before the coming of the Portuguese. An elaborate stone carving has been uncovered, with many coins and a quantity of good-quality Chinese porcelain. The carving is reminiscent of Muslim India. But there are also signs that suggest to me the survival of African animism and of human sacrifice even among the rich merchant class at Kilwa; bowls buried intact beneath the flooring, a severed human skull, possibly the litter of scattered bones.

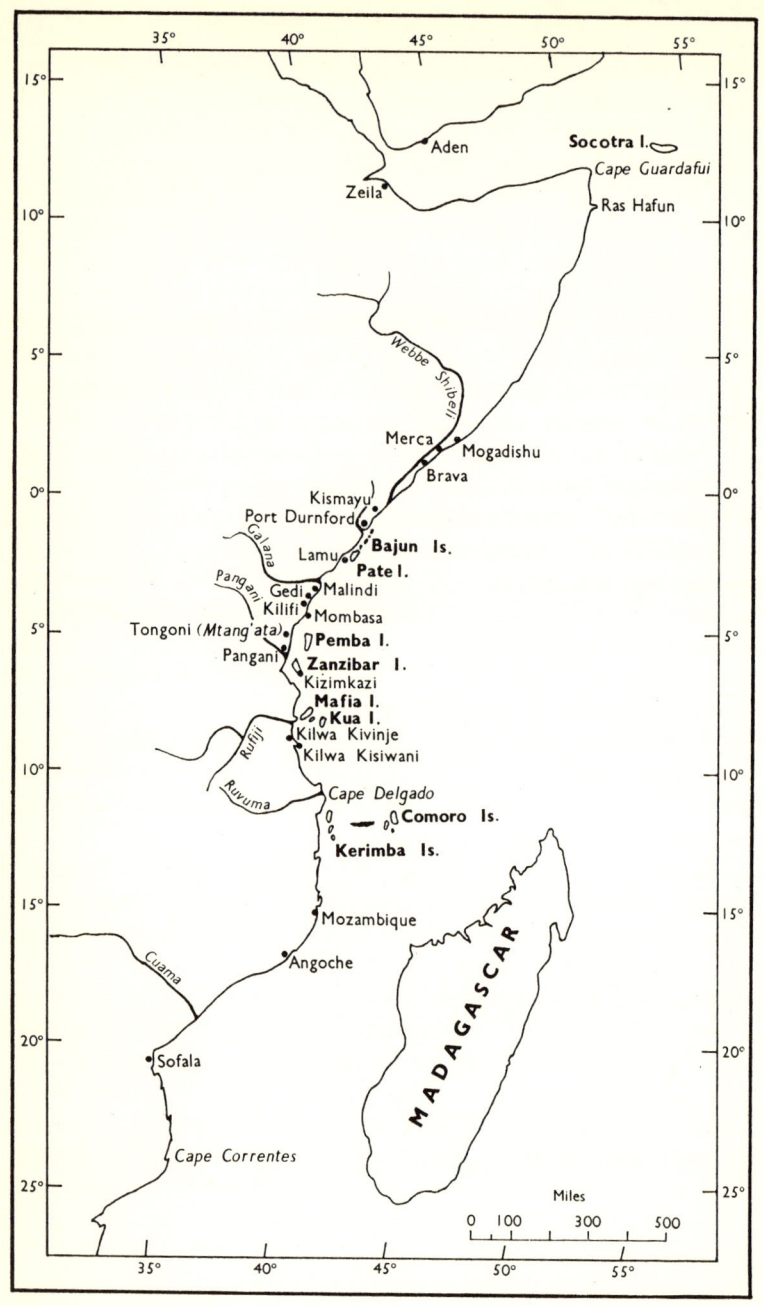

MAP 7. The East African coast, showing the principal places mentioned in the text

V

THE COAST
1498–1840

G. S. P. FREEMAN-GRENVILLE

Two grand climacterics bound the history of the East African coast between 1498 and 1840—the first incursion of Europeans into the Indian Ocean and the final removal of the Court of Oman to Zanzibar. Within this period there are other, more crucial, more formative, themes. New unities from which a Swahili culture now grew and flourished replaced those of medieval times. At the end of the sixteenth century the savage, cannibal Zimba wrought a massive destruction on the population from the Ruvuma to Malindi, and, in response to other pressures, new tribes percolated into the coastal hinterland. They mingled with the relics of the earlier population and with fresh migrations from South Arabia and the Persian Gulf to which many family traditions attest, and were absorbed on the coast into a linguistic unity with a common ground not only in religion but also in commerce and culture. Their struggle, evoked by the Portuguese, paved the way for Sayyid Said's attempt to realize a political unity in the nineteenth century. The period is thus essentially one of the fusion of hitherto diverse elements: and we shall not understand the nature of the coastal towns, about which we are best informed, without some conception of the tribal hinterland.

Its population in the early sixteenth century cannot be described with precision. The only tribe the Portuguese mention by a recognizable name is the Segeju, who are recorded in 1571 near Malindi. It is probable that Swahili had long been spoken in Mogadishu; but no clear idea is possible of the situation in the extreme north. Presumably the coastal Bajuni, who now inhabit the 'five hundred islands' north of Lamu and who claim Arab origins, represent in their present amalgam the remainder of the population which, in the late sixteenth and early seventeenth centuries, was expelled or destroyed by the incursions of the Galla, who themselves suffered attack from the Somali.[1]

A common thread of tradition links many tribes in present-day Kenya and northern Tanganyika with the legendary Shungwaya, probably rightly identified[2] with the vast, deserted site of Bir Gao or Port Durnford. The *Kitab al-Zanuj*[3] and their own oral traditions claim Shungwaya as the dispersal point for the Kikuyu, Meru, Kamba, Taita, Asu, Chaga, Shambaa, and Digo, if not the Segeju, together with some tribes whose names are no longer identifiable. If these tribes had indeed sprung from a town with the institution of kingship and with Islam as its religion, one would have expected to find both among these tribes at an early period. Yet of these only the Shambaa coalesced to form a kingdom, in the eighteenth century: and none, according to the surviving genealogies and traditions, show any trace of Islam before the end of the nineteenth century. These traditions are largely those of the ruling families, and one may conclude that many of the alleged tribal movements are rather those of ruling clans whose traditions have been absorbed by others, and who reached their present habitats, as the traditions in general agree, during the sixteenth and seventeenth centuries. These movements were rather those of loose groups of kindred or even simply of individuals who were absorbed into earlier, small, and unorganized populations. Such a view is supported by the absence of political organization of a dynastic kind amongst tribes claiming Shungwaya origin: rather, these are characterized by loose, informal assemblies of elders, gathered as local convenience, need, or interest might dictate, and occasionally by 'secret' societies in which religion and politics merged. A common general origin for all the northern people, extending also to the Bondei and to the Zigua of Handeni and Morogoro District, in spite of the superficial diversity of their languages, is suggested by their wide common or similar vocabulary. An excavation in Handeni District in 1955 suggests the survival there of a race like the Bushmen into the eighteenth century; moreover, throughout the coastal area

---

[1] Cf. E. Cerulli, *Somalia, scritti vari editi ed inediti*, 1957, vol. i; A. H. J. Prins, *The Coastal Tribes of the North-Eastern Bantu*, 1952.   [2] Prins, ibid., p. v.
[3] See Chapter IV, p. 103, above, and bibliographical note to this chapter.

great variations of physical type among the so-called coastal tribes, including not infrequent steatopygia and smallness of stature, suggest the absorption of earlier peoples of this type.[1] Farther south, between the Pangani and the Rufiji, the traditions again are of loose-knit family groupings. An African author of a history of the Doe in the Bagamoyo hinterland states categorically that his is not a tribe, but a group of families in free association. None the less, in the course of time, they have developed a common dialect and common, specialized, agricultural techniques. His view is no doubt accurate for all the so-called tribal groupings as far as the Rufiji and for the earlier elements in Kilwa District. Farther south, the mass of the inhabitants, among whom the Makonde predominate in the coastal hinterland, are confessedly descendants of nineteenth-century arrivals pressed thither from beyond the Ruvuma.

From the height of Singino Hill, behind Kilwa Kivinje, a splendid view illumines the contrast between the trade of the Middle Ages and that of the following centuries. Behind the medieval island settlement of Kilwa Kisiwani lies a long stretch of creeks and marshes, which makes contact with the mainland impossible without transhipment and a long detour by land. Seventeen miles north of Kilwa Kisiwani is its eighteenth-century replacement, Kilwa Kivinje, on the mainland, necessitated by the new direction of trade. In the sixteenth century the sea trade was the preoccupation of the coastal towns: the loot taken by the Portuguese from Kilwa and from Mombasa was gold and ivory largely from the southern part of the continent, payable against cloth and beads imported from India and elsewhere. It is curious, in view of the large quantity of sherds found, that imports of Chinese porcelain find no mention in local literature before the eighteenth century. Based upon its possession of Mozambique and Sofala, and deriving its main wealth from them, the range of the economic dominance of Kilwa seems demarcated by the area within which its coinage has been recovered. It runs no farther north than Mafia, for fifteenth-century Zanzibar not only had political independence but its own coinage.[2] It is significant that

[1] I am grateful to Mr. J. G. Stephenson for a note on his as yet unpublished excavation in 1955.
[2] Cf. G. S. P. Freeman-Grenville, 'Coinage in East Africa before Portuguese Times', *Num. Chron.* 17, 1957, pp. 151–75.

coinage, save for a few small finds, is absent on the mainland north of the Rufiji. There are differences, too, in architecture: especially in the cemeteries, which, unlike those in other parts of the Muslim world, lie close to the mosques. A notable feature in the north is the pillar tomb unknown in the south.

In the north there seems to have been a series of small rulers. Pangani has been thought to be the ancient Rhapta, but it has no tradition of settlement anterior to the late sixteenth century. At Mtang'ata, (the modern Tongoni), where Vasco da Gama found the oranges so much to his liking, no traditional history remains in spite of a fifteenth-century mosque of considerable size. Vumba, among the creeks of the present Tanganyika–Kenya border, had a sultanate, but its sway was no more than local.[1] Opposite, in Pemba, the traditions are of autonomous settlements under at least five 'kings', while in Zanzibar it is probable that more than one sultanate contested the island. Mombasa is said to have been ruled by a 'Shirazi' dynasty, related to, if not descended from, that of Kilwa. Its population, so say the Portuguese, numbered 10,000 against 4,000 in Kilwa. Seventy miles north of Mombasa was its rival Malindi, also with a 'Shirazi' dynasty, with Gedi,[2] perhaps a Windsor in relation to the Westminster of Malindi, nine miles to the north. If sixteenth-century tradition is to be trusted, Mombasa included Vumba and the Mtang'ata, but the territory between Mombasa and Malindi was in dispute. Yet farther north, with several other petty sultanates, including Lamu and Siu, was Pate; beyond them, again with some small states intervening, were the important cities of Brava and Mogadishu, the latter with her own coinage. In a letter to the King of Portugal, written on 20 November 1506, Diogo de Alcaçova reported, with many palpable exaggerations, that Mombasa collected customs dues for Kilwa from ships which had not put in there on their northward journey. This might imply vassalage, but there is no other reliable evidence for this.

Thus, if religion and the Swahili language[3] gave a semblance of unity to the area, it was divided politically amongst

[1] A. C. Hollis, 'Notes on the History of Vumba, East Africa', *J. Roy. Anthrop. Inst.* 30, 1900, pp. 275–97.
[2] Cf. J. S. Kirkman, *The Arab City of Gedi*, 1954.
[3] Cf. G. S. P. Freeman-Grenville, 'Medieval Evidences for Swahili', *J.E. Afr. Swahili Committee*, 29, 1959, pp. 1–14.

a number of rulers. Kilwa looked southwards, and controlled the sea traffic from Mafia far away to Sofala. Mombasa, and to a lesser extent Malindi and other larger towns, participated in this traffic and controlled the intermediate marts. Farther north the only centre of first-class importance was Mogadishu, with the Webi Shebelle behind. As to the larger islands, Pemba, Zanzibar, and the Mafia group were primarily agricultural, although Kisimani Mafia was important as a watering station. Pemba was chiefly a purveyor of rice, and the rest of coconuts and their by-products. They had not the wealth of the mainland, which indeed was confined to a small number of merchant cities: between these lay a great number of coastal villages which served not only for fishing and agriculture but also for the collection of ivory. It was ivory which lent importance to places such as Vumba, Lamu, and Pate. Their relations with the interior, on which their wealth depended, will be explored later in this chapter.

It is not known to what extent fifteenth-century Portugal was informed about East Africa and the Indian Ocean. Fra Mauro Camoldese's map of 1460 shows Sofala and a number of other places. It was only six years after the expulsion of the Moors from Spain when Vasco da Gama set out on 5 July 1497. Christendom and Islam had long been in contact in the Iberian peninsula. A Portuguese, Pedro da Covilha, had travelled in the Indian Ocean collecting information on behalf of the King of Portugal between 1487 and 1491.

Since the early fifteenth century the main aim of Portuguese policy, under Henry the Navigator's inspiration, had been to seek some means to reach the eastern markets and to outflank Islam. The struggle to free the Iberian peninsula from Islam had been a national struggle, and thus far more bitter than the Crusades of which it was preached as a revival. The discovery and exploitation of the East and of India was itself a Crusade in a new aspect as well as an opportunity to gain untold riches. This religious aspect of what is primarily a commercial matter must not be passed over: it is the counterpart of the Islamic dogma of the *Dar al-Harb*, and to the Portuguese and to all their Christian contemporaries it justified the bringing of kings under tribute, the seizure of their lands, and the control of their trade.

Da Gama's mission of 1498 was as an explorer. His achievement was the discovery of the sea route to India and the proof that at Sofala there was 'infinite gold'. He made no conquests or trade agreements, save an understanding with the King of Malindi. The latter was on bad terms with his neighbour of Mombasa, and this disagreement provided the Portuguese with a foothold. Even if Vasco da Gama had been more tactful, it is doubtful whether he could have been more successful. The sultanates, with an unbroken sequence of independence, saw no attraction in the proffered protection of the Portuguese king nor reason in the payment of tribute. Looking back in about 1520 on the events that followed the Portuguese arrival, the chronicler of Kilwa wrote: 'Those who knew the truth confirmed that they (*sc*. the Portuguese) were corrupt and dishonest men, who had come to spy out the land in order to seize it.' But unfortunately for the sultanates, they had no tradition of cohesion in the face of a common enemy, such as the King of Mombasa urged on Sayyid Ali of Malindi in 1502. The Portuguese could deal with them piecemeal.

So, in 1502, on his second voyage, da Gama was able to bring Kilwa under tribute and, in 1503, Ruy Lourenço Ravasco received tribute at Zanzibar. But the grand design, in which Pedro Alvares Cabral failed in 1500, was the establishment of a factory at Sofala, and this was one of the main objects of the largest fleet yet sent, under Francisco d'Almeida, in 1505. Sofala, in the possession of Kilwa since the twelfth century, was occupied by Pedro d'Anaya with little bloodshed, because the aged, blind Shaikh Yusuf had already declared himself independent. He calculated too, and rightly, that fever would take more Portuguese lives than resistance. As to Kilwa and Mombasa, it is a matter of opinion whether the attacks upon them and their sacking were premeditated. Both resisted and both were laid waste. In the following year Lamu, Pate, Oja, and Brava were attacked. The first two capitulated. Oja declared that its only allegiance was to the Caliph in Egypt, and with Brava, likewise defiant, was sacked and burnt. Only the lateness of the season saved Mogadishu; but, within eight years of their arrival, the Portuguese could claim to have seized the whole of the coast and the trade route to India and beyond.

This apparent success was costly, for it had been wholly destructive. The gold of Sofala had been the life-blood of Kilwa, and its benefits were shared likewise by Mombasa and Malindi, whose inhabitants had enjoyed special privileges there. The Portuguese had too few men to hold so great a stretch of coast, nor had they reckoned on the capacity of the conquered for dissension and intrigue. At Sofala the compliant shaikh was soon murdered, and another shaikh named Mokondi attacked the fort, and was driven off only with appalling loss. In Kilwa the puppet monarch was soon dethroned and killed, and a large number of the townsfolk migrated northwards, leaving the rest to dynastic squabbles. Trade in the town was at a standstill. In 1506, only a year after the building of the fort which was also to be the factory for Kilwa, the Portuguese found it necessary to permit the merchants of the city to trade in every commodity and in every direction, as they had done before the Portuguese arrived. This served as some slight palliation, but absolute control of Sofala was lost, and Kilwa never wholly revived. If there was quiet farther north, it was because the Portuguese at this stage made little attempt to interfere.

The Portuguese, indeed, in their inability to co-operate either with Arab or African, were their own greatest enemies. Strandes calculated that Sofala may once have produced as much as £600,000 a year:[1] it was never so profitable to the Portuguese and barely paid its way. Both Arabs and Africans adopted a policy of judicious non-cooperation combined with smuggling: the Portuguese had not their partial immunity to malaria, and also had corruption within their ranks. Even an honest factor, like Soares, could scarce make the Sofala factory meet its debts.[2] He ventured to tell the king that the effort was hardly worth while. To the home government his arguments were unintelligible. The gold of Sofala must be held and developed, while Mozambique, where a fortress was built in 1507, was of undoubted use as a refreshment station on the way to India. It was the search for gold which, during the rest of the

[1] J. Strandes, *Die Portugiesenzeit von Deutsch- und Englisch-Ostafrika*, 1899, pp. 99–100, and 47. (English translation, trans. J. F. Wallwork and ed. J. S. Kirkman, 'The Portuguese in East Africa', *Transactions of the Kenya History Society*, vol. ii, 1961.) Gold values have trebled since he wrote.
[2] Cf. E. Axelson, *South-East Africa, 1488–1530*, 1940.

sixteenth century, led to attempts to open up the hinterland which is still Portuguese East Africa; in the area with which we are concerned, if there were unofficial Portuguese settlers along the coast, there was never any serious organized attempt to penetrate the interior. An insufficient fleet attempted to dominate the sea route, and it was this which led to the revolt, and second sack, of Mombasa in 1528, supported by Mtang'ata and the petty sultanate of Utondwe, north of Bagamoyo.

For the next forty years there is scanty information. In 1569 Fr. Monclaro, of the Society of Jesus, accompanied an expedition led by Barreto, whose object was to quell a revolt in Pate. He found Kilwa virtually deserted, and Mafia quiet and rich in agricultural produce. Zanzibar was in confusion from an invasion of 'Kaffir rebels' from the mainland, whom Barreto put down, receiving the gift of the island from the king. There were a few Portuguese and a chaplain, and trade in cloth, beads, iron, ivory, and some ambergris, and very much fruit. He found Mombasa large and populous but was disappointed to find that Christianity was not yet established among the people, while Malindi, as St. Francis Xavier observed when he passed in 1542, was partly in ruins: it was much pressed by the barbarous Segeju. Farther north was Cambo, 'a large city with several fine edifices', which was ruled by a queen. Her country had suffered from Turkish raiders, and even she had been captured, only escaping by throwing herself into the sea from the poop of a galley; she was glad to have Barreto's protection. At Pate the people had fled, and the king's appeal for mercy was accepted on payment of 12,000 cruzados.

Monclaro's disappointment at the failure to implant Christianity, by contrast with the success already won in Goa, is easily intelligible. The earliest fleets had carried Franciscan friars as chaplains rather than as missionaries. This was, perhaps, because there were reports of Christian communities on the coast, and formal missionary activity might seem unnecessary. Nevertheless, in a letter to the King of Portugal dated 31 August 1506, the Captain of Kilwa reported that forty men and women wished to become Christians there, and that the sultan was angrily refusing permission. In 1542 St. Francis Xavier passed along the coast, touching at Mozambique, Malindi, and Socotra, on his way to Goa as Vicar Apostolic of the Indies.

There was no immediate increase in missionary activity. Jesuit missions began in the south in 1560, and were shortly followed by Dominicans, while houses of Augustinian canons were set up at Mombasa, Faza, Pate, and Zanzibar. The southern missions had considerable success, but they worked in largely pagan areas. In the north the small number of clergy was faced by a Muslim population. Converts were not numerous, but by 1599 the Augustinians had baptized 600 persons, including the King of Pemba. At Faza in 1624 a congregation of seventy attended the ceremonies of Holy Week. The number of clergy was indeed too small for the vast work undertaken.

In sum, the years between the Portuguese conquest and the year 1580, when Spain asserted dominion over Portugal, were of steady decline rather than of progress, as recent excavation has displayed at Gedi and elsewhere.[1] The fact was, whatever may have been the profits of medieval prosperity, there was an insufficient margin for another sharer in them. There was no room for both the Arabs and the Portuguese, who, in the north, had done no more than seize the Arab colonies. If lack of fighting skill and unity and the possession of inferior weapons handicapped the Arabs, events were now in train which were to thrust back the Portuguese to the only area they had effectively tried to settle.

The Turks had conquered Cairo in 1517 and by 1547 their power extended beyond the Hijaz to the Yemen and Aden. At this point the impulse of conquest seemed spent; they only raided along south Arabia, gaining no permanent footholds. Had they done so, East African history might have told a different story. Their raids as far as the northern neighbourhood of Malindi before 1569 have already been mentioned. There are only stray literary mentions of these raids and no information as to the policy, if any, which may have directed them. Certainly about 1580 they were most troublesome to the Portuguese in the Persian Gulf as well as in the Red Sea.

In 1585 an Amir Ali Bey came in a single galley, and visited Mogadishu, Brava, Kisimayu, Faza, Pate, Lamu, and Kilifi, telling the people he had been sent by the Turkish sultan to free them from the Portuguese. No document warrants that this was official Turkish policy, but the northern coast

[1] Cf. Kirkman, *Gedi*, p. xiv.

rose as one in revolt. Mombasa, the most southerly point he reached, asked for a permanent Turkish garrison. In Mogadishu a recently rediscovered coinage was struck, in imitation of the Turkish mints.[1] Faza, at this time or shortly after, acquired a Turkish dynasty, which ruled until the end of the nineteenth century, and whose descendants still survive. If in 1586 Amir Ali sailed back with plunder worth £50,000, and left behind promises he was unlikely to fulfil, he had at least crystallized a desire both for independence and for unity which it was to take a century to attempt to realize. One galley was insufficient force. When, in 1587, loyal Malindi summoned a fleet from Goa, the coast meekly submitted, awaiting, with Arab caution, the Turkish return. In 1588 Amir Ali came again with five ships and once again the northern coast revolted, Pemba this time joining in, massacring most of the Portuguese on the island and driving out their ruler who had been friendly with them.

Another fleet was dispatched from Goa, and, having called at Brava, Lamu, and Malindi, where the king joined it, it stormed Mombasa, overpowered the Turks, and drove them, with the Arabs and Swahili, into the walled town. The Portuguese were unaware that a worse menace was approaching from the south.

No writer of this period reads more graphically than the Dominican missionary, Fr. João dos Santos: he was an eyewitness of much he describes. The Zimba tribe, he tells us, was first heard of in history near the modern Sena, and had long eaten their prisoners of war and slaves who were past labour. Like the Nguni some centuries later, they abandoned without apparent reason a peaceable, simple existence for a war economy of swift-moving migration and plunder. One band suddenly moved northward, overwhelming the Portuguese and Arab colonies at Sena and Tete and committing the most revolting atrocities. Another band, reckoned at 5,000 men, moved along the coast (in the words of Samuel Purchas) 'killing and eating every living thing, Men, Women, Children, Dogs, Cats, Rats, Snakes, Lizards, sparing nothing . . .'. At Kilwa in 1587 a 'treacherous Moore' guided them over the

---

[1] G. S. P. Freeman-Grenville, 'East African Coin Finds and their Historical Significance', *Journal of African History*, 1, 1960, pp. 31–43.

mud flats to the island at the low ebb of the spring tides, and 3,000 of the inhabitants were taken and slain and the city sacked. Since the population was reckoned at only 4,000 in 1505, and some of the inhabitants are stated by dos Santos to have escaped, this was an unimaginable slaughter of the population. Many, too, had migrated earlier in the century, and this was probably the death blow to Kilwa. Unfortunately this is the only figure dos Santos gives, but it is reasonable to suppose that a similar destruction attended the Zimbas' further passage up the coast. It would explain too the apparent ease with which new elements seem to have settled in what the traditional histories depict as virtually unpopulated coastland areas in the period that follows.

The Zimba arrived in Mombasa at the moment that Thomé de Sousa had driven the Turks and their co-religionists into the walled town. Saying that 'the Portugals were Gods of the sea, and hee of the Land', the Zimba captain sent to de Sousa professing friendship, 'requesting that seeing they had honourably ended their enterprise, he might beginne his, namely to kill and eate every living thing in the Iland . . .'. Thus for the third time in this century Mombasa suffered a destruction of the utmost horror. The Zimba, however, were unsated and proceeded to Malindi, where the king, loyal to Portugal, had a European garrison of thirty. On their way the Zimba certainly passed Gedi, but excavation has not displayed any evidence that it was sacked. But at Malindi, by good fortune, the warlike tribe of the Segeju had been moving down the coast, and they now took the Zimba in the rear as they besieged the city, and killed all but the captain and a hundred men who escaped. The Portuguese were indeed lucky in this unexpected ally.

From this destruction by the Zimba and from the Turkish raids two results ensued. The first was a partial depopulation along the coast: the older populations were destroyed, and in the seventeenth and following centuries new tribal elements entered in. But the most immediate result was that the Portuguese were compelled to attend more to the northern coast and, in particular, to reorganize their administration and to make Mombasa a captaincy separate from that of Mozambique. They feared further Turkish raids, but none took place. Already in 1569 Monclaro had reported that a fort had been begun at Mombasa to guard the port, but that it had not got beyond foundation level. In 1593, with masons from India, labour from Malindi, and an Italian architect, Fort Jesus was begun, and occupied by a garrison of a hundred men the following year. The Captain of Mombasa set up as sultan al-Hasan ibn Ahmad, King of Malindi, as a reward for his loyalty; the last of the 'Shirazi' rulers, Shah ibn Mishham, had opportunely died and apparently there was no obvious successor. If the Portuguese had enjoyed a century of friendship with successive kings of Malindi, it was now to be lost by ignorant arrogance and grasping commandants at Mombasa. In 1614 al-Hasan ibn Ahmad went to Goa to state his complaints, refused to accept the assurances of the viceroy and, on his return, fled to the Nyika tribesmen, by whom he was murdered for a bribe offered by the Portuguese. Such treachery was disapproved at Lisbon and his son, Yusuf, was ordered to be taken to Goa to enjoy the advantage of a Christian education, while Muhammad, an elder kinsman, acted as regent.

Yusuf was then only seven years old, and only returned to Mombasa in 1630, by now a grown man and a Christian under the name of Dom Hieronimo Chingulia. He had taken a Portuguese wife. Such a sultan could hardly be pleasing to the people of Mombasa, while he was not less mistrusted by the Portuguese. The situation inevitably led to friction, and Yusuf, influenced perhaps more by family ties than by his new religion, turned to Islam. On the Feast of the Assumption, 15 August 1631, his own dagger-thrust in reply to the commandant's greeting gave the signal for revolt, and his supporters threw themselves upon the Portuguese who had gathered for the festival. Only four priests and one layman escaped to Pate as survivors of the massacre. Many, including numerous converts to Christianity, refused to abjure their religion and died as martyrs, while 400 were deported as slaves to the market of Mecca.

Messages calling on the whole coast to rise met with little response. There was no supporting force such as the Amir Ali had brought, but the kings of Utondwe, Manda, Luziwa, Chwaka, Pate, Siu, and Pemba refused tribute, and there were risings at the little towns of Tanga and Mtang'ata. At the end of the year a small fleet was sent from Goa to punish the rebels, but Fort Jesus was turned to good account against its own builders. Next year another fleet was dispatched. It found the town in ruins and the fort dismantled. It is not clear what had happened, but Yusuf had apparently been repudiated by his followers. He had certainly left Mombasa, and until 1637 he stirred up a series of petty revolts. In that year he died at Jiddah, no doubt endeavouring to enlist Turkish support. An inscription of the utmost complacency over the gateway of Fort Jesus boasts that Francisco Seixas de Cabreira had reduced the coast into subjection to the King of Portugal after the revolt led by 'the tyrant king'.

When the Portuguese had established themselves in East Africa, they had also taken Muscat in Arabia and Hormuz in the Persian Gulf. The revival of Persia had already taken the latter from them in 1622: in 1650 the Omani rose under the Imam Sultan ibn Saif, and threw them out of Muscat. Mombasa promptly sent a delegation to the Imam to complain of 'the iron yoke and the injustices . . . that weighed upon them,' and the evil deeds of all sorts committed' there. 'So far', they said, 'injustice and the law of the strongest had prevailed.'[1] So in 1652 the Imam sent a small number of ships to raid Pate and Zanzibar, and wiped out the Portuguese there. It was the beginning of the Omani intervention in East Africa. Francisco Seixas de Cabreira—he of the complacent inscription—was sent back to govern Mombasa. In 1660 the Imam returned again at the invitation of Pate, attacked Faza, and also took the town, but not the fort, of Mombasa. In 1669 he pushed south as far as Mozambique and almost succeeded in seizing it. In 1678 the viceroy, Pedro d'Almeida, came in person to punish Pate, whose king had gone to Muscat to invite the Imam. The Portuguese arrived first, beheaded the rulers of Pate, Lamu, Siu, and Manda, with many others, but in January 1679 the Muscat fleet arrived and the Portuguese hurried down to Mozambique with their booty. In 1686 Pate again rebelled, and the king was seized and carried captive to Goa. There he was killed on Christmas Day 1688 with twelve of his counsellors.

The scene was thus set for Muscat to attack Mombasa and in 1696 Saif ibn Sultan sailed there with a large fleet. He had

[1] Anonymous, 'A History of Mombasa', *c.* 1824, apud G. S. P. Freeman-Grenville, *The East African Coast: Select Documents*, 1962.

not less than 3,000 men. The defenders had warning of his approach, and the fort was held by some 2,500 persons. A siege began, and by January 1697 the defenders had been reduced to not more than twenty by a disease similar to bubonic plague. The survivors, commanded for a time by a scion of the house of Faza, held on, and in September a small relieving force succeeded in entering the fort. It did not fall until December 1698, when an insufficient relieving fleet arrived to find the red flag of Oman flying. In 1728, after a long interval, the Portuguese profited by a quarrel between the Omani governors of Mombasa and Zanzibar to seize the fort once more, but in 1729 they were easily expelled. Thereafter the Portuguese made no active attempt to assert their power north of the Ruvuma.

Their departure was no miracle: rather it was remarkable that their occupation had lasted so long. It is hard to estimate the coastal population they controlled after the ravages made by the Zimba: south of Mombasa it was greatly reduced but, counting the whole area from the Ruvuma as far as Brava, it could hardly have been less than 100,000. The Mombasa establishment of 1606 was little over 100 men, 80 of whom were at other stations: in 1634 there were about 200, including 80 sailors and some customs officers. Counting non-official settlers, the Portuguese population at its very greatest never exceeded 900 persons.[1] Such a small personnel was too slender to resist revolt, and it was never possible to deal with trouble before it arose. Nor was there, in fact, any profit, but rather the contrary: there was a deficit of 40 per cent. over revenue. Much of this may well have been due to dishonesty and incompetence, facts which Portuguese historians make no attempt to conceal. There was another enemy too, malaria, for which we have no statistics, but which was to take heavy toll of the British there during their short occupation between 1824 and 1826. Among the Portuguese there were individuals capable of genuine heights of personal heroism. Yet it is not by these that the state stands or falls.

In all this period, even if the traditional *History of Pate* runs continuously throughout it, it is difficult to reconstruct the

[1] J. M. Gray, 'Rezende's Description of Africa in 1634', *Tang. Notes*, 23, 1947, pp. 2–28.

life of the coastal cities in any detail. It is perhaps of greater utility to survey some general tendencies.

The monarch of Kilwa whom Monclaro saw in 1569 was black. The shaikh of Sofala with whom Pedro d'Anaya had negotiated in 1505 had been of a similar hue: he was of a family which, in the fifteenth century, had provided a succession of *amirs* for Kilwa. In Pate the founder of the dynasty which began in 1204, Sulaiman ibn Sulaiman, had married the daughter of the preceding king: how a rapid Africanization of the Arab colonists took place is discernible from the royal titles of Pate—Swahili, and not Arabic, titles being used for the heir apparent, the princes and princesses of the royal house.[1] de Barros speaks of black and white Moors at Kilwa and Mombasa, the former preponderating. We cannot tell how closely the Arab and African became knit, but knit they were. A consideration of the names mentioned in the *Mombasa History*, or of those in the traditional histories collected by Velten, all of whom were leaders of the people, shows a far higher proportion of Bantu than Arabic names. The process of Africanization may be illustrated from the Ndagoni manuscript in the Zanzibar museum, which gives a list of the members of a 'Shirazi' family settled there from about 1500 to 1840. The founder of the line has a Persian name, Darhash, and the next five generations common Muslim names. In the following nine generations, although three of the names are common among Muslims, no less than six, Shaham, Makame (three times), and Mbwana, are peculiar to the Swahili and are of Bantu origin. These arguments, perhaps, are dangerous, but the facts can be explained on no other basis. It would be wrong to regard the coast towns as more than conventionally Arab; rather the Arabs, ever swelling the number of immigrants, were not less speedily being absorbed. These facts are difficult to explain and difficult to follow, for we have not the necessary demographic material. Nevertheless they must not be forgotten: they are an essential ingredient in the background of any consideration of the life of the coastal sultanates during this time.

We have already noted the effect of the Zimba. There were

[1] A full text of the *History of Pate* is in C. H. Stigand, *The Land of Zinj*, 1913, pp. 29–102.

also Galla raids which reached as far as Malindi at the end of the sixteenth century, while the excavation of Gedi has suggested that there were natural difficulties as well. The water-table was falling, and perhaps rainfall too was diminishing. The Portuguese found great abundance of provisions at Kilwa, Mombasa, and Malindi, which certainly does not exist in their immediate vicinity now. It is remarkable how many ruined mosques throughout the whole coast became disused in the course of the sixteenth century, and that, with the exception of Mboamaji, where an inscription gives the date of foundation as 1608, there are apparently hardly any buildings of this type attributable to the seventeenth century. Perhaps the town of Pujini in Pemba, where a mortar partly compounded with mud replaces the earlier mortar of ground coral and lime, may belong to this period. None of this decadence could be attributed to the Portuguese; they, not less than the coastal folk themselves, were the losers by it. All these factors form part of the background to any consideration of the coastal cities during this period.

It is not possible to give any consecutive account of the extreme north, Mogadishu and Brava, and the settlements of Somalia. In the *History of Pate*, however, there is consecutive information, even if its chronology is hard to interpret. The Nabhani dynasty continued to rule until the end of the nineteenth century. If in the fourteenth century Pate had had some brief hegemony over the coast, it was now but one among several principalities within the same group of islands, with Siu, Manda, Lamu, and others as equals. Towards the end of the sixteenth century Pate received a new group of Arab settlers with the name Hatimii, immediately from Brava, but who claimed they had originated in Andalusia. Another group of the same family, as we shall see, were to be the founders of Dar es Salaam, now capital of Tanganyika. Pate only grudgingly accepted the Portuguese for a brief period in the first half of the seventeenth century, the period of greatest Portuguese strength on the northern coast following upon the building of Fort Jesus; the burden of the whole story is one of intrigue and fighting to be rid of them. In this the Nabhani are in sharp contrast with Kilwa which, torn with dissension within its ranks, speedily succumbed.

As to Lamu, the *Chronicle*[1] is unfortunately defective until towards the end of the eighteenth century; its dynasty is best considered together with the Mazrui governors of Mombasa who, appointed first in the early eighteenth century, ruled there until, attempting a dominion of their own, they were forced out by Sayyid Said. Apart from the *History of Mombasa* itself two curious traditional histories, the *Origin of the Jomvu* and the *Story of the Changamwe*,[2] give an account of two of the nine tribes of the town. The latter makes the very interesting statement that many Africans had taken to Christianity during Portuguese times, but regrettably it is weak in its chronology, and does not say when. Taken together, they portray a city with an Arab upper class which, to maintain its power, was obliged to rely upon the goodwill and support of the Bantu majority, with whom many of them intermarried.

Most interesting is the *History of Vumba*, which survives only in an account written by Sir Claud Hollis sixty years ago, in which it appears that a most complicated compromise had been worked out by which both African and Arab customs were embodied in the investiture of the *Diwan* of Vumba.[3] The dynasty claimed an origin dating back to 1204, but the earliest sultans appear to be known by Digo names. The chronology is most suspect and it is hard to credit that the dynasty antedates the earlier sixteenth century. The members of the royal family were sharifs, descendants of the Prophet, and as such held in the deepest veneration both by the Digo and the Segeju. The *Diwan* was formally invested with sandals and a turban covering a worked skull cap, placed upon his head by a leading member of the Ba-Amiri clan. The *Diwan* then announced the throne name by which he would be called, and was carried in procession upon a native bedstead. On this and other formal occasions he was attended by a band of music, and even on informal occasions he was attended by slaves carrying the royal horn, or *Siwa*, an umbrella, and his chair of office. He was addressed as *Mwenyi*, lord, owner, a Swahili title of respect. Such, or rather similar, ceremony attended the shaikh of Mogadishu. Ibn

[1] 'Shaibu Faraji bin Hamed al-Bakariy al-Lamuy, Chronicle of Lamu' (ed. W. Hichens), *Bantu Studies*, 12, 1938, pp. 1–33.
[2] H. E. Lambert, *Chi-Jomvu and Ki-Ngare, sub-dialects of the Mombasa area*, Dialect Study III, East African Swahili Committee, Kampala, 1958.
[3] Hollis, *J. Roy. Anthrop. Inst.* 1900.

Battuta describes him formally attending the Friday Prayers at Mogadishu in 1331: he had a band of music and guards, a canopy carried over his head, and he and the *Qadi* alone returned from the mosque wearing sandals. Pate, Lamu, and Kilwa likewise had their royal horns, and at Kilwa in the late eighteenth century the sultan, even in his pitifully reduced state, was attended by a sword-bearer, musicians, and one or two executioners.[1]

Of Mtang'ata, which Vasco da Gama saw as a flourishing town, and where he burnt and abandoned the storeship *San Raphael*, little is known, but culturally and linguistically it was closely connected with Vumba. Of Tanga we have only the names of a few of the rulers. *The Ancient History of Pangani* is almost wholly a genealogy,[2] which appears to start about 1550 and continues to about 1900. The earlier settlement was north of the present town at Muhembo, at the top of a high cliff and, although Pangani is shown on early eighteenth-century maps, the *History* dates the present town as only *c*. 1827. These dates accord well with the ruined mosque at Muhembo and the graves near it, as well as with the clearly nineteenth-century character of the earlier buildings in the town itself. The first seven generations of the genealogy are shown by descent in the female line, and all the names appear to be Zigua: it is not until the seventh generation that a daughter of the house takes a husband with an Arab name, and thereafter the reckoning is patrilineal; yet very shortly after Swahili names replace the Arabic. Thirty-four miles farther south at Mkwaja the genealogy of the *Diwans* claims that the dynasty was founded about 1300, and that the ruling family was 'Shirazi', albeit that the founding ancestor hailed from Baghdad.[3] The genealogy says nothing of the fortunes of the dynasty, but local tradition remembers a revolution which split the area under several *jumbes* about 1789, in which year a dated inscription records the building of the mosque at Kipumbwi Mkuu, one of these *jumbeates*. If the mosque is in the Arab coastal tradition, the names of the genealogy display a fundamental Africanization, as do the ruins of the former traditional capital of Uzimia.

[1] Rhodes House MS. Afr. r. 6, see below, pp. 148, 155.
[2] C. Velten, *Prosa und Poesie der Suaheli*, 1907, pp. 306–9.
[3] I am obliged to the late E. C. Baker for the use of the manuscript and for the notes made by him at Mkwaja in 1923.

The History of Former Times in Bagamoyo[1] tells a similar story, that the first settlement was made by an Arab. Decorated tombs, similar to those which can be dated at Mkwaja and in its vicinity, near the Friday Mosque of Bagamoyo, show that there was a new influx in the early seventeenth century. The story, of little intrinsic interest, is of constant friction between the part-Arab, part-African settlers and the people of the interior. Farther south *The History of Dar-es-Salaam*[2] relates how the Barawi, who had settled at Pate in the late sixteenth century, occupied the coast between Kunduchi and Mboamaji. This is confirmed by a number of epitaphs at Kunduchi, from which unfortunately some dates have been erased, but the tombs themselves can be dated to some extent by references to those of known date at Mkwaja. It is interesting to compare the mosques of Kunduchi and Mboamaji, the former following the traditional pattern of the Middle Ages, the latter with lateral arches and intricate decorations which perhaps derive from India. Its lintel is dated 1608.

The Ancient History of Kilwa Kivinje[3] relates that the foundation of the town was by two tribes with Bantu names, the Ngarwe and the Bumura, neither of which now survives: they were joined by Nyasa who had first lived at Mchinga, some 120 miles south, and who had opened up a trade route between the coast and their homeland. Apparently this trade was at first exclusively in the hands of one Nyasa family, who brought others in their wake. The tradition of this route, from Kilwa Kivinje to the south of Lake Nyasa, still survives, and at the end of the nineteenth century Eustace, a British naval officer, recorded that it was defended by a series of stone forts, set conveniently for marching distances. It was perhaps a revival of the route to which Ibn Battuta refers in the early fourteenth century. The story involves economic questions to which we shall shortly return. The real building and development of what was to become an important town by the mid-nineteenth century is ascribed to a Sultan Yusuf ibn Hasan, a 'Shirazi', who came from Persia with his brother Ismail. He was received and accepted as sultan by the local people. The chronology is obscure, and much of the story is plainly myth: the foundation of the town

[1] Velten, *Prosa und Poesie*, pp. 300–65.
[2] Ibid., pp. 289–99.   [3] Ibid., pp. 253–64.

may tentatively be placed in the earlier eighteenth century, but it was still of no importance when Morice, a French speculator, visited Kilwa Kisiwani about 1776. Its development, as its earlier surviving buildings bear witness, belongs to the early nineteenth century.

The Swahili *Ancient History of Kilwa Kisiwani*, stated by Prins[1] to be derived from the same ultimate source as the Arabic and Portuguese versions, both of which end in 1502, differs so much from them in structure and content that his view is difficult to maintain. It gives only five generations of rulers, of whom two in the final generation died shortly before the death of Sayyid Said in 1856. The earlier part may represent a variant tradition of the events of the tenth century, and reminiscences of the reforms of the fifteenth,[2] but the major part unquestionably refers to the eighteenth century. It is unlikely that another dynasty had replaced that which ruled Kilwa in medieval times, for in 1812 the sultan told Prior, the surgeon of a visiting British warship, with a perhaps pardonable exaggeration, that his family had ruled there for 1,200 years. Family pride is the last refuge in disaster, and already in 1589 João dos Santos had seen 'the ruines of their sumptuous Mezquites and Houses'. The essential historicity of this Swahili tradition is affirmed, however, by its reference to the visit of a Frenchman there in the late eighteenth century, for the purpose of trading in slaves. A curious manuscript in Rhodes House, Oxford, describes the plan of this Frenchman, Morice, to start an establishment at Kilwa.[3] Like Dallons in 1804, Morice found the place much reduced and poor. The sultan, Prior remarked in 1811, is only 'rich in the pride of ancestry'. Apart from the fort there were only two or three stone houses: the rest were of bamboo, twigs, and thatch. Prior saw too the ruins of the once prosperous town of Songo Mnara, which has no documentary history, by then wholly abandoned to nature. The people he describes, though inaccurately, as negroes.

As to *The Ancient History of Lindi*, what was to become a small German port frankly claims Makonde, not Arab, origins,

---

[1] A. H. J. Prins, 'On Swahili Historiography', *J.E. Afr. Swahili Committee*, 28, 1958, pp. 33–35; Velten, *Prosa und Poesie*, pp. 243–52.
[2] See Chapter IV, pp. 112–13, above.
[3] Rhodes House MS. Afr. r. 6; Mauritius Archives GA 11, no. 119, 1804.

although the petty rulers of neighbouring villages proudly bore the title sultan. One sultan bought the country from another for what are described as 'all his goods', a curious miscellany: 'a female slave, a mortar for grinding snuff . . ., lengths of cloth and the head and liver of a wild pig'.[1] It is hard to assign dates to this simple state of affairs, which is plainly non-Islamic. It occurred again farther south at Sudi, Mikindani, and Kionga, whose traditional histories Velten also preserves. Even if there was occasional intermarriage with the Arabs, the African predominated and Arab influence was at a low ebb.

All this seems a far cry from two accounts of the trade of the east coast as seen in the early sixteenth century from two different vantage points, that of Tomé Pires in Malacca from 1512 to 1515, and that of Duarte Barbosa in Cananor on the Malabar coast, about 1517–18. In four brief paragraphs Tomé Pires, who was subsequently Portuguese ambassador to China, summarizes Indian Ocean trade from the Malacca point of view. It is particularly interesting that, like the *Periplus of the Erythraean Sea* 1,400 years before, he notes a slave trade exclusively from the Somali ports of Zeila and Berbera, and nowhere south of this area. For the western Indian Ocean Aden was the pivot which connected Cairo with Cambay, and from which the fleets set out to East Africa. The chief articles taken south were rice, wheat, soap, indigo, butter, lard, oils, carnelians, coarse pottery 'like that of Seville', and all kinds of cloth. Northward and eastward were taken the slaves from the extreme north of the coast: from the southern area came gold and ivory, and we may suppose copal and ambergris. At Aden, too, the products of Italy, Greece, and Damascus could be bought, manufactured goods and jewellery, beads, cloth, and weapons of various kinds. At Malacca there were gathered the agents of many nations, concerned no doubt to sell their wares farther east: Cairenes, Arabs (chiefly from Aden), Ethiopians, Persians, Turks and Turcomans, Armenians, and men of Kilwa, Malindi, and Mombasa.

All this serves to explain the riches Duarte Barbosa describes, the cottons and silks, and many kinds of beads which the ships of Kilwa, Malindi, and Mombasa brought to Sofala

[1] Velten, *Prosa und Poesie*.

to exchange 'at such a price that these merchants departed well pleased'. The 'Moors' of the interior came there laden with gold, which they exchanged for the cloth without weighing it. A great store of ivory was also brought, and sold in Cambay, not only long the traditional ivory mart for India, but on the route al-Masudi described 600 years before to China. There were minor marts of a similar sort on the coast at Cuama and Angoxa, which, with Sofala and Mozambique, swelled the wealth of Kilwa, with its 'fair houses of stone and mortar', its Moors, some 'fair' and some black, clad in gold and silk and cotton, their women gay in gold and silver and jewelled earrings. (It is doubtful whether the veil was worn in East Africa at this time.) Mombasa was 'a place of great traffic', while at Malindi the people were 'great barterers', dealers in cloth, gold, ivory, and wax. It is significant that the city was rich enough to import food, rice, millet, and some wheat from Cambay. In Pemba, Mafia, and Zanzibar there was trade in cereals, flesh, fruit, and vegetables of all kinds, some of which were sold to the mainland: as elsewhere the women were notable for their finery, and the men dressed in very fine silks and cottons, bought in Mombasa from Cambay merchants.

How far these descriptions represent an idealization of the past, or Portuguese expectations, or Arab exaggeration, is difficult to gauge. What is clear, however, is that they are based upon what is essentially a carrying trade deriving from the gold of Sofala, with Mozambique, Cuama, and Angoxa as lesser centres, and from ivory from these and in all probability other collecting points upon the coast. It is thus salutary to turn to the actual figures given by Soares, factor of Sofala between 1512 and 1515. It must be recognized that everything which the Portuguese had taken in the south could only result in a lessening of the carrying trade, and thus of the wealth, of the cities of the north. Soares's figures represent Sofala alone, but Sofala, it must be remembered, had been the basis of the prosperity of Kilwa and the rest of the coast so far as gold was concerned since the twelfth century. On an average he collected annually 12,500 miticals of gold during his period of office, a total of 130 pounds only, which, at the present-day price of about £7 an ounce would seem to be a mere £14,600.[1]

[1] Cf. Axelson, *South East Africa*, pp. 6–7, n. 2.

To reach its early-sixteenth-century value we should, perhaps, regard this as equivalent to half a million pounds. There were 51,000 lb. of ivory, 137 branches of amber, 4 grains of seed pearls (which are still obtained with little difficulty near Mikindani, but small and irregular in size), 20 lb. of coral, 2,000 lb. of copper, and some lead. There were only six slaves.

To pay for the gold and ivory Soares imported beads in great quantity, more than 7,000 lb. weight, together with some cloth.[1] These are the figures for but one centre; we have none for Mozambique, Cuama, and Angoxa. Nor do we know to what extent the volume of trade had diminished by reason of the Portuguese intervention. Soares had told the king that the factory did not cover its expenses, and it may reasonably be assumed, apart from some smuggling which doubtless continued among the Arabs, that the volume of trade had been far greater at Sofala and elsewhere when it was still in local hands.

Having lost their carrying trade, Kilwa and the cities to the northward were thus thrust back upon their own resources and such trade as they might attract by contact with the interior. They no longer had their profits as middlemen. Monclaro noted in 1569 that Kilwa was still trading in ivory, which was bought from the 'Kaffirs' of the interior, and also in honey and wax. Mafia, as now, was rich in coir, and also produced an abundance of pitch, both of which, however, were bought by the Portuguese, who had established a small agency on the main island. At Mombasa iron goods were imported for resale to the people of the interior, and ivory-handled daggers were manufactured, clear evidence not only of the ivory trade but also of interior contact. Monclaro saw Malindi in a ruinous condition, and this had probably been the case for some time, for St. Francis Xavier relates that all but three of its seventeen mosques were in ruins.

Pate carried on trade with India, and by then it was the only town on the coast reported to be importing silks, which the Portuguese exchanged for iron ware, beads, and cottons. Sir James Lancaster, the first Englishman known to visit East Africa, who watered at Zanzibar on his way to India in 1591, only mentions the excellence of the timber and pitch 'or rather

[1] Ibid., pp. 126-7.

a kind of gray and white gumme like unto frankincense, as clammie as turpentine . . .'.

In 1634 we have some figures for the revenues of the Portuguese customs at Mombasa and Pate:[1] they are highly suspect, for one cannot believe that collection was either efficient or uncorrupt. Moreover, it was the first year of collection. There were duties on opium, tobacco, and ivory, while all ambergris was appropriated to the crown. Pate, Faza, Lamu, Mombasa, and especially Pemba, with its rich rice crop, abounded in produce of every kind. Zanzibar produced ships' timber, and Mafia pitch. The principal revenues were from ivory and ambergris, which, together with civet-musk, is an important basis for perfume. It is remarkable that in all these descriptions slaves are scarcely mentioned, and never as an export. While, at this time, slaves were an export from Zeila and Berbera, there is, indeed, no evidence to suggest that they were exported at all from the coast farther south during the sixteenth or earlier seventeenth century, and the contention of Coupland, that the slave trade was continuous from earliest times, rising to a peak in the nineteenth century, cannot be substantiated. Certainly there was domestic slavery, but it was only after the Omani had begun to intervene that slaves began to be exported.

Most of the evidence for internal contacts is disappointingly recent. Most of it derives from traditions along the trade routes of the nineteenth century and the histories collected along the coast by Velten c. 1900, which, even if they reflect late-nineteenth-century ideas, may give some indication of the occupations of small centres in earlier times. Unfortunately the record is far from complete, and it is necessary to deduce the trade of the many from what little we know of a few. Looking at it from north to south, we may first note that mosque-building in the coastal villages on any permanent basis in stone was not resumed until the eighteenth century. All these mosques are on the coast, and there is no trace of Islam inland. It may perhaps be fairly assumed that until then the peoples of the interior came for the most part to the coast, and that caravans did not set out westwards. For Mogadishu we have no account of trade conditions, nor for Brava: geographically the Webi Shebelle speaks for itself as a channel of trade. Pate,

---

[1] J. M. Gray, *Tang. Notes*, 1947, pp. 2–28.

according to its history, looked north to Egypt, and, with Lamu and doubtless the lesser surrounding kingdoms, east to India. But these were rather entrepôts with but immediate inland connexions, for behind them lay the most pitiless bush. Slightly farther south the Tana river behind Kipini, the Sabaki behind Malindi, and a small system of streams behind Mombasa could give access by forced marches to the uplands, and to the mountains of Kilimanjaro and Kenya. That such a route was known in the seventeenth century seems shown by the traditional expedition or raid of the Segeju to what is now the Kamba country. Again, behind Tanga, and some thirty miles farther south, is the system of streams which feed the Pangani river, which skirts and reaches the farther side of Mount Kilimanjaro. Burton knew this route in the mid-nineteenth century, and himself used it to reach the Usambara, but, although the Nyamwezi were already using it, it may have been comparatively recent in his time. Farther south still the earliest that is heard of any central route is that by which, according to Buganda traditions related by Sir Apolo Kagwa, Kyabugu (who ruled Buganda from *c*. 1763–1780) imported cups and plates, and his son Semakokiro, *c*. 1797–1814, exchanged ivory for *kaniki*, the coarse Indian blue calico, cowries, and copper bracelets.[1] This route was presumably through the Nyamwezi country to Lake Victoria, and thence, via Karagwe, to Uganda. It may then only have been of very recent date. South of the Rufiji only one route is indicated by tradition, that already mentioned leading from Kilwa Kivinje to Lake Nyasa; but later, in the mid-nineteenth century, Livingstone went inland from Mikindani almost due west. It is doubtful how old this route is, but it would have crossed that taken by Gaspar Bocarro in 1616, when he made an adventurous journey from Tete on the Zambezi to Kilwa. There were villages all the way, and one may deduce some rough tariff from his interchanges of gifts and food.[2]

Largely, then, we may postulate not an organized trade based upon well-trodden and recognized routes, but rather the passage of goods from hand to hand and their eventual sale on

[1] See Chapter VI, pp. 190–1, below.
[2] J. M. Gray, 'A Journey from Tete to Kilwa in 1616', *Tang. Notes*, 25, 1948, pp. 37–47.

the coast. The organization of caravans belongs rather to the late eighteenth century or the early nineteenth. Before this time the more normal pattern seems well depicted in a passage in the *History of Sudi*, a village of trifling importance between Lindi and Mikindani:

> The occupation of the people was growing food, rice, maize, vegetables, sesame, millet, beans, peas, bulrush millet and cassava. After they have reaped them, they sow again, and so they have food for the year. They sell part of their food, and buy trade goods or silver. When this silver has accumulated, they buy slaves with it. They use these slaves in their plantations, and cultivate with them. Another of their occupations is hunting elephant. If they get ivory, they sell it for silver or trade goods. And with this silver they buy slaves and place them in their plantations and cultivate with them.[1]

Here there is no hint of inland journeying, and according to the *Ancient History of Kilwa Kivinje* the first founder of its trade was a Nyasa, who travelled inland because 'the inland folk were fools and did not know the price of things here on the coast'. It is in the eighteenth century that this change begins to take place, and no doubt its pace was accelerated by the intervention of Oman, with a consequent greater demand for goods and produce. This tendency is in strong contrast with what Duarte Barbosa recorded of Sofala in the sixteenth century, where the people of the interior came laden with goods for sale. They obtained what they wanted, and, as he smugly remarks, the merchants were well content with the price.

Thus, taking the traditional histories of the coast as a whole, the principal emphasis, in agriculture, is upon millet, rice, vegetables, beans, peas, cassava, ground nuts, maize, cattle, and fishing, and for trade up-country, upon cloth, beads, salt, cowries, and some silver with which to pay for ivory, copal, ambergris, and, at Kilwa Kivinje, copper. Slaves are mentioned incidentally at several places south of Kilwa Kivinje; at Mikindani it is said that the Makonde came to sell themselves in times of famine.

It is hard to determine when the slave trade, in the proportions it assumed in the nineteenth century, began. If the Portuguese had made any serious profit from it in the sixteenth and seventeenth centuries, we should surely have heard of it.

[1] Velten, *Prosa und Poesie*, p. 281.

Their accounts provide no evidence. In the mid-seventeenth century Oman had won its independence from Portugal, and had made itself the principal trading state of southern Arabia, and the entrepôt for India. In the next century, after the appointment of La Bourdonnais as governor of the Île de France in 1735, the French colonies began to undergo a rapid development. Their Dutch predecessors had had a slave trade with Madagascar and with Mozambique, and it was to this area that he originally turned for his needs. By the end of the century the French had over 100,000 slaves as against 20,000 whites and coloured free persons. The needs of further development, of replacement through disease, old age, and death, and the failure of slaves sufficiently to reproduce themselves in captivity, required an optimum recruitment of some 10,000 slaves a year. It was with this as a background, for we have no information on the side of Oman, that the French attempted to enter into what had hitherto been an Arab preserve towards the end of the eighteenth century. Morice, a trader from the Île de France, wished, as a result of his experience at Kilwa Kisiwani between 1775 and 1779, to persuade the French Government to start an official slave-trading centre there. By then Oman had been in possession of the coast for ninety years, and each year vessels from India were arriving in Zanzibar, which served already as an entrepôt for the whole coast from Pate southwards. Ivory and provisions precede slaves in the list of articles of commerce given by Morice. Slaves went to Surat, either directly or indirectly, some to China, but the greatest number to Malabar. We are not told how many, but clearly the French were entering into competition with an important Arab trade. On 14 September 1776, Morice signed a treaty with the sultan of Kilwa allowing him a monopoly of slaves at 20 piastres each, the sultan receiving a duty of 2 piastres. He contracted to supply Morice with a thousand head a year. By 1790 this figure had risen to a total for the three previous years of 4,193, with seven ships visiting Kilwa from the Île de France, two of them making the voyage twice. The price had also risen to 40 piastres, of which the sultan received half.[1]

In 1804 the French Captain Dallons was at Zanzibar, and

[1] Rhodes House MS. Afr. r. 6; de Curt, *Reports*, 29 Jan. 1791.

wrote to the Île de France to complain of the treatment he received there. A eunuch of Ethiopian stock commanded the garrison and a Hindu was in charge of the customs. The French were at the mercy of an unscrupulous interpreter, who raised the price of slaves to an impossible height, over and above which a duty of 11 piastres a head was extracted by the governor in defiance of the correct rate paid by the Arabs of a piastre only. The Arabs sent their purchases to Muscat, the Red Sea, and the Persian Gulf—India is not mentioned—so that Dallons felt it unfair discrimination, especially since the French had no desire to compete with the Arabs in the markets in which they were selling. Regrettably he gives no precise figures except that the governor of Kilwa paid dues to Muscat of 6,000 piastres a year. Probably this represented the dues on more than 6,000 slaves, for some no doubt remained in his pocket.[1]

In 1812 the sultan complained to the captain of the English frigate *Nisus* of his loss of trade with the French, which he attributed to English activities in the Indian Ocean, presumably since the Anti-Slavery Act of 1807. Prior, the surgeon, was told of the activities of Morice, and also of the enormous profits made on the coast by the imports of European and Indian wares. He noted too that, whereas many thousands had formerly been taken away as slaves, by then the Arabs 'do not take away many'. It was a situation which was not to endure for long.

The political history of the Omani in East Africa provides the reasons. According to Salil ibn Raziq, they took possession of Pemba and Kilwa shortly after the capture of Fort Jesus in 1698. At an early date in the seventeenth century there were Omani governors, called *Liwali* in Swahili. By 1710 the viceroy in Goa was told that Kilwa had a garrison of fifty: in 1724 the sultan of Kilwa applied to Mozambique for help to rid himself of the Arab intruders. In 1745 Mozambique reported to Goa that, because of civil war in Oman, Pate, Malindi, Pemba, Zanzibar, Mafia, and Kilwa had virtually recognized the Portuguese as their legitimate overlords. No doubt they had welcomed Omani intervention to rid themselves of the Portuguese and found that the Omani were no less grasping and

[1] Mauritius Archives GA 11, No. 119, 1804.

oppressive. In 1759 Sultan al-Hasan ibn Ibrahim of Kilwa wrote to Mozambique, reporting that Mombasa and Pate were at war with Oman, and assured the governor of his friendship towards the Portuguese. In 1765 he sent again, with some messengers from Mombasa, to complain of Omani oppression, and in 1769 an abortive Portuguese mission was sent to Mombasa. At some time about 1772 there was a palace revolution in Kilwa, and we may judge from Morice's account that Kilwa already regarded herself as independent, as indeed the Mazrui governors of Mombasa had virtually been since the appointment of Muhammad ibn Uthman al-Mazrui before the overthrow of the Yorubi dynasty of Oman in 1744.

It was not until 1784, after the death of Ahmad ibn Said al Busaidi (1744–84), the founder of the Omani Busaidi dynasty, that any direct control of Oman's African possessions was attempted. In that year the rival claimant to the succession, Saif ibn Ahmad, came south with his son Ali with a view to carving out an independent domain. Kilwa surrendered to Ali, but Zanzibar held out under its governor: in the following sailing season Sultan, another brother of the Imam, and the Imam's eldest son, Ahmad, arrived with a relieving force. Said was banished to Lamu, while Zanzibar, and Kilwa under his son Ali, surrendered to the legitimate ruler. These events made it clear that if the Imam was to make good his African claims he must perforce exercise more control. The real author of the new system, and the most potent personality in the history of the East African coast in the nineteenth century, now succeeded, Sayyid Said ibn Sultan (1804–56), who obtained absolute power after the murder of his brother in 1806.

It was under these circumstances that Zanzibar became the centre for Omani activity in East Africa, and developed as the main pivot of the slave trade. In the late winter of 1811 Captain Smee sailed down the coast on the orders of the Bombay Government.[1] Mogadishu had virtually no trade and was remarkable only for the minarets of four mosques. Merca and Brava were equally unimportant, while at Pate trade was at a standstill in a town torn between two rival sultans. The monsoon carried him past Mombasa, but at Zanzibar he came on the same chief of customs, Yaqut, whom Dallons had so disliked

[1] *Transactions of the Bombay Geographical Society*, 6, 1844, pp. 23–61.

in 1804. He found him devoted to the interests of Sayyid Said, and extorting large sums from the inhabitants, from a considerable Indian community, and from French slave traders. Already there was a good number of stone buildings, and the town was a mart for slaves, ivory, bees-wax, and tortoise-shell, largely from the mainland. He considered the Indians excessively taxed, and questioned the right of Sayyid Said to tax them since they were British subjects. He left the brig *Sylph* to protect their interests. So it was that Indians forged the first link between England and East Africa. As to slaves, he considered that about three-quarters of the population were servile, the island exporting some 6,000 to 10,000 annually to Muscat, India, and the Mascarene islands. It might seem hardly necessary to inquire farther for reasons for Sayyid Said's removal to Zanzibar when there was so valuable an asset to be developed and protected. But however prosperous the interests of Oman in Zanzibar, the situation was far from secure. The writ of Oman ran in only part of the island, and except for Kilwa, and perhaps that part of the mainland immediately opposite Zanzibar, the coast as a whole was virtually independent, while Mombasa was making a determined bid to regain the hegemony it had enjoyed in the Portuguese times.

The Mazrui family have a story common in oriental annals, of several generations of ability and leadership as governors, whose consistent aim was sovereignty. The first Mazrui became deputy governor of Mombasa in 1727, and treated the inhabitants with such ferocity that they revolted. Another member of the family, Muhammad ibn Uthman, was appointed there *c.* 1740; and, when Ahmad ibn Said al Busaidi seized power in Oman in 1744, himself proclaimed: 'The Imam is an ordinary citizen like myself. He has usurped Oman: I have usurped Mombasa.'[1] Two years later, when Ahmad had time for East African matters, he had him murdered. Mombasa opinion did not support the assassins, and after a siege of Fort Jesus they were put to death, and Ali ibn Uthman al-Mazrui seized the reins of power. Shortly after he was able to seize Pemba, whose rice served Mombasa as a granary, and in 1753 he led an expedition to Zanzibar. Here, however, a family quarrel caused

[1] J. M. Gray, *The British in Mombasa, 1824–1826*, 1957, p. 9, 1, gives an alternative version.

his own nephew to murder him, and the Busaidi in Zanzibar, already half defeated, enjoyed an unexpected victory. The next half-century was uneventful, but in 1807 Ahmad ibn Said al-Mazrui led an expedition to Pate, where there had been a dispute over the succession, and installed his own candidate as a vassal of Mombasa. It was not until 1813 that he overreached himself in an expedition to seize Lamu, which ended in the disaster of the battle of Shela. In 1814 his son Abdullah succeeded, and sent Sayyid Said in Oman a defiant message stating his independence. He then appealed to Bombay for British protection, but without success. It was not to be the only appeal of its kind.

Sayyid Said could do no more than take note of these matters, for his control over Oman was not yet complete. In 1822 he was able to send a kinsman, Hamid ibn Ahmad, with a fleet, in response to an appeal from Pate. Brava and Lamu acknowledged the sovereignty of Oman, and after some negotiation Fumoluti, the Mazrui puppet in Pate, withdrew. In Pemba Hamid captured the Mazrui fleet, and the Mazrui had to make their way home as best they could. Abdallah ibn Ahmad al-Mazrui died of chagrin, but not before he had sent another appeal to Bombay for protection, claiming he was the ruler of an independent state that Sayyid Said sought to annex. He was told that 'it was contrary to our policy' and that fidelity to British engagements with Muscat forbade it. But, before the arrival of this communication, a letter came requesting that the British be permitted to survey Mombasa Island and to purchase cattle. No one could read it. The Mazrui and the people of Mombasa took it as an affirmative answer, and prepared to hoist the British flag.

The British on their part had shown no disposition to possess themselves of an unimportant city on the African coast which, it was plain, offered neither commercial nor strategic advantages, and which would be the occasion not only of expense to Britain but also of offence to Muscat. But on 3 December 1823 there arrived at Mombasa H.M.S. *Barracouta*, under Lieutenant Boteler, part of Captain Vidal's command engaged in a survey of the hydrography of the western Indian Ocean. The Mazrui were importunate, but for a time Boteler and Vidal managed to fend them off. They sailed on 7 December. On 7 February

Captain Owen arrived in H.M.S. *Leven,* at the very moment that a fleet sent by Sayyid Said was attempting to bombard the Mazrui into submission. Believing that acceptance of the Mazrui offer would enable Britain to use Mombasa as a centre from which to destroy the slave trade, he acceded to their demand for protection. The Union Jack was at last hoisted. He explained his action in a moving dispatch to the Admiralty, and no one can doubt either the sincerity of his patriotism nor the ardour of his religious desire to end the slave trade.

Thus he left a party on shore under Lieutenant Reitz, who shortly died, the command passing to Midshipman Phillips and later to Midshipman Emery. None of these young men—Reitz was only twenty-one—understood that what the Mazrui had really wanted was the mere protection of the British flag and to continue in the old way. The clause of the treaty forbidding the trade in slaves was unintelligible to them. Moreover, on the other side of the Indian Ocean, the Bombay Government could not concede that Mombasa or an unexplored possibility of its exploitation to end the slave trade could be worth a disruption of relations with Sayyid Said in the Persian Gulf. It was thus, after prolonged negotiations, that the British withdrew in October 1826.

Sayyid Said's action was prompt, and he immediately sent an ultimatum ordering the Mazrui to submit to him. They replied that they were prepared to submit to his formal sovereignty and even to pay taxes, but that they were not prepared to evacuate or to hand over Fort Jesus. Said had no choice, and in 1827 set sail with a fleet to Mombasa. The show of force and his personal diplomacy secured a temporary submission, but after a short visit to Zanzibar he was compelled to return to Muscat to dispose of a rebellious nephew. The Mazrui then rose again, but it was not until 1829 that Said could attend to them. His attack on Fort Jesus failed, but the Mazrui, exercising discretion in their success, agreed to negotiate peace. On this occasion Said spent three months in Zanzibar in a house four miles from the town, which he subsequently confiscated from the owner for a contravention of the Moresby Treaty of 1822. It was this that he later enlarged, and where he spent the latter part of his life.[1]

Once again he had to return home in haste to deal with a rebellion. He came back to East Africa in 1832, and in 1833 sent

---
[1] *A Guide to Zanzibar,* Government Printer, Zanzibar, 1952, based on reliable

an embassy to Madagascar to ask for the hand of the widowed queen and 2,000 soldiers. The offer of marriage may have been a diplomatic gesture, for a proffered princess was ignored. The soldiers were promised, and were to have been used against the Mazrui, but none ever materialized. But in these events there is a hint at least of a claim to an Indian Ocean dominion. Three years elapsed before Said came to Africa again, and, on this occasion in 1837, the Mazrui themselves were engaged in a family quarrel as to who should succeed as governor. Rashid ibn Salim had seized Fort Jesus, and had gained the support of the majority of the townsfolk: by judicious bribery Said won their opinions for his opponents within his family. Eventually his failure had become a foregone conclusion, and at this point Said changed sides, supporting him on condition that Fort Jesus should be conceded to Zanzibar. Rashid had no choice. Said then, having failed by negotiation to persuade him to leave Mombasa, so that a nominee of Said's own choosing could be appointed, arrested the whole family piecemeal, and deported them to Bandar Abbas on the Persian Gulf. He thus became undisputed master of the East African coastline: and it was thither to Zanzibar, without any announcement of permanent removal, that he sailed from Muscat in 1840. Thereafter, save for one brief visit to Muscat the year before his death, he was to stay in Zanzibar until the end of his reign in 1856.

Thus 1840 is a climacteric in the history of the East African coastline, since it saw the setting up in Zanzibar of a court whose ruler had foreign diplomatic connexions—the first in tropical Africa—with France, with Britain, and with the United States. From the United States he accepted a consul in Zanzibar in 1836, and from Britain a consul, Atkins Hamerton, in 1840. This, so far as the British were concerned, was mainly to confirm and support the Moresby Treaty of 1822, by which Said had agreed to make illegal throughout his dominions the sale of slaves to the subjects of Christian powers, and to confirm trading conditions already existing. For the Americans it was a similar confirmation.

These changes, which brought Europe and the United States into permanent formal contact with Zanzibar, if slight in

local traditions which contradict R. Coupland, *East Africa and its Invaders*, 1938, p. 295. Cf. Gray, *British in Mombasa*, pp. 178 sqq.

immediate appearance, were the precursors of a new phase in the history of the coast. It is thus a convenient moment to look back, and to attempt some account of the social and cultural history of these three and a half centuries. The comparative paucity of local documents, the disinterest of foreign visitors—it is significant that no Portuguese writer so much as mentions even the existence of the Swahili language!—and the fact that the study of the ruined buildings of this period along the coast is still in its infancy, all combine to make this a most difficult task. As yet it precludes the tracing of stages of transition, and it is only possible to point certain contrasts between the beginning of the period and its close. It is essential to recollect that throughout this time not only were there new African elements reaching the coast; there was also a constant infiltration of new adventurers from Arabia, in particular from Oman and the Hadhramaut. In addition there was from about 1800 a considerable influx of settlers from India, especially in Zanzibar and Kilwa Kivinje; Indians had been settled in petty numbers at least by the end of the fifteenth century.

In the early sixteenth century, when the unknown chronicler was ordered by his sultan to write the *History of Kilwa*, he did so in Arabic. He used, however, a number of words and also a system of dating which would not have been understood by any but a Swahili reader.[1] At the end of the eighteenth century, or in the early nineteenth, when the *History of Mombasa* was written, Swahili was used: it was only translated into Arabic in 1824 for Captain Owen's interpreter.[2] From the sixteenth century until the *History of Mombasa* there is no prose writing known. It is possible that certain elements of the *History of Pate* may have been composed at different periods, as likewise the late-nineteenth–early-twentieth-century compilation in Arabic, the *Kitab al-Zanuj*. Recent research along the coast by J. W. T. Allen has shown that there is a considerable body of manuscripts yet to be known which are in private hands, and it is possible that their recovery may enable us to know when Swahili prose came first to be written.[3]

[1] G. S. P. Freeman-Grenville, 'Chronology of the Sultans of Kilwa', *Tang. Notes*, 50, 1958, pp. 88–89.
[2] Gray, *British in Mombasa*, p. 70.
[3] J. W. T. Allen, 'The Collection of Swahili Literature and its Relation to Oral Tradition and History', *Tang. Notes*, 53, 1959, pp. 224–7.

As to poetry, the earliest known manuscript of the Swahili poem *Utendi wa Tambuka* is preserved in the library of the Hamburg Seminar für Afrikanischen Sprachen, and dated 1728. A manuscript of the poem *al-Hamziya* in the library of the School of Oriental and African Studies is dated 1749. Poetry in Swahili is known as *mashairi*, and includes a great variety of types, epic, lyric, and didactic. Some of the *mashairi* are apophthegmatic, and written *ad hoc* to illustrate a point in an incident. The earliest datable poem, as may be argued from its relevance to its context, is embedded in the *History of Pate*, and would appear to belong to the fourteenth century.[1] All these poems are in a strict, formal style with rules which derive ultimately from Arabic poetry, by which it has certainly been inspired, but from which it does not wholly derive. Among many similar works we may select for mention *al-Inkishafi*, The Soul's Awakening. It is in as grand a manner and diction as Dante or Milton. The economic decline of Pate, once glorious in its elegance and culture, affords a symbol of man's neglect of God, and the Judgement man must thus endure. Sin, the theme, enables the writer to paint a portrait of the slippered ease of the merchant nobles, rich from the Indian trade, their houses elegant with glass and porcelain, their beds inlaid with silver and ivory, their women a foretaste of the Islamic paradise. On such the Divine Judgement is ineluctable.[2]

In the early sixteenth century Duarte Barbosa and others could describe the grandeur of the houses of late medieval Kilwa, Mombasa, and Malindi. But, except at Kua, on Juani Island in Mafia, which has a palace, at Mtitimira, a few miles north of Kilwa, a late walled city, in the grand palace and buildings of Songo Mnara, and near by at Sanje Magoma, together with Mogadishu and the places already mentioned, the towns were of one-storied buildings only. Sixteenth-century Gedi has a small, provincial elegance, but outside such centres there were few stone houses: at Pujini, in Pemba, built probably mostly in the seventeenth century, the decline of architecture is displayed by the use of mud in place of the earlier mortar of ground

---

[1] Freeman-Grenville, *J.E. Afr. Swahili Committee*, 1959, p. 19.

[2] Abdallah bin Ali bin Nasir, *al-Inkishafi*, ed. W. Hichens, 1939. Cf. E. Dammann, *Dichtungen in der Lamu-Mundart des Suaheli*, Berlin, 1940, and id., 'German Contributions to Swahili Studies in Recent Decades', *J.E. Afr. Swahili Committee*, 26, 1956, pp. 9–17.

coral, traditionally mixed with a proportion of white of egg, together with lime, which dries with the hardness and consistency of cement. The use of stone declined, to be replaced generally by mud and wattle, and a thatch of *makuti*, plaited palm branches.

There is a curious change in mosque architecture which cannot yet be explained. Mecca lies almost due north from the East African coast, and, since the Muslim must turn thither when he prays, it was logical that the arcades of the medieval mosques should run north and south. Only one mosque, Mboamaji, which has an inscription over the lintel of the entrance dated A.H. 1017–A.D. 1608/9, can be ascribed with any certainty to the seventeenth century. Its arcades run laterally, and, although it cannot be said to have created a precedent, for many of its spectacular details of decoration are not imitated elsewhere, all the mosques ascribable to the eighteenth and nineteenth centuries follow a similar ground plan. There is a marked contrast, too, with the medieval *mihrabs* of carved, faced coral masonry: the later *mihrabs* are invariably of plastered rubble.

The chronology of fine plaster work is at present a problem which can only be noted. It was certainly known in the Middle Ages, and is illustrated best in the numerous domes and small barrel vaults of the Friday Mosque of Kilwa Kisiwani. The Mboamaji mosque, however, displays a type of plaster work which is wholly different and more fanciful of character, as do a series of tombs of a small local dynasty which ruled in and near Mkwaja, some thirty miles south of Pangani. These tombs would appear to be largely of the eighteenth and nineteenth centuries, but the genealogy of the dynasty has yet to be applied to the local traditions of the names of the occupants of the tombs.[1] It is clear, however, that variations in style were not great: features of the Mboamaji mosque appear upon them and again in a mosque at Kilwa Kivinje which can hardly be earlier than c. 1800. The patterns are closely allied to those of the carved wooden doors, whose beauty Duarte Barbosa admired in the sixteenth century. The earliest dated surviving example is of 1698, but they are still currently

---

[1] Some preliminary views as to dating are given in the writer's *Medieval History of the Tanganyika Coast*, 1962.

manufactured in the traditional manner both in Zanzibar and elsewhere.

Not only mosque architecture, but also that of domestic buildings, underwent a change. On the Kenya coast the only medieval stone houses so far excavated are at Gedi and at one or two small sites. Although they, and some houses which survive above ground at Kilwa, had elaborate sanitary arrangements, they were of an extreme simplicity. At Kua, of whose surviving houses some thirty would appear to belong to the late eighteenth century, there is no trace of the earlier sanitation. Some appear to have been arranged to include mere huts of wattle and thatch within a containing wall. No feature of the domestic architecture of the Portuguese such as survives within Fort Jesus appears to have had any influence on coastal habits or taste. On the contrary, the two alleged Portuguese houses at Mvuleni in Zanzibar appear to have followed local building tradition. Houses of the northern Arab type built round an internal courtyard, of which there are a number at Kilwa Kivinje, were most probably not built until the later nineteenth century: certainly not one can be assigned within the present period.

It can only be imagined that the furnishings were comparatively simple. At Kua certain houses still possess the wooden pegs for hanging carpets or silken hangings on the walls. The poem *al-Inkishafi* witnesses to beds inlaid with silver and ivory at Pate, and the Mogadishu museum preserves an ivory inlaid bed of uncertain date. A Portuguese account of 1505 describes beds at Kilwa of similar type to those in use on the coast today, a crude wooden frame with roughly made legs and woven palm twine stretched across and supporting an *mkeka*, a sleeping-mat. Chairs were rare, and ordinary people sat cross-legged on a *baraza*, a raised platform in front of the house, sheltered by the overhang of the roof. The Zanzibar museum possesses a superb chair inlaid with ivory, but such chairs were surely the mark of royal status.

Some domestic vessels have survived. Cooking, as the middens of many sites bear witness, was done in clay pots. The profusion of Chinese porcelain, in Swahili *bakari-sini*, shows that the more comfortable classes ate from plates and bowls imported from China, and that the import was continuous throughout the period. It is curious that local literature makes no reference to it save a solitary reference in *al-Inkishafi*. The porcelain plates and bowls were carried by the domestic slaves on brass *sinia*, richly chased trays, from the Arab lands and from India. The same areas provided elegant brass-bound chests which served to keep clothes and articles of value. The best of these, as also boxes for writing implements and for paper, were of camphor, which is proof against insects, and especially against the white ant, which is to be found everywhere along the coast.

Duarte Barbosa speaks of the rich dress and jewellery of the well-to-do. There is ample sixteenth-century evidence of the import of silks and brocades, as well as of the scarlet cloth which the nineteenth-century travellers were to find so popular. The Zanzibar museum possesses a quantity of silver jewellery of coarse appearance and little elegance. It appears to be of Indian inspiration, and there is no strong evidence to suggest that it was manufactured earlier than the nineteenth century, although it may have followed earlier traditional patterns. In a strict Islamic society there is no pictorial art to aid such identification. If the quantity exhumed at Gedi and which litters the beaches at Kilwa and elsewhere be witness, beads had a continuous popularity from medieval times onwards. In early Portuguese, as in medieval, times, they came from Cambay, but also, via Aden, from Venice, and many no doubt from farther afield, as far as Bohemia and Java.[1] There are many of an extremely small calibre, which would suggest that they were sewn on to garments like sequins, but no clothing of this period survives. The frequency of poetical references to pearls would suggest that they were in common use, but we should not, perhaps, draw any inference from their profusion in the portrait in the National Galleries of Scotland, which depicts a Princess of Zanzibar, who married Sir John Henderson in 1604, in company with her personal attendant, for their dress is indisputably European.[2]

[1] W. G. N. van der Sleen, 'Ancient Glass Beads with Special Reference to the Beads of East and Central Africa and the Indian Ocean', *J. Roy. Anthrop. Inst.* 88, 1958, pp. 203–15.
[2] A. G. Mathew, 'The Culture of the East African Coast in the 17th and 18th Centuries in the Light of Recent Archaeological Discoveries', *Man*, 56, 1956, p. 61: the princess is portrayed as white, and her attendant as African.

It is notable that a considerable number of epitaphs of this period commemorate women. During this period there were 'queens' of Zanzibar, Siu, Mombasa, and Kua, while at Kilwa for a time there was a woman regent, an appointment for which there was a medieval precedent. We should perhaps be wrong to regard women as a suppressed class. In general we hear little of them, or of children. It is not until the end of the nineteenth century that there is any description of education. Velten's *Desturi za Wasuaheli*, if written down by a Bagamoyo *mwalimu* at the end of the nineteenth century, is probably representative of all this period. It describes a set and formal rule of life and custom which it may be presumed successive generations scarce altered: the pious ceremonies which surrounded birth and circumcision; the elaborate initiation into life in the Koran school, both as to the Islamic religion and as to the observance of the ancestral customs of courtesy of which the aim is *heshima*, a single untranslatable word which subsumes self-esteem, dignity, and the public respect; the formalities of marriage which permitted both human inclination and the deliberation of family arrangement; and the solemnities of death and succession. Just as the coastal people absorbed Islam, so the continuance of certain ceremonies of pagan, animistic origins was permitted. Men, women, and children participated in the *Nau Roz*, or *Kuoga Mwaka*, the Swahili New Year ceremonies in which Islam is intertwined with rituals both of Persian and of African reference, and whose object seems to be the purification of the worshippers before cultivation and the hallowing of both the land and the sea. In this, in the cult of the ancestral spirits (albeit their bodies were buried orthodoxly in the cemetery beside the mosque), and in certain ritual formulae of witchcraft which include Islamic invocations, Arab and African elements are inextricably intertwined.

To this amalgam the Portuguese made comparatively little contribution. Of greatest importance was the introduction of certain new crops and trees, cassava or manioc (Swa. *muhogo*; Port. *mandioca*) and the cashew tree (Swa. *mbibo*; Port. *bibo*), both of which originated in Brazil; maize or Indian corn (Swa. *muhindi*; Port. *maiz*); the fruit of the cucumber tree, *malus indica* (Swa. *bilimbi*; Port. *bilimbinos*); *mpea*, the avocado pear tree (Port. *pera*), with *mpera*, the guava, from the same

Portuguese origin; and for cultivation the use of dung, *mboleo* (Port. *boleo*, dropping). There were also important improvements in the rig of vessels and their cut, and many useful household articles, handkerchiefs (Swa. *leso*; Port. *lenço*), the screw (Swa. *parafujo*; Port. *parafuso*), and wine (Swa. *mvinyo*; Port. *vinho*). They introduced snuff-taking (Swa. *tabakelo*, snuff-box; Port. *tabaqueira*), while almost all the names for playing cards have a Portuguese origin, and perhaps too the name for a brothel (Swa. *danguro*: ?Port. *dançador*). Altogether some sixty words of Swahili have a Portuguese origin, and in one way or another they are mostly objects and innovations of genuine utility. But the presence of the Portuguese for 200 years and more had little deep effect, and contributed nothing to art, building, or literature, and nothing lasting to politics. Their passing was as a dream that is lost, but, if nothing else, they had at least brought to the coast what are now the staples of maize and cassava.

The coastal civilization was something unique, neither wholly Arab nor exclusively African, eclectic but not cosmopolitan. In Swahili, civilization is spoken of as *ustaarabu*, being like an Arab. Until the late eighteenth and early nineteenth centuries the immigrants, absorbed in trade, had their faces turned towards the homelands to which they exported goods, just as their religion turned their faces to Mecca. They were essentially trading communities, and were contented to let the trade of the interior flow to them. It is only about 1780 that the first traders seem to have gone inland. As they intermarried, and employed African serviles, or enfranchised African *surias*, concubines by whom they had had children, so they unwittingly and unconsciously absorbed African ideas, African religious practices enshrined in the witchcraft of the coast, and turned to their own use African words in the creation of the Swahili language. It is some measure of the strength of African influence that this language, in absorbing foreign words, has perfectly preserved a Bantu form of grammar. Inevitably there were differing degrees of Africanization: they were the very essence of the Swahili civilization.

VI

DISCERNIBLE DEVELOPMENTS IN THE INTERIOR

c. 1500–1840

ROLAND OLIVER

THE 350 years from the beginning of the sixteenth century to the middle of the nineteenth are, for the interior of East Africa, the age of tradition. During the first two centuries of this period the Portuguese were active along the coast, but the only eyewitness account of any part of the interior that has come down to us is the bare narrative of one overland journey from Tete to Kilwa in 1616.[1] During the last century of this period Arabs were certainly penetrating beyond the coastal fringe, but of their journeyings at this time no contemporary written record has yet been discovered. No outside observer can therefore be called in evidence. All that we know, and that only in part, is what the peoples of the East African interior have themselves handed down or imagined about their own past. A grave or a grove, a ditch or a bank, may here and there add some corroboration. In time the archaeologist's spade will doubtless add much more, but as yet it has been sparingly applied. For the present, tradition holds the field.

Traditional evidence in Africa is not by any means evenly spread. There are communities which remember little beyond the time of their grandparents, and others whose oral history extends over fifteen or even twenty generations, that is to say, over four or five centuries. Wealth or poverty of tradition seems to be in part a product of the social organization of the people concerned. Decentralized societies have few and short traditions. Monarchic and oligarchic states, on the other hand, tend to be rich in traditional lore, because it is one of the instruments

[1] J. M. Gray, 'A Journey by land from Tete to Kilwa in 1616,' *Tang. Notes*, 25, 1948, pp. 37–47.

of centralization, taking the place of archival and statistical services in modern, literate societies. Oral history is best preserved in kings' courts, where there is a hereditary leisured class of chiefs and councillors, and where a knowledge of precedent is constantly in demand. But even granted favourable conditions of society, the survival of unwritten history depends upon the society's historical survival. When a literate state is overthrown, most of its records will probably survive. When a non-literate state meets disaster, its living repositories of learning will be scattered and dispersed, their knowledge will become irrelevant, and there will be no occasion for handing it on to the younger generation, and it will wither and die with its last possessors. In East Africa, therefore, the bulk of our information comes from the relatively organized states, and, more narrowly, from the organized states that were still in existence at the time of the European contacts of the late nineteenth century.

There is a further difficulty. The European contact and the introduction of literacy did not of itself ensure that the oral tradition surviving at the time was fully or accurately or objectively recorded in writing. Explorers, such as Burton, Speke, and Stanley, were assiduous in reporting what they could find of traditional historical information during their brief sojourns in particular tribal areas. Later, often, however, much later, missionaries and administrators started, as a spare-time interest, to search out the history of the peoples among whom they worked. Inspired by their example, literate Africans pursued their own inquiries and set down the results in vernacular publications of varying merit. Anthropologists, entering the field later still, added historical chapters here and there to primarily sociological monographs. The evidence from all these sources adds up to a formidable literature. Glaring gaps nevertheless remain. Peoples as crucial for the history of the whole region as, for example, the Nyamwezi have been virtually ignored. Even where the evidence is thickest, the information to be derived from it is painfully inadequate and imprecise about the most important matters. What emerges is the barest outline only, and much of that still speculative. But if a history of East Africa is to be anything more than a record of the most recent immigrants, such speculations must have their place. If the sum of finally ascertained fact is small, a discussion of the problems may at least induce some proper orientation towards the local scene.

The first fact which becomes plain from an examination of the traditional evidence is that East Africa has seen within the last 500 years population movements on a scale which have not occurred in Europe since the Dark Ages, in the Middle East since the coming of the Turks, in South Asia since the Moghul invasions. It is not, of course, that the land was previously empty, although sparsity of population was doubtless the largest single cause of political instability and economic backwardness. But the largest linguistic frontiers—those separating the different families of languages—seem to have moved only slightly within this period, indicating that most of the movement has been more like the ebb and flow of tides in the sea than like the shunting of a goods train in which distinct units push each other bodily to and fro following the impulsions of a distant engine. Migrants into or within East Africa have usually been absorbed, linguistically at least, by the pre-existing peoples of the areas where they have settled. The principal exception to this rule has been in the north-west corner of the region, which has suffered within the period under review an influx of peoples speaking languages of the Nilotic family, who have arrived in sufficient force to impose those languages over wide areas where languages of the Sudanic, Nilo-Hamitic, and Bantu families had previously been in use.

These Nilotic invasions were spread certainly over two, perhaps over three centuries. They brought into northern Uganda and to western Kenya a large proportion of the peoples known collectively as Lwo, who had probably evolved as a single coherent group, occupying most of the south-east of what is now the Republic of the Sudan. It is a reasonable inference from the ethnographic evidence that this area received during the period in question a series of substantial population movements from the Nilo-Hamitic region immediately to the east of it, causing the Lwo group to disperse southwards and northwards, that is to say both up and down the line of the White Nile. Though lying farther back in time than the earliest traditions, it seems the most reasonable hypothesis that it was this earliest stage of the dispersion that sent the first Lwo invaders southwards into Uganda, and the ancestors of the Shilluk northwards to Fashoda. The subsequent migration of the Anuak up the Sobat river from Shillukland is authenticated by tradition, as are still later movements from the Anuak country reaching northern Uganda as late as the early eighteenth century.[1]

It can be accepted as certain that the spearhead, and, indeed, more than the spearhead, of the early southward immigrations crossed the Somerset Nile, between Lake Albert and Lake Kyoga, into the southern or Bantu-speaking half of Uganda. There they settled in considerable strength, adopting the Bantu speech of the country, while their leaders, having ousted the previous rulers, founded a number of related Bito dynasties, some of which survive to this day, including, notably, those of Bunyoro and Buganda.[2] It is from the genealogical evidence of these dynasties that the arrival of the first Lwo can be roughly dated, for a large number of distinct and collateral lines show between eighteen and twenty generations from the foundation of these dynasties to the present day. Even at twenty-five years to the generation,[3] this is a period which must extend to the very beginning of the sixteenth century and is likely to have started a little earlier still.[4]

[1] Cf. D. Westermann, *Geschichte Afrikas*, 1952, p. 317. J. P. Crazzolara, *The Lwoo*, 1950, pp. 31–32, places the Lwo cradleland to the west of the Nile, on the basis of doubtful etymological evidence. He also (pp. 33–58) believes that all the Lwo who entered Uganda came round via the Shilluk and Anuak. The traditional evidence does not demand this interpretation, the acceptance of which also creates serious chronological problems as between the Shilluk traditions, extending through fourteen generations only, and those of the southern Lwo extending through eighteen or nineteen.

[2] J. P. Crazzolara, 'The Lwoo People', *Uganda J.* 5, 1937, pp. 1–21.

[3] A 'generation' in this context may be defined as the average difference in age between a king and the son or nephew who will succeed him, ultimately if not immediately. The kings of these states were of course polygynous, and there was no rule of primogeniture, indeed only princes born during their father's reign were eligible. In these circumstances a dynastic generation was probably seldom less than twenty-five years and usually more.

[4] e.g. the genealogies of (i) the royal house of Buganda, in J. Roscoe, *The Baganda*, 1911, which shows 20 generations to Kimera, the founder of the present dynasty; (ii) the Mpewo clan of Buganda, whose head was chief steward of the Kabaka's enclosure, and whose genealogy recorded in A. Kagwa, *Ekitabo kye Bika bya Buganda*, 1949, pp. 73–78, shows a sequence of 18 generations covering the reigns of all the Kabakas back to Kimera; (iii) the Bito dynasty of Bunyoro, as recorded by J. W. Nyakatura, *Abakama ba Bunyoro-Kitara*, 1947, shows 18 generations to Rukidi, the brother of Kimera; (iv) the Bito dynasty of Kiziba, recorded in E. Césard, 'Le Muhaya', *Anthropos*, 32, 1937, pp. 32–57, showing 17 generations to Kibi, the great-grandson of Rukidi; (v) the Hinda dynasty of Ankole, recorded

It is unlikely that any of the Lwo settlements in northern Uganda were substantially earlier than those in the south. Some of them were certainly later. There are, however, at least two of the score or so of small tribal groups under Lwo ruling clans today collectively called Acholi who have traditions extending back to the Lwo invasion that established the Bito dynasties in the south.[1] There is also traditional evidence that a strong concentration of Lwo-speaking people, including the ancestors of at least six more of the surviving Acholi tribal groups, built itself up on both sides of the Somerset Nile, in an area now usually called Chope but formerly Pawir, lying just to the north of the Bantu linguistic frontier, and subject politically to the Bito dynasty of Bunyoro.[2] Pawir would seem in fact to have been the area settled by the rearguard of the original migration, which arrived too late to join in the conquest of the south, and which therefore escaped the Bantuization undergone by the van. The first three Bito kings of Bunyoro are believed to have been brought back into this region for burial:[3] this affords some indication of how long it took before the conquerors of the south began to feel themselves more Bantu than Lwo.

So far as northern Uganda is concerned, the picture of Lwo migration moving steadily up the Nile Valley towards a destination in Pawir would seem to hold good for the whole of the sixteenth century. In the seventeenth century, however, a new pattern developed. Fresh contingents continued to arrive from the north, but Pawir was by now overcrowded. As early as the seventeenth generation back there is a reference in tradition to the emigration of a Patiko chief into the country to the west of the Albert Nile.[4] The main Lwo occupation of this area, however, began about the thirteenth and fourteenth generation

in A. G. Katate and L. Kamugunguru, *Abagabe b'Ankole*, 1955, showing 19 generations to Ruhinda, the contemporary of Rukidi; (vi) the Nyiginya dynasty of Rwanda, exceptionally well recorded by A. Pagès, *Un royaume hamite . . . le Ruanda*, 1933, showing a major invasion of Rwanda by the Nyoro in the 18th generation back. The Nyoro ruler is remembered as Chwa lya Nyabongo and is probably to be identified with Chwa, the son of Rukidi: certainly, however, the reference is to a Bito ruler.

[1] The traditions of the Patiko group, extending through nineteen generations, have been recorded in some detail by Crazzolara and are particularly circumstantial: *The Lwoo*, pt. ii, pp. 223–56.
[2] Cf. R. M. Bere, 'An Outline of Acholi History', *Uganda J.* 11, i. 1947, p. 3.
[3] Nyakatura, *Abakama*, p. 88.
[4] Crazzolara, *The Lwoo*, pt. ii, p. 226.

back, when members of the Atiak and Uchibu ruling clans crossed over the Nile a little to the north of Lake Albert and

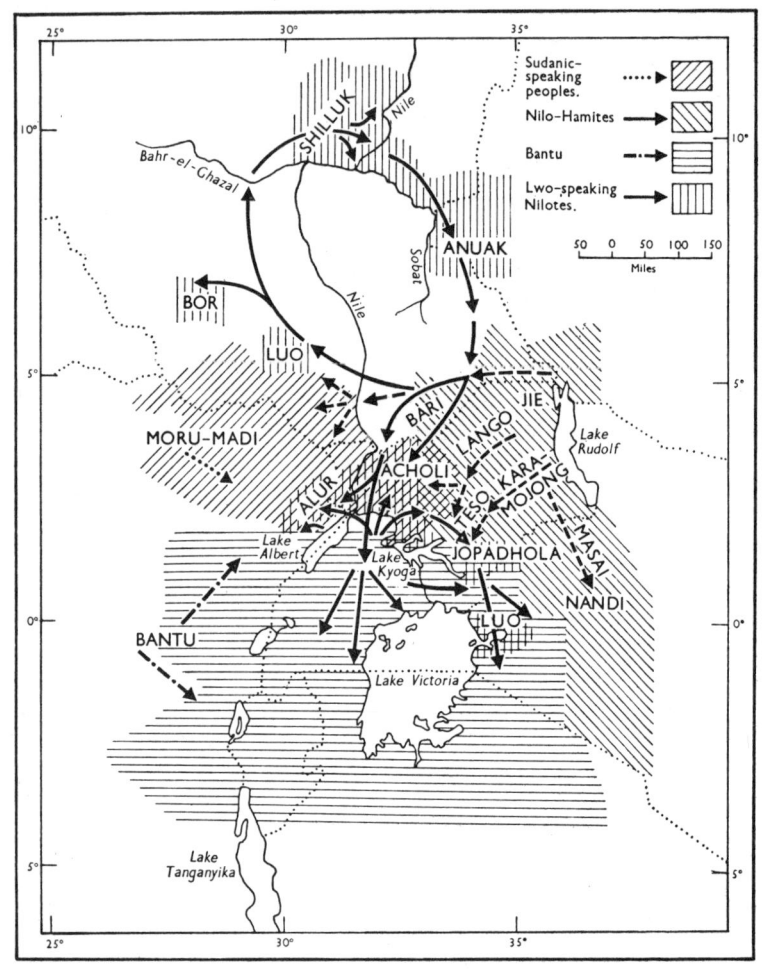

MAP 8. Approximate movements of Lwo-speaking Nilotic peoples and Central Nilo-Hamites, *c.* 15th–18th centuries

proceeded in the course of a century or so to establish a whole series of Lwo chieftainships in the country now called Alur.[1] Meanwhile, to the east of the Albert Nile, Pawir was expanding

[1] A. W. Southall, 'Alur Tradition and its Historical Significance', *Uganda J.* 18, 1954, pp. 144–5.

fast, not as a unitary state but as a succession of Lwo groups, who colonized new areas in southern Acholiland and in the present Lango country to the north of Lake Kyoga, and then sought recognition and chiefly insignia from the Bito rulers of Bunyoro.[1] Finally, and still at the same late-seventeenth-century period, there arrived a definite new influx from the Anuak country, which founded the political structure of most of northern Acholiland. The passage of these groups through the south-east corner of the Sudan would seem to be clearly remembered in Bari tradition.[2]

So much for the Nilotic newcomers into northern Uganda. Who, then, were their predecessors and what became of them as a result of the Nilotic settlements? Broadly, the answer is that they were people of two types, the majority of whom were absorbed politically and culturally by the Lwo invaders, but some of whom have survived as small independent groups up till the present day. In the western half of the area, in the valley of the Albert Nile and in the hill country to the west of it, were peoples of the Moru–Madi type, speaking 'Sudanic' languages, with ethnic and cultural affiliations extending northwards and westwards into the south-west of the modern Republic of the Sudan. In the eastern half of the area, to the north of the Somerset Nile and Lake Kyoga, were peoples speaking languages of the Nilo-Hamitic family, whose cradleland lay eastwards astride the present Kenya–Uganda frontier to the west of Lake Rudolf.

The evidence on these submerged layers, which is far more conclusive in the case of the Moru–Madi than of the Nilo-Hamites, comes from the sociological analysis of the modern Acholi and Alur, which also throws considerable light upon Lwo methods of colonization. It shows that most political units consist of a ruling clan of real Lwo origin and of a number of clans of commoners, who are the descendants of the people absorbed by the Lwo at various stages of their migration.

> On the march [writes Crazzolara], as still in big hunts, the tribal, clan and family groups kept closely united. . . . Female prisoners were absorbed and became completely submerged. . . . For male

[1] R. M. Bere, *Uganda J.* 1947; cf. A. W. Southall, *Alur Society*, 1956, pp. 218–19.
[2] Crazzolara, *The Lwoo*, pt. i, pp. 52–55; cf. A. C. Beaton, 'A Chapter in Bari History', *Sudan Notes*, 17, 1934, pp. 169–200.

captives the case was different in theory, but scarcely in practice. Prisoner slaves were allocated to a family or clan-group, and treated as blood-relations, and even given wives, or cattle as dowry. But with their children and descendants they started their own sub-clans and social life, as a clan-segment, related and hence exogamous, to the main Lwo clan.[1]

Members of the ruling clan, whose head was the chief or *Rwot*, were called Jo-*kal*, Jo-*bito*, or Jo-*ker*[2] and generally had no more particular name. The non-Lwo clans, however, were known as the *Lwak* (herd) or *Lobong* (slaves) and were called by their special clan-names. It is from the analysis of these subject clans in the several tribal groups of the Acholi and Alur that the incorporation of large bodies of Madi, Okebo, and Lendu can be detected. It would seem clear that these, together with the ancestors of the still-independent Madi and Lugbara tribal groups living to the north of the Alur in the West Nile District of northern Uganda, formed the basis of the pre-Lwo population in the north-west. The ancient population of the north-centre is more problematical since no descendants of these survive as independent groups. Crazzolara, however, whose views deserve respect, is of the opinion that several of the subject clans of the southern and eastern Acholi tribal-groups are descended from ancient Nilo-Hamitic peoples of the same general type as the Didinga of the Imatong mountains. It is at least a reasonable hypothesis.[3]

The final stage in the Lwo settlement, and the last major episode in the history of the peopling of northern Uganda, seems to have been due to a marked expansion of the central Nilo-Hamitic peoples—the ancestors, that is to say, of the Lango and the Jie, the Teso and the Karamojong. We have already seen that the first westward explosion of this group of peoples occurred a little to the north of the present Sudan–Uganda frontier, probably causing the original Lwo dispersion, and bringing some of the ancestors of the Fajelu and Kakwa,[4]

---

[1] Crazzolara, *The Lwoo*, pt. i, p. 47.
[2] *Kal*: the reed-fence, surrounding the residence of the chief E*kikali* which means the same in Lunyoro. *Ker*: chiefship. The meaning of *Bito* is uncertain, but the Bantuization of this word as Ba-*bito*, the family name of the royal house of Bunyoro, is highly significant.
[3] Crazzolara, *The Lwoo*, pt. i, pp. 81–82, 333–4, 349–50, 475, &c.
[4] Who now speak Sudanic languages.

and, later, those of the Bari into their present positions along this line. The final settlement of the Bari may perhaps be placed in the late sixteenth or early seventeenth century.[1] The movements of the following century and a half were neither so sudden nor so far-reaching. They involved merely a steady pressure of warlike, mainly pastoral groups from the north-eastern corner of modern Uganda south-westwards towards Lake Kyoga.

According to tradition it was twelve generations ago, towards the mid-seventeenth century, therefore, that the ancestors of the Lango separated from those of the Jie in the region of Mount Otukei, in the north-west corner of the present Lango country. It seems that the first Lango thrusts went westwards into an Acholiland where Lwo and Madi groups were still living a comparatively separate existence. It is said that the Lango at first avoided the Lwo and concentrated their attacks on the Madi, but that the Lwo later made common cause with the Madi. The Lango invasions may therefore have been one of the largest factors leading to the integration of Lwo and Madi groups into the modern Acholi. It would seem also that this integration was successful in diverting the expansion of the Lango southwards into their present home, the occupation of which took most of the eighteenth century. The Lango are now primarily a nation of cultivators, but traditionally this is a recent development. They entered the country as pure pastoralists, living on milk and the blood of their cattle, as the Jie do to this day.[2]

Just as the Lango were a section of the pastoral Jie of Karamoja District who hived away and gradually changed their mode of life in a new environment, so the Teso, the south-easterly and even more strongly agricultural neighbours of the Lango, claim descent from the pastoral cousins of the Jie, the Karamojong. The Teso settlement probably began almost contemporaneously with that of the Lango, but proceeded more slowly. According to Lawrance, the great-grandfathers of the present generation were mostly still living in Usuku, the part of the country nearest to Karamoja.[3] It is a significant claim of

---

[1] A. C. Beaton, *Sudan Notes*, 17, 1934, p. 200, lists the reigns of fourteen generations of rain-makers between Yuyok, the first to settle in Shinduru, south of Juba, and Lugor, who died there in 1885.

[2] A. Tarantino, 'Notes on the Lango', *Uganda J.* 13, 1949, pp. 145–53; and 'Lango wars', ibid., pp. 230–4.

[3] J. C. D. Lawrance, *The Iteso*, 1957, pp. 10–11.

both Teso and Lango traditions that, with the exception of the very primitive Tepeth groups of Nepak Mountain, they found the country open before them, though there were broken pots and the remains of hearths and homes. Both peoples emphasize that the occupation was completely peaceful.

This claim touches the last and most mysterious problem of the expansion of the Lwo, who, it is agreed on all hands, must have passed in considerable numbers along the northern shores of Lake Kyoga on their long road south-eastwards to the Kavirondo Gulf at the north-east corner of Lake Victoria, where their descendants today number nearly a million people. Chronologically, it is perfectly possible that the Lwo should have passed by while the Lango and Teso were still evolving a settled life on the north-eastern fringes of their present sphere. It is difficult to believe, however, that the Lango, at least, did not conquer some of them and speed the rest upon their way. All the Lango today speak a Lwo language. So do the Kumam, who were probably the vanguard of the Teso expansion, now settled on the Kaberamaido peninsula west of Soroti. Tarantino holds the view that the Lango adoption of Lwo speech is the result of recent neighbourhood contacts with the Acholi.[1] Crazzolara challenges this in the strongest terms, pointing out that many Lango names are of Lwo and Madi origin, and are common among both the Acholi and the Alur. They point, he says, to a population of much the same composition as Acholi, which was gradually absorbed by the three main Lango clans when they arrived.[2]

It would seem, therefore, that the southern Lango country, including the whole Kaberamaido peninsula, formed part of the great concentration of Lwo peoples which piled up against the northern frontier of the Bantu world during the century or so after the Bito vanguard had burst through and become Bantuized. The buffer area was Chope, the Lwo-speaking enclave in northern Bunyoro. From Chope the main stream of immigrants moving up the Nile Valley rebounded in the late seventeenth and early eighteenth centuries, westwards into Alur, northwards into Acholi, eastwards into Lango country. With the arrival of the Lango people, these eastern Lwo moved still

[1] Tarantino, *Uganda J.* 1949, p. 145.
[2] J. P. Crazzolara, *The Lwoo*, pp. 564-5.

farther east, passing rapidly through Teso with little settlement, but pausing at the eastern end of Lake Kyoga, where the Jopadhola form a Lwo wedge between the Bantu Soga and the Bantu Gisu of Mount Elgon and point the route to their Lwo relatives still farther southwards in Kavirondo.

Like the Arabs in North Africa and Spain, the Lwo must have ended their great dispersion as a tiny minority among those whom they ruled. Like the Arabs, they were extremely successful in spreading their language and certain aspects of their culture. Unlike the Arabs, they do not seem to have been outstanding warriors; and there is little suggestion in tradition that their material equipment was in advance of that of their predecessors and later subjects. 'The impression they made', writes Southall of the Alur, 'must be attributed to their superiority in the scale of their social and political organization, and the self-confidence which gave them their reputation as rain-makers, arbitrators and administrators, and preserved their sense of moral destiny to dominate.'[1] This is an interesting verdict in that it has specially in mind the smaller Alur chiefdoms, many of which had as few as 500 subjects. Size, therefore, was not the main element in this political advance. Rather it was the transition from purely kinship polities to those which were in some sense territorial. In purely kinship polities the blood-feud was an insuperable obstacle to prosperity and progress. With territorial chieftainship, on however small a scale, arbitration and compensation could have much fuller play.

The chief was continually initiating cycles of production and consumption . . . in which the realisation of services was balanced by the satisfaction of needs. He was accumulating girls paid to him in fines, and providing wives to the destitute. . . . He was receiving supplies of grain and livestock, and organising feasts which evoked sentiments of loyalty, and intensified social relationships leading to a sense of unity. He was marrying girls he fancied from all the subject clans, so giving them a renewed stake and interest in the chieftainship . . . and from this heterogeneous stock he himself begot an ever-growing population observing common values . . . .[2]

Without doubt, factors such as these were at least as important as the mere movement of people. The Lwo occupation amounted to a reorganization as well as a migration. This

[1] Southall, *Alur Society*, p. 231.   [2] Ibid., p. 82.

aspect of the matter becomes even more relevant as we move from the north of Uganda to the south.

In southern, Bantu-speaking, Uganda, even more clearly than in the north, the Lwo invasion of the late fifteenth or early sixteenth century was the outstanding event that marked the beginning of a new age. It was not, in this case, that it brought a higher civilization. On the contrary, it would seem that in southern Uganda the Lwo were the barbarian conquerors of a culture more advanced than any they knew themselves, and which absorbed them far more truly than they absorbed it. Nevertheless, like the Anglo-Saxon invaders of romanized Britain, they gave this area the first outline of its modern shape. They established new political nuclei themselves, and their presence caused other nuclei to form in defence against them. Some of these new units were to be more successful than others. Some were to grow, others to disappear. But from the time of the Lwo invasions all the main ingredients of the early-nineteenth-century scene were present—in north-west Tanganyika and Rwanda-Urundi as much as in Uganda. Only the pattern of their interaction remained to be evolved.

The Lwo invasion is not in southern Uganda the beginning of traditional history. What comes before is hard to interpret, but at least it is clear that this section of the Bantu world already enjoyed a social and material culture very different from that of the pre-Lwo peoples of the north, very different also from that of its own Bantu neighbours in Kavirondo and the Kenya highlands. No doubt this was to some extent the result of the specially favoured climate and well-distributed rainfall of the north-western periphery of Lake Victoria, as also of the remarkable properties of the banana plantain as an almost labour-free source of food, drink, and building material. No doubt in such an environment population was always, by East African standards, comparatively dense, and possibly mere density is apt to be a factor making for social coherence and economic specialization. Early traditions, however, leave little doubt that, long before the arrival of the Lwo, immigrants had had an important part to play.

The pre-Lwo traditions of southern Uganda fall broadly into two streams. There is an eastern stream associated with central

Buganda and Busoga, which tells of a culture hero remembered as Kintu, 'the marvel', who entered the region from the northeast, from beyond Mount Elgon, and who in the course of an incredibly long and peripatetic career united the ancestral clan-heads and attracted them into ceremonial positions about his person in a way which suggests the evolution of some kind of centralized monarchical state. Kintu did not die, he just moved on. He was a period, not a person. And the same may well be true of Chwa, his successor in Buganda tradition, who was succeeded after an interval by Kimera, the first of the Bito (Lwo) dynasty.[1] The other stream comes from the western half of the region, where, in Bunyoro tradition, the Bito dynasty was preceded by three others, the Gabu, the Ranzi, and the Chwezi, all of whom ruled over a kingdom called Kitara, with its centre a little to the south of modern Bunyoro, in the downlands of eastern Toro, western Buganda, and northern Ankole. These western dynasties are clearly more real, for they are the names of clans which still exist among the pastoral aristocracy known collectively as Hima in Uganda and Tutsi in Rwanda. On the whole it seems likely that the Ganda kingdom of Chwa was identical with that of the Chwezi, and that what the Bito conquered was already, in a very loose feudal sense, a single political unit dominated by Hima pastoralists under kings of the Chwezi clan. According to tradition, the last capital centre of the last of the Chwezi kings was in the celebrated entrenched earthwork site at Bigo, on the Katonga river.[2] Recent archaeological investigations have tended to confirm the traditional evidence that Bigo was the seat of a Hima dynasty. It has been established that the centre of the site, as it now appears, is the result of a very considerable reconstruction, which probably took place in early Bito times. The original layout of the central embankments was almost certainly similar to that of the

[1] For Buganda traditions see Roscoe, *The Baganda*, pp. 133–85 and 214–30; and A. Kagwa, *Basekabaka be Buganda*, 4th ed. London, 1953, and *Ekitabo kye Bika bya Buganda*; cf. J. M. Gray, 'The Early History of Buganda', *Uganda J.* 2, 1935, p. 260. For Busoga, C. L. Bruton, 'Some Notes on the Basoga', *Uganda J.* 2, 1935, p. 291; D. W. Robertson, *Historical Considerations Contributing to the Soga System of Land Tenure*, Entebbe, 1940; Y. K. Lubogo, *A History of Busoga*, Jinja, 1960.

[2] For Nyoro traditions, see K. W., 'Abakama ba Bunyoro-Kitara', *Uganda J.* 3–5, 1935–7; and Nyakatura, *Abakama ba Bunyoro Kitara*; for Ankole, A. G. Katate and L. Kamugungunu, *Abagabe b'Ankole*; cf. P. J. Gorju, *Entre le Victoria, l'Albert et l'Édouard*, Rennes, 1920.

typical *orurembo*, or royal town, of the early Hinda kings of Ankole.[1]

It is significant of their sense of cultural inferiority that the Bito have felt it necessary to claim in their traditions a genetical relationship with their Chwezi predecessors. One consequence, however, is that we are very ill informed about the actual process of the conquest. According to the official accounts the Chwezi royal family, having decided to depart, sent messengers across the Somerset Nile to conduct their Bito 'cousins' safely into the kingdom. They were also obliging enough to leave intact their buildings and their regalia, their herds, and even the royal wives, who were thus able to instruct the newcomers in all the pastoral ceremonies of Hima kingship. The first Bito monarch, Isingoma Mpuga Rukidi (the father of the drums, the spotted one, from the land of the naked people), whose Lwo name seems to have been Labongo (the commoner), built his capital on the hill Eburu, a few miles to the north of Mubende; and a 'division of the kingdom' is next reported, by which sub-dynasties of the Bito family were planted out as tributary rulers of the central provinces of the former Chwezi kingdom, among them the nucleus of the modern Buganda, which was allocated to Kato Kimera (the twin, the germinator), said to have been the twin brother of Rukidi. In Buganda tradition, as in that of Bunyoro, the new dynasty represented itself as the near relative of the old. Kimera is described as the grandson of Chwa, the offspring of a secret love-affair between Chwa's son and a daughter of the great king of Bunyoro. Only in the traditions of Rwanda do we meet the stark reality, where, in the seventeenth generation back, early in the sixteenth century therefore, we find the report of a major invasion by 'the Banyoro', a strange and terrible people 'as numerous as the grains of millet in a good harvest', whose spears were of iron 'and all of one piece' and whose shields were so strong that the arrows of the Rwanda were useless against them. These barbarians not only ravaged the cows but committed the supreme outrage of killing them for food; they also marched, driving in front of

---

[1] Cf. R. Oliver, 'Ancient Capital Sites of Ankole', *Uganda J.* 23, 1959, pp. 51–63. The results of a preliminary expedition to Bigo are briefly summarized in P. L. Shinnie, 'Excavations at Bigo', *Uganda J.* 24, 1960, pp. 16–28. The more recent and very significant excavations of Dr. M. Posnansky at Bweyorere (1959) and Bigo (1960) are as yet unpublished.

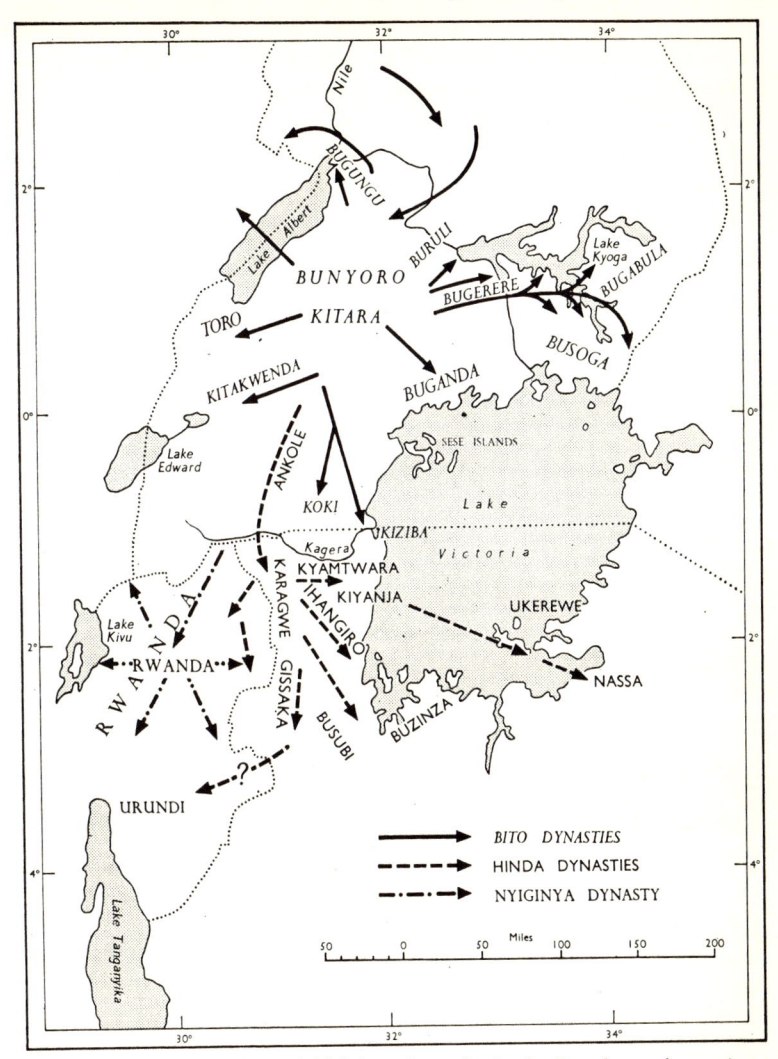

MAP 9. Extension of Bito and Nyiginya dynasties in the Interlacustrine region, 15th–18th centuries

their cohorts the women and children whom they had already enslaved.[1] Here there would seem to be a more convincing picture of an as yet un-Bantuized Lwo horde of a type which was no doubt all too well known all over southern Uganda but

[1] Pagès, *Royaume hamite*, pp. 79–80 and 558–74.

of which there is no record in the Uganda traditions because these are so largely the traditions of Lwo dynasties, which even when Bantuized did not care to remember themselves in such a role.

The first effect of the Lwo invasion of southern Uganda, therefore, was the substitution of a Lwo, Bito dynasty for the Hima, Chwezi dynasty in the central kingdom of Kitara, henceforward known to its neighbours as Bunyoro, a term which referred, so it seems, specifically to the new Lwo element in the population, the bulk of which tended to form an intermediate class between the Hima pastoral aristocracy and the ordinary peasant cultivators.[1] The centre of this kingdom can be roughly fixed by reference to the royal tombs which lie along a belt of country parallel to, but well to the south of, the Kafu river, the oldest being in north Singo, a mere fifty to eighty miles north-west of Kampala.[2] From Singo right across Mubende District into northern Toro was the real heartland where the royal herds were pastured. But all around this centre, still within easy reach of the royal armies, was a ring of what were probably nearer to tributary states than regular provinces. Most, though not all, of them were ruled by sub-dynasties of the Bito family, who were not disturbed while they remained loyal, and who on the whole remained loyal as long as they were within reach of punishment.

Of these peripheral states Buganda, to the east, was originally one. To the south-east lay Buddu, with a Fumambogo sub-dynasty, and southwards Bwera with a Mori sub-dynasty; south-west and west lay Kitakwenda and Kyaka, both under Bito; and, to the north-west, Bungungu with a sub-dynasty of Chwa. On the north and north-east Buruli and Bugerere both had Bito rulers. Beyond this inner circle raids from Bunyoro often resulted in the establishment of Bito dynasties still farther afield. An early example of this was Kiziba, the northernmost of the Haya states of Bukoba District, which received a Bito dynasty in the sixteenth or seventeenth generation back.[3] Koki and Toro followed in the eighth and fifth generations respectively. But the classic direction in which Bito princelings found thrones and fortunes and virtual independence was eastwards

---

[1] Cf. Gorju, *Le Victoria, l'Albert et l'Édouard*, p. 42; Roscoe, *The Bakitara*, pp. 2–3.
[2] Cf. K. Ingham, 'The *Amagasani* of the Abakama of Bunyoro', *Uganda J.* 17, 1953, p. 138.   [3] E. Césard, 'Le Muhaya', *Anthropos*, 1937.

in Busoga. Here all the princely houses called Baisengobi, which held sway right over the north and east of the country, were of Bito origin. The oldest of them, Bukoli, for example, and the little state of Bugwere in Mbale District, would seem to be contemporary with the Bito of Bunyoro themselves. Others, such as Bulamogi and Bukono, would seem to have been founded about the thirteenth generation back. Others, like Bugabula and Bugweri, have only been in existence for nine to ten generations.[1] It was these outlying sub-dynasties, whether in the far east or the far south, whose relations with the parent house were the briefest and most tenuous. Living beyond the reach of easy vengeance, they quickly came to lead their own lives and to form alliances outside the family circle.

Quite apart from their own conquests, however, the coming of the Bito seems to have had secondary results of the greatest importance, and especially in the area immediately to the south of greater Bunyoro. The populations here—both in Ankole and in the Bukoba District of Tanganyika—are essentially similar to those of Bunyoro and Toro. Many of the same clans occur. The languages are closely related. Northern Ankole, at least, had formed part of the Chwezi kingdom and it looks very much as though that kingdom's defeat by the Bito in the north led to an attempt to resurrect it farther south. We hear no more of the Chwezi themselves: in their place there arose another clan, the Hinda, which became almost as prolific in ruling houses as the Bito. The eponymous founder of the clan, Ruhinda, is traditionally described as a sort of illegitimate son of the last of the Chwezi rulers, Wamara, by a slave woman, Njunaki. This is of course as suspect as the relationship claimed by the Bito to the Chwezi, and the question must at least be asked as to whether Ruhinda was not also a Lwo. Probably, however, he was not. He was, more likely, the leader of a band of Hima cattle-owners, and there must have been many such who emigrated southwards with their herds to avoid the Lwo invasions.[2] His

---

[1] D. W. Robertson, *Soga Land Tenure*; cf. L. A. Fallers, *Bantu Bureaucracy*, 1956, pp. 30–32.

[2] Cf. P. H. van Thiel, 'Businza unter der Dynastie der Bahinda', *Anthropos* 6, 1911, p. 501, according to which Kayango, founder of the Ngo clan, married in Koki a slave-girl of the Hima, Njunaki of the Bogo clan, Ruhinda being the issue. According to this version, Ruhinda bought the first (Karagwe) drum, Nyabatama, from its Sita keeper during a famine.

Uganda base was in the high hills of Isingiro, overlooking the Kagera Valley. His Tanganyika home was in Karagwe on the south side of the river. There is nothing improbable in such a dualism: Hima herdsmen still pass to and fro in seasonal migrations to this day. From Karagwe his descendants, armed each with a royal drum of high significance, went forth, west to Gissaka and east to Kyamtwara, the parent of three more of the Haya principalities; north to Ankole, and south to Ihangiro, Buzinza, Busubi, and even to Ukerewe and Nassa on the southeast coast of Lake Victoria.[1] South-westwards the Hinda are said in some traditions to have founded the dynasty of Burundi.[2]

Unlike the Bito, Hinda dynasties as they divided preserved no sense of seniority as between founder and offspring; and their penetration lacks the military emphasis which seems to characterize the Bito states.

The Bahinda had two main ideas ... the idea of chieftainship and that of cattle-breeding. ... As cattle-breeders they were not at first interested in land under cultivation. However the introduction of cattle on a large scale produced an economic revolution. ... Up to that time the Bahaya had lived only partly on bananas. But the usual practice of African cattle breeders, which the Bahinda also followed, of farming out their cattle [to peasants] made it possible to cultivate bananas on a far larger scale because manure ... was now available. Thus bananas gradually became the staple food of the tribe. ... On the political side, the new unit, the chiefdom created by the Bahinda, was called the Ihanga. The Bakama [chiefs] assigned certain duties about their court to every clan. This meant that some members of a clan were always in attendance at the court since the clan sent men by turn to fulfil their special functions. ... This practice ... was of advantage both to the chief and to the clans. To the chief because he had a large body of his subjects living permanently at the court who could defend him in case of necessity, to the clans because these duties placed the well-being of the chief to a great extent in their hands.[3]

There is of course a natural tendency, which must be resisted,

[1] Kiyanja, Bugabo, and Maruku. Cf. J. Ford and R. de Z. Hall, 'The History of Karagwe', *Tang. Notes*, 24, 1947, pp. 3–27. There are further valuable unpublished accounts in the District Books of Bukoba, Biharamulo, and Mwanza (Ukerewe).

[2] e.g. E. Simons in *Bulletin des juridictions indigènes et du droit coutumier congolais*, 1944, p. 144.

[3] H. Cory and M. M. Hartnoll, *Customary Law of the Haya Tribe*, 1945, pp. 259–68.

to telescope the origins of chieftainship in any given area into a single epoch-making episode. No doubt the Hinda did introduce into north-west Tanganyika a more sophisticated and complex type of chieftainship than was there before. But the traditions of Ankole, Kiziba, and the Haya states all tell of an earlier form of chieftainship associated in every case with the Sita clan, of which, however, we know nothing beyond the name. We have seen that the Chwezi, too, had their predecessors. And of course the Chwezi, though their kingdom was probably the largest one of pre-Lwo times, were not the only ruling clan of southern Uganda. For example, the two Chwezi royal wives who were captured and taken over by the first Bito king of Bunyoro were of the Lisa clan, another Hima clan, which had the rule over two little kingdoms of Buwheju and Buzimba, lying to the west of the historical Ankole and incorporated into it only in British times. To the south of the Lisa there were the Shambo, rulers of Mpororo and Ndorwa, on the frontiers of modern Uganda and Rwanda. South of the Shambo were the Hondogo of Bugessera, and the Nyiginya of Buganza, who were to end by ruling all Rwanda. Thus, while the coming of the Bito was indeed an epoch-making event, the sense in which it was so must not be misconstrued. They planted a number of new dynasties, of which two were to become of outstanding importance. They precipitated the Hinda move southwards and therefore the political reorganization of north-west Tanganyika. By attacking Rwanda and stimulating the Nyiginya of Buganza to active defence of their country they helped to set in motion a process of assimilation, consolidation, and conquest in Rwanda which has been likened to the growth of France from its first Parisian island. In northern Uganda Lwo political know-how was new enough to justify an emphasis on the *diffusion* of culture. In southern Uganda the emphasis should fall on the stimulus it gave to the political *evolution* of the area.

The outstanding example of such a process of evolution is provided by the history of Buganda. In the early sixteenth century it was, as we have seen, merely one of a number of peripheral sub-kingdoms of the Bito kingdom of Bunyoro. Like Kiziba, or Koki or Kitakwenda, it was ruled by a hereditary sub-dynasty of the main Bito house. Its extent would have been

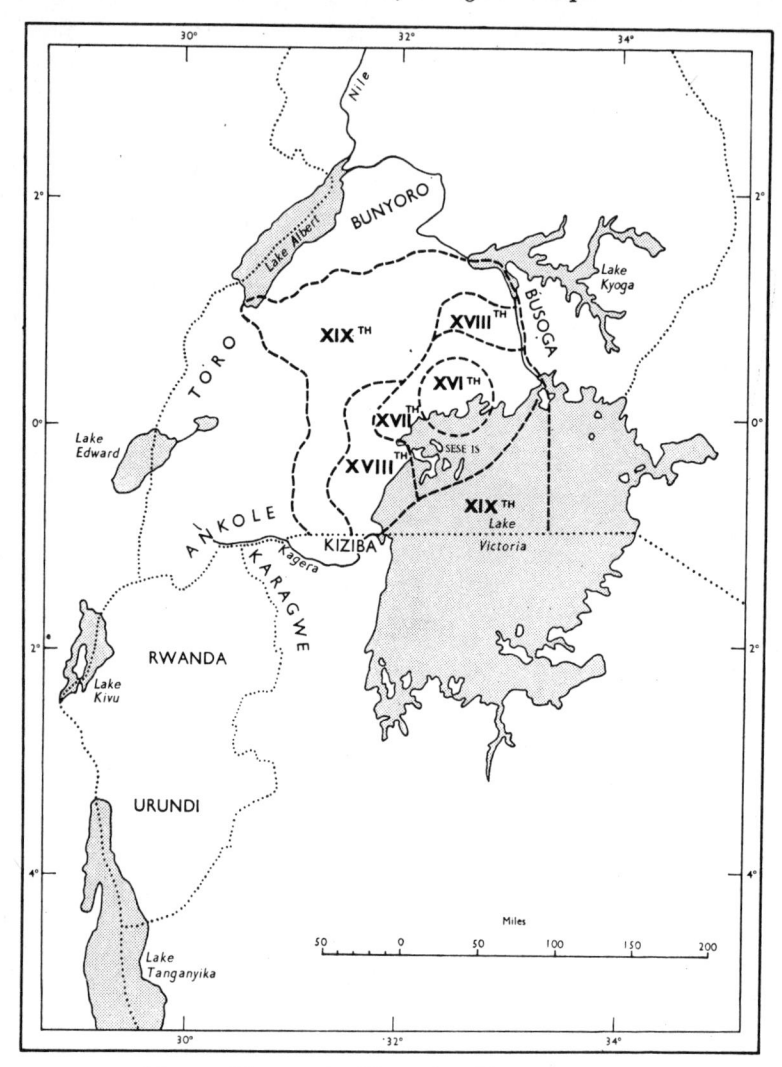

MAP 10. The growth of Buganda, 16th–19th centuries

easily comprised within a twenty-five-mile radius drawn from Kampala.[1] Through the whole of the sixteenth century Buganda remained inconspicuous. The limelight of tradition is all upon Bunyoro, featuring Nyoro raids into Ankole and Rwanda and

[1] i.e. to the counties of Busiro, Busuju, Kyadondo, and part of Kyagwe.

the installation of Nyoro sub-dynasties in Busoga to the east and Kiziba to the south.

In the fifteenth generation back, perhaps therefore towards the turn of the sixteenth and seventeenth centuries, Buganda, apparently with the aid of the Sese Islanders, just beat off a heavy onslaught from Bunyoro;[1] but still for two more generations it was Bunyoro that held the initiative, leaving Buganda alone, but raiding far and wide, north across the Nile, east to Busoga, south as far as Karagwe. Then suddenly in the twelfth generation back, in the middle of the seventeenth century, Bunyoro overreached itself with a disastrous campaign through Ankole into Karagwe and Ndorwa, and Buganda moved on to the offensive. In the reign of Kabaka Katerega, Buganda more than doubled its size, driving westwards into the countries of Mawokota and Gomba, Butambala and Singo.

The first climacteric in the history of southern Uganda had come. It was not merely that Bunyoro was declining and Buganda taking the lead. It was that during the years of its obscurity Buganda had been slowly evolving a much more homogeneous and centralized form of society than Bunyoro was ever to achieve. Perhaps just because of its small beginnings, the king and court of Buganda were intimately linked by bonds of matrimony or of office with every single clan in the country. The slight Hima and larger Lwo element in the population had been completely absorbed: they did not even persist as separate, more or less endogamous classes as they did in the grazing lands of the west. When Buganda expanded, therefore, it also assimilated. The territorial 'county' chieftainships were still hereditary within particular clans. The two new chieftainships which were created in Katerega's reign for Gombe and Butambala were likewise to be hereditary, the first for thirteen, the second for eighteen reigns. But they were hereditary within Ganda clans which were themselves far more closely integrated within the Ganda state and the Ganda monarchy than the various collateral branches of Bito ruling groups upon which greater Bunyoro depended for its cohesion.[2]

In the next phase of Buganda's development this centralizing

[1] Cf. J. M. Gray, *Uganda J.* 1935, p. 260.
[2] Cf. A. H. Cox, 'The Growth and Expansion of Buganda', *Uganda J.* 14, 1950, pp. 153–9.

tendency became still more marked. It is related that the early-eighteenth-century kabakas, of the tenth and ninth generations back, instituted and built up the strength of the royal bodyguard, one result of which was that a number of rulers are remembered chiefly as bad and bloody tyrants.[1] Without doubt, however, strong government was an aid to expansion, as also was the fact that Buganda, as it pushed outwards from its centre, had the increasing advantage of interior lines as against a greater Bunyoro strung out in two long lines to the west and north. The eighth generation back saw the consolidation of the country north-westwards to the borders of Busoga, the southern part of which had already become a Ganda sphere of influence. In the seventh generation back Buganda, allying itself with the newly founded Bito sub-dynasty of Koki, occupied the large southern province of Buddu, thus completing its control of the Lake Victoria coastline from the mouth of the Kagera to the exit of the Nile.

It is most significant that the chieftainship of Buddu after the Ganda annexation never became hereditary. Buganda was in fact moving from a feudal to a bureaucratic system, in which subordinate administrators held office during the king's pleasure. From the end of the eighteenth century all but the very oldest-established posts began to be filled in this way. Buganda was also, from the late eighteenth century at least, becoming conscious of a world outside the little circle of the Interlacustrine states. It is said that already in the eighth generation back plates, cups, saucers, and glass imported from overseas had reached the court of Kabaka Kyabagu. Tradition is emphatic that Kyabagu's son, Semakokiro, was an extremely wealthy king, who employed an army of hunters to obtain ivory, which he traded to the south of Lake Victoria in exchange for imported articles, especially cotton cloth. It seems quite clear that this early long-distance trade was a royal monopoly and that its effect was to carry political centralization to new lengths, because it was now possible for the monarchy to reward its servants in goods as well as in land. The effect on the military forces at the king's disposal was certainly immense. The growth of long-distance trade equally naturally created new motives for territorial expansion. It was important to close the land routes

[1] e.g. Kabakas Kagulu and Kikulwe.

leading to Bunyoro round the west of Lake Victoria: hence from the earliest years of the nineteenth century there developed a regular pattern of intervention in the affairs of the little states on Buganda's southern border. Kiziba and Koki were the first to come under Ganda influence. Kyamtwara was next, followed soon by Karagwe and Ihangiro. North-eastwards Busoga was increasingly penetrated, not so much for control of the approach-routes as to exploit its resources in ivory and slaves.

An interesting comparison could be drawn between the achievement of Buganda during these three and a half centuries and that of Rwanda, which had also from small and slowly matured beginnings developed a political and cultural coherence never reached by Bunyoro, Ankole, Karagwe, or Burundi, the other main states of the region. But Rwanda, though rising to heights unattained in Buganda on the ceremonial, social, and poetic sides of life, never reached either the administrative efficiency of the Ganda concept of officialdom, or Buganda's alertness to the outside world and readiness to respond to its challenge. It was these two qualities, whose origins lie far back in history, that were to be the secret of Buganda's triumphant adaptation to the coming of Europe in the later nineteenth century.

South of this so-called Interlacustrine region of southern Uganda, Rwanda-Urundi, and north-west Tanganyika, there lies another vast region, comprising almost the whole western half of the modern Tanganyika, which throughout the period of traditional history was populated entirely by Bantu-speaking peoples, patrilineal like the Interlacustrine Bantu,[1] and having, though in embryonic form, the same notions of chieftainship and social organization as their northern neighbours. Almost throughout this region a chief was, or could be, known as *Ntemi*[2]—he who cuts short the discussion by giving judgement.[3] Almost everywhere, and even if he only ruled a thousand

---

[1] Though the succession to Nyamwezi–Sukuma chieftainships is matrilineal, suggesting that patrilineal succession is a more or less recent introduction.

[2] *Ntemi* or *Mtemi* is used by the Sukuma–Nyamwezi, Gogo, and Bena, and also on occasion, and most significantly, by the Ngonde of the Kyungu—see Monica Wilson, *The Peoples of the Nyasa–Tanganyika Corridor*, 1958, p. 48. Ntwa, or Mutwa, is used by the Hehe, the Kinga, and Nyakyusa.

[3] The noun is derived from the verb *ku-tema* 'to cut'. It has been suggested that

subjects, the chief could be described as a 'divine king', the possessor of special insignia, and of royal fire from which all fire in the kingdom must be kindled. Everywhere the king's death and burial was the subject of special rites. In principle the king could not die a natural death;[1] he must be buried in a completely special way and usually to the accompaniment of human sacrifice. All this, and on the ritual side much more, the typical chieftainship of the West Tanganyika region shared with the chieftainship of the Interlacustrine region. Only in the south it so happened that no centralizing, state-building tendency developed on anything like the scale of the larger kingdoms of the north. The normal limit on the size of a political entity was what a ruler could administer personally, with the aid of his own close family. As populations grew and cut deeper into the uninhabited bush, so kingdoms divided and sub-divided. As the agricultural resources of the more favoured areas were taken up, segments of ruling houses migrated with their followers to the marginal lands which form the dry eastern fringe of the region, where previously only hunting peoples had wandered in search of game. The multiplicity of small units has not been favourable for the preservation of traditional history, nor has its content often been of more than parochial significance. One broad trend is, however, clear. It is that whereas the general pressure of population has been from west to east—and more accurately from south-west to north-east—the institution of chieftainship has spread from north to south, from the western shores of Lake Victoria to the northern shores of Lake Nyasa.

Along the north-western border of the West Tanganyika region, marked today by the southern frontier of Burundi, there lies a belt of territory occupied by the Ha, Jiji, and Vinza peoples, where the existing chiefly institutions have spread south from Rwanda-Urundi within the last one hundred and fifty years. Before this invasion of Tutsi ruling families the Ha and Jiji were organized under rulers called Bateko, whose descendants survived as a priestly class under the Tutsi, but about

originally 'the cutter' may have signified 'the clearer of the bush', 'the axe-man', just as the word *mufumu*, used for sub-chiefs among the Shambaa and elsewhere, would seem to mean 'the spear-man'.

[1] This does not of course imply that violence was necessary, but merely that natural death must be hastened or assisted. Cf. F. Bösch, *Les Banyamwezi*, 1930, pp. 81–86 and 487–8.

whose origins no traditions have been reported. Again, along the south-western border of the region, the Mambwe, the Rungu, the Fipa, the Mpimbwe, the Bende, all have chiefs and ruling clans which look for their origins to Congo territory west of Lake Tanganyika.[1] East of this border country with its comparatively modern intrusive elements, however, there stretches the large territory of the Nyamwezi–Sukuma people, whose political institutions appear to come from the north rather than the west, and have certainly been operative in the region for much longer. The explorer Richard Burton, who was usually a reliable reporter, declared that native tradition of the 1850's supported the hypothesis that Unyamwezi was formerly

a great empire, united under a single despot, whose tribe was the Wakalaganza, who still inhabit the western district of Usagozi. According to the people, the last of the Wanyamwezi ancestors died in the days of the grandfathers of their grandfathers. His children and nobles divided and dismembered his dominions, further partitions ensued, and finally the old empire fell into the hands of a rabble of petty chiefs.[2]

Subsequent inquiry has not, however, substantiated this well-known and often-quoted passage. It is true that the two or three hundred *machalo* or small independent chieftainships, into which the Nyamwezi and Sukuma peoples are today divided, can to some extent be classified into groups whose ruling families are related. Thus, the Kanga clan bears rule over twenty-one small chiefdoms in the Nzega District. In Tabora District the Sagali clan has nineteen chiefdoms and the Songo clan fourteen.[2] These groups of chiefdoms doubtless descended each from a single parent territorially larger than any of the surviving subdivisions. This is not, however, to say that the parent states were stronger or more important than their descendants: indeed

---

[1] Kibondo, Kasulu, Kigoma, Sumbawanga, and Mbeya District Books. Cf. A. Lechaptois, *Aux rives du Tanganyika*, 1913. Lechaptois believed that the ruling family of Ufipa were Hima from Burundi: and even tried to derive the family name *Watwaki* from *Batutsi*. The traditions reported by the late Sir John Lamb in the Sumbawanga District Book, however, make it clear that the Watwaki came from the western side of Lake Tanganyika. Moreover, the Twaki chiefs were known as Mwene, the royal stools were made in Rungu, and the probability would seem to be that they were the offshoots of some dynasty of the Luba–Lunda of the south-east Congo.

[2] R. F. Burton, *The Lake Regions of Central Africa*, 1860, ii. 4.

[3] Bösch, *Les Banyamwezi*, pp. 320–2, 493–4.

it seems more likely that population growth has been the main cause of subdivision.

A superficially stronger argument in favour of the original political unity of the Nyamwezi–Sukuma peoples is to be found in the near uniformity of the chieftainship customs throughout the area. But in the opinion of the most competent observers, this uniformity is due more to a common origin of the political institutions in the Interlacustrine region to the north than to any former political unity in Nyamwezi–Sukumaland itself.

While the part of the Lacustrians living west and north of Lake Victoria has for many years formed and retained a number of kingdoms . . . the part living south of Lake Victoria has never built an empire. But the immigrants from the north, mainly from Bunyoro, being well acquainted with the idea of powerful and extensive kingdoms, fundamentally influenced the political structure of the area. The immigrating families established authority over sections of the existing population and came to rule over them. The Hamitic element has been wholly absorbed, however, and only faint tradition remains of the historical events of the invasion.[1]

The Interlacustrine origin of Sukuma–Nyamwezi chieftainship is agreed by all authorities. The chronology of the movement which this implies is, by contrast, a question of the utmost difficulty, and one which tradition alone is unable to solve. The genealogies of the ruling houses which are profusely recorded in the District Books of Tabora and Kahama, Shinyanga, Mwanza, and Nzega frequently show successions of twenty or twenty-five rulers. These genealogies are notoriously suspect as the conscious response to a political inquiry by the Tanganyika government into the seniority of chieftainships in connexion with Sir Donald Cameron's policy of Indirect Rule. Nevertheless, if Sukuma–Nyamwezi chieftainship did come south from the Interlacustrine region, there are strong arguments for supposing that it did so at a fairly remote period, when Interlacustrine chieftainship was probably small, segmented, and matrilineal in succession. It is significant that the long-horned cattle which Hima and Tutsi conquerors introduced into the northern kingdoms stop short at the most southerly of the Hinda kingdoms. This in itself suggests that the chiefly groups of the Nyamwezi–Sukuma must have left southern Uganda before the

[1] H. Cory, *The Ntemi*, 1951, pp. 1–2.

coming of the Hima–Tutsi. Moreover the very difference in scale between the smallest Hinda or Bito kingdom of the north and the typical Ntemiship of the south suggests that the Ntemi is probably the surviving representative of an older layer of Uganda chieftainship, corresponding perhaps to the Gabu rulers of Bunyoro tradition or to the Kintu period of Buganda.

The significance of the chronological problem in Nyamwezi chieftainship appears above all in relation to the history of the peoples to the south. There have been extensions of Sukuma chieftainship eastwards and north-eastwards into the Singida, Maswa, and Musoma Districts to the east of Lake Victoria.[1] And there have been corresponding Nyamwezi extensions eastwards into Ugogo.[2] But the traditions of these groups are short and clear. They tell of the expansion of the last 200 years, into country always marginal for cultivators, which had been previously occupied by hunting peoples on the one hand and pure pastoralists on the other. It would seem that the older extensions of Nyamwezi–Sukuma culture ran south-eastwards to the Iringa highlands, the country of the Hehe, the Bena, and the Sangu, and next, beyond them, to the Pangwa, the Kinga, the Nyakyusa, the Ngonde, the Safwa, the Nyika, and the Namwanga, spread in a crescent from the Livingstone mountains north-east of Lake Nyasa to the Mbeya highlands south of Lake Rukwa.

In spite of tribal peculiarities, all these peoples present, as Professor Monica Wilson has observed,[3] a certain broad cultural homogeneity. All of them are cattle-owners as well as cultivators. All of them reckon descent in the male line, which distinguishes them sharply from their southern and western Bantu neighbours—the Makua and the Yao, the Bemba and the Lunda. There is, moreover, a common element in all their traditions, according to which the big event of all their histories was the establishment of chieftainship by cattle-owning strangers, who were also the possessors of many new techniques. It is clear that

[1] i.e. the people of Ushora and Isanza in Singida District and the Shashi in Musoma District. The Maswa District Book shows that the Sukuma element, here mostly under Binza ruling families, entered the district *later* than the Hinda dynasties of Ukerewe and Nassa.

[2] H. Claus, *Die Wagogo*, Leipzig, 1911; T. Schaegelen, 'La Tribu des Wagogo', *Anthropos*, 33, 1938, pp. 195–217 and pp. 515–67.

[3] *The Peoples of the Nyasa–Tanganyika Corridor*, pp. 46–49

these traditions refer to one broad stream of settlement which spread through the Iringa District and the Livingstone Mountains, and thence down to the Nyasa lakeshore and up again on to the Mbeya plateau. The Namwanga, whose traditions form the final link in the chain, place their arrival in the sixteenth generation back, therefore probably in the late sixteenth century.[1] Ngonde traditions, which claim a close relationship between their own royal family and that of the Namwanga chiefs, show a succession of eighteen rulers since the descent from Bukinga in the Livingstone Mountains. Nyakyusa chieftainship has a peculiar system of succession whereby generations succeed each other at fixed intervals of thirty years irrespective of the life-span of individual chiefs; and the descent of the founders from Bukinga can therefore be assigned, with considerable confidence, close to the turn of the sixteenth and seventeenth centuries. Kinga traditions confirm the connexions with Nyakyusa and Ngonde and add to their list of relationships the chiefly groups of the Nyika and Safwa, as well as those of the Bena and Pangwa. They refer these relationships to an earlier dispersal centre called Ilongo, on the borders of the Hehe and Bena country, which appears also in the early traditions of the Hehe.[2]

The missing link in the chain of traditional evidence is any firm connexion between the Hehe–Bena group on the one hand and the Nyamwezi–Sukuma group on the other. On the showing of Nyakyusa, Ngonde, and Namwanga traditions the movement of the cattle-owning chiefly ancestors must have reached the Kinga country at least twenty generations ago, and probably more. It is unfortunately the case that neither Hehe nor Nyamwezi traditions have kept so long a record. As in the case of Nyamwezi connexions with the Interlacustrine peoples, therefore, the only evidence is the always inconclusive one of cultural similarity. Nevertheless a chain of evidence with only two such weak links is one that must be taken seriously, and as a hypothesis which fits the sum of present knowledge it may be concluded that, before the advent of the Lwo and probably before that of the Hima–Tutsi, a highly ritualistic form of territorial chieftainship associated

---

[1] Article by C. Chisholm Richards in Mbeya District Book; cf. J. C. C. Coxhead, *The Native Tribes of North-eastern Rhodesia*, London, 1914.
[2] Monica Wilson, *Nyasa–Tanganyika Corridor*, pp. 12–13.

with immigrants who combined cultivation and stock-rearing, appears to have spread round the western shores of Lake Victoria, influencing first the earlier Bantu inhabitants of the Interlacustrine region and later those of western and southwestern Tanganyika. The western limits of this sphere of influence are provided by the matrilineal Bantu peoples of the eastern Congo and Northern Rhodesia, whose chieftainship traditions point to the north-west. The southern limits are provided by the matrilineal Bantu of southern Tanganyika, among whom chieftainship and historical traditions are almost non-existent. The eastern boundary is the most difficult to define, since this area seems to have been the scene of much secondary expansion. Nyamwezi-Sukuma expansion northwards and eastwards has already been mentioned. There has equally been much northward and eastward expansion from the Hehe country. Half the ruling class of Ugogo came from Uhehe during the second half of the eighteenth century. Sagara and Kaguru traditions of Kilosa District point to a late-eighteenth-century origin in Uhehe. The Luguru people of Morogoro District tell of an eighteenth-century movement from Ubena.[1] It cannot be quite accidental that Mbega, the late-eighteenth-century founder of the Kilindi dynasty of Usambara, is credited with the marvellous skill in hunting wild pig (*nguruwe*) attributed to the founders of ruling houses from Ugogo all the way south to the Nyakyusa. Nor is it without significance that among the Chaga chiefs of Kilimanjaro at least three of the ruling houses claim an ancestor called Ntemi. It would appear, therefore, that the primary southward movement of Ntemi chieftainship from the Interlacustrine region through western Tanganyika, which had reached Lake Nyasa by the sixteenth century, was followed during the seventeenth and eighteenth centuries by a series of secondary movements from central Tanganyika eastwards and north-eastwards into the region between Morogoro and Kilimanjaro.

It is easy to lose the significance of chieftainship in the dry recital of its spread from one little-known part of East Africa to another. To recover some sense of what it could involve we need only remember how, in the *Habari za Wakilindi*, Mbega's predecessor, Turi,

[1] Dodoma, Kilosa, and Morogoro District Books.

set out with the men of Vuga, a very great army, young men and elders, with two war-horns and his flute and his signal-horn. . . . And they came to that place with song and dance, and every man

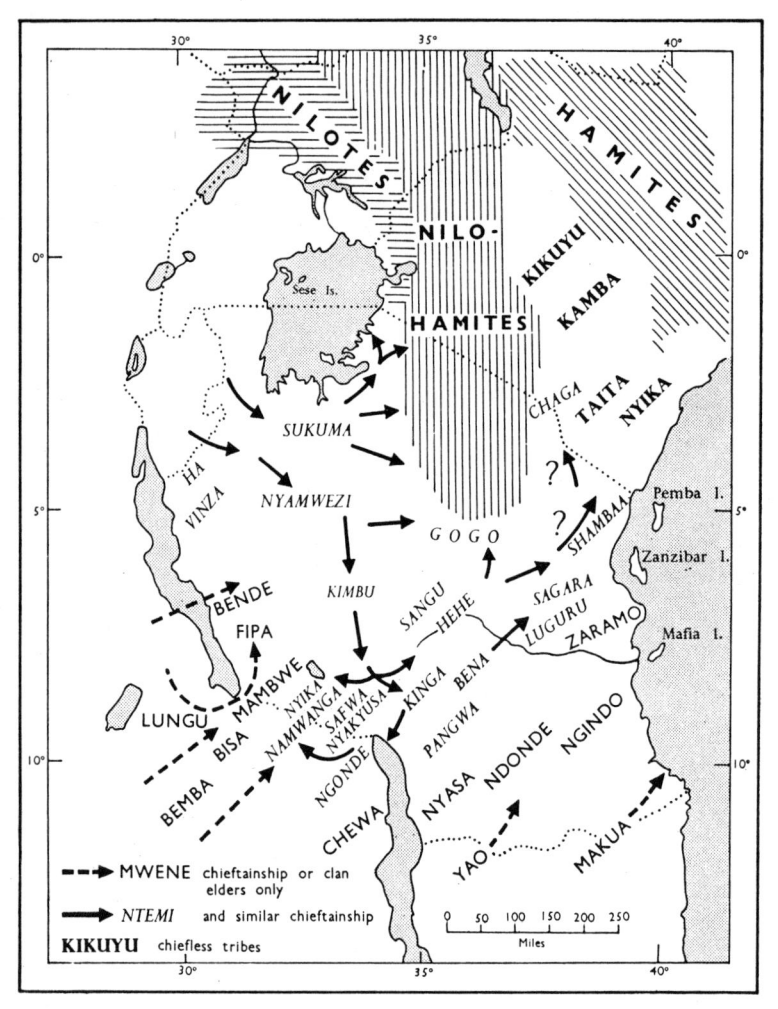

MAP 11. Probable spread of Ntemi chieftainships from Nyamwezi–Sukuma area

vaunted his prowess and glorified his country and boasted of his pre-eminence and rank. And Turi, their headman, was received with loud shouts of applause. And he proclaimed his skill in his craft, and glorified the fire of his forge, and boasted how he slew the men of Pare with his spear, and how he beat out arrows and axes and hammers and knives. . . . And he boasted that he was rich in oxen and goats and sheep, and he claimed that it was he who protected all the people and cared for them with gentle kindness. And at every word he asked his people, 'Is this that I say true or false?' and his people responded with one voice, 'It is true.'[1]

What we have been mainly considering so far has been the impact of Nilotic people and influences[2] upon the Bantu-speakers of southern Uganda and western Tanganyika. When we move east, to Kenya and eastern Tanganyika, the primary interaction which we have to consider is that between Bantu-speakers and Nilo-Hamites. Unlike the first, this second interaction did not throw up as its central theme the formation of states. In some ways it was a stronger and more pervasive influence than the first. The Bantu peoples who live in the neighbourhood of Nilo-Hamites have been visibly changed by the contact in a way that the Bantu who abut on Nilotic areas have not. They tend to be distinctly lighter in colour, to practise initiation rites with circumcision at puberty, and to have taken over in varying degrees the weapons, ornaments, hair-styles, and other social customs characteristic of the Nilo-Hamites. Like the Nilotes, the Nilo-Hamites seem to have been responsible for the spread of patrilineal ideas about marriage and inheritance.

All in all, it looks as though there has been a much more fundamental mixture of populations in and around the Nilo-Hamitic area than between Nilotes and Bantu. Mainly, no doubt, this is because the Nilo-Hamites have been in the area for much longer than the Nilotes. They have probably been there longer than the Bantu. For despite the apparent lesson of the linguistic map, which would seem to show that a Nilo-Hamitic invasion has bent and battered a Bantu line that once stretched tidily along the second parallel of north latitude from Lake Albert to Mogadishu, it is more likely that it was on the contrary a northward- and eastward-flowing Bantu tide which has lapped around both sides of the Nilo-Hamitic bulge, eroding its frontiers and even breaking in to form enclaves within them.

[1] 'Habari za Wakilindi', tr. Roland Allen, *Tang. Notes*, 2, 1936, p. 82.
[2] Including, of course, influences which though not Nilotic in origin were nevertheless transmitted southwards through the Nilotic sphere.

The centre of Nilo-Hamitic disturbances and displacements was, as we have already seen, the Rudolf basin in the far north. Probably it was the proximity of the Galla which made it so. At all events it was from the northern end of the Rudolf basin that the Bari and their predecessors had burst out westwards along the Sudan–Uganda frontier to disperse the Lwo.[1] And it was from the western side of the basin that the Lango and the Teso later spread out into north-eastern Uganda, cutting off the Acholi from the Jopadhola and from the Luo of Kavirondo.[2] It was equally from the Rudolf basin, and this time from its southern end, that there broke out the most spectacular and virile of all the Nilo-Hamitic eruptions—that of the Masai.

The Masai, like other Nilo-Hamitic peoples, are a nation only in the sense that they speak one language, follow a common way of life, and observe the same customs and beliefs. They have scanty, though fairly homogeneous traditions of immigration, beginning with an ancestor called Maasinda who made a gigantic ladder which enabled the forefathers of the tribe to climb up the escarpment from the Rudolf basin to the Uasin Gishu plateau north of Kitale. From there they deployed in six main sub-tribes over the central part of the Kenya highlands and southwards, down the line of the Rift Valley, into Tanganyika. None of these units formed a regular territorial state, but the young men of each were initiated together and thereafter progressed in an elaborate system of age-classes through warriorhood to eldership and old age. The sequence of these age classes forms a fairly accurate system of chronology as far back as the last quarter of the eighteenth century. For earlier times there is only the genealogy of the Ngidong'i family of rain-making priests or laibons, originating in the late seventeenth century, when the Masai expansion had already reached the Ngong Hills in the neighbourhood of Nairobi. They are presumed to have entered the Kenya highlands perhaps a century earlier.[3]

The rapid southward movement of the Masai should not, however, be taken as typical of the Nilo-Hamitic peoples as a whole. The Masai were late-comers who broke through from

---

[1] See above, pp. 171–2.

[2] See above, p. 176. The term 'Luo' is here used to distinguish the group settled in Kavirondo from the Lwo-speaking peoples as a whole.

[3] H. A. Fosbrooke, 'An Administrative Survey of the Masai Social System', *Tang. Notes*, 26, 1948, pp. 3–12.

the east centre of the Nilo-Hamitic world towards its southern periphery. In doing so, they pushed through and past the Nandi group of tribes—the Sapei, Pok, and Kony of Mount Elgon; the Suk (Pokot), Keyo, and Tuken of Baringo; the Nandi proper, and the Kipsigis, farther south towards the Mau. All these peoples were already established in the highlands when the Masai came up, and while the Nandi have traditions of an ancient southward movement from the mountains to the north of Elgon, their occupation of the Nandi Hills can be placed by reference to their age-class system at least as far back as the turn of the sixteenth century.[1]

Moreover, the Nandi themselves were comparative newcomers among the southern Nilo-Hamites. To the south of them lived the widely dispersed group of Tatoga tribes, pastoralists like the Nandi and the Masai, of whom the Barabaig and Kisamajeng of Mbulu District are two of the best-known sections. The traditions of these two sections and of others living in Singida District begin with the account of a southward migration from what is now Masai country to the south of the Mau range, some ten generations ago. But this was a local movement only. The very wide distribution of Tatoga elements, from Musoma and Maswa Districts on the west, right round to Manyoni and Dodoma Districts on the south, suggests that they were once the principal inhabitants of a large region to the south of the Nandi tribes, from which they were only slightly shifted and dispersed by the advent of the Masai. The antiquity of their occupation of this general area is attested by the considerable differences in language and social organization that have grown up between them and the Nandi and Masai groups.[2] Finally, driven south in turn by the Tatoga, there were those mysterious peoples the Goroa or Fiomi; the Iraqw or Mbulu, the Alagwa or Burungi, living nowadays between Mbulu and Kondoa, whose languages are so exceptional that they have hitherto defied classification.[3]

[1] G. W. B. Huntingford, *Nandi Work and Culture*, 1950, pp. 2-11.
[2] Mbulu, Singida, Musoma, Maswa, Manyoni, Dodoma District Books; G. McL. Wilson, 'The Tatoga of Tanganyika', *Tang. Notes*, 33, 1952, pp. 34-47.
[3] According to O. Reche, 'ein noch wenig abgeschliffenes rein hamitisches Idiom'—*Zur Ethnographie des abflusslosen Gebietes Deutsch-Ostafrikas*, 1914, pp. 103-11. The latest research on these languages now being conducted by Dr. W. H. Whiteley does not justify so simple a verdict.

As the Masai spread out across the Kenya highlands, they passed to the east of the Nandi peoples, dividing them from the scattered hordes of Dorobo hunting peoples on the eastern side of the Rift, who though very different in culture and physique from the other Nilo-Hamites, nevertheless speak an unmistakably Nandi language. Farther south, down the Tanganyika stretch of the Rift Valley, the Masai again passed eastwards of the Tatoga, the Goroa, and Iraqw. There were Bantu peoples to the east again—the Kikuyu, the Kamba, the Taita, the Chaga, the Asu, the Shambaa, the Zigua, the Ngulu, to whose history we shall return. Ahead of them, to the south, though a few primitive Bantu-speaking hunters like the Sonjo fled from their path, the first definite Bantu resistance they encountered came from the Gogo, who, as we have seen, were mostly eighteenth-century colonists from Unyamwezi and Uhehe. Who therefore preceded the Masai in the Rift Valley and on the so-called Masai steppe to the east of it? There can be no certain answer to this question: three hints are, however, worth following.

The Bantu-speaking Meru people on the eastern side of Mount Kenya preserve the memory of cattle-keeping Hamites, whom they called the Mwoko, who used to live in the plains below them, and who used to bury their dead in a sitting position in stone-built cairns like the Galla. Clashes between the Meru and the Mwoko continued until the time of a Meru age-class which would have been in the warrior stage in about 1760.[1] Again in the Rift Valley in northern Tanganyika, at Engaruka some twenty miles south of Lake Natron, three miles of stone terracing on the steep sides of the Rift wall mark the site of a town, occupied less than 300 years ago, which still shows the ruins of more than 6,000 rudely built stone houses. Elderly Masai questioned by Dr. Leakey expressed the opinion that the settlement had been built by the ancestors of the Iraqw.[2] Farther south still, Gogo tradition of about ten generations back speaks, on the one hand of small hunting peoples in the western part of the country, on the other hand of cattle-people

---

[1] H. E. Lambert, *The Systems of Land Tenure in the Kikuyu Land Unit*, 1950, pt. i, pp. 12–13.
[2] L. S. B. Leakey, 'Preliminary Report on an Examination of the Engaruka Ruins', *Tang. Notes*, 1, 1936, pp. 57–60.

in the east, who dug dew-ponds and carved draught-boards in the rocks, and again, of people with red and white skins who built clay houses, which they fired like pottery.[1]

There is no need to try to conflate these isolated scraps of evidence. What they amount to is simply that what appears on the linguistic map of today as a Nilo-Hamitic bulge sweeping into what looks as if it ought to be Bantu territory, is in fact a natural, ecological bulge of dry highland steppe, abounding in wild game for the hunter and offering good pasture for cattle, but of distinctly marginal use to the cultivator. Consequently the Bantu cultivator avoided it for as long as possible. He moved up each side of it, but left it what it had been since time immemorial—a funnel of pasture-land down which Hamites and Nilo-Hamites moved on their way to absorption in the Bantu south.

Of that absorption the Bantu peoples on all sides of the funnel show plentiful traces. On the western side particularly, the other great stream of influence coming down through Uganda, which we have already examined, is a source of some confusion. Yet it is significant that the two streams have not always or everywhere overlapped. The Gisu of Mount Elgon and all but one[2] of the Bantu tribes of Kavirondo are examples of people living on the borderland between the Interlacustrine region and the Nilo-Hamitic region, who have been quite uninfluenced by the political institutions of the first, but who have taken things as fundamental as patrilineal institutions and male initiation rites from the second. On the eastern side of the funnel the Kikuyu, the Kamba, the Taita, and the Nyika all show, in their initiation customs and age-classes as well as in their appearance and adornment, the clearest signs of influence from the Nilo-Hamitic and Hamitic sources, but none at all of the chiefly institutions of the Interlacustrines. It is farther south that the two influences begin to be overlaid. In the west the Nyaturu, the Iramba, and the Isanzu of Singida District are said to be Tatoga who have been Bantuized by immigrants under Ntemi chiefs from the eastern and southern districts of Unyamwezi. On the east, the Chaga, the Shambaa, the Ngulu, the Sagara,

---

[1] Claus, *Die Wagogo*, p. 64; Schaegelen, *Anthropos*, 1938, p. 199.
[2] The Hanga (Wanga) or Abahanga, who had a ritual kingship, derived in all probability from Bunyoro.

while exhibiting all the typical signs of Nilo-Hamitic contact, have received in addition, though in attenuated form, the notion of ritual kingship, which seems to have reached them from Uhehe round the southern end of the funnel.

Chaga genealogical material is copious and contradictory, but there is a curious consistency about the ancestry of the ruling houses of Machame, Kilema, and Marangu, in each of which a ruler called Ntemi, and coming apparently from Usambara, appears about the ninth generation back. On the other hand the clan history of the chiefs of Moshi appears to show thirteen generations to one Makilo, who 'came from Usambara, but was not a true Shambaa, being of the Kilindi tribe which afterwards founded the kingdom of Vuga in Usambara'.[1] If true, this would take the history of the Kilindi family back into the seventeenth century; but in view of the fact that the Kilindi kingdom of Vuga was founded by the great-grandfather of Kimweri, who was a contemporary of the mid-nineteenth-century missionary Johann Krapf, the Moshi chronology seems highly suspicious.[2] Disappointingly little confirmation comes from the genealogy of the original Kilindi family of Ngulu, which is evidently much foreshortened.[3] On the whole the evidence perhaps suggests that Ntemiship reached north-eastern Tanganyika towards the middle of the eighteenth century—a conclusion which would tally well enough with the chronological evidence for this kind of chieftainship in Ugogo.

In the face of all these indications—for the evidence is far too slight to call them proofs—of an important secondary dispersion of chieftainship systems, spreading from western Tanganyika towards the east in the course of the eighteenth century, it becomes necessary to speculate briefly on the possible causes of such a movement. Probably there is no one simple answer. Expansion of the population of Nyamwezi–Sukumaland may account for the beginnings of the eastward movement. Expansion of the chiefly families, as distinct from that of the population as a whole, would be an hypothesis which has an evident parallel in the expansion of the Alur ruling class over the West Nile District of Uganda. For such an expansion there are many possible

---

[1] Moshi District Book, cf. C. C. F. Dundas, *Kilimanjaro and its People*, 1924.
[2] J. L. Krapf, *Travels, Researches and Missionary Labours . . . in Eastern Africa*, 1860, pp. 382–3.   [3] Handeni District Book.

explanations. Southall has attributed the success of the Alur basically to their success as judicial arbiters in superseding the endless and destructive process of the blood feud. Cory, writing of the Hinda, has emphasized an economic reason—the value of cattle farmed out in twos and threes to peasant cultivators in need of milk and manure. The *Habari za Wakilindi* makes it quite clear that Turi, the predecessor of the first Kilindi ruler of Vuga, was the descendant or successor of a line of blacksmith chiefs. 'Whoever comes to this country of Usambara must admit that he is Turi's man. That clan holds the country because God gave them the gift of working iron, and skill in war. ... None but they have ornaments, none dares to boast himself a son of the country.'[1] The frequency of the spear and the hoe among the insignia of African chieftainship is a potent witness to the political significance of the blacksmith's art.

By the time we reach the eighteenth century, however, there is another possibility—the trading chief, with a mobile force of hunters at his command, and subjects willing to undertake long journeys in organized caravans to the east coast. Long-distance trade of any kind in country where head porterage was the only means of transport, and the ivory trade in particular, was a cooperative venture requiring the kind of political and economic infrastructure which chieftainship could most easily provide. It may be therefore more than a coincidence that the Nyamwezi who seem to have been the prime exporters of Ntemiship became also the acknowledged leaders and caravaneers of East Africa's long-distance trade. Long before the coast-men ventured into the interior, the Nyamwezi ventured to the coast. About the chronology we cannot be precise, but on the one hand there is the expulsion of the Portuguese from Mombasa in 1699 and again in 1730, and on the other hand there is the reported arrival of the first long-distance imports in Buganda in the reign of Kabaka Kyabagu in the middle years of the eighteenth century. It is a curious symbol of the connexion between trade and kingship in the Nyamwezi area that the chiefly headgear which had formerly carried the horns of the *mpala* antelope was changed after the opening of the ivory trade for one incorporating the *kibangwa*, a portion of the shell of the giant clam, a rarity much prized by those who first

[1] Translation by Roland Allen, *Tang. Notes*, 2, 1936, p. 86.

trudged the long miles from western Tanganyika to the Indian Ocean.

Our survey so far has shown us a Bantu world covering roughly the southern two-thirds of the East African interior, abutting on a world of Nilotic peoples in the north-west and enveloping on both sides a world of Nilo-Hamitic peoples in the north-centre. Both of these northern worlds have projected their influence southwards into the Bantu world, so that right down the western side of East Africa we find a zone of Nilotic influence, while down most of the eastern part of the interior we find a zone of Nilo-Hamitic influence. There is a third or coastal zone, whose Bantu inhabitants were influenced, partly from Hamitic (Galla and Somali), but mainly from overseas sources. This zone forms the subject of the previous chapter, and perhaps the most remarkable thing about East African history in the light of our present knowledge is how slight a part the coast appears to have played in the life of the other two zones until late in the eighteenth century. Even when the contact did begin, it seems clear that it was, for a long time further, the interior which came to the coast, rather than the coast which went to the interior. We have now, finally, to consider the southernmost region of East Africa, lying between the Rufiji and Ruvuma rivers, to which neither Nilo-Hamitic nor Nilotic influences seem to have penetrated to any appreciable extent. Here again, though it formed the direct hinterland of Kilwa, the influence of the coast seems to have been negligible until the nineteenth century, and even then the most important changes which affected it came not from the east, but from the south.

North of the Rufiji it is clear that in the three centuries preceding the European penetration we have been dealing for the most part with peoples who had already developed a certain ethnic stability and character. It is obvious that a people like the Ganda had, by the seventeenth century, if not by the sixteenth, a real sense of tribal nationalism, capable of absorbing and moulding into its own pattern most new elements that were likely to find their way into its orbit, whether by migration or conquest. The same could hardly be said of the hundreds of little states to the south and the north of the Interlacustrine group: here there were no tribal institutions, but a considerable homogeneity was nevertheless maintained by the interplay of small units of like origin. Even among the Nilo-Hamites there was the cultural solidarity promoted by pastoralism—the most specialized and all-absorbing of human occupations. So long as a Masai kin-group retained its herds it was in no danger of losing its cultural identity: only if it lost its cattle through warfare or sickness would it seek incorporation in a neighbouring society of cultivators. Here and there, in a swamp or on a mountain-top or in a stretch of desert bush, some few descendants of the ancient hunting cultures would survive, proud in their individualism, their demography to be reckoned in square miles to the man and not in men to the square mile. During the period under consideration most of southern Tanganyika was only just beginning to emerge from this stage. Many of its people were primarily hunters or fishermen. Most of the rest were shifting cultivators of the most diffuse kind, scattered, chiefless, largely anonymous. Ndonde, Ngindo, Ndendeule, Matambwe, Pogoro, Nyilila, Mawindi, Nindi, Matengo, Nyasa—terms such as these denoted merely people of the bush or the valley, the savannah, the forest, or the lakeshore.

Gaspar Bocarro, whose overland journey from Tete to Kilwa in 1616 has already been referred to, reported that the lands to the north of the Ruvuma were at that time under the lordship of 'Manhanga'. This, as Sir John Gray has pointed out, means 'the people of the lake' and doubtless refers to the ancestry of one of these loose linguistic and cultural groupings. Bocarro marched for seven days through country laid waste by the Zimba, whose depredations in the coastal belt were described in the preceding chapter, and he next spent eight days in crossing country that was just inhabited. In exchange for Bocarro's present of 800 copper bracelets the chief whom he misnamed Manhanga gave him a tusk of ivory. Seven days farther on, he received another, much smaller, tusk from Manhanga's 'brother' Chiponda. Along the inhabited sections of the route there were good and cheap provisions—millet, rice, fruits, hens, sheep, cows, and goats— but between Kilwa and the Ruvuma not much was inhabited.[1]

Such, it seems, were the conditions prevailing over most of the Songea, Tunduru, Masasi, and Mahenge Districts until little more than a century ago. Towards the east, sections of the

[1] Cf. J. M. Gray, *Tang. Notes*, 1948, pp. 44–46.

large, loose group of Makua peoples from Mozambique had probably been drifting northwards in small numbers for a long time. Farther west the Yao, who were to become so important during the apogee of the slave trade and so remarkable later for their wholesale adoption of Islam, were still living in their homeland to the south of the Ruvuma. It was indeed only in the present century that any substantial number of them migrated northwards into what was by then German East Africa. The continuous history of the interior of southern Tanganyika, therefore, begins only with the Ngoni invasions, the last large-scale immigration into East Africa in pre-colonial times.

The Ngoni were in origin close relatives of the Zulus. Their leaders formed one of several groups which moved away from Natal in order to avoid forcible incorporation in the empire of the tyrant Shaka. The Matabele of Southern Rhodesia and the Shangaans of Mozambique were of the same general dispersion. The Ngoni group, however, moved faster and farther than the others. Crossing the Zambezi in 1835, they were in Nyasaland the following year, and by 1840 the main body was temporarily settled on the Fipa plateau above the south-east corner of Lake Tanganyika.[1] What is especially suggestive in relation to earlier African migrations, such as that of the Lwo, is that what had started as the flight of a few thousand true Ngoni had snowballed by including the survivors of every people plundered or defeated during the 2,000-mile exodus. At this stage the true Ngoni were still the aristocrats who occupied all the offices of the military and civil hierarchy. But such was the impact of Zulu tactics with shield and stabbing spear, and such was the power of Ngoni military training and social discipline, that by the next phase of the dispersion all those who had been incorporated during the migration were accounted Ngoni and fit to be employed in all but the very highest posts.[2]

Already during the year or two after the Zambezi crossing some sections had broken away from the main body of the migration, of which one under Induna Mputa had crossed the Shire river and moved up through the hills to the east of

---

[1] D. G. Lancaster, 'A Tentative Chronology of the Ngoni', *J. Roy. Anthrop. Inst.* 67, 1937, pp. 77–90.
[2] Margaret Read, 'Tradition and Prestige among the Ngoni,' *Africa*, 9, 1936, pp. 453–84.

the lake, incorporating many Yao groups, and finally settling as a ruling class in the Songea District of southern Tanganyika, north of the upper Ruvuma. These were known to Livingstone and other early travellers by their Yao nickname of Magwangwara. It was, however, in Ufipa, on the death, in about 1845, of Zongendaba, the original leader and paramount chief of the Ngoni, that the main body of the tribe broke up. Mpezeni and Mombera, the rival heirs, went south with their followings to Nyasaland. Smaller groups broke away and raided north-westwards into the Congo and north-eastwards to Ujiji and Unyamwezi. Two important Indunas, Zuru and Mbonani, hived away south-eastwards through the country of the Safwa, the Kinga, and the Pangwa until they came into contact with the kingdom of Mputa in Songea. In the clash between these two Ngoni hordes, the newcomers prevailed. Mputa was killed. His heir Chikuse retreated across the Ruvuma with his close followers and lived to found a new nation, called after his son Gomani, centred to the south-west of Lake Nyasa. The victors on the eastern side of the lake, Zuru and Mbonani's successor, Chipeta, divided Songea and western Njombe between them, Chipeta taking the northern half and Zuru the south.[1]

The effect of these Ngoni invasions on the peoples of East Africa was threefold. On the one hand there was the effect upon the small, disorganized tribal groups, who now found themselves incorporated into one or other of the new Ngoni states of the south-west. It is perhaps the fact most illustrative of the chaos out of which these states were created that in them, alone of all the Ngoni states north of the Zambezi, the Ngoni language has prevailed over the languages of the conquered. Another and more sinister effect of the Ngoni invasions on East Africa was the loosing of hordes of raiders and refugees on the regions all round the sphere of direct Ngoni settlement. These were the *ruga ruga* of the early travellers' accounts, bands of detribalized brigands living by ivory-hunting and plunder, but ready to follow any ambitious chief in need of soldiers of fortune. And, finally, among the effects of the Ngoni invasions there were the significant reactions which they provoked among the near-by

---

[1] Songea District Book; G. W. Hatchell, 'The Angoni of Tanganyika Territory', *Tang. Notes*, 25, 1948, pp. 69–71; P. W. Gulliver, 'History of the Songea Ngoni', ibid. 41, 1955, pp. 20–23.

tribal polities, above all among the Hehe, the Bena, the Sangu, and on the Nyamwezi empire of Mirambo.

The first of these peoples to unite were the Sangu, whose country bordered on Ufipa and was laid waste by the Ngoni in the early 1840's. Mwahawangu, the *Mtwa* of Mapunga, one of the foremost of a group of small chieftainships, retreated into Uhehe until news came of the Ngoni dispersion following Zongendaba's death. Then he led his people back, reconquering other small units as well as his own. Mwahawangu's grandson and successor was the first Merere, whose fortified settlement at Utengule was such a landmark in the great days of the Arab penetration.[1] The next group of small chieftainships to become united under one of their own number were the Bena. This was a short-lived amalgamation, for it was overtaken by the still more successful unification of the thirty-odd Hehe chieftainships by Muyugumba the father of Mkwawa. It was Muyugumba who, by adopting the military training and weapons of the Ngoni, checked the expansion of the more northerly of the two Songea Ngoni kingdoms—that of Chipeta. In the course of these wars the Hehe absorbed most of the former Ubena, whose ruling family, the Manga, retreated into the Ulanga valley, and there carved out a new supremacy over the indigenous Ndamba and Pogoro of Mahenge.[2] The rise of Mirambo in Unyamwezi is dealt with elsewhere.[3] Here it need only be remarked that it was when the two northward-moving sections of Ngoni under the Indunas Mtambalika and Mtambarara, after raiding all the eastern coast of Lake Tanganyika from Ufipa to Ujiji, finally settled under Mirambo's overlordship near Kahama, that the Ngoni invasions of East Africa came to be more or less absorbed.

In a real sense the Ngoni had set the scene for the Arab penetration that was to follow. For the Arab contact was to prove mainly destructive in some areas and mainly constructive in others. Which it was to be in any particular area depended more than anything else upon the type of political organization achieved before the Arabs came. The tendency was for the strong to grow stronger and for the weak to be annihilated.

[1] Mbeya District Book.
[2] Iringa and Mahenge District Books; F. Fülleborn, *Das Deutsche Njassa- und Ruvumagebiet*, 1906, pp. 179, 201; A. T. and G. M. Culwick, *Ubena of the Rivers*, 1935, pp. 20–21.
[3] See Chapter VIII below.

The Ngoni both by the political units they formed themselves and by the reactions which they provoked in others provided a kind of inoculation against what was to follow. They did so in what were previously some of the most backward and defenceless areas of East Africa, which might otherwise have witnessed the scenes of holocaust and massacre which occurred on the upper Congo and its tributaries. In pre-colonial Africa migration and warfare, conquest and settlement, were not always the unmitigated evils they have been painted. Sometimes, like later and larger manifestations of imperialism, they were the means of communicating the achievements of the strong and successful to the weak and the outpaced.

VII

ZANZIBAR AND THE COASTAL BELT
1840–1884

J. M. GRAY[1]

BY the year 1840 Zanzibar had become for all practical purposes the capital of the Omani realm, and Said ibn Sultan's permanent residence.[2] The move had been foreshadowed by the increasing amount of attention which the ruler had paid to his East African dominions in the previous decade, and was the culmination of the long process of establishing Omani control over the islands and coastlands which had been going on since the expulsion of the Portuguese. Said himself had played a crucial role in this development. During the latter part of the eighteenth century local governors in East Africa had been able to make themselves virtually independent rulers; only the island of Zanzibar remained consistently loyal to Oman, and even there the hold was tenuous.[3] In 1804 the Frenchman Dallons reported that the dominant power on the island was the Harthi clan, who constantly interfered in affairs of state, and regarded the Busaidi dynasty as mere parvenus.[4] For several years after his accession to power Said had been unable to relax his attention from affairs in Oman. In 1814,

[1] Sir John Gray wishes to acknowledge the help he has received from Dr. J. E. Flint of King's College, University of London, who, at a time when he was unable to do so himself, prepared the final draft of this chapter.

[2] R. Coupland, *East Africa and its Invaders from the Earliest Times to the Death of Seyyid Said in 1856*, 1938, p. 295. This work will be cited hereafter as Coupland, *Invaders*.

[3] Sultan ibn Ahmad, who had ruled Oman until his death in 1804, had evidently attached little importance to Zanzibar. In 1798 the sultan's garrison in Zanzibar had been under the command of a certain 'bahadur' who would appear to have been of Baluchi or Indian origin. The person in charge of the customs was an Abyssinian eunuch named Yaqut ibn Ambar, who by 1811 had become hakim or magistrate. *The Transactions of the Bombay Geographical Society*, 6, 1844, pp. 42–57 contains descriptions of Yaqut and of the government of Zanzibar Island by an English naval officer who visited the island in 1811.

[4] For the early origins of the Harthi see Mrs. C. E. B. Russell, *General Rigby, Zanzibar and the Slave Trade*, 1935, p. 348.

when Abdullah ibn Ahmad Mazrui succeeded to the governorship of Mombasa, Said had demanded the customary presents due to him as overlord. Abdullah's reply was to send presents which were openly contemptuous in form—one of them was an empty corn-measure.[1] Internal dissensions in Oman compelled Said to bide his time, but the insult was not forgotten. In the meantime he realized that it was incumbent upon him to consolidate his position in Zanzibar. A succession of Arab governors were appointed, on whom Said felt he could rely.

In 1822 Said began the long struggle with the Mazrui by capturing first Pate, and then Pemba, the granary of Mombasa. After failing to recapture Pemba, the Mazrui hoped to bring British power to their aid, and Owen's 'protectorate' did give them a brief respite.[2] After 1828 Said made continuous attempts to control Mombasa by force, each of which was concluded by a temporary truce, the terms of which were violated almost as soon as they were made. Finally, in 1837, Said took advantage of a succession dispute among the Mazrui, and by a combination of force, guile, bribery, and treachery he succeeded in seizing the most prominent Mazrui, who were exiled to the Persian Gulf, where they all died in captivity.[3]

The elimination of all the leading members of the Mazrui family had made Said ibn Sultan finally supreme not only in the islands of Zanzibar and Pemba, but also along the whole of the coast from the Juba river to Cape Delgado. There were still one or two pockets of resistance which required to be mopped up and there were destined to be a number of rebellions to be suppressed, but both Said and his immediate successor were eventually able to stamp out all resistance of this description. For nearly half a century no really formidable rival ever appeared to dispute the sovereignty of the rulers of Zanzibar in the regions just mentioned.

All the same Said ibn Sultan soon realized that even in Zanzibar strong measures had from time to time to be taken in order to maintain his authority. The Harthi clan ill brooked

[1] Coupland, *Invaders*, p. 220.
[2] See Chapter V, pp. 158–60, above.
[3] Coupland, *Invaders*, pp. 217–94, contains a full account of the destruction of the Mazrui power and of Owen's protectorate. See also M. Guillain, *Documents sur l'histoire, la géographie, et le commerce de l'Afrique orientale*, 1856, i. 584–9, and ii. 26–27; J. M. Gray, *The British in Mombasa, 1824–1826*, London, 1957.

the interference of the newly arrived Sultan of Oman in the affairs of Zanzibar, where hitherto they had always had very much their own way. Said was, however, strong enough to assert his own personal authority in Zanzibar whenever he was in residence there. Whenever affairs of state made it necessary for him to return to Oman, he always took the precaution of insisting that the leading member of the Harthi clan should accompany him to Muscat as a hostage for the good behaviour of his turbulent kinsfolk in Zanzibar.

Hitherto the administration of the interior of the island of Zanzibar had been left in the hands of local rulers. By the time of Said ibn Sultan's first arrival in Zanzibar the number of these had been reduced to two. The northern part of the island and the adjacent islet of Tumbatu was administered by a ruler known as the Sheha, who resided on the islet itself and whose subjects were, and are still, known as the Tumbatu. The remainder of the main island was administered by a ruler variously known as the Jumbe or Mwenyi Mkuu, whose subjects came to be known as the Hadimu. The title of Mwenyi Mkuu was hereditary in a family of mixed Arab and African blood. The name of the holder at the time of Said ibn Sultan's arrival in 1828 was Hasan ibn Ahmad Alawi. We know that on this occasion he and Said ibn Sultan met, and some sort of *modus vivendi* was established between them, though its terms are not altogether clear.

The Mwenyi Mkuu, Hasan ibn Ahmad, appears to have been living in the town of Zanzibar in 1828, but apparently found Said ibn Sultan too close a neighbour for his liking. Accordingly he moved out to Bweni in the middle of the island where he died in 1845. His successor was his *wazir* (adviser), named Muhammad ibn Ahmad Alawi. It would appear that, early in the reign of Said ibn Sultan's son and successor Majid, the Mwenyi Mkuu and sultan fell foul of one another with the result that the Mwenyi Mkuu was temporarily deported to the mainland, but eventually in deference to public opinion was allowed to return to Zanzibar. The subsequent attitude of the Mwenyi Mkuu to the sultan is best evidenced by the fact that he built himself a fortified house at Dunga, some twelve miles from Zanzibar town, and had armed sentries constantly posted outside.

Muhammad ibn Ahmad Alawi died in 1865. He left an only son surviving him, named Ahmad, who at the time of his father's death was only twelve years old. The boy left Dunga and came to reside in Zanzibar, where he died in 1873 of the smallpox. He left no male successors and with him the title and post of Mwenyi Mkuu came to an end.

Up to the time of the death of the last Mwenyi Mkuu the administration of each village or group of villages had been in the hands of officers called Shehas. The Shehas combined the functions of village constables and tax collectors as well as the performance of many other multifarious duties. Until 1873 instructions from the Sultan of Zanzibar to the Shehas had passed to the latter through the Mwenyi Mkuu. It was not until the passing in 1897 of a Local Government Decree that any attempt was made to place the Shehas under the direction and control of a superior local government official. The result was that for many years the Hadimu territory remained more or less a native reserve.[1]

Like the Hadimu, the Tumbatu had a hereditary ruler, called the Sheha. Whilst his influence was more or less supreme on the islet of Tumbatu, his influence on the main island appears never to have been as extensive as that of the Mwenyi Mkuu. The post of Sheha was not confined to males. The earliest recorded Sheha was in fact a woman called Mwana wa Mwana, who married Hasan, the Mwenyi Mkuu of the Hadimu who died in 1845. This marriage did not, however, lead to a union of the two tribes, as her son Ali succeeded her as Sheha but did not succeed to the post of Mwenyi Mkuu. But the dynasty of hereditary Shehas came to an end some time before 1865, when Sayyid Majid appointed one Msellem as Sheha and placed him under the orders of the Mwenyi Mkuu. When Muhammad ibn Ahmad died in 1865, Sayyid Majid issued a letter announcing that Msellem was to remain Sheha under the orders of the new Mwenyi Mkuu.[2]

Another problem confronting the early-nineteenth-century

---

[1] The above account is based on Zanzibar Archives; W. H. Ingrams, *Zanzibar, its History and its People*, 1931, pp. 30-32, 147-53; C. P. Rigby, *Report on the Zanzibar Dominions*, 1861, pp. 19-20 (reprinted in Russell, *Rigby*, Appendix II); R. F. Burton, *Zanzibar, City, Island and Coast*, 1872, i. 272 and 410.

[2] W. H. Ingrams in *Zanzibar, An Account of its People, Industries, and History*, 1924, p. 64.

ruler of Zanzibar was that of the people commonly called in contemporary correspondence the 'northern Arabs'. From very early days vessels from Arabia and the Persian Gulf had taken advantage of the north-east monsoon to proceed from their home ports to East Africa with cargoes to exchange for the products of that continent. Unfavourable winds compelled them to delay their return voyage until the south-west monsoon set in some four months later. Zanzibar was their principal port of call. During their stay there and at other places on the coast their crews ran riot and the local inhabitants had to submit to all the evils that could be inflicted upon them by a horde of piratical invaders. Sayyid Said, who had a strong fleet at his disposal and who was himself a person of great strength of character, had, generally speaking, been able to curb the nefarious activities of these visitors whenever he was resident in Zanzibar, but his son and successor, Majid, lacked his father's strength and tried to temporize with them by gifts of money. Their activities declined after matters had come to a head in 1861 and the assistance of a British man-of-war was invoked to expel a piratical horde which had run riot in Zanzibar for several weeks. There were from time to time subsequent attempts to assert their immunity from interference with their misdeeds, but with the aid of British men-of-war these were generally suppressed. After 1873, when Sayyid Barghash ibn Said concluded a treaty whereby the carriage of slaves to and from his dominions was severely limited and extended powers were given to British men-of-war to search vessels suspected of engaging in the slave traffic, their activities were almost entirely limited to surreptitious attempts to evade British naval patrols by means of smuggling in canoes and other small vessels and in small numbers at a time.[1]

A large number of Sayyid Said's compatriots followed in his train to Zanzibar. Some came to serve in his garrison and others to take up civil appointments under him. Others were attracted by the economic and commercial potentialities which his African dominions appeared to offer. Sayyid Said himself appears to have realized very quickly that tropical Zanzibar could produce economic crops which could not be produced in Oman. Cloves had been introduced to Zanzibar in about 1818 by one

[1] Coupland, *Invaders*, p. 168.

Salih ibn Haramil Abri. With extraordinary intuition Said ibn Sultan quickly grasped the potentialities of cloves as an economic asset for Zanzibar and by his personal example and in many other ways did everything in his power to encourage their cultivation in his African islands. His foresight stood Zanzibar in good stead. In the course of a very few years the revenue yielded by the clove crops of Zanzibar and Pemba became a mainstay of those two islands. Said also encouraged the cultivation of many other economic crops. Though his experiments were by no means always successful, his realization of the extreme desirability of developing agriculture in his African dominions showed a remarkable insight.[1]

These innovations in the agricultural economy of the two islands of Zanzibar led to a number of other social changes of a sweeping character. In the first place the newly introduced crops, such as cloves, were grown for the purpose of export and called for planting on an extensive scale. Many of the Arabs already settled in the islands, and many more who followed in Sayyid Said's train, began to acquire extensive areas of land in the western and more fertile parts of the two islands for conversion into plantations. This led to the gradual retreat of the aboriginal African population with its more primitive forms of husbandry into the northern, eastern, and southern parts of the islands, where for many years they held themselves very much aloof from the inhabitants of the more fertile belts. Furthermore, plantations called for an increased labour supply, firstly, for clearing the land for planting, and, secondly, for keeping the plantations in a state of good husbandry and, lastly, for harvesting the crops in due season. In this connexion it is not inapposite to mention that cloves are harvested twice in each year and coconuts four times in each year. One of the many reasons for the retreat of the aboriginal inhabitants from the fertile belts was to avoid being compelled by the new landlords to work under servile conditions on these plantations. Consequently plantation owners had to look to the mainland for their labour supply. African slavery was a centuries-old institution. The inevitable consequence was that the East African slave trade received a fresh impetus in order to meet the demands

[1] Ibid., pp. 313–14; F. B. Pearce, *Zanzibar, the Island Metropolis of Eastern Africa*, 1920, pp. 295–7.

of landowners in Zanzibar and Pemba. The Arabs thus established themselves not only as a trading and administrative class, but as a planter aristocracy, whose interests were vitally concerned with the maintenance of the slave trade. They began to develop many of the attitudes traditionally associated with a planter class, a certain aloofness towards other groups in the society, and the cultivation of pursuits made possible by leisure. Some planters began to acquire houses in the town, where they could entertain and perhaps obtain access to the high society of the sultan's palaces. Political groups integrated themselves into the older rivalries of the island, some composed merely of social climbers and place seekers; others of men determined to preserve and extend the influence of the planter class. Indian merchants and traders made the planter's life easier by providing credit; the price to the Arabs was a mounting burden of indebtedness.[1]

The Indian money-lender was not much liked by the Arab planters, but Said was well aware of his importance, and did all he could to encourage the inflow of capital and the expansion of trade which the Indians produced. Said gave the Indians complete religious toleration, listened to their advice on commercial and even political affairs, and gave them control of financial administration, including the collection of customs. As a result the Indian population grew from about a thousand in 1840 to nearly six times that number twenty years later, many of whom, particularly the Muslim element, had become permanent settlers. Their control over the commercial life of Zanzibar was virtually complete.[2]

At the time of Said's first arrival in Zanzibar he had realized its potential importance as an emporium for trade with the interior of Africa. By 1839 he had become in the habit of sending trading expeditions some hundreds of miles into the interior.

[1] Rigby's *Report*, p. 8, contains a rather hostile, but illuminating, description of the Arab planter's way of life.
[2] W. C. Palgrave, *Narrative of a Year's Journey through Central and Eastern Arabia*, London, 1865, ii. 369–70, for an account of Said's attitude to Indians; also Guillain, *Documents*, ii. 80; Rigby, *Report*, p. 4. Said himself belonged to the Ibadhi sect of Islam, which had the general reputation of being extremely puritanical. He was none the less very tolerant of other faiths. When he discovered that the public slaughtering of animals during the festivities at the close of Ramadhan offended the religious susceptibilities of his Hindu customs master, he gave orders that the practice should cease.

Other traders might join these caravans for protection. Owing to the example set and the encouragement given by the rulers of Zanzibar this trade with the interior grew steadily and brought in constantly increasing revenues in the shape of customs duties.[1] Before 1828 the total revenue yielded from Zanzibar had been about £10,000 per annum. By 1834 it was twice that figure, and by 1859 it had risen to £50,000 per annum.[2] In 1859 Consul Rigby could report that the total trade of Zanzibar had reached £1,664,577. Eighty ships of a total of 23,340 tons had used the port of Zanzibar, and the sultan's realm had exported ivory worth £146,666, cloves worth £55,666, and gum-copal worth £37,166. Cowries valued at £51,444 had been shipped to West Africa and sesamum worth £20,800 to France. In return Zanzibar bought £93,744 worth of American, £53,777 worth of Indian, and £37,711 of British cottons; she imported rice costing £38,444, Venetian beads for the interior trade costing £21,879, and lesser values of brass wire, gunpowder, muskets, and provisions. Total imports were £908,911, total exports £755,666.[3]

In a report which Consul Hamerton sent in 1855 to the Bombay Government he stated that Said ibn Sultan's government was

positively of a purely patriarchal nature. There are no establishments of any kind similar to those existing in the states of native princes in India. All things are in a most primitive state and I suppose just as they have been from a very remote period.... There are but two men who can be fairly said to be the principal men of the court who interfere in affairs of government. One is at Muscat... and the other at Zanzibar, Seyyid Suleiman ibn Hamed.... All the respectable men in Muscat and Zanzibar generally attend daily at the durbar, or when the Iman generally holds his assembly, three times a day, but they do not interfere in the government.... The Imam has not any ministers or secretaries for different departments.[2]

This description fairly accurately represents the form of government which lasted not only throughout Said ibn Sultan's reign, but also throughout those of his two immediate successors.

[1] Coupland, *Invaders*, pp. 304–13, 316–20.　　　　[2] Ibid.
[3] Russell, *Rigby*, Appendix II.
[4] *Selections from the Records of the Bombay Government, No. XXIV*, N.S., Bombay 1856, pp. 237–8. (Cited hereafter as *Bombay Records*.)

The sultan was his own lawmaker and was also the fountain of justice. Writing eleven years earlier on the subject of the administration of justice in Said's dominions, Hamerton denounced the venality and maladministration of the Kadhis, but said that 'from His Highness the Imam alone a poor man can obtain justice . . . who is most truly every man's friend; he wishes to do good to all'.[1] In those last words lay the secret of Said's long and successful reign.[2]

Said's East African empire outside Zanzibar and Pemba was still a tenuous affair. On the Somali coast there was little Zanzibar influence outside the environs of the fortified towns of Mogadishu, Merca, Brava, and Kismayu.[3] The island of Lamu had been loyal since 1813, when Said had supported its people against the Mazrui and had installed a governor who built a fort.[4] Such was not the case with the neighbouring island of Pate. That island was divided into three petty sultanates, each separated from its neighbour by a swamp which very nearly intersected the island. Of these that of Pate, fairly well controlled by Said since its capture in 1822, was the largest and most extensive. The others were Faza and Siu.

Faza was ruled for close on sixty years by Shaikh Mzee ibn Saif Stambuli. He succeeded an uncle in about 1836 and during the period under review outwardly remained consistently loyal to the Sultans of Zanzibar, though he was upon more than one occasion strongly suspected of playing a double game. He also had a great influence over the people on the mainland. As Vice-Consul Holmwood reported in 1874, it was tactless to suggest to the people on the mainland that Sayyid Barghash was their ruler. 'They hinted with some irritation of manner that their own chief had the whole district under his foot.' Nevertheless, the fact remains that it suited the purposes of their chief to acknowledge the Sultan of Zanzibar as his overlord, and it may be fairly said that the sultan exercised indirect rule over the mainland territories of Mzee ibn Saif.

---

[1] A. Hamerton, *Report on the Affairs of the Imam of Muscat*, 1844, *Bombay Records*, 238, p. 5.
[2] For Said's administration in Zanzibar see Coupland, *Invaders*, pp. 321–8; Rigby, *Report*, pp. 6–8; Guillain, *Documents*, ii. 237; Burton, *Zanzibar*, i. 263.
[3] R. Coupland, *The Exploitation of East Africa, 1856–1890*, 1939, pp. 452–3.
[4] A. Werner, 'A Swahili History of Pate', *J. Af. Soc.* 14 and 15, 1914–15, pp. 148 sqq., 392 sqq.

The description of Siu as a sultanate is somewhat of a misnomer. Until about the middle of the nineteenth century the town was subject to a curious form of dual administration. During the seventeenth century the townsmen had invoked the aid of the Somalis against the aggression of their neighbours in Pate. The Somalis responded to the call and helped to drive the people of Pate out. Thereafter a government was established consisting of a shaikh chosen from the Famau family (the descendants of the original Asiatic settlers in Siu) and a shaikh chosen by the Somalis. In the 1820's the government of the place had become concentrated in the hands of a Famau shaikh named Mataka ibn Mbaraka, who sided with the Mazrui of Mombasa in their attempts to make their own nominees rulers of Pate. Even after the Mazrui had been compelled by Said ibn Sultan to abandon all claims to interfere in the affairs of Pate, Mataka persistently refused to recognize Said ibn Sultan as his overlord.

Mataka's first challenge to Said ibn Sultan had occurred when he came to the aid of one Fumuloti ibn Shaikh, who in 1822 had been installed by the Mazrui as Sultan of Pate but was driven out by the inhabitants after having 'reigned from morning till after the hour of afternoon prayers'. Fumuloti had subsequently established himself on the mainland near the mouth of the Ozi river. In 1825 he came back, landing at Siu and invoking the aid of Mataka in an attack on Pate, but failed to evict Said ibn Sultan's nominee from that place. Thereafter Mataka continued to intrigue against the Sultan of Zanzibar.

Five expeditions were sent by Said against Siu, but each of them ended disastrously. After the ignominious defeat of the last of them in 1845,[1] in the words of Colonel Playfair at a later date, Said 'found it expedient to adopt a temporizing policy, which had the effect of smoothing over difficulties for the time, only to make them most unmanageable afterwards'.

From a very early date in the nineteenth century there were between the Pangani and Rufiji rivers a number of coastal towns

[1] The 1845 campaigns are documented in Werner, *J. Af. Soc.* 1914–15; C. H. Stigand, *The Land of Zinj*, 1913, pp. 91–93; Salil ibn Razek, *History of the Imams and Seyyids of Oman*, trans. G. P. Badger, 1871, pp. 355 and 360; Guillain, *Documents*, iii. 99–102; Coupland, *Exploitation*, pp. 338–9.

and villages from which caravans, which had been equipped and financed in Zanzibar, set out for, and returned from, the interior. The main points for the arrival and departure of these caravans extended from Saadani about forty-five miles to the south of Pangani, to Mboamaji, about fifteen miles to the south-west of the modern town of Dar es Salaam.

As trade between the coast and the interior increased, Arabs, Swahilis, and Indians began to cross over and settle at various points between Saadani and Mboamaji. This phase of coastal settlement was of a peaceful nature and its success would appear largely to be attributable to the tactful handling of the local inhabitants by Said ibn Sultan's kinsman, Sulaiman ibn Hamid Busaidi. From what we can learn of Sulaiman it is evident that he preferred diplomacy to the use of force. It would appear that it was largely due to his influence that, when Sayyid Thuwain, ruler of Oman, threatened to invade Zanzibar in 1859, some 20,000 to 30,000 armed men from the mainland were mustered to repel the invaders.

The erstwhile independent sultans of Kilwa Kisiwani claimed to exercise dominion over the whole of the coast from the Rufiji delta down to Cape Delgado and the Sultans of Zanzibar claimed the like authority as their overlords. Nevertheless the sultans of Kilwa still dreamt of the day when they could shake off the Omani yoke. When in 1812 H.M.S. *Nisus* called at Kilwa, Sultan Yusuf ibn Hasan had been secretly importing arms and ammunition in the hope that one day he might be able to expel the Arab Governor and garrison.[1]

According to a local chronicle[2] Sultan Yusuf died shortly after a raid by the Sakalava of Madagascar on Máfia, which took place in about 1822. Thereafter the prosperity of Kilwa Kisiwani began to wane. There was a dispute as to Yusuf's successor, which led to the imprisonment at Muscat of Muhammad ibn Sultan, one of the rival candidates. Muhammad was eventually released and set himself up as sultan at Kilwa Kivinje, about fifteen miles to the north of Kilwa Kisiwani. His rival remained at Kilwa Kisiwani, but the trade of that place

[1] J. Prior, *Voyage along the Eastern Coast of Africa . . . to St. Helena*, London, 1819, pp. 66–69.
[2] Swahili histories of Kilwa Kisiwani and Kilwa Kivinje in C. Velten, *Prosa und Poesie der Suaheli*, Berlin, 1907, pp. 249–52, 253–61.

was gradually transferred to Kilwa Kivinje. Local chronicles inform us that, 'since Sultan Muhammad and Sultan Hassan ruled, there has been no proper kingdom. . . . After the death of Sultan Muhammad the power was in the hands of Said ibn Sultan himself. Sultan Hassan had no power but only the honour of the title of Sultan.'

The jurisdiction of the sultans of Kilwa had at one time been exercised as far south as the Ruvuma river and it would appear that Said ibn Sultan's claim to sovereignty along this stretch of the coast was never seriously contested. At Lindi, Sudi, and Mikindani the Sultans of Zanzibar appear in the first instance to have exercised indirect rule through the hereditary local chiefs, but during the reign of Sayyid Barghash garrisons were installed at Lindi and Mikindani under the command of Arab governors.

After their expulsion from Mombasa the Portuguese exercised from their headquarters at Mozambique a somewhat precarious hold upon the coast to the south of Cape Delgado. Until the latter part of the reign of Said ibn Sultan no serious attempt was made from Zanzibar to contest their sovereignty in those regions, but with the gradual increase in trade between Zanzibar and the interior of the continent the situation began to change. Caravans began to set out from Tungi Bay, just to the south of Cape Delgado, for the interior of the continent, and Arab and Swahili traders began to establish depots at the various villages south of the Ruvuma river. In 1837 a certain Amadi succeeded as local ruler with the title of Mwenyi Mkuu. In 1853, he allowed the Sultan of Zanzibar to establish a customs post in Tungi Bay. The Portuguese thereupon sent an expedition to expel the Zanzibar officials but failed in the attempt. Various abortive attempts were made by the Portuguese to settle the dispute by treaty.

Sayyid Said's ambition was not to create a territorial, but an economic empire in East Africa. His position was that of a great middleman,[1] controlling the intercourse between East Africa and Europe, Arabia, India, and America. 'I am nothing

---

[1] The missionary Krapf fell out of favour with Said, because he unwisely furnished the names of European and American firms to Kimweri of Usambara, so that the king could trade directly with them. C. P. Groves, *The Planting of Christianity in Africa*, ii. London, 1954, p. 116.

but a merchant', he once confessed to a French visitor.[1] Apart from his private income, much of which came from trade on his own account, Said was almost entirely dependent upon customs duties and upon taxes on the slave-trade. His wealth and effective power were thus determined by the extent to which his subjects could expand trade.

The coastal towns on the mainland were the focal points through which trade with the African interior was channelled. Sayyid Said's reign witnessed a remarkable expansion of this trade; the trading population of the coast increased in number, new routes to the interior were established, penetration extended farther westward, and the caravans increased in size and frequency. The basis for this expansion lay in the growing demand of Europe for ivory, gums, and vegetable oils, in the expanding demands of the island and coastal plantations for slave labour, and the constant demand for slaves in Arabia. But Said's policies quickened the pace. Whilst reserving to himself the monopoly of ivory and gum-copal between Pangani and Kilwa,[2] the rest of the coast and all other commodities were left open to all comers. The chaotic local variations in customs duties were gradually consolidated into a simple 5 per cent. duty on all imports. There were no export duties or other taxes, licences, or monopolies upon trade. The customs were placed in the hands of an Indian farmer, who paid a fixed sum in return, which was periodically increased as trade expanded.[3] Said also did much to facilitate exchange, importing pice from India as a small-denomination currency.[4] Said's encouragement of Indian immigration allowed Indians to settle in the coastal towns, where their function was to finance, through credit, the expeditions to the interior.

Each coastal settlement had its governor, customs officials, a few Baluchi soldiers, its Indian merchants counting their riches by the debts owed to them, and the Arabs, organizing expeditions, setting off, or returning. Usually these people could build stone houses, often with plantations and provision grounds attached, whilst the African population lived in the traditional

---

[1] A. de Gobineau, *Trois ans en Asie, de 1855 à 1858*, Paris, 1905 ed., p. 99, quoted by Coupland, *Invaders*, p. 299.
[2] Coupland, *Invaders*, pp. 303, 386, 423, 481.
[3] Guillain, *Documents*, ii. 251 (details for 1847); Rigby, *Report*, p. 19 (for 1859).
[4] Burton, *Zanzibar*, ii. 405–6.

clusters of huts. Each settlement had its own hinterland and routes to the interior, and its own pattern of trade with Zanzibar and the rest of the world.

Just as the ports of the Benadir coast maintained a considerable political autonomy from Zanzibar, so their economic life was virtually an independent one. The northern Arabs traded directly with them, and most of their trade with India did not pass through Zanzibar. In 1846 Mogadishu and Brava could each boast a population of 5,000 souls, and Merca had 3,500. Each of these towns had its Arab quarter of stone, but Warsheikh at the same date consisted of only 'twenty-three miserable huts'.[1] Mogadishu was still a centre of cotton manufacture, though the industry declined throughout this period in face of the competition of cheap American imports. The cotton was imported from Kutch to be spun by the women of Mogadishu, whilst the weavers were mostly men. The occupation was lowly in rank, and often performed by slaves. In mid-century the town produced about 300,000 woven pieces each year, which were consumed locally, or exported southwards along the coast, or even to Zanzibar, the Red Sea littoral, and Egypt.[2] Cotton-spinning and weaving also took place at Merca, but on a very much smaller scale.[3] The Benadir coast imported its slaves by sea, finding very few in its own hinterland. Working slaves were bought from Patta, or from the Mrima—the coastland opposite Zanzibar and Pemba; Mogadishu imported 600 in 1846.[4] A few Abyssinian slaves, highly priced and well regarded, were obtained from the northern Arabs. Often the working slaves were sent inland to work the ivory trade, amounting to some 50,000 lb. annually.[5] Camels were frequently used on trading expeditions inland, bringing back—besides ivory—gums, hippopotamus teeth, and hides in exchange for beads, cottons, tobacco, and sometimes even re-exported slaves. The Africans themselves often sent trading parties into Brava.

Mombasa undoubtedly suffered economically and commercially from the long struggle between the Mazrui and Zanzibar. Geographically it was well placed to tap the trade of the Chaga, the Kamba, or even Usambara, and its harbour was

---

[1] Guillain, *Documents*, iii. 117.   [2] Ibid., pp. 531–2.
[3] Ibid., p. 148.   [4] Ibid., p. 537.
[5] Estimated from Guillain, *Documents*, iii. pp. 174, 310, 533.

the finest in East Africa. But internecine strife weakened Mombasa's hold on the hinterland and depopulated the country. Guillain estimated that its population fell from 5,000 in the 1820's to less than 3,000 in 1846, of whom some 220 were Arabs and 50 or so Indians. Most of this fall was caused by emigration to Pemba, though many of these people began to return in the 1840's, and in 1860 Burton could estimate the population at between 8,000 and 10,000.[1] Mombasa, though a distribution centre for the slave-trade, obtained only a few slaves from its hinterland at this time. The Nyika would sometimes sell their criminals, or enslave their own people in time of famine, as in 1840. But slaves were generally imported from Zanzibar or direct from Kilwa. From Mombasa they were resold to Pemba and Lamu, where they were again resold to Patta, which sold them again to the Benadir ports.[2] Mombasa had no significant industry, and her agriculture was insufficient for her needs; subsistence came from trade, and the Nyika and the interior peoples supplied provisions. Inland the chief product sought was again ivory, provided by the Chaga and Kamba. About 70,000 lb. were bought annually in the 1840's in exchange for American and British cottons, beads, and brass wire. At this time much of the trade was brought to the coast by Africans, but by the 1860's large and well-armed Arab expeditions were common.[3] Besides ivory the interior supplied rhinoceros horn, hippopotamus teeth, and, nearer home, gum-copal from the Nyika. Mombasa itself produced good quality coconut oil and some sesamum.[4]

The coast opposite Zanzibar and Pemba was remarkable for a greater emphasis on commercial penetration inland, and for the larger quantity and better quality of its ivory. Tanga in 1857 was a town of 4,000 or 5,000 people, with twenty banyans, sending out trading parties of 400 or 500 to Chaga and the Masai for ivory.[5] By 1843 Arab traders had reached the capital of the kingdom of Buganda, and they were moving west of the Great Lakes after 1851.[6] Pangani had a similar population, and sent out parties of up to a thousand men, who sometimes carried

[1] Burton, *Zanzibar*, ii. 75.   [2] Guillain, *Documents*, iii. 305.
[3] Burton, *Zanzibar*, ii. 61 describes such a party.
[4] Ibid., pp. 45–46; Guillain, *Documents*, iii. 315–17.
[5] Burton, *Zanzibar*, ii. 116–17.
[6] See Chapter VIII, below.

goods worth £4,000 to Chaga, the Masai, and Ngulu. Pangani exported about 35,000 lb. of ivory at this time, and lesser quantities of rhinoceros horn and hippopotamus teeth. The banyans were very prosperous, and Burton knew of one who was owed $26,000.[1] Saadani was of less importance, a town of 700 or 800 huts, trading mainly in provisions to Ngulu, and in copal.[2] Bagamoyo, perhaps because of its position directly opposite Zanzibar, was a centre for penetration both south-westward towards Lake Nyasa and also north-westward to Tabora and on to Ujiji.[3] The entire Mrima was closely linked with the commercial life of Zanzibar Island. Sea communication was maintained throughout the year, and many of the Indian merchants were agents for Zanzibar merchants.[4]

The Kilwa region was the slave-trading centre of the East African coast. British naval patrols seem to have made little impression on the trade, and the French 'free' labour immigration scheme, in operation from 1843 to 1864, actively stimulated the Kilwa traffic. Kilwa flourished on the chaos and anarchy around Lake Nyasa, and the Yao were the main intermediaries who supplied slaves and ivory to the coast. The operations were financed by Indians, who numbered about fifty in 1843, and had increased to over 150 in 1857. After a harrowing journey from Nyasa, in which the weakest usually perished, the slaves were stowed in tiers on the dhows at Kilwa for the two- or three-day journey to Zanzibar, where they often arrived sick and starving. The trade needed to be profitable, for Kilwa was an unhealthy place. Cholera struck the region in 1857 with terrible effects. Burton, who had seen cholera in India, declared that he had seen nothing so terrible as the Kilwa outbreak; there were corpses everywhere, in the streets, in the countryside, and floating in the sea. In the 1870 outbreak 200 people died every day, and slaves could not be sold at a dollar a head for fear they were infected.[1]

As the result of his making Zanzibar his second capital Said ibn Sultan had been brought into closer touch than he had

[1] Burton, *Zanzibar*, ii. 136–47.   [2] Ibid., pp. 267–79.
[3] R. F. Burton, *Lake Regions of Central Africa*, London, 1860, ii. 57.
[4] Guillain, *Documents*, iii. 370–2.
[5] Burton, *Zanzibar*, ii. 344–7. For Kilwa's trade see ibid., pp. 342 sqq.; and Coupland, *Invaders*, p. 302.

hitherto been with European traders and European governments. In 1833 he entered into a treaty of amity and commerce with the United States of America[1] and four years later an American consul took up his residence in Zanzibar.[2]

Another treaty of amity and commerce was concluded with France in 1844 and a French consul was posted to Zanzibar in the same year.[3] Relations between the Sultans of Zanzibar and the French representatives were never quite so cordial as those with the representatives of other European states. The trouble principally arose out of French opposition to the sultan's efforts to conform to the wishes of the British Government in regard to the suppression of the slave-trade. There was constant friction between both Said and Majid and the French consuls over the exportation of slaves from the African coast for plantation work in Réunion, which the French endeavoured to conduct under a so-called 'free emigration' scheme. It was only after strong representations from the British Government that the French agreed to abandon this scheme in 1862.[4]

One other bone of contention, which caused for some time further friction between the French government and the Sultans of Zanzibar, was the carriage of slaves by the sultan's subjects in vessels flying the French flag. Some of the most notorious slave-dealers were allowed to sail with impunity under the tricolour. At times an energetic French consul at Zanzibar tried his best to collaborate in putting down the traffic under the French flag, but, as often as not, his good intentions were entirely thwarted by the refusal of the court in Réunion to condemn vessels seized by him or to deal with the offenders on the pretext of lack of evidence. It was further notorious that French passes were issued by the authorities in that island on the flimsiest of claims to French nationality. What can only be described as this prostitution of the tricolour was to be allowed

[1] For full details see E. Roberts, *Embassy to the Eastern Courts of Cochin China, Siam and Muscat, 1832–4*, New York, 1837; and C. T. Brady, Jnr., *Commerce and Conquest in East Africa*, 1950, pp. 89–97; J. M. Gray, 'Early Connections between the United States and East Africa', *Tang. Notes*, 22, 1946, pp. 55–86.

[2] Brady, *Commerce & Conquest*, pp. 102–14.

[3] Guillain, *Documents*, ii. 213–14, text of treaty in iii. 459–64; R. H. Crofton, *The Old Consulate at Zanzibar*, 1935, p. 16.

[4] Coupland, *Invaders*, pp. 421–58, provides a full analysis of French activity in Zanzibar between 1817 and 1856; Guillain's *Documents* is virtually a history of the French in Zanzibar.

to continue until 1888 when France agreed to give British and German men-of-war the right of search.[1]

The first German ship to call at Zanzibar would appear to have been the schooner *Alph*, which called there in 1844. During the next twelve years four German firms set up business in Zanzibar.

The second European state to appoint a consul in Zanzibar was Great Britain. This appointment was made by virtue of a commercial treaty of 1839.[2] The first consul to be appointed was Atkins Hamerton, who had a dual role to perform. He was answerable to the Foreign Office in London in his consular capacity and to the Government of Bombay as agent of the East India Company's interest in Said ibn Sultan's Asiatic as well as his African dominions. At the time of his appointment in 1840 Said was in Muscat and Hamerton took up his duties there, but in 1841 in consequence of reports of threatened French aggression in his East African dominions Said betook himself to Zanzibar and Hamerton was directed to follow him. Though for some years more Hamerton was to be an itinerant consul between Muscat and Zanzibar, from 1841 until his death in 1857 he spent most of his time in Zanzibar. During sixteen years of close association there were inevitably times when he and Said had differences of opinion—more particularly on those occasions when on instructions from England Hamerton tried to spur Said to adopt stronger measures for the suppression of the slave-trade, but such rifts were only momentary.

As his correspondence in the Zanzibar archives shows, Hamerton appreciated the difficulties which confronted Said in regard to the slave-trade, as he was working more or less single-handed without reward and at considerable sacrifice of his personal revenue. Hamerton pointed out that, if he were to introduce the more drastic measures for the suppression of the slave-trade for which abolitionists in England were constantly clamouring, their introduction might well cost Said his throne, if not his life, and thus entirely defeat the very object which the abolitionists had in view. In addition, whilst Hamerton insisted on British subjects' being given the full facilities for trade which had been bestowed upon them by the treaty of 1839, he declined to accede to the demands of British trading

[1] Coupland, *Exploitation*, pp. 134-6 and 218-19.
[2] For text see *Bombay Records*, no. 24, 1856, pp. 251-2.

firms that they should be allowed to transgress those rights. In course of time Said came to appreciate the attitude of the British consul and the differences of earlier years were buried in oblivion.

On 16 April 1854 Said ibn Sultan left Zanzibar for Muscat for the last time. As Hamerton reported to the Bombay Government, there was at the time 'an exceeding great fear of the French'. Five days before his departure Said went to the British Consulate, accompanied by his son Khalid and all the principal Arabs. In the presence of them all he placed his son's hands in those of Hamerton and desired him in all difficulties to be guided by the British consul's advice. The Arabs then rose and kissed Hamerton's hand, saying, 'We are now satisfied through the favour of the Almighty and the powerful destiny of Her Majesty Queen Victoria all will go well with us.' Said then intimated to Hamerton that he wished to be alone with him. The two accordingly went into another room. Said then told him that Khalid and the chief Arabs had for the last three days, both night and day, been with him and declared that they would not remain at Zanzibar, if His Highness and Hamerton both left them in the time of difficulty.[1]

Affairs in and around the Persian Gulf had recalled Said from Zanzibar to Muscat. He had hoped to settle matters speedily and to be able to return to Zanzibar at an early date, but affairs in Oman did not go as well as he wished and his sojourn in his Asiatic dominions was prolonged. A few months after Said's departure Khalid died. In due course orders arrived from Muscat appointing Said's fourth son, Majid, as acting governor of Said's East African dominions. Thanks to the intervention of Hamerton and the prompt measures of precaution taken by Din Muhammad, the Baluchi commander of the sultan's troops, Majid's appointment was not contested. At long last Said ibn Sultan was able to set sail for Zanzibar in September 1856. Three days after leaving Muscat he fell ill. The voyage proved a slow one and he grew steadily worse. On 19 October he died whilst his ship was passing an outlying island of the Seychelles group. The vessel reached Zanzibar six days later, when his son Barghash had the body taken secretly ashore and buried near the palace beside the body of Khalid.

[1] Quotation from Coupland, *Invaders*, p. 455.

Barghash was then about nineteen years old and had already formed plans to seize the throne of Zanzibar. By conveying the body secretly by night to Zanzibar he had hoped by a sudden *coup d'état* to gain possession of the palace and proclaim himself sultan. But the vessel carrying the dead sultan's body had dropped anchor overnight off Chumbe Island, about seven miles from Zanzibar. As the ship and its escort flew no flags, those on shore realized that something was amiss and the Baluchi commander once more took active measures to prevent a *coup d'état*. It was known in Zanzibar that after the death of Khalid Said had nominated Majid as successor to his East African dominions and for the moment Said's wishes were accepted without dispute.[1]

Sayyid Majid ibn Said was about twenty-one years old when he became ruler of Zanzibar. Ever since his childhood he had suffered from a mild form of epilepsy, which at times was to prove a serious handicap when he was at the wheel of the ship of state. At the same time he was not wanting in physical courage and, despite a very limited and narrow upbringing, was capable of a far wider outlook on affairs of state than most of his compatriots. About a year after his appointment as his father's deputy, he had been approached to denounce his father's treaties and edicts for the suppression of the slave-trade, but had refused to do so. Behind the scenes the reactionaries began to lay plans for a complete reversal of Sayyid Said's policy. So long as Hamerton, the British consul, was alive, it was realized that any attempt at a revolution had but a small chance of success, but Hamerton died in July 1857, and his loss deprived Majid of his chief mainstay. More than twelve months were to elapse before Hamerton's successor arrived. During that time Majid's grip on the reins of government became gradually weakened. Fearing assassination, he eventually took refuge on one of his men-of-war. Rigby arrived on 28 July 1858, and a few days later Majid returned to his palace. But it soon became evident that his opponents were preparing for a rising. During his lifetime Said ibn Sultan had given directions that after his death his second son Khalid should succeed to his African

[1] *Proceedings of the Commission on the Disputes between the Rulers of Muscat and Zanzibar* (cited hereafter as *Disputes Commission*), Bombay, 1861, pp. 58, 94, 116; Russell, *Rigby*, ch. vii; Coupland, *Invaders*, pp. 455–6 and 551–5.

dominions and his third son Thuwain to his Asiatic dominions. As already mentioned, after Khalid's death Majid was nominated by his father to succeed him in East Africa. Thuwain resented being deprived of the more prosperous portion of his father's dominions and resolved to gain possession of them. An expedition was accordingly equipped and actually set sail from Muscat in February 1859, but was intercepted at sea by a British man-of-war and induced to return to Muscat.

The rebellion was, however, merely postponed. The Harthi clan and other malcontents began plotting to set Barghash on the throne in place of Majid. In October 1859 matters came to a head. Barghash had been ordered by Majid to leave Zanzibar and proceed to Muscat. He agreed to do so, but then began to procrastinate. He was thereupon placed under house arrest, but with the help of certain of his half-sisters managed to escape into the country to a country house known as Marseilles. Members of the Harthi clan and a large following of armed slaves rallied there to his support. As he was unable to rely entirely upon his own local troops to cope with the situation, Majid invoked the assistance of the British Navy. Marseilles had been put in a state of defence and was bombarded by the guns of the Navy. Thereafter the place was evacuated by the rebels and the followers of the Harthi dispersed. Barghash fled back to the town of Zanzibar, where he was eventually compelled to surrender. He was deported to Bombay, where he spent the next two years in exile. The Harthi made their submission to Majid and, with the exception of some of their principal chiefs who were deported to Lamu, were leniently dealt with.

Back in Muscat Thuwain still insisted that Zanzibar and East Africa formed part of his dominions. In support of his claim he urged that during the reign of Said ibn Sultan part of the revenues collected in Zanzibar had been regularly remitted to Muscat in order to supplement the meagre revenues of Oman. After Said's death Majid had remitted at Thuwain's request a sum of 40,000 German crowns to Thuwain and had expressed himself willing to continue making similar remittances annually. Thuwain insisted that this payment was in reality tribute money and that he was sovereign of all the dominions once ruled by his father. Majid, who had discontinued the payment after

Thuwain's abortive expedition against Zanzibar, strenuously denied this. Eventually the two sultans agreed to submit their dispute to the arbitration of Lord Canning, Governor General of India. Canning's award is dated 6 April 1861. The terms of his decision were, firstly that Majid should be declared the ruler of Zanzibar and the African dominions of his father; secondly that the ruler of Zanzibar should pay annually to the ruler of Muscat a subsidy of 40,000 crowns. The award expressly stated that this annual payment was not to be understood as a recognition of the dependence of Zanzibar upon Muscat, but was to be regarded as a final and permanent arrangement, compensating the ruler of Muscat upon the abandonment of all claims upon Zanzibar and adjusting the inequality between the two inheritances derived by each sultan from Said ibn Sultan, which two inheritances were thenceforward to be regarded as distinct and separate. On 13 March 1862 the British and French Governments issued a joint declaration engaging reciprocally to respect the independence of the two sultans, each of whom had formally accepted it and had agreed to abide by the terms of the Canning award.[1]

Though it was a heavy drain upon his resources, Majid paid the annual subsidy until 1866, when Thuwain was murdered by his son Salim. Thereafter Majid announced that he would not pay the subsidy to a parricide. For two years he remitted the money under protest to the Governor of Bombay but ceased to do so after Salim had been deposed by his kinsman Azzan. Thereafter all claims of Zanzibar for the recovery of the moneys due in respect of the subsidy were written off as a bad debt. In addition to being a heavy drain on the revenue of Zanzibar, the payment had been a galling one. By whatever other name it might be called it looked like tribute money and its payment conveyed the impression that Zanzibar was still a dependency of Oman.[2]

As already mentioned, the outrages of the visiting northern Arabs increased beyond all bounds at the beginning of Majid's reign. His troops proved utterly unreliable, and he had weakly

---

[1] Coupland, *Exploitation*, pp. 14–37; *Disputes Commission*, 1861; R. N. Lyne, *Zanzibar in Contemporary Times*, 1905, pp. 53–56. The text of the Anglo-French Declaration can be found in *British and Foreign State Papers*, 57, 1866–7, p. 785.

[2] Coupland, *Exploitation*, pp. 72–75.

tried to buy the intruders off by money presents.[1] News of their open flagrant violation of all anti-slave-trade treaties and laws roused public opinion in Great Britain and the British consul was instructed to bring pressure to bear on Majid to tighten up measures for the suppression of the traffic.

Hamerton had died in July 1857 and the new consul who arrived twelve months later was Christopher Palmer Rigby.[2] It was Rigby who had been responsible for invoking the assistance of the British navy against the rebellion of Sayyid Barghash and the Harthi. Reading between the lines of General Coghlan's report,[3] one can see that it was very largely on Rigby's recommendations that the Canning award severed Zanzibar once and for all from Oman. He put down with a strong arm the holding of slaves by British Indians residing in East Africa. Rigby tried his best to spur the reluctant Majid to take drastic action against the slave-trading activities of the northern Arabs, but here he had little success.

Rigby left Zanzibar owing to ill health in 1861. The consuls who followed him were no more successful in persuading Majid to strike at the northern Arabs. Lewis Pelly stayed only a short time before being transferred to the Persian Gulf, and was followed in rapid succession by Robert Lambert Playfair (1863–5) and Henry Adrian Churchill (1865–70). Both these last-named suffered severely in health and during their illnesses their duties were undertaken by two Agency surgeons in succession, G. E. Seward and John Kirk. Kirk frequently acted as consul between 1868 and 1873, and in the latter year he was definitely appointed as consul general. Latterly relations between the British consuls and Majid were tense and strained. Partly because his health was beginning to decline and partly because he knew that any drastic measures against the slave-trade and the northern Arabs would rouse further hostility against himself, Majid kept on procrastinating and refusing to co-operate with the British proposals.

In 1859 a treaty of amity and commerce was concluded between Sayyid Majid ibn Said and the Hanseatic Republics of Lübeck, Bremen, and Hamburg. The first consul to be

---

[1] Russell, *Rigby*, pp. 328–9.
[2] Russell's *Rigby* contains considerable documentary material culled from Rigby's journals and dispatches.   [3] *Disputes Commission.*

appointed under the terms of that treaty to Zanzibar was John Witt, the local representative of O'Swald & Company. By 1870 German trade with Zanzibar had outstripped that of America and France and was second only to that of British India.[1]

It was during Majid ibn Said's reign that the Sultan of Zanzibar first asserted his authority in many places in the coastal region and came to be recognized by the local chiefs as their overlord. Matters in Siu came to a head in the reign of Majid. Muhammad ibn Shaikh, brother of that Fumuloti whose exceedingly brief rule over Pate has already been recorded, after having gone through the usual phase of obedience to the Sultan of Zanzibar and of assertion of independence, had in 1848 been expelled by Said ibn Sultan and taken refuge on the mainland. He died in exile in about 1860. During Muhammad's six years' rule of Pate Said had managed to construct a fort at Siu and to install a garrison there. This fort had remained in the hands of the Sultan of Zanzibar, but in 1861 Ahmad, the nephew of Muhammad and son of Fumuloti, who will hereafter be referred to by his nickname Simba (lion), joined with Mataka in a successful attempt to expel the Busaidi garrison from Siu and to destroy the fort. After an abortive attempt by his kinsman, Sulaiman ibn Hamid, to recover the place, Sayyid Majid ibn Said personally led an expedition which speedily restored order. The fort was then reconstructed in a better strategic position and a garrison installed. In 1866 Muhammad and Omar, the sons of Mataka, were seized when on embassy to Zanzibar and thrown into prison at Mombasa, where they shortly afterwards died. After this there was no disputing that the Sultans of Zanzibar were absolute in the island of Pate.

The position on the mainland adjoining the island of Pate was, however, very different. When driven out of Siu in 1862 Simba managed to escape to the mainland. He first of all established himself at Kau a few miles from the mouth of the Ozi river. After a short time owing to pressure from the Sultan of Zanzibar's troops he retreated about a dozen miles to the north of Kau and constructed a strongly fortified post at Witu in what was fairly dense forest country which was difficult to

---

[1] H. Brunschwig, *L'Expansion allemande outre-mer du XVIème siècle à nos jours*, Paris, 1957, pp. 76–79; M. E. Townsend, *The Rise and Fall of Germany's Colonial Empire*, New York, 1930.

approach. In 1866 a force about a thousand strong was dispatched against him from Lamu, but failed to dislodge him. In the following year a German explorer, named Richard Brenner, visited Simba, who, anticipating further hostilities on the part of the Sultan of Zanzibar, requested Brenner to use his good offices with the Prussian Government to procure for him a treaty of friendship and a Prussian Protectorate. Brenner conveyed Simba's wishes to Berlin, but the paper remained pigeon-holed in the Wilhelmstrasse for another eighteen years.[1] Nevertheless, Simba was able to maintain his hold upon Witu and to gain a considerable influence over the neighbouring Somali and African tribes. Though Simba was cut off from access to the coast, Witu was something more than a cave of Adullam. It is difficult to estimate the extent of his influence on the hinterland, but at Witu and in the country adjacent thereto he brooked no equal or overlord.

In 1862 the attention of Majid ibn Said was drawn to the landlocked harbour, which is known today as Dar es Salaam. He was quick to realize its potential value as an outlet for trade with the interior. In 1866 he started to build on the banks of the harbour and announced his intention of transferring his capital there. Majid, however, died in 1870 before his plans had reached completion.

Majid's reign witnessed further commercial growth. Between 1859 and 1871 the tonnage of American and European shipping using Zanzibar harbour rose from 18,877 to 27,626. The growth of British shipping was most noticeable; in 1859 it had been insignificant, in 1871 it was larger than that of any other nation. By this latter date British and British Indian trade was twice that of Germany and almost three times that of the United States. But Germany's trade with Zanzibar had doubled, and the tonnage of her shipping almost doubled, in the same period.

The pattern of trade remained the same, growing in value and quantity. In 1867 exports of ivory were worth $663,500, of cloves $321,000, and exports of gum-copal, cowries, and sesamum seed were each valued at about $100,000. Cloth and beads for the inland trade were still the largest imports.[2]

---

[1] Coupland, *Exploitation*, p. 114.
[2] Ibid., p. 77, based mainly on Kirk's *Administrative Report* for 1870, P.P. 1872, liv. 783, and P.P. 1867, lxvii. 299.

Accurate figures for the slave-trade are almost impossible to compile, but the explorers were all agreed that the Arab slavers were expanding the area of their activities, and bringing more and more slaves to the coast, despite British naval patrols and the treaties. Kilwa maintained its predominance as a slave-collecting centre, and by this time its hinterland was showing signs of the appalling effects of the trade. Consul Rigby was told by Indian residents of Kilwa that areas ten or twelve days' march inland, once thickly populated, were now ruined and empty of people.[1] But this did not dry up the flow of slaves. Political troubles in the interior could have their effect on the coastal trade; slaves poured into Pangani for a period coinciding with the disintegration of Usambara after Kimweri's death, before his son re-established control.[2] This expanding slave-trade contributed in no small measure to the sultan's finances. Consul Kirk calculated that the revenue from taxes on the slave-trade in 1871 amounted to £20,000,[3] about twice the yield of 1845.

Majid left no male heir surviving him and the only one of his half-brothers residing in Zanzibar who had attained majority was his former rival Barghash. Since his return from exile Barghash had given no further trouble. Experience had taught him that the Harthi and other malcontents only wanted him as a mere cat's-paw, caring no more for him than they did for Majid and being ready to discard him when he was no longer useful to them or no longer ready to fall in with their wishes. Moreover, whilst in exile at Bombay he had gained a wider outlook on life. He had occupied his time in reading and study and had come back to Zanzibar a well-read man and far better informed as regards social and political problems than any of his compatriots resident in Zanzibar. He was therefore the obvious choice as Majid's successor. When it was clear that Majid's days were drawing rapidly to a close, Barghash had been approached by H. A. Churchill, the British consul, and had assured him that, in the event of his becoming sultan, he would not only observe all the treaties entered into by his two

[1] Coupland, *Exploitation*, p. 140. But Ngoni incursions may also have been a factor. See Chapter VI, pp. 208–11, above.
[2] Coupland, *Exploitation*, pp. 137–8.
[3] *Report of the East African Slave Trade Committee*, P.P., 1871, XII, questions 99–104.

predecessors and all laws enacted by them for the suppression of the slave-trade, but would also undertake to carry out the measures which the British government had proposed to Majid. The result was that on Majid's death he was recognized as Sultan of Zanzibar without incident.

Churchill had acquiesced in Barghash's succession without prior consultation with the Bombay government and was rebuked for doing so. In actual fact the situation called for prompt action by the man on the spot and the delay caused by prior reference to Bombay might have led to a repetition of the troubles occasioned on the death of Said ibn Sultan. Nevertheless, it looked at one time as if Churchill had merited a rebuke. No sooner was Barghash on the throne than he proceeded to disavow every one of the promises which he had made to Churchill during Majid's lifetime.[1] Ill health forced Churchill to leave Zanzibar two months after Barghash's succession and the British consulate was left in charge of Dr. Kirk, the Agency surgeon. Fortunately Kirk had known Barghash for some time previously and their relations had been of the friendliest. Back in Oman, Barghash's half-brother Turki had recently seized the throne. It was clear that he had his eyes on Zanzibar. Barghash soon realized that, if he was to hold his own in East Africa, he must look, like Majid before him, to the British government for support. Within three months of taking over the duties of consul, Kirk was able to report that there had been a complete change in Barghash's relations with the British consulate.[2] Kirk finally left Zanzibar in 1886. During their sixteen years of close association the same friendly relations and the same mutual confidence existed between him and Barghash as had existed in the latter days of Said ibn Sultan and Hamerton.

Within eighteen months of his accession Barghash was confronted by a major disaster. Though Zanzibar lies well to the north of the normal cyclonic zone, on 15 April 1872 a hurricane devastated the island. All the ships of the sultan's fleet, which were in the harbour, became wrecks and some two-thirds of the clove- and coconut-trees in the island were destroyed. The damage to his ships deprived the sultan for the time being of his

---

[1] Coupland, *Exploitation*, pp. 87–95, quotes most of the documents in the Zanzibar Archives relating to Barghash's pledges.
[2] Kirk to Bombay Government, 9 Mar. 1871, quoted ibid., pp. 99–100.

most effective means of asserting his authority over the seaports on the mainland. As new trees would not bear fruit for many years, the losses to landowners were incalculable. Fortunately only the fringe of the hurricane caught the southern shores of Pemba and the havoc wrought in Zanzibar led to increased planting of cloves in the other island.[1]

Nine months later a special mission, headed by Sir Bartle Frere, arrived in Zanzibar to negotiate a more stringent treaty for the suppression of the slave-trade. Barghash was placed in a difficult position. The rehabilitation of the clove plantations in Zanzibar and the opening up of fresh plantations in Pemba had called for an increased demand for labour and the only means of supplying that demand was by the recruitment of slaves. This and the extremely conservative attitude of many Arabs at once aroused the strong hostility of most of his compatriots. Barghash felt that he could not run counter to such opposition without risking the loss of his throne. Eventually Frere had to leave without achieving his purpose. But Kirk was left behind with instructions to inform Barghash that, if he did not agree to the treaty which had been submitted to him, Zanzibar would be subjected to a blockade by the British Navy. It was an extremely delicate situation calling for very careful handling, but by the exercise of supreme tact on Kirk's part Barghash was induced to sign.[2]

The treaty prohibited the export of slaves from the main continent whether destined for any other part of the sultan's dominions or for foreign parts. Under this treaty the sultan engaged to close all public slave-markets in his dominions. He further bound himself 'to the best of his ability to make an effectual arrangement throughout his dominions to prohibit and abolish' the export of slaves from his dominions. Thereafter he loyally carried out in so far as he was able everything he had undertaken. Within twenty-four hours of his signature to the treaty the public slave-market in Zanzibar was closed and messages had gone out to all his officers on the mainland to enforce the provisions of the treaty.[3]

---

[1] Ibid., pp. 56–57; Crofton, *The Old Consulate*, p. 52.
[2] For the full story of the events leading to the treaty of 1873 see Coupland, *Exploitation*, pp. 182–216; P.P. 1873, lxi, *Correspondence respecting Sir Bartle Frere's Mission to the East Coast of Africa*.
[3] *British and Foreign State Papers*, lxi, 1872–3, 173–4. Besides the severe loss to the

In 1875 Sayyid Barghash paid a visit to England and was the guest of Queen Victoria at Windsor.[1] Shortly after his return he was persuaded by Kirk to issue two proclamations to prevent evasions of the treaty. It had been discovered that many slave-traders, especially in the vicinity of Kilwa, were collecting slaves near the coast and then marching them up the coastal belt to sell in batches at other towns along the line of march.[2] Many of the slaves who were thus disposed of were subsequently smuggled in canoes and other small vessels into Zanzibar and Pemba. By the two proclamations which Barghash issued in 1876 the conveyance of slaves by land was prohibited and slave caravans were forbidden to approach the coast from the interior.[3]

Kirk quickly realized that these proclamations would be no more than dead letters unless the sultan had an efficient and reliable force to ensure that the law was obeyed. On his advice Barghash decided to raise a small army to be trained under the supervision of a European officer. An ideal man was found for the post in the person of Lieutenant William Lloyd Mathews[4] of H.M.S. *London*, who was destined to serve under five successive Sultans of Zanzibar until his death in 1901. Mathews began drilling his first recruits in 1877 and within a remarkably short time had raised a force which proved a useful supplement on land to the work of the British naval patrols at sea.[5]

The status of slavery was not to be abolished until nine years after Barghash's death. Until this was brought about, surreptitious dealing in the slave-traffic was bound to continue. Nevertheless, though execution of the sentence might be postponed for some years to come, and although slaving in the interior might persist and even increase, so far as the coast was concerned when Sayyid Barghash signed the Treaty of 1873, he signed the death warrant of both the status and the traffic.

The proclamations of 1876, and their enforcement, set the

Zanzibar Arabs, the sultan himself lost about a quarter of his total revenue, which had come from the tax on slave imports.

[1] Lyne, *Zanzibar in Contemporary Times*, pp. 92–95; Pearce, *Zanzibar*, pp. 265–6.
[2] J. F. Elton, *Travels and Researches among the Lakes and Mountains of Eastern and Central Africa*, London, 1879, pp. 72–106.
[3] For text see *British and Foreign State Papers*, lxvii, 1875–6, pp. 455–6; the Arabic texts can be found in FO 84/1453, enclosed in Kirk to Derby, 28 Apr. 1876.
[4] R. N. Lyne, *An Apostle of Empire, being the Life of Sir Lloyd William Mathews*, K.C.M.G., 1936.
[5] Coupland, *Exploitation*, pp. 240–3; Lyne, *Apostle of Empire*, pp. 40–48.

seal on the future commercial development of the coast. Riots in 1876 in Mombasa were provoked by the proclamations, though directed against the freed-slave work of the missions. But the sharpest effects were seen at Kilwa, against whose slave-trade the proclamations were openly directed. Kilwa at this time made £120,000 a year in profits from the trade; 6,000 slaves were hidden near the coast, and about £40,000 worth of goods were in stock for future trade. But Barghash, supported by Kirk, determined to enforce his will. Two hundred more Baluchi soldiers were sent by steamer to put down any opposition; these were supported by a British warship, and Kirk made a personal visit. In 1877 slaves were still being smuggled out of Kilwa, but the Indians were refusing credit for the trade. When the governor's participation was exposed, and the sultan threw him into prison, the trade virtually collapsed.[1] Within two years Kilwa had found compensation in the development of a rubber trade exporting over 500 tons at £250 a ton, and by exploiting more intensively the local supplies of gum-copal, sesamum, and cereals.[2]

The strangulation of the slave-trade to the coast after 1876, following upon the hurricane of 1872, might have been expected to cause serious commercial depression. Yet the economy passed through the crisis with great resilience. The hurricane seems to have had a worse effect on trade than the measures against slave shipment. From an initial decline after 1872, growth began again after 1875, and by 1879 the value of trade had reached £2 million. Customs revenue rose from £65,000 in 1869 to over £100,000 per annum after 1876. Though the pattern of imports remained much the same, exports were rapidly changing. Ivory declined to third place, with a value of £160,000, on the list of exports for 1879. Rubber exports of £250,000 headed the list, with cloves worth £170,000 in second place. The growth of British shipping at Zanzibar was very striking. Already predominant in 1871, the tonnage of British shipping increased sevenfold in the next eight years, whilst that of other nations declined or remained stationary.[3] The opening of the Suez

[1] Coupland, *Exploitation*, pp. 225–31.
[2] See the very interesting report on Kilwa and environs by Holmwood enclosed in Kirk to Salisbury, 6 Mar. 1880, in FO 84/1574.
[3] Coupland, *Exploitation*, pp. 319–22.

Canal in 1869 brought Zanzibar 2,000 miles closer by sea to Britain, and the British were quick to exploit this advantage. In 1872 Sir William Mackinnon's British India Steam Navigation Company began a monthly mail service to Zanzibar with a subsidy from the British government, and in 1877 the Eastern Telegraph Company joined Zanzibar to Aden by cable.[1]

Another indication of the growing connexions between Zanzibar and Europe was the increasing activity of Christian missionaries. The first contacts in the nineteenth century were made almost by accident. Johann Krapf, prevented by Ethiopia from working among the Galla, determined to reach the same people by way of Mombasa. Under the aegis of the Church Missionary Society he arrived in Zanzibar in 1844. Said gave him a most friendly reception and, when he expressed a wish to start work at Mombasa, he sent him on his way with a letter of introduction to his officers and subjects, describing him as 'Dr. Krapf, the German, a good man who wishes to convert the world to God. Behave well to him and be everywhere serviceable to him.'[2] Krapf and his compatriot and colleague, Johann Rebmann, were to penetrate farther inland and to be the first Europeans to set eyes on Mounts Kenya and Kilimanjaro, but their missionary labours were confined to the vicinity of Rabai on the mainland near Mombasa. Krapf left East Africa owing to ill health in 1853, and in 1873 Sir Bartle Frere found a solitary and blind Rebmann at Rabai with about half a dozen converts. A year later, largely through Frere's influence, the Church Missionary Society acquired a large property on the mainland just opposite to Mombasa for the establishment of a settlement for freed slaves.[3]

The next British mission in the field was that of the United Methodist Free Churches. The first party arrived under the leadership of Krapf in 1862 and was given a very friendly reception by Sayyid Majid. On the advice of Krapf their first mission station was established at Ribe on the mainland near Mombasa. Two members of the society—Thomas Wakefield

[1] Ingrams, *Zanzibar*, p. 166.
[2] J. L. Krapf, *Travels, Researches and Missionary Labours during an 18 Years' Residence in Eastern Africa*, London, 1860, p. 127; R. Oliver, *The Missionary Factor in East Africa*, 1952, pp. 5–9.
[3] Coupland, *Exploitation*, pp. 359–60.

and Charles New—made valuable contributions to geographical knowledge as the result of their journeys into the interior.[1]

The Methodists were quickly followed by the Universities' Mission to Central Africa. After the failure of the society to establish a permanent station in the Shire Highlands, Bishop Tozer withdrew the mission to Zanzibar with the intention of using the island as a base for the renewal of missionary labours on the mainland. Bishop Tozer and his party reached Zanzibar in 1864. Like the Methodists, they met with a very friendly reception from Sayyid Majid. In 1868 the attack was renewed on the mainland by planting a station at Magila in Usambara. Tozer resigned the bishopric in 1873 and was succeeded by Bishop Edward Steere. Meanwhile work on the mainland was gradually extended. Further stations were opened in Usambara and in 1875 Bishop Steere made a journey from Lindi into the Nyasa country. Two years later in preparation for a further advance into those regions he opened a station at Masasi about a hundred miles inland from Lindi and another at Newala about fifty miles south of Masasi. In 1878 Herbert Clarke at Newala took on the role of peacemaker between the Maviti, one of the Ngoni sections, and the Makua. He induced the two warring tribes to agree to abstain from raiding each other for slaves and to submit their disputes to a representative of the Sultan of Zanzibar and not to interfere with the construction of a road which Sayyid Barghash was contemplating building from the coast to Newala. In 1882 Masasi was attacked by a band of raiders from the south and a number of Christian converts were enslaved. Thereafter Masasi was temporarily abandoned as a mission station.[1]

In point of time the second mission to arrive in the dominions of the Sultan of Zanzibar came from Réunion in 1860. Monseigneur de Maupoint, bishop of that island, sent his Vicar General, the Abbé Fava, to Zanzibar with two secular priests and six sisters of the order of Filles de Marie to found a mission in Zanzibar itself.

Soon after the arrival of the missionaries they were reinforced from Réunion by two Fathers of the Congregation of the Holy

---

[1] Ibid., p. 361; Oliver, *Missionary Factor*, p. 8 and p. 54.
[2] Ibid., pp. 12–15, 38–39; G. H. Wilson, *History of the Universities' Mission to Central Africa*, 1936, pp. 56–58.

Ghost. One of these was Antoine Horner, who had been in charge of a leper settlement in Réunion. Very shortly after Horner's arrival Bishop de Maupoint found that his limited resources both of personnel and of money would seriously handicap the mission's work and decided to hand over the management to the Missions of the Congregation of the Holy Ghost and the Sacred Heart. Accordingly the Abbé Fava and the secular priests withdrew in 1863 and Horner was appointed Vice-Prefect Apostolic. Much of the missions' labours were among liberated slaves and Horner was quick to realize that Zanzibar itself was too restricted a place to allow of the expansion of that work to its full extent. Having acquired a large area of land on the outskirts of Bagamoyo on the mainland, Horner in 1868 established there an agricultural colony for liberated slaves. During the next five years the organization of the mission on a wider scale began to take effect. As their converts grew up and became self-supporting, they were settled in charge of a missionary in self-contained Christian villages farther inland. In 1877 Father Horner had reached the Uluguru Mountains and had established one such station about a hundred miles inland at Mhonda near Morogoro.[2]

Christian missionaries in Zanzibar and on the coast were working in a peculiar, and sometimes difficult, environment. From the first they had looked to the interior of the continent as their field of work, but circumstances dictated that a considerable amount of their energies should be expended on mission stations and settlements on Zanzibar Island and the coastal region. In early days these settlements had been essential as bases for later moves inland, and they continued to exist as centres for the training of the later evangelists of the interior. Placed as they were in the centre of an Islamic slave-trading and slave-holding society, it was natural that the missions should provoke tensions and sometimes active resentment from the Muslim Arabs and Swahili. Christianity, whether Catholic or Protestant, was now deeply convinced of the sinfulness of slavery; and the missionaries, whatever their instructions, or however well they understood political realities, could not help at times reacting strongly against it. Missionaries assumed the

---

[1] Oliver, *Missionary Factor*, p. 18, pp. 21–23; Coupland, *Exploitation*, pp. 33–34, pp. 355–7.

task of caring for and educating the slaves freed by British naval patrols. It was a natural extension of this work to take responsibility for refugees from inland wars, and for the misfits and unwanted of tribal society near the coast, who in earlier times would almost certainly have become enslaved. To a slave-holding society such activities could only appear disruptive and subversive, designed to upset the proper ranks and orders of the community.

Runaway slaves were a particularly acute problem for the Christian missionaries. To harbour, conceal, and care for runaways was clearly illegal and a violation of property rights by Zanzibar law, yet to the missionary it often seemed a positive duty to assist runaways, which it was sinful to avoid. Moreover, the mission stations soon became known to slaves as places where runaways might find help and sympathy. When a fugitive arrived who had been cruelly treated, few missionaries could turn him away, still less denounce him. In 1880, during Ramadhan, the settlement at Freretown was openly attacked by Mombasa Arabs and Swahili, and the missionaries defended themselves with arms. It was later admitted that the missionaries possessed a white flag, with 'Freedom' written in Swahili characters upon it, which was to have been used to provoke a slave-rising. In 1888 an inquiry by an official of the Imperial British East Africa Company revealed that over 1,400 runaway slaves were harboured at the Mombasa freed-slave centres.[1]

Sayyid Barghash continued the process of expansion into the coastal regions of the mainland, particularly after 1875. The attempt of the Egyptian Government, moving across from the Lakes region, to obtain a foothold on the East African coast, had exposed to Barghash the vulnerability of Zanzibar's position. British diplomatic help in removing the Egyptian threat, Barghash's visit to England, and British assistance in equipping and training Mathews's forces convinced the sultan that the modernization and expansion of his dominions could best be achieved in alliance with Britain, and with the British commercial and humanitarian group headed by Sir William Mackinnon.

[1] Oliver, *Missionary Factor*, pp. 55–56. The above analysis of the missionaries' situation is largely based on Oliver's discussion, pp. 15–26 and pp. 50–56. See also Groves, *The Planting of Christianity in Africa*, ii. 283–90.

The most serious local challenger to Zanzibar authority on the coast was Mbarak, a younger son of Rashid ibn Salim, the last independent Mazrui ruler of Mombasa. After the deportation of the leading members of the Mazrui family from Mombasa in 1837, Sayyid Said had nevertheless permitted two members of the family to remain as local Walis in Gazi and Takaungu. Mbarak had taken up residence at Gazi. The long-standing feud between his family and that of the rulers of Zanzibar, intensified by the recollection of his own father's fate, filled him with a burning sense of wrong and fierce desire to recover what his ancestors had lost. In 1850, when he had only just reached man's estate, he raised a force of his own and proceeded to Takaungu, where he attacked and expelled Rashid ibn Hamis, a member of the younger branch of the family who had just been appointed as Wali of that place. Said ibn Sultan speedily dispatched troops to the aid of his nominee and Mbarak was driven back to Gazi. He was, however, allowed to remain there unmolested and, on the death in 1860 of his kinsman, Abdullah ibn Khamis, was appointed Wali of Gazi.

Mbarak did not remain quiet for long. Complaints reached Majid ibn Said that he was abusing his powers and his subsidy was stopped. It was also found necessary to send a force from Mombasa to attack him, but the death in 1870 of Sayyid Majid led to a suspension of hostilities. Majid's successor Barghash made peace with Mbarak. But Mbarak did not mend his ways. In 1872 he was once more in open rebellion, but eventually surrendered and swore allegiance to the Sultan of Zanzibar as his overlord. A year later he was once again in revolt. On this occasion his stronghold at Mwele was stormed and he fled into the interior, where for the next eight years he remained in a state of outlawry 'in the bush'.[1]

In 1882, however, he managed to return. In February of that year he succeeded in raising a formidable force of two thousand men recruited from the local tribes, including the Masai. He then made a sudden raid on the coast town of Vanga. Sayyid Barghash thereupon sent part of his recently raised army under their trainer and leader, Lloyd William Mathews, to deal with the situation. After several days of close investment Mathews took Mbarak's stockaded post at Mwele

[1] Coupland, *Exploitation*, pp. 248-9.

by assault. Mbarak, however, succeeded in cutting his way through his assailants and making his escape. Eventually he surrendered to Lloyd Mathews, who promised him his life and obtained from him a promise that he would never again engage in hostilities against the Sultan of Zanzibar. Thereafter he was allowed to return to Gazi and in the meantime for a few years the land enjoyed comparative peace.[1]

During the period of the decline of the fortunes of the Mazrui, Kimweri, a local chief in Usambara, began to expand the Kingdom of Vuga, which extended from the slopes of Kilimanjaro to the sea coast between Vanga and Pangani. Having stepped into the shoes of the Mazrui, Said ibn Sultan also claimed suzerainty over this strip of the coast. His and Kimweri's conflicting claims were settled by a compromise. Tanga and its environs remained under the administration of a governor appointed by the Sultan of Zanzibar, but elsewhere along the coast Kimweri appointed his own officers, whose appointments had, however, to be confirmed by the Sultan of Zanzibar. In 1852 or 1853 the caravan route from Usambara to the coast was interrupted by a raid of the Zigua. Kimweri found himself unable to cope with the situation. It was eventually agreed that the Sultan of Zanzibar should undertake the protection of the coastal end of the trade route by establishing garrisons of his own. Accordingly one such post was established about ten miles up the Pangani river at Chogwe and another some ten miles farther to the north-west on a hill called Tongwa. In his latter years Kimweri's hold on the outlying portions of his dominions appears to have weakened. After his death in about 1870 his kingdom declined in power during the strife and anarchy which arose from disputes regarding his successor. These disputes brought the previous condominium to an end and thereafter the Sultans of Zanzibar acquired the sole control over this coastal strip.[2]

Barghash displayed little interest in Majid's project to develop Dar es Salaam. The buildings which had been erected, or were being erected, soon fell into disrepair through neglect, and trade which had begun to concentrate there was diverted to other

[1] Ibid., pp. 253-5.
[2] Coupland, *Invaders*, pp. 345-60; Krapf, *Travels*, contains a great deal of material descriptive of Kimweri's kingdom.

places along the coast. There was a slight revival of prosperity in 1877 when Mackinnon and Buxton inaugurated work on the construction of a road for wheeled traffic between Dar es Salaam and Lake Nyasa. But this work proceeded very slowly and in 1881 was eventually abandoned after the road had reached its seventy-third mile. Thereafter Dar es Salaam relapsed once more into its former state of decay. At the end of Sayyid Barghash's reign Bagamoyo, situated some thirty-five miles distant from Zanzibar, with its well-protected, broad, sandy shore, which was uncovered at low tide and was suitable for beaching dhows, had become the principal starting-point for caravans proceeding into the interior.

It is difficult to give any estimate as to the extent of the influence of the Sultans of Zanzibar beyond the immediate vicinity of their garrisons and customs posts on the coast. It varied considerably at different points. On the Somali coast it would appear never to have reached any more than a few miles beyond their walled towns. On the mainland opposite to the Bajun Islands, whilst the Sultans of Zanzibar had effective control of a narrow strip of the coastal belt, Simba exercised quasi-independent rule over the inhabitants farther inland.

Between Ras Ngomeni and the Pangani river the Sultans of Zanzibar had succeeded to the former dominions of the Mazrui, who had considerable influence over the tribes of the hinterland. Despite the sporadic insurrections of Mbarak ibn Rashid, who more than once induced large numbers of the local tribes to make common cause with him, the Sultans of Zanzibar had much the same influence over the local tribes as had their predecessors.

As already mentioned, Said ibn Sultan had military posts more than ten miles inland in the region of the Pangani river. After the Mackinnon Road had been pushed some distance inland, a military post was established some miles inland at a place which is referred to in British official correspondence as 'the Bungalow'; but it appears to have had very little military value. In 1873 a British naval officer reported that the sultan's Baluchi soldiers were afraid to penetrate more than thirty miles into the interior and that their relations with the local inhabitants 'savour somewhat of armed neutrality'.

In 1880 an expedition sent out by Leopold of Belgium's

International Association was attacked and its leaders murdered in the Nyamwezi country. The incident was followed by a demand from the Belgian authorities that a punitive expedition should be sent by Sayyid Barghash against the assailants. As Sir John Kirk said at the time, such an expedition was impossible and beyond the sultan's power, but he managed to persuade Barghash to establish a fortified post at Mamboya, about 120 miles from Bagamoyo on the trade route to Tabora.[1] The sultan's troops remained at Mamboya until 1885, when they undertook a punitive expedition into Usagara against certain tribesmen who had attacked a party led by two Germans bent on treaty-making.

South of the Rufiji delta the hinterland suffered for many years from constant raids of various sections of the Ngoni, known locally as the Maviti and Magwangwara.[2] In 1868 a band of these marauders appeared in the neighbourhood of Kilwa Kivinje and blocked its trade with the interior. When a force comprising troops of the sultan and some armed slaves attempted to dislodge them, they were ignominiously put to flight. When in 1873 Sir Bartle Frere visited Lindi, he learnt that the outskirts of the town had been more than once raided by these marauders with impunity.[3] In about 1878, following the example of Buxton and Mackinnon at Dar es Salaam, Sayyid Barghash contemplated the construction of a road from the coast to the U.M.C.A. mission station at Newala, but, unlike the Buxton–Mackinnon project, that of the sultan was stillborn. In 1884 Lieutenant Charles Smith, vice-consul at Kilwa, made a journey through the interior behind Kilwa and Lindi and reported that in the belt of country extending from the immediate neighbourhood of Kilwa to the upper reaches of the Ruvuma river, a distance of 250 miles, there were absolutely no inhabitants, though the ruins of more than one village indicated that the district had been fairly well populated at a recent date. The abandonment of this large area was largely attributable to an extensive raid carried out by the Magwangwara in about 1866 and partly to a wholesale migration to places of greater safety due to that and other raids.

[1] Coupland, *Exploitation*, pp. 263–6. For the background of this incident see Chapter VIII, pp. 290–1, below.
[2] See Chapter VI, pp. 208–11, above.   [3] Coupland, *Exploitation*, p. 247.

In 1882 the same tribe raided Masasi, where the U.M.C.A. had established a station five years before.[1]

The evidence given to the Delimitation Commissioners[2] in 1886 shows that by this time Sayyid Barghash had 120 customs posts established on the coast between Warsheikh and Tungi Bay in addition to garrisons at all the principal ports. Whilst the posts on the Somali coast were somewhat few and far between, there were over ninety separated from each other by short intervals along the 600-mile stretch of coast between the Tana river and Tungi Bay.

If the Sultan of Zanzibar had appointed a Kadhi to officiate in a garrison town, the local inhabitants living in close vicinity to the town would bring their cases before him. If there was no Kadhi in the place, they would take their disputes to the Liwali or Governor. Sometimes Africans living at a distance of several days' journey from the town would likewise submit their cases to the arbitration of the Kadhi or Liwali, but in such cases the jurisdiction of the Kadhi or Liwali would appear to have been only with the consent of all the parties to the dispute.

Sir John Kirk's correspondence with the Foreign Office shows that he had hopes that the Sultan of Zanzibar would one day be in a position to bring even such far distant lands as Unyamwezi under his effective administration, but the Foreign Office was not ready to back Kirk's plans for the pacification of the interior as far as he wished. In 1881 Kirk was told that he could continue to use his discretion in the matter but that he 'should avoid committing Her Majesty's Government to any policy entailing a definition of the extent of the Sultan's territory inland'. If Kirk's plans were to be carried out, they required the active co-operation of Sayyid Barghash. Barghash indeed was not unwilling; at one time he had sent a force of several thousand Baluchi soldiers to Unyamwezi to try to restore peace there and to keep open the trade route through Tabora. This venture proved costly and unsuccessful, and he subsequently—with Kirk's approval—sought instead to secure the pacification of the interior by giving support and recognition to the powerful Nyamwezi chief Mirambo. The failure of this policy also and the reaction of Barghash in embarking on a yet more ambitious scheme of political expansion are traced in the following

[1] Coupland, *Exploitation*, p. 362.   [2] Ibid., ch. xix.

chapter. In fact both Barghash's scheme and Kirk's more modest plan of consolidation were doomed to failure. It was not only that ill health dogged Barghash in his latter years. If the costs of administering the country to the east of the Lakes were not to be met, as Barghash came to hope, by the even further expansion which would secure to Zanzibar the ivory of the Congo, they must be met by increased taxation at the coast. As Barghash knew only too well, the imposition of increased taxes would have been most unpopular with his subjects and the revenue might have been exceedingly difficult to collect. Moreover finance was not the only problem which the sultan had to face. If the hinterland was to be not only secured, but efficiently and effectively administered, the tasks of securing and administering it would have to be entrusted to men upon whose initiative, loyalty, probity, and tact in dealing with the local inhabitants Barghash could confidently rely. Such qualities would in any case have been severely tested in the strenuous and troubled period of international rivalry that was impending, and a civil service combining them was not to be created in a day. Preparations for such plans as those envisaged by Kirk and by Barghash alike required time, and the sands of time were running out.

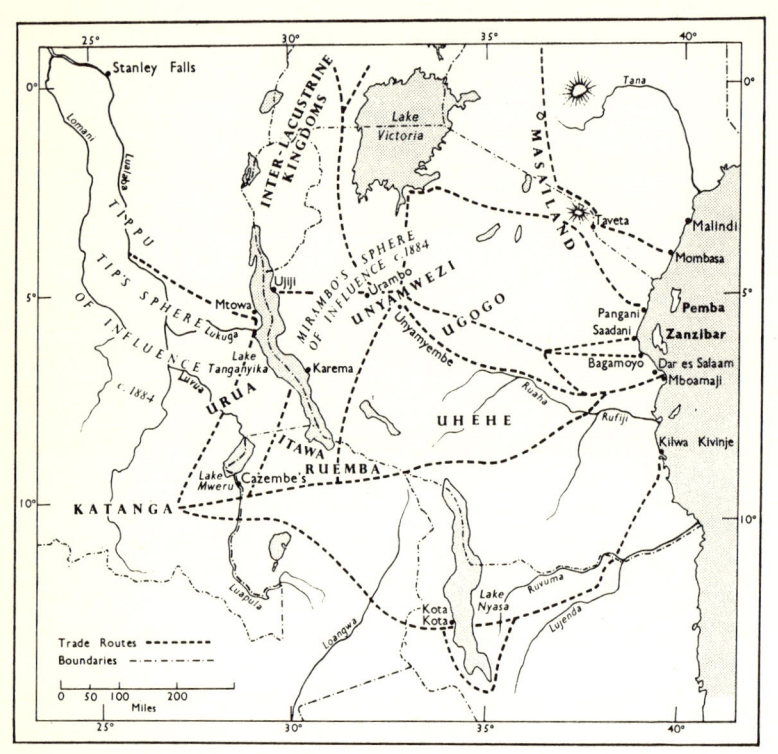

MAP 12. Approximate lines of the main trade routes to the interior based on Zanzibar, 1840–1884

VIII

THE SOUTHERN SECTION OF THE INTERIOR
1840-84

ALISON SMITH

UP to the end of the first quarter of the nineteenth century the hinterland of East Central Africa, within the rim of its interior plateau, was still much as it had been for centuries. By contrast with the Muslim coast it was a pagan world: a world not of cities, connected with one another and with the other lands bounding the Indian Ocean by a sea-borne trade, but of self-sufficient villages, where even the largest centres of population, linked only by the porterage of human head or shoulder, hardly knew anything approaching an exchange economy. It contrasted also with the lands farther north. There a small number of peoples, characterized by a more massive sense of tribal coherence, were pre-eminent, and were recurrently engaged in hostilities with their neighbours; whereas this was a very heterogeneous, and also a more static world. In the 1830's, it is true, the pastoral and predatory Masai made an unsuccessful bid to extend their grazing lands southward;[1] and within a few years the irruption of the Ngoni from the south was to set in motion considerable changes in political and economic structures, as was the quickening penetration of influences from the coast: but during the previous 200 years, movements of population seem to have been slight and political evolution undramatic. Thus the picture is one very diverse, very indeterminate.

Yet, perhaps because their life was reasonably stable, the lot of the inhabitants of this region, although they lacked the amenities of the coast-dweller, was not necessarily one to be pitied. Their material culture was that of the Iron Age. They used iron-tipped weapons to hunt with, and often iron axes, hoes, and knives to cultivate the sorghum, root crops, and pulses

[1] J. Thomson, *Through Masai Land*, 1885, p. 414.

which were the mainstay of their diet. They kept goats and fowls, and in those areas—more extensive than today—where the tsetse permitted it, they herded cattle. Pastoralism was the dominant economic pursuit only of a minority. Most were agriculturalists, and their economy was a subsistence one, uncomplicated by writing, unaided by the wheel. But the explorer Burton, comparing their condition with that of other peasantries with whom his wide travels had made him familiar, found village life in East Africa a tolerably comfortable affair:

> The African is in these regions superior in comforts, better dressed, fed and lodged and less worked than the unhappy Ryot of British India. His condition, where the slave trade is slack, may, indeed, be compared advantageously with that of the peasantry in some of the richest of European countries.

His wants were simple, revolving principally on food: he found leisure, aided by draughts of *pombe* or some other home-brewed intoxicant, for plenty of sociability and hospitality.[1]

Finally, over most of its extent this part of the African interior was a sparsely peopled world. Across the wide central belt of what is now Tanganyika, rainfall is too low to mask the essential poverty of the soil, and this arid land of scrub and granite can support only the lightest population. There are, however, some more favoured zones, where the rainfall is more bountiful or where volcanic activity has enriched the earth, or both: around and to the south of the shores of Lake Victoria and away along the foothills of the mountains of Rwanda-Urundi; on the slopes of the Kilimanjaro range; in the neighbourhood of Lake Nyasa. Here the first explorers and missionaries came upon more or less thickly settled populations.

In one or two localities of high natural or artificial fertility, the land was intensively cultivated, but over the great bulk of the country the classic slash-and-burn methods of agriculture necessarily prevailed. In some areas, notably among the Gogo of the arid central zone and their neighbours to the south, pastoralism outweighed agriculture in importance; and in a few, as in the forested hills inhabited by the Safwa and the Kinga, even the Bantu-speaking inhabitants were primarily hunters, rather than farmers or herdsmen.

[1] R. F. Burton, *The Lake Regions of Central Africa*, 2 vols., 1860, ii. 278–80.

Along most of the southern part of the Sultan of Zanzibar's dominions the inland concentrations of population were separated by many days' march of desert and scrub from the coast: but between the Indian Ocean and Lake Nyasa there was no such absolute break. The traveller bound for the lake from Kilwa or its vicinity might pass all the way through fairly fertile and inhabited country, the dry lightly forested land near the coast gradually giving place to stream-watered uplands as he climbed inland. Here, by about 1840, the main peoples he would have encountered would have been the Makonde, and the Yao and Makua. The Yao and Makua were folk of small groups, loosely amalgamated through headmen under petty independent chiefs. After the middle of the nineteenth century, with the advent of more dangerous times, large settlements became common; but it seems that in earlier days the villages were always small ones. Their inhabitants, matrilineal and matrilocal, were cultivators first and foremost, with 'gardens' of maize, cassava, sorghum, and tobacco, but at least in Livingstone's day they had some cattle also, while their own tradition stresses the importance of hunting—'they set traps for small game, and dug game pits for the larger kinds; hunting with dogs and nets was in full swing. All this was the work of the men....'[1] They wove their own bark-cloth and they smelted iron—Livingstone everywhere observed, even in districts no longer inhabited, the signs of an iron-working population.

The picture of Yao life as it was lived in the earlier years of the nineteenth century is, moreover, so far as it can be reconstructed, an essentially peaceable one. The U.M.C.A. missionary, Archdeacon Johnson, recounts meeting an old man whose

face would light up as he described the good old days when he was a boy, when each of the many streams we had crossed, and were to cross, was the site of a village, when strangers could pass from one end of the country to another, and Indian corn, rice, and goats abounded: he dwelt on the social life, the free hospitality, the ready transit, the giving food and receiving news in turn that then prevailed....[2]

[1] Y. B. Abdallah, *The Yaos* (trans. and ed. M. Sanderson), 1919, p. 4.
[2] W. P. Johnson, 'Seven years' travels in the region east of Lake Nyasa' in *Proc. R.G.S.*, n.s. 6, 1884, p. 518. It is true of course that old men's stories such as this need to be taken with a grain of salt.

A similar description comes from Yao written tradition:

> In the old times, long long ago, in their old homes, the Yao were in accord and united. If a quarrel arose they used to fight without rancour, avoiding bloodshed. . . . If strangers came to a village, would they have to pay for their food? No, it was bestowed on them gratis; directly a man heard that a stranger was at his door, he would rejoice and say 'Verily, I have the plant of hospitality at my door, bringing guests'.[1]

The Yao were at this stage, it may be surmised, not very obviously to be distinguished from their Makonde and Makua neighbours, and all three groups were considerably to the south of their present positions; the Yao had not as yet spread beyond the small triangle of land between the Lujenda and Ruvuma rivers. Within the next fifty years they were, even more than the Ngoni invaders or their own Makua kindred, to dominate the history of the area between Nyasa and the coast.

Crossing the watershed and descending the shores of Lake Nyasa, one enters a potentially more productive region. This is less perhaps because the lakeside soil is in general particularly fertile than because the resources of the water have been available to supplement those of the land. At all events, the first Europeans to record their visit to the lake were chiefly struck by the numbers of the population they found along its margins:

> Never before in Africa have we seen anything like the dense population on the shores of Lake Nyasa. In the southern part, there was an almost unbroken chain of villages. On the beach of well-nigh every little sandy bay, dark crowds were standing, gazing at the novel sight of a boat under sail; and wherever we landed we were surrounded in a few seconds by hundreds of men, women, and children.[2]

In origin, these Maravi tribes—the Nyanja, Tumbuka, Chewa, Nsenga—inhabiting the greater part of the lake shore were related to those just described farther east. Like them, they were matrilineal and matrilocal, had elaborate initiation rituals for girls as well as boys, and lived in villages, small groups of which owned a loose allegiance to independent minor

---

[1] Abdallah, *The Yaos*, p. 11.
[2] D. and C. Livingstone, *Narrative of an Expedition to the Zambesi and its Tributaries*, 1865, p. 372.

chiefs. The tribal units, however, were smaller, and the crops more various. In tsetse-free areas they kept cattle, and they also dug and smelted iron and grew and wove their own cotton. As a supplement to vegetable diet, fish naturally took the place of game, and the sedentary, pacific life implied by a fishing community was perhaps one reason why they subsequently fell an easier prey than the Yao and related peoples to the era of raiding that was impending.

For while there is every reason to believe that the lakeside scene was in the early part of the nineteenth century one of relative peace and stability, it was shortly to be disrupted by Ngoni forays from the plateau to the west of the lake and by Yao slavers from the east. At its north-western corner, however, one enclave of people succeeded, almost to the end of this period, in defending themselves from attack, and remained little touched by outside influences of any kind up to the time when they came under the observation of the first European visitors. The kindred Ngonde and Nyakyusa people, distinct in origin from the Maravi folk, inhabited the pocket of superbly fertile country lying at the foot of the Kipengere Mountains; and while the Ngonde shared more or less in the effects of marauding by their more warlike neighbours, as well as being themselves early participants in the ivory trade, the Nyakyusa afford an outstanding example of the level of comfort to which, given a sufficiently favourable environment, African life could attain even within a simple Iron Age culture. Possibly it was this very degree of social consciousness and material well-being which, together with their numerical concentration and their remoteness, enabled them for long to repel attackers.

The most distinctive feature of Nyakyusa society was the age-village, whereby male age-groups, together with wives and children, formed villages until the children became herd-boys and in turn hived off to form settlements of their own. They also had chiefs, who, unlike the main body of the people, traced their origin to the Kinga tribe—hunters and ironworkers—of the Livingstone Mountains in the east; and to these is attributed the introduction of cattle, which were the Nyakyusas' chief pride and care. They hung them with bells, and stalled them in cleanly byres, separate from their own dwellings. But they were also skilled agriculturalists, whose carefully ridged and manured plots climbed far up the valley slopes, planted with maize, yams, and sweet potatoes, but above all with bananas, the staples of their ample diet. The stability of an intensive cultivation, moreover, together with the need for protection from a heavy rainfall, caused them to be expert builders, and their conical huts, set about with banana groves from which the fallen leaves were swept daily, made an unforgettable impact on the African traveller.

It seemed a perfect Arcadia, about which idyllic poets have sung, though few have seen it realized. Imagine a magnificent grove of bananas, laden with bunches of fruit, each of which would form a man's load, growing on a perfectly level plain, from which all weeds, garbage, and things unsightly are carefully cleared away. Dotted here and there are a number of immense shady sycamores, with branches each almost as large as a separate tree. At every few spaces are charmingly neat circular huts, with conical roofs, and walls hanging out all round with the clay worked prettily into rounded bricks, and daubed symmetrically with spots. The grass thatching is also very neat. The *tout ensemble* renders these huts worthy of a place in any nobleman's garden.[1]

Joseph Thomson, whose description this is, might well be astonished and delighted. He had just come down through the mountains that separate this lakeside paradise from the bare and inhospitable uplands which largely comprise what is now the Southern Highlands region of Tanganyika. On and around these uplands lived, thinly scattered, the patrilineal Hehe, Bena, and Sangu peoples. By the time that they first came under European notice, they had already been consolidated into sizeable political units, each vying for supremacy: but it seems likely that this was a recent development due to Ngoni pressure, for by their own accounts they still formed, well into the nineteenth century, a multiplicity of independent chiefdoms. No doubt Ngoni influence also helped to crystallize their military organization—introducing for instance such features as the regimenting of the boys and young men of the Hehe kingdom into military colonies (though it is worth noting that the Hehe had, even before the coming of the Ngoni, taken part together with the Gogo in repulsing the Masai from the north); at all events they had in earlier days been more peaceable than they

[1] J. Thomson, *To the Central African Lakes and Back*, 1881, vol. i. 267.

later became, for until the 1850's the main trade route from the coast ran through their territory, and the accounts of early travellers stress its safety. 'In travelling in the country from the East side to the Lake [Tanganyika] there is no danger.'[1]

Yet their way of life was rough and raw by comparison with that of the Arcadian Nyakyusa. Like them, they were cattle-keepers: but these upland peoples were more nearly true pastoralists, to whom, as hungry travellers found to their cost, their plots of grain and vegetables were of very secondary importance, and whose villages, 'small, ragged and low', were redolent of the proximity of man and beast. To quote Thomson once more:

> Mwhanna formed one continuous line of flat-roofed houses, divided longitudinally into two parts, the front and back. The front part is occupied by the cattle, and remains undivided throughout the greater part of its length,—all the cattle being housed in common during the night. . . . The back part of the house forms the dwelling-place of the Wahehe, and there they store the milk. . . . The entrance is through the cattle's portion, by which they also get the little light they require, and as there is no other means of ventilation, the natives are kept warm by the crowding and hot breathing of the cattle,—not a very wholesome way, certainly; but the natives are by no means fastidious.[2]

To the west these highlands dip gradually north of Lake Rukwa into a rocky and very probably then already tsetse-infested land: but along the savannah and fly-free ridge to the south of that lake, the 'corridor' between Lakes Tanganyika and Nyasa, the population is somewhat more plentiful, only to thin out once more in the range of hills that fringes the eastern shore of the Tanganyika lake. This again was a region which lay in the path of the Ngoni northward progress and which, being a cattle-rearing one, was much subject to Ngoni raiding: but its peoples failed to coalesce under the Ngoni impact into more solid political units. Their tribal traditions are fragmentary, and it is difficult to arrive at any clear picture of conditions as they prevailed in the early nineteenth century. We may, however, envisage a cluster of patrilineal, agricultural folks who

---

[1] 'The visit of Lief ben Saeid to the Great African Lake' communicated by James Macqueen, *J.R.G.S.* 15, 1845, pp. 371–6.
[2] Thomson, *Central African Lakes*, i. 214.

spoke languages distinct but mutually intelligible—the Namwanga, the Mambwe, the Lungu, and the Fipa. But adjoining these, and in places mingled with them, there were other elements of a western or north-western origin, hailing from across Lake Tanganyika. Sometimes these elements constituted the main body of the population, as in the case of the Holoholo and the Tongwe close to the lake shore. More commonly they provided an aristocracy or a line of chiefs. Thus the aristocracy of Ufipa and the chiefs of Ulungu both trace their origin to the western side of the lake, while Ufipa is distinctive in having also a chieftainship of possible Tutsi origin. To this may probably be ascribed the fact that Ufipa society was more centralized, more authoritarian than its neighbours, with the typically Tutsi institutions of district chiefs, court officials, and royal pages. Yet Thomson described the people as, although ready to defend themselves at need, 'more of a purely agricultural race than any other tribe I have seen'.[1] Despite an abundance of game, they were not hunters. The London Missionary Society representative, Swann, found their neighbours the Lungu, at the southern end of the lake, to be 'a cheerful, peaceable folk', industrious since the work of forest clearance had made them so; cotton cloth was still made (this was in 1880) in almost every village.

The Tutsi chiefs who established themselves in Ufipa seem to have reached that country circuitously by way of the south end of Lake Tanganyika. In the densely settled districts to the north of the Malagarasi river a similar element, here infiltrating direct from Rwanda-Urundi, was even more readily apparent: among the Jiji, the Ha, and the Vinza the Tutsi strain is still said today to be physically well marked in the chiefs and leading headmen. At this period, indeed, the arrival of these Tutsi chiefs from the hills must have been a fairly recent event—perhaps no more than forty or fifty years old: their rule over the local Bantu population seems to have been achieved peacefully.

All these were patrilineal peoples; all were cultivators, growing manioc, holcus, millet, and vegetables in wide variety, while the Jiji, living by the lake, were skilled boatmen and fishermen. The immediate lakeside was unsuited to cattle: but elsewhere these were important. The Ha in particular Burton described as inhabiting, before the depredations of the Ngoni, a 'fertile

[1] Thomson, *Central African Lakes*, ii. 222.

country, well stocked with the finest cattle'; and probably the Ha, who remained out of the main stream of commercial influence, most nearly represented the condition of this country in the early years of the century. Burton, however, who with Speke was the first European to record this part of the country, had little to say either of them or of their neighbours the Vinza, reserving his chief descriptions for the Jiji. These were 'a burly race of barbarians', rough and uncivil in manner to strangers, although showing elaborate courtesies among themselves; almost amphibious; much addicted to toddy and to bhang, and given to violence in their cups. Some of these characteristics they may have accentuated in consequence of the coming of traders; but the conduct of their public life had very likely long been much as the explorer painted it.

The affairs of the nation are settled by the *mwami*, the chief, in a general council of the lieges, the *wateko*, presiding. Their intellects, never of the brightest, are invariably fuddled with toddy, and after bawling for hours together and coming apparently to the most satisfactory conclusion, the word of a boy or of an old woman will necessitate another lengthy palaver.[1]

In all the broad stretch of East Africa between the Great Lakes and the sea, however, there was no region which so absorbed the attention of travellers as the wide plains, interspersed with granite hills, of Unyamwezi, merging in the north with Sukumaland. Although without rivers and poor in soil, it has sufficient rainfall to sustain a fairly thickly settled population, and, at least in the wet season, it struck the eye as a scene of comfort and prosperity.

The Land of the Moon, which is the garden of Central Intertropical Africa, presents an aspect of peaceful rural beauty which soothes the eye like a medicine after the red glare of barren Ugogo, and the dark monotonous verdure of the western provinces. The inhabitants are comparatively numerous in the villages, which rise at short intervals above their impervious walls of the lustrous green milk-bush, with its coral-shaped arms, variegating the well-hoed plains; whilst in the pasture-lands frequent herds of many-coloured cattle, plump, round-barrelled and high-humped, like the Indian breeds, and mingled flocks of goats and sheep dispersed over the

[1] Burton, *Lake Regions*, ii. 62–71.

landscape, suggest ideas of barbarous comfort and plenty. There are few scenes more soft and soothing than a view of Unyamwezi in the balmy evenings of spring.[1]

It was a land of villages and hamlets: even in Burton's time there was nothing that could be called a town—although later in the century much bigger settlements became common. The characteristic dwelling was the *tembe* noted by Thomson in Uhehe, the hollow square or oblong of stable-like buildings of wattle and daub, with mud, clay, and grass roof over low rafters supported by the long walls, and partitioned according to the diverse needs of families and household purposes. A particular feature was the *iwanza* or common club-house, one for each sex, which was the central social institution of every village. Here again an intermixture of populations was easily to be perceived, a small chiefly class unmistakably traceable to the Interlacustrine region to the north-west being superimposed on longer-established Bantu peoples, 'the ryots or subjects . . . collectively styled Wasengi'. Besides these there was a slave-class; the Nyamwezi were early noted as large slave-owners.[2]

The Tutsi of Rwanda-Urundi were also represented, not as an aristocracy but simply as professional herdsmen, tending, and purveying the products of, the cattle which were so conspicuous a feature of the Unyamwezi landscape. Almost certainly these cattle were more abundant in Unyamwezi and Sukumaland than they were towards the end of the century; the Nyamwezi chief Mirambo later recalled how since his youth war had caused depopulation, and how depopulation had brought in the tsetse.[3] Twenty years before the coming of the first Europeans the Nyamwezi were famed at the coast for their herds; and in 1858 Chief Fundikira of Unyanyembe was said to have no less than 2,000 head of cattle. There were the humped Zebu beasts, characteristic of eastern Bantu Africa in general; and in the north-western part of the region there were also the long-horned Ankole variety, corresponding to the latest, the Hinda, wave of penetration from the north-east.[4]

[1] Burton, *Lake Regions*, ii. 7.
[2] *Missionary Herald*, 87, 1840, pp. 118–21. Letters from Mr. Burgess, a member of the American missionary party which visited Zanzibar in 1839. See p. 270 below.
[3] E. J. Southon, *The History, Country and People of Unyamwezi*, manuscript in Central Africa, Box III, L.M.S. Archives.
[4] See Chapter VI above.

The men had care of the livestock; to the women fell in the main the business of agriculture—the chief crops being probably then, as they are now, millet, cassava, and sweet potatoes. But, in contrast to the situation in the Interlacustrine region, the division of economic functions did not reflect the political and social structure—there was no sharply defined gulf between an agricultural peasantry and a cattle-owning aristocracy.

When Burton came to the country he found it divided, as it is at the present day, into a mosaic of petty chiefdoms, each with its *ntemi* or *mwami*, its councillors, and elders. Each had also its court slaves; and he noted in particular the characteristic rituals with which the chiefs were buried—in sitting position, dressed in skins, and provided with weapons, wine, food, and often human companions to ease their passage to the future world. He was told, however, of a time, not more than 150 years back, when there had been a united Nyamwezi empire. Genealogical evidence lends no support to this claim; yet it is noteworthy that accounts brought down to the coast even in the earlier part of the nineteenth century hint at fewer and more important political units.[1] Probably the explanation of the latter at least is that these accounts were traders' accounts; and that the chiefs who figured so notably in them were those who had come to play a significant part in trading transactions with the coast.

It is possible that trade was at all events one of the factors responsible for another phenomenon apparently recent at this time—the spread of Nyamwezi influence eastwards into Ugogo. The Gogo, the last of the peoples of whom it seems essential to take account in setting the scene for the developments traced in this chapter, were basically a pastoral folk, who had in places adopted some of the characteristics of the Masai to the north—their language and their weapons. But by the time they were first visited by Europeans, they were already conspicuously impenetrated by influences deriving from Unyamwezi. A rude, brawling, and boundlessly inquisitive race, whose harsh manners and fierce looks almost all early travellers described at length, they were physically much intermixed with slaves apparently brought from the west; they used Nyamwezi knives, spears, and hoes; and their *ntemis* were of Nyamwezi origin.

[1] See Burton, *Lake Regions*, ii, ch. i.

They were moreover to some extent sedentary and agricultural, so far as the poor physical resources of their bare, hot land allowed it. Of course this Nyamwezi penetration may represent simply migration, either of whole groups or more probably of families, but it is also possible that commercial factors entered into it. The dialects spoken on the borders of Ugogo were often not those of the nearest adjoining Nyamwezi Districts, but of areas considerably to the west and south, and these were the districts where the accounts of coastal traders located the chiefs they regarded as important.

What is certain is that by the time this period opens the Nyamwezi were known, far distant from their homeland, not only as prosperous agriculturalists, with slaves and extensive flocks and herds, but also as traders. For this Bantu world, static and self-sufficient although in the main it was, was already broken by some currents of intercourse, some features which proved the African cultivator, within the limited opportunities open to him, not less capable than other men of responding to commercial stimuli. Slaves and ivory, the immemorial produce of the Dark Continent, found their way down to its coast for export; and within it, the return goods which went inland helped to lubricate a thin flow of exchange which was stimulated by other articles of trade that never had much more than an internal circulation—bark-cloth, salt, and copper. Of these, copper was perhaps the most significant. Little reached the coast; but in the form sometimes of bars and crosses, sometimes of already made-up ornaments, it travelled widely in the interior. To trace its part in the economic development which forms so large a part of the subject-matter of this chapter, it is necessary to go somewhat beyond the limits of the regions bounded by the Great Lakes, to the basin of the Upper Congo.

The copper deposits of the Katanga region to the south-west of Lake Tanganyika were undoubtedly being worked by the medieval period and possibly much earlier still;[1] and it was almost certainly through the control which they established over this area and its adjacent salt-pans that the Lunda Kazembes built up a power which came to rival and eventually to overshadow the parent Luba empire of Mwata Yamvo in the Upper

[1] See Chapter IV, p. 117, above.

Congo. Their capital, in the Luapula Valley, became in the late eighteenth and early nineteenth centuries the objective of a number of Portuguese-sponsored expeditions. These brought back impressive accounts of its display of foreign cloth, glassware, and other luxuries, and of the prosperity and orderliness of the realm over which the Kazembes ruled. Katanga and the Kazembe's dominions to the north and east of it seem indeed to have been something of a pivot of Central African trade at this time, to some extent determining its patterns throughout the eastern half of the continent. For to the east the Bisa, a tribe related to the Lunda, developed a marked propensity for commerce, their caravans often making their way right down from the Kazembe's capital to the Indian Ocean. At other times they sold their slaves and ivory to the Yao, east of Lake Nyasa, who by the latter half of the eighteenth century at least had also made a name for themselves at the coast as a trading people. In the 1790's the Portuguese explorer Dr. Lacerda was told how the goods taken inland by the Yao to be traded with the Bisa, and bought from the Arabs of Zanzibar and its vicinity, were displacing those formerly supplied by the Portuguese; and early in the nineteenth century the Yao were bringing down wax and ivory, as well as slaves, to the port of Kilwa, by a route which ran north and east from Lake Nyasa.[1]

Other traders came to Katanga from farther north. It seems more than likely that, at any rate by the beginning of the nineteenth century, caravans were going thither from Unyamwezi. Two Portuguese 'pombeiros' or literate African itinerant traders, visiting the Kazembe's court in 1806–10, found there companies of 'Tungalagazas'. These can probably be identified with the 'Garaganza' (the Wakalaganza of Burton)—the inhabitants of one of the districts of Unyamwezi—who were to play a significant part in that region's future history. Certainly it seems that copper at this period was finding its way northward both through Urua and across Lake Tanganyika near Ujiji, and around the southern tip of the lake, to furnish ceremonial

[1] The chief source for the above accounts of the Kazembe's court and of the trading activities of the Yao and the Bisa is R. F. Burton, *The Lands of Cazembe*, 1873, which gives accounts of three expeditions thither, including Lacerda's, between 1798 and 1831. See also F. T. Valdez, *Six Years of a Traveller's Life in Western Africa*, London, 1861; and I. Cunnison, 'Kazembe and the Portuguese', *Journal of African History*, 2, 1961, pp. 61–76.

spearheads and ornamental bracelets among the peoples of the Interlacustrine kingdoms. Certainly also, the Nyamwezi had by this time established trade connexions to the eastward, passing along the main east–west highway which, throughout the nineteenth century and into the twentieth, formed the chief artery of traffic from the Zanzibar coast to the interior. Yearly they came down in great caravans—it was said in thousands—to the coast, and there settled and cultivated for two or three months before returning to their own *shambas* in Unyamwezi when the rains fell there in November and December.[1]

These two main routes, the one running south-west from Kilwa, to cross or round Lake Nyasa and so on westward to the lands of Kazembe and Katanga, the other westward from Zanzibar to Unyanyembe—in the neighbourhood of present-day Tabora—and thence bending south-westwards towards the Katanga region in its turn (but throwing off an important branch also to the Interlacustrine kingdoms in the north), were and remained pre-eminent. Upon their existence the achievement of the European explorers depended. Lacerda and his Portuguese successors followed the Bisa trade route beyond the Loangwa River. Livingstone traversed a section of it to the west of Lake Nyasa in 1863, and three years later joined it again, near the outset of his final journey, between Lake Nyasa and the coast. But the Kilwa route remained far less well known than the more northerly one, first pioneered by Burton and Speke to the shore of Lake Tanganyika in 1857–8, and followed again two years later by Speke on his way to the Interlacustrine kingdoms north and west of Lake Victoria, which thereafter became a regular highway for explorers, missionaries, and traders. There were of course other routes also. One, comparatively little known to European travellers, diverging south-westwards from the main central line, struck down towards Ubena and the country north of Lake Nyasa: from there the enterprising might yet again pursue their way westwards towards the lands of Kazembe. To the north of Zanzibar, from the coastline between Pangani and Mombasa, a series of lesser tracks ran inland to the Chaga and the Masai country around

---

[1] Burton, *Lake Regions*, especially ii, 29–30; and information obtained by a party of American missionaries in 1839 and reported in *Missionary Herald*, 86 and 87, 1840, pp. 61–62 and 118–21. See p. 272 below.

and beyond Mount Kilimanjaro, and thence west towards Lake Victoria and eventually north-west of Mount Kenya—although, since these never became a channel of political penetration, they too remained fairly obscure.[1]

It would be misleading, indeed, to picture these 'routes' in any very clear or tangible terms. At their best no more than single-file trails through long grass or bush which had to be broken afresh each season, often nothing beyond the experience of the guide from one village or one landmark to another, they varied their course from season to season, from year to year. The road to Unyanyembe, for instance, which in Burton's and Speke's day ran through Ugogo, not far from the present-day course of the Tanganyika central railway, had up to about the middle of the century made a wide detour to the south to follow the Ruaha Valley before bending back north-westwards. Its starting-point might be anywhere from Saadani, north of Zanzibar, down to Mboamaji, near the site of the present Dar es Salaam.[2] There is, finally, no doubt that the main initiative in building up this thin network of trade that threaded east central Africa came from the peoples of the hinterland themselves. Whatever the factors that impelled one people rather than another to supplement the hoe and the spear with prolonged excursions for the trafficking of salt and copper, slaves and ivory, the early caravans were composed of the members of interior tribes—the Bisa, the Yao, the Nyamwezi, and farther north, the Kamba of present-day Kenya. These indigenous caravans continued to handle the bulk of the trade up to the middle of the nineteenth century and even later.

By at least the second and third decades of the century, however, a new element had begun to enter into the picture—a handful of Arab and Swahili merchants were to be found operating far inland. One of them, Juma bin Rajab, whose grandson was to become famous under the name of Tippu Tip, was a figure of some importance in Unyamwezi.[3] The accounts of two or three others who had journeyed as far as Lake

[1] See Chapter IX below.
[2] Burton, *Lake Regions*, i. 302–3, cf. the account of Lief bin Said in *J.R.G.S.* 1845, pp. 372–3; and of Nyamwezi informants interviewed at Zanzibar in 1839, *Missionary Herald*, 86 and 87, 1840; pp. 61–62 and 118–21. See p. 270 below.
[3] H. Brode, *Tippoo Tib*, London, 1907.

Tanganyika formed the basis of the first serious speculations in Britain on the geography of the Great Lakes region. Meanwhile, far down to the south-west, the father of one Muhammad bin Salim, an Arab whom Livingstone in 1867 found at the Kazembe's court, was, so his son claimed, the first Arab ever to open up trade in those parts,[1] although Swahili traders had evidently penetrated some distance up the Kilwa route—possibly as far as Lake Nyasa—a good deal earlier.[2] These early pioneers seem to have been mostly men of the coast (Mrima) rather than pure-bred Arabs of recent Muscat origin. Also, as one would perhaps expect, they were individual and isolated adventurers, who from the first married into the country and became involved in tribal politics. Thus Juma bin Rajab was said to have set up the local Nyamwezi chief of the small territory of Ugowe, a chief whose grandson Mirambo was to make history in the region thirty or forty years later, while Juma's son Muhammad married the daughter of a second Nyamwezi chief, the powerful Fundikira of Unyanyembe.

For centuries the Arabs and Swahili of the coast had been settled there, apparently largely indifferent to all beyond the thin coastal fringe with its towns and plantations. Why now did they seek to push inland for themselves? Almost certainly the answer is to be found in the rapid expansion during the nineteenth century of the demand at the coast for the two ancient products of the interior, slaves and ivory. There is a time-honoured formula about the inter-connexion between these two trades. The slave, it is said, obviated the cost of transport by carrying down the ivory: at the coast he could be sold along with his burden. But there must always have been a good deal of myth about this. For one thing, both the slaves and the ivory that came down to the coast had to be paid for—and they were paid for in bulky goods—cotton, beads, and hardware, particularly brass and iron goods—which had to be carried back to the

[1] Livingstone, *Last Journals*, (ed. H. Waller), vol. i. 277.
[2] Morice MS. Morice was a French speculator who in 1770–5 made a systematic collection of information on inland trade prospects at Kilwa. This information, in the form of a questionnaire and answers thereto, is preserved in a manuscript at Rhodes House, Oxford. See Bibliography. Krapf also speaks of Swahili traders, armed with muskets, as engaged, despite personal risk, in commerce with the Masai and the Kwafi. J. H. Krapf, *Travels, Researches and Missionary Labours in Eastern Africa*, 1860, p. 363.

interior; moreover, porterage, both of the incoming trade-goods and of the coast-bound ivory, seems often to have been a business of skill and prestige. Nevertheless, there *was* a link between the two, in the sense that, time and again, the demand for the one commodity seems to have been the precipitating factor in stimulating commerce in the other. If it is possible to pick upon a single starting-point for the nineteenth-century burst of expansion, it is probably to be found in the need of the French Indian Ocean islands for labour on their plantations. This it was that set the French speculator Morice inquiring about the possibilities of developing Kilwa as a French trading centre—he was looking towards a regular annual supply from the East African coast of some 10,000 slaves for the French islands alone.[1] The early years of the nineteenth century therefore saw a gradually growing European interest in East Africa, while by its second and third decades ivory was entering upon an unprecedented phase of popularity in both Europe and America.

It was at this point that the Imam Sayyid Said of Muscat, the Omani Arab ruler of Zanzibar, determined to devote his main energies to the exploitation of his Zanzibar rather than his Persian Gulf possessions. Undoubtedly his decision in itself gave an important stimulus to the trade of the coast. He successfully promoted the cultivation of cloves, and the labour requirements both for the clove and for the other plantations which multiplied along the coast on the profits of the ivory trade brought in turn an increased demand for slaves.[2] Said also encouraged the settlement within his dominions of Indian traders, a sprinkling of whom had been on the coast since the early Portuguese times or before, and it is likely that the influx of these, all along the East African littoral, was a direct as well as an indirect encouragement to the Arabs and Swahilis, who had previously managed the coastal trade, to push farther inland. On the one hand the commercial acumen and intense industry of the Indians made it difficult to compete with them: on the other hand they were able and willing to provide the credits upon which long expeditions to the interior depended. Their opportunities were further enlarged during the 1840's, when a combination of circumstances caused the

[1] See Chapter V above.   [2] See Chapter VII above.

transatlantic European slave-trade temporarily to require a substantial recruitment from the East African coast. In 1846 Britain announced her intention to abolish sugar preferences, and the next three years saw a spectacular increase in the rate of slave shipments to Cuba and Brazil. In accordance with the Anglo-Portuguese Agreement of 1817 almost all these slaves were procured from regions south of the equator, and during the 1840's there is evidence of a very considerable traffic round the Cape of Good Hope. It is true that most of the slaves shipped to the Americas came from the Portuguese sector of the coast, but there, as in the non-Portuguese territories to the north, the Indian trader, with his connexions in Zanzibar, was by this time in the ascendant, and it was probably Zanzibar that chiefly profited by the shipment of natives of East Africa to Cuba and Brazil.

It seems possible to discern a real contrast between the commercial penetration of the interior deliberately sponsored by Sayyid and the sporadic venturings which had preceded it. It happened that in July 1839 a party of American missionaries called at Zanzibar on their way to Bombay, and were anxious to collect, during their short stay, all the information they could about conditions in the East African interior. They learned not only that the Imam was systematically exploring the prospects of trade along the central route, but that he was at that moment negotiating with an embassy of leading Nyamwezi a treaty for the security of his up-country caravans.[1] It may even have been on this occasion that an agreement was arrived at whereby Arabs residing in Unyanyembe were to be exempt from the payment of any dues to the Nyamwezi chief of that district.[2] It would seem certainly that at about this time a regular Arab 'colony' was consciously established in Unyanyembe (or a little to the north of it). It was in some sort under the authority of a recognized leader, Abdullah bin Salim, who had 200 armed slaves; and forty years later a venerable and respected Muscat Arab called Sultan bin Ali was pointed out to a Belgian visitor as its sole surviving member.[3] There may be detected, moreover, a degree of estrangement, if not of jealousy, between these apparently more pure-bred Arab newcomers and the earlier

---

[1] *Missionary Herald*, 86, 1840.   [2] See p. 279 below.
[3] Burton, *Lake Regions*, i. 327, 329; J. Becker, *La Vie en Afrique*, 1887, ii. 29.

traders from the Mrima coast. It is noteworthy for instance that Burton makes no mention of Muhammad bin Rajab, although his son Tippu Tip claimed that he was at this time the foremost of the Unyanyembe Arabs. Burton, on the other hand, does describe a settlement at Msene, a few miles to the west, as 'the capital of the Coast Arabs and Waswahili, who having a natural antipathy to their brethren of Oman, have abandoned to them Unyanyembe and its vicinity'.[1] The lines of this division, between the coast-man and the Zanzibarite, the half-caste and the pure-bred Arab, were often blurred; nevertheless in one form or another the mutual distrust between the two groups was to be a persistent feature of Arab society in the interior.[2]

The character of the Arab settlement in Unyanyembe, at least as it existed ten or fifteen years later, has been vividly described by Burton:

this 'Bandari-district' contains villages and hamlets, but nothing that can properly be termed a town. The Mtemi or Sultan Fundikira, the most powerful of the Wanyamwezi chiefs, inhabits a Tembe, or square settlement, called 'Ititenya' on the western slope of the southern hills. A little colony of Arab merchants has four large houses at a neighbouring place, 'Mawiti'. In the centre of the plain lies 'Kazeh', another scattered collection of six large hollow oblongs, with central courts, garden plots, store-rooms and outhouses for the slaves. Around these nuclei cluster native villages—masses of Wanyamwezi hovels, which bear the names of their founders. . . .

The Arabs live comfortably, and even splendidly, at Unyanyembe. The houses, though single-storied, are large, substantial and capable of defence. Their gardens are extensive and well planted; they receive regular supplies . . . from the coast; they are surrounded by troops of concubines and slaves, whom they train to divers crafts and callings; rich men have riding asses from Zanzibar, and even the poorest keep flocks and herds.[3]

The aspect of this, as of other settlements, changed from time to time with political or trading conditions. On revisiting it two years later Speke noted with surprise how, as a consequence of the disturbances which had followed the death of the Nyamwezi chief Fundikira, the Arabs, instead of appearing as merchants,

[1] Burton, ibid., i. 395–6.
[2] Although for simplicity's sake the coast-men are frequently designated in this chapter as 'Arabs', this distinction should be borne in mind.
[3] Burton, ibid., i. 326–8.

as formerly, 'looked more like great farmers, with huge stalls of cattle attached to their houses'.[1]

At Ujiji by Lake Tanganyika, second in importance of the early settlements, changes mainly reflected its varying importance as a trading centre. When first visited by Burton and Speke in 1858, it gave the impression of being a place which had seen better days. The Arabs, they were told, had found that the air did not agree with them, and now visited the lake only in flying caravans during the dry season. Livingstone, however, in 1869, and two years after him Stanley, found it a thriving and active place; but afterwards, as the main tide of Arab exploitation flowed on to the west of Lake Tanganyika, it again fell into decay, as described by Joseph Thomson ten years later:

> but for the appearance of the inhabitants, it might have been supposed to be a deserted village, so ruinous did the houses appear. These represented almost every style of African architecture—the huge-roofed Indian bungalow, the flat-roofed tembe, the quadrangular hut of the Waswahili with baraza in front, and the beehive-shaped hut of most of the natives, with composite forms of every description.

There was still, however, an atmosphere of bustle:

> the frequent appearance of Arabs in their flowing garments . . . bands of Waswahili, strings of slaves laden with grain or ivory, flocks of sheep and goats, and small herds of cattle, together with the canoes on the water, all gave the place an appearance not unlike a coast village on the Mlima.[2]

It was not until the 1870's that any of the other Arab settlements in the interior came to rival Unyanyembe or Ujiji; but meanwhile Arab merchants, individually or in small groups, continued to extend their activities. In the north, the first visit by an Arab to the court of Buganda was paid by one Ahmad bin Ibrahim in 1844. Four years later Arabs or Swahili equipped with firearms were aiding the Ganda in a raid on Busoga; and Burton was told in 1858 that the Ganda ruler Suna, who had just died, 'greatly encouraged, by gifts and attention, the Arab merchants to trade at his capital'. Those who did so evidently

---

[1] J. H. Speke, *Journal of the Discovery of the Source of the Nile*, 1863, p. 91.
[2] Burton, *Lake Regions*, ii. 57; Livingstone, *Last Journals*, ii. 7–11; H. M. Stanley, *How I found Livingstone*, London, 1895 edn., pp. 330–1, 387–8; Thomson, *Central African Lakes*, ii. 88–91.

prospered. Burton's host Thiney bin Amir, who had first journeyed to Buganda six years earlier, had become one of the wealthiest of the colony at Unyanyembe. In the south-west, activity seems to have been renewed chiefly in the 1860's, when a handful of Arab traders, some with semi-permanent establishments there, were trading in Itawa and the neighbourhood of Lake Mweru. In the same decade it was apparently possible for a trader far up-country in lesser-known Uhehe to borrow locally 4,000 to 5,000 dollars' worth of trade goods to pursue his operations into Ufipa and Ruemba,[1] while on the Kilwa–Nyasa route at least one Arab merchant, Said bin Habib, was even before 1860 trading, via Kazembe's and Katanga, across the whole breadth of the continent.[2]

Both in the regular Arab settlements and in the scattered establishments farther afield stockaded depots contained, for months and sometimes years, the accumulated stocks of the traders—barter goods on the one hand, ivory and occasionally copper on the other. But the trade had its own distinctive patterns. It has already been suggested that the thesis that the slave eliminated transport costs by carrying the ivory cannot be generally sustained; and indeed slaves and ivory did not by any means always go together. No doubt the slaves did sometimes carry tusks, especially in the latter (and more publicized) days when the whole trading system had run amok, but in its heyday the main slave and ivory routes were distinct. The great majority of the slaves brought to the coast hailed from the tribes towards and around Lake Nyasa—the Makua, the Yao, the Ngindo, the Nyanja; and although phases of war and famine would from time to time throw up brief bursts of human merchandise elsewhere along the coast, Kilwa remained throughout the period the principal port for slave shipments. The ivory that came by the same route consisted mainly of

[1] Livingstone, *Last Journals*, see especially i. 271–3, 321–2; *Maisha ya Hamed bin Muhammed al Murjebi yaani Tippu Tip*, trans. W. H. Whiteley, Supplement to *East African Swahili Committee Journals*, 28, ii, 1958 and 29, i, 1959, p. 43.

[2] Rigby to Anderson, 20 Mar. 1860, in Zanzibar Archives. There were also several others trading on the Nyasa route as far as Katanga (D. and C. Livingstone, *Narrative of an Expedition to the Zambesi and its Tributaries*, p. 393); and the journey is recorded of three 'Moors' who traded successfully between the Kazembe's and Benguela in about 1851–2 (Pakenham to Malmesbury, 8 Aug. 1852, in *Slave Trade Correspondence*, P.P. 1853, A. 343). See also a reference in Krapf, *Travels*, p. 430, to a Swahili who had journeyed via Kilwa and Nyasa to Loanda on the west coast.

smaller tusks. The biggest and most valuable tusks came down from the neighbourhood of Lake Tanganyika to the Mrima coast. Some ivory too, but very few slaves, came from Chaga- and Masai-land to the ports north of Zanzibar. The slaves brought from elsewhere than the Nyasa region, moreover, were for the most part for specialized markets—Galla boys as house servants, girls from Unyamwezi, and occasionally Hima women from the Interlacustrine kingdoms for the harems of wealthy Zanzibar Arabs. The produce arriving at the coast does not indeed by any means tell the whole story. Slaves might be bought in one district, to be traded elsewhere for ivory. The readiness of the Gogo, for instance, to give tusks in exchange for slaves was probably one reason why their inhospitable country became established as a main trade-route; and indigenous copper, as already described, constituted one of the chief lubricants of the trade of the interior.

It is difficult to generalize about the size or other characteristics of caravans. The early accounts of Nyamwezi caravans on the central route suggest almost mass movements, with hundreds, if not thousands, taking part. Later descriptions give the impression that even at mid-century they were generally larger, while faster moving and more leanly provided, than the parties under Arab leadership. Here, as along the Kilwa slave route, it was common for two or three parties to join together, for mutual protection or to secure some mitigation of the *hongo*, or passage tolls, demanded along the road. On the other hand, away from the main highways comparatively small parties of from a dozen to thirty or forty members, whether under African headmen or the leadership of Swahili or Arab traders from the Mrima coast, were often to be met with.

Size depended much, of course, on the length of time that the expedition had been absent from the coast. When that of Tippu Tip headed back towards Zanzibar after no less than twelve years of exploitation and trading, it numbered some 2,000 porters carrying a similar number of tusks and guarded by 1,000 *askaris*—and one may be sure that its size and prestige attracted hundreds of further followers. Such mammoth processions, returning after years of absence, spoke not only of the enterprise and fortune of their leaders, but also of the extensive credit facilities which were available to back them at the coast.

Tippu tells how at the outset of this, his third and longest venture to the interior, Sultan Majid of Zanzibar himself was ready to intervene to ensure that he obtained the 50,000 dollars' worth of trade goods which he required;[1] and although this was certainly beyond the ordinary scale, it gives some idea of the amount of Indian capital which might be locked up in the interior at any moment. But it was not only at the coast that credit was to be had. There were many accidents, delays, and opportunities which might detain an expedition far beyond its original intention, and the general trading system inland could never have extended as it did had it not also been possible to borrow on a large scale locally. At Unyanyembe the principal financier until his death in 1861 was the Indian Khoja, Musa Mzuri (Musa the 'handsome' or the 'good'), whom Burton described as having his fortune wholly tied up in outlying debts, so that he could not, if he had wished it, have returned to the coast. But Musa was an almost isolated phenomenon. In general, lending in the interior must have been in the hands of the Arabs themselves. That it could nevertheless be on an impressive scale is shown by the occasion, already referred to, when Tippu Tip was able to borrow more than 4,000 dollars' worth of trade goods at Mtengera's, far away in the highlands of Uhehe.

It will be seen, from the complexity of the political and economic scene into which the coastal traders penetrated, that the process by which the trading 'frontier' moved 'forward' in East Africa may easily be oversimplified. Observers in Zanzibar were apt to visualize a kind of regular fanning-out process, and to attribute this to the exhaustion and depopulation of successive regions nearer the coast. What actually happened was a good deal more complicated. It is true that at certain periods one may clearly discern fresh areas being opened up by Arab traders (although often these areas had been more indirectly involved in commerce long before), and such thrusts may be to some extent correlated with economic incentives on the coast. Thus a sharp rise in ivory prices in 1856/7 seems to have caused a rush of Arab traders into Urua on the far side of Lake Tanganyika—in which the young Tippu Tip was involved, and which was commented on by Burton. The speculative surge

[1] *Maisha ya Hamed bin Muhammed*, p. 59.

forward which followed another ivory boom in 1869–70 made possible Livingstone's explorations into Manyema, while another may have prompted the first successful penetration of the Upper Congo forests seven years later. Equally, a cause for going farther afield may often have been the overstocking of nearer areas with trade goods to such an extent as to raise the price of ivory to unattractive levels. Again, it is possible that it was the demand for labour for the French sugar plantations on the island of Réunion which set in motion the first major slave-raiding incursions into the Shire District south of Lake Nyasa in 1859–60; moreover, high ivory and clove profits, bringing increased prosperity on the coast, seem generally to have brought an enhanced slave demand in their wake. But elephants' tusks, far less human beings, are not commodities to which crude or rigid economic laws can be applied. It was of little moment that high ivory prices ruled on the coast, if wars in the interior blocked the road thither; while in seeking for the circumstances which determined the volume of the slave traffic, or the sources from which it was supplied, it is virtually impossible to distinguish between economic and political elements. Slaves were not only an article of trade: they were also a commodity sought for their own sake by some tribes who by adopting the girls and children recruited their own strength. Even when sold they could well be the by-product, rather than the principal object, of African wars.

Such wars became more destructive and more extensive as the use of firearms spread over the interior. From very early in the century there is evidence that African chiefs on the mainland wanted, and occasionally obtained, muskets and powder. But then, and for another fifty years, the goods which found their way up-country in exchange for the down-traffic of slaves and ivory consisted almost entirely of cloth and beads. In 1859 and 1860 what seems to have been an unprecedented quantity of firearms passed through the Zanzibar customs house, and was practically all shipped across to the mainland. Thereafter, firearms constituted a steadily growing proportion of Zanzibar imports, until in 1880 they seem to have been at least a third of the total value. Three years later it was estimated that the Nyamwezi chief of Unyanyembe alone had 20,000 guns at his disposal. There was a cannon in Unyanyembe in 1860 and cannon were being transported across Lake Nyasa in 1861.[1] The increase reflected a change in kind as well as quantity. Muzzle-loading muskets gave way to breech-loading rifles, although up to the end of this period the old 'tower' musket was still the most characteristic weapon. It was as much as anything the growing traffic in arms which conditioned the shape of political change in the interior—and with it, in large measure, the rate and direction of fresh exploitation, and the availability of slaves, whether at the coast or elsewhere. It altered by degrees the character of the Arab infiltration. In the areas originally exploited this had been in the first instance wholly peaceful, in those penetrated later it partook more and more of the nature of open conquest. And, since the Arab trading community never had the capacity, if indeed it had had the will, to reserve the means to power in its own hands, the infiltration also steadily transformed the political landscape on and beyond the fringes of direct Arab influence.

The people most immediately and most deeply affected by the Arab incursion were the Nyamwezi. The earliest accounts—though something must perhaps be allowed for the mellowing effect of distance—speak of them as a peaceful, prosperous, and relatively united people. They owned many slaves and beasts of burden. Trade, and even porterage, was amongst them a matter of status and pride—there is good evidence that several Nyamwezi *ntemis*, or chiefs, learnt of their elevation to that dignity while engaged in the caravan traffic. One chief who in youth took part in it was Mirambo, whose career is traced below. Another was Kalasa, the father of Msiri of Katanga; a third, Fundikira, Sultan of Unyanyembe at the time of Burton's and Speke's visit.[2] But it seems certain that one effect of Arab penetration was to compete with the older Nyamwezi caravans at all events to the coast. In Burton's time—and later—the latter were still common, but he refers to them somewhat disparagingly; and it is likely that Arab traders (who were also debtors) would be given the greater encouragement by the

---

[1] D. Livingstone, *The Zambezi Expedition of David Livingstone* (ed. J. P. R. Wallis), 1956, i. 210.
[2] Stanley, *How I found Livingstone*, p. 426; Burton, *Lake Regions*, ii. 29–32; Livingstone, *Last Journals*, ii. 180.

Indian merchant houses upon whom all trade ultimately depended. But on the more remote interior routes the advantage of the Arab traders was less marked, and one reaction of the Nyamwezi was to infiltrate, economically and politically, along one of these routes, that from Unyamwezi to Katanga; in very much the same way as the Arabs had done along the road to Unyanyembe and Ujiji.

Whether or not it can be shown that 'Garaganza' traders—Nyamwezi so called from the name of a district just south of Unyanyembe—were already operating south-west of Lake Tanganyika at the beginning of the century, they were certainly finding their way to Katanga by the 1820's. Livingstone met at Tabora in 1871 an old chief who as a small boy had made the journey thither with his father, who was 'a great trader'. At one time it must have seemed that here as elsewhere they would be cut out by Arab rivals. But in the event the Nyamwezi held and consolidated their own position. One of them was Ngelengwa, son of a subchief in Usumbwa (just west of Tabora), who, apparently profiting by the experience of an apprenticeship in Arab caravans, established himself, with a small armed following, in the neighbourhood of Katanga itself. There he steadily entrenched his position until he was strong enough to defy successfully his nominal overlord, the Lunda Kazembe, and to make himself effective sovereign, under the name of Msiri, 'the mosquito', of the copper region.[1] It seems to have been from this time (between 1865 and 1870) that the Nyamwezi gradually acquired the virtual monopoly of the Katanga route, settling also in formidable colonies along the road thither. By 1880 the traveller Thomson reported that the Arabs had largely abandoned the Katanga trade to them.[2]

But Arab competition was also meeting with an effective Nyamwezi challenge nearer home. It seems likely that the

[1] A. Verbeken, *Msiri, roi du Garenganze*, Brussels, 1956, gives a general account of Msiri's origins.

[2] Livingstone, *Last Journals*, i. 308, 310, 312; Thomson, *Central African Lakes*, ii. 46–47. The decision of two Arab merchants to leave off trading in this area and go northward is as it happens recorded by Livingstone. It followed the disasters arising (*a*) from the operations of Msiri and the Garaganza on the west, (*b*) from Tippu Tip's bid for power in the Bemba country to the east (see p. 288 below), and (*c*) from a serious incursion of one section of the Ngoni which in this year was raiding north of Lake Bangweolo. *Last Journals*, i. 276, 321, 330, 349–51; D. G. Lancaster, 'Tentative Chronology of the Ngoni', in *J. Roy. Anthrop. Inst.* 67, 1937, pp. 77–90.

Arab colony in Unyanyembe had been founded on a treaty with the principal chief there whereby its members were to be exempt from all taxes on merchandise. The next chief of Unyanyembe, Manua Sera, coming to power at a moment when trade was particularly thriving, not unnaturally sought to revise the contract, and announced his intention of imposing a tax on incoming merchandise, only to be driven out by the Arab residents. They, of course, had nothing but abuse for Manua Sera, but Speke depicts him as an attractive young man, a dashing *condottiere* chief, who contrived even in adversity to command a remarkable loyalty among his followers. He finally met his death in 1865, but not before he had played serious havoc with Arab trade by harassing the main coastal road, thus setting a fruitful example to his countrymen.[1] At Unyanyembe the Arabs replaced him by a series of more compliant puppets, but some forty miles distant, at the time of his death, another and more formidable young Nyamwezi chief was already in the ascendant.

This young man, Mirambo, claimed descent on both sides from the last ruler, alleged to have held authority seven to nine generations back, over a united Nyamwezi kingdom. Starting from his father's small patrimony of Ugowe, he first added to it his mother's family territory of neighbouring Uliankuru. From this nucleus he built up a power which by 1870 was strong enough to take up directly a challenge from the now jealous Arabs of Unyanyembe. Five years of desultory skirmishing left the honours pretty well on Mirambo's side. Jealousy and indecisiveness sapped the Arabs' efforts to destroy him, despite their at first infinitely greater resources; and although Mirambo and his followers were reduced to skins for their backs, and stones, copper wire, and ornaments for their guns, they came near on at least one occasion to sacking the whole Arab settlement. By the time peace was concluded in 1876, Mirambo's capital at Urambo was fairly established as a rival market to Unyanyembe. Meanwhile he was extending his territorial influence. Between 1876 and 1880 he gained control of the main roads leading north-westward to Karagwe and Buganda, and by the latter year he dominated the Ujiji road to Lake

[1] *Maisha ya Hamed bin Muhammed*, pp. 41–43; Speke, *Journal*, pp. 77–78. Manua Sera's death is recorded in Stanley, *How I found Livingstone*, p. 267.

Tanganyika. Only the mountain kingdom of Urundi on the west, and the wild half-nomadic Nyaturu on the east, brought him up short. His last years were spent in extending his power southward.[1]

In human terms, Mirambo owed his achievements to the clearness of his objectives and to his personal courage and boundless energy in realizing them; in material terms to his success in holding together his force of *ruga-ruga* and in supplying them with firearms. The importance of the latter needs no emphasis. Throughout his life Mirambo's interest in firearms was paramount, and certain of his campaigns were evidently undertaken with the special aim of replenishing his stocks of guns and ammunition. The *ruga-ruga* require more explanation. Very early in his career Mirambo had made use of Ngoni warriors—he himself spoke their language and according to some accounts had been brought up amongst them. These Ngoni, a splinter group of the main body whose raiding progress northward is traced in an earlier chapter,[2] had reached Unyamwezi about the middle of the century, and there had more or less settled down. They retained, however, their warlike disposition, and seem to have formed the inspiration and sometimes the nucleus of the bands of mercenary warriors which were now becoming a conspicuous feature of the scene in the region between the great lakes. *Ruga-ruga* were drawn from all the most restless elements in a fluid society—runaway slaves, deserting soldiers and porters, professional elephant-hunters, young men impatient of the tedium of tribal life. Yet at their best they were not lacking in a certain panache and *esprit de corps*. Distinctive with red cloak, feather head-dress, ivory and copper ornaments, they are somewhat reminiscent of the predatory companies of the European Hundred Years' War— attaching themselves to whichever leader held out most hope of plunder, condescending to no manual labour save that of building his house and stronghold.[3]

So outstanding a 'war-lord' as Mirambo was naturally able to attract into his train the best and the largest force of such

---

[1] Southon, *History of Unyamwezi*; Becker, *Vie en Afrique*, ii. 173-6; Stanley, *Through the Dark Continent*, 2 vols., 1878, i, pp. 131-7, 489-94; and correspondence between the Zanzibar Consulate and the Foreign Office in the F.O. 84 (Slave Trade) series.   [2] See Chapter VI, pp. 208-11, above.

[3] See Becker, *Vie en Afrique*, ii. 507-8.

soldiers of fortune. He was also unusual amongst African leaders in that he looked beyond the immediate rewards of successful campaigning towards more peaceful and statesmanlike ends. Yet he was in some ways only the greatest of the series of 'war-lords' to which the coming of gunpowder was giving rise all over the region. A recurring feature of the travel literature of the late 1870's and early 1880's is the description of the *bomas*, or strongholds, of chieftains of this kind, ranging from a few dozen huts within a palisade to the headquarters of the Sangu chief Merere in Urori, which was said to have 300 gates and to shelter (apart from its human population) some thousands of cattle.[1] Mirambo's capital was thus described by a member of the London Missionary Society:

a large square enclosure the sides of which are composed of a substantially built wall against which houses are built all round. It is nearly half a mile square and encloses nearly two square miles of ground. In the space thus enclosed about two hundred round huts—well built and some of them fifty feet in diameter—give habitation to about 10,000 inhabitants; quite another 5,000 live in the houses built against the wall . . . the industrious agriculturalists and enterprising *wapagazi* live at a distance from the capital; the former supplying food for the army, the latter carrying the plunder to Zanzibar and returning with guns, powder and the like.[2]

Generally the *bomas* were defended by stout hedge-palisades of prickly, poisonous, euphorbia with loopholes for firing from. These in turn might surmount high embankments and be surrounded by one or more ditches, while the gateways were commonly adorned by human skulls grinning on the top of long poles. Within, the chief's quarters, strengthened to serve as a final citadel in case of attack, would, together with the armoury, form the most prominent buildings.

Sometimes the 'war-lords' who grasped these new opportunities were, like Manua Sera and Mirambo, the natural leaders of existing tribal groups—in which case the new developments shifted, but without disintegrating, the old patterns of

[1] For some such descriptions see J. F. Elton, *Travels and Researches among the Lakes and Mountains of Eastern and Central Africa*, 1879, p. 368 (Merere's); Thomson, *Central African Lakes*, ii. 233 (Simba's); *Maisha ya Hamed bin Muhammed*, p. 129 (Kasanura's); and V. Giraud, *Les Lacs de l'Afrique équatoriale*, 1890, pp. 253-5 (the Chitimukulu's) and 369-70 (the Kazembe's).

[2] Southon to L.M.S., 8 Sept. 1879, Central Africa Box III, L.M.S. Archives.

tribal politics. This seems to have been so, for instance, with the Vinza, a Hima-ruled tribe of some importance between Unyanyembe and Ujiji. Around the southern end of Lake Tanganyika it was so among the Bemba, who under their *Chitimukulus* or paramount chiefs rose with the advent of firearms to an unprecedented degree of power in that region; also, although less successfully, with their neighbours the Lunda of the Luapula Valley, who diminished in power indeed, but remained united under the authority of their Kazembes.[1]

It is interesting to speculate why some peoples did, while others did not, take to firearms. For the dispersal of these did not follow as closely as might be expected the opportunities of trade. It is true that the Nyamwezi and the Yao, the latter also among the earliest traders to the coast, were the first to adopt the new weapons. But inland of the Yao the Bisa, although in earlier days by no means militarily incompetent, and also the foremost traders of a wide region, had by the 1880's lost the political initiative to their Bemba neighbours, among whom the advent of firearms ushered in a brief period of what might almost be called imperialist expansion. Much here may be due to the outstandingly strong organization of Bemba chieftainship; but west of Lake Tanganyika the paradox is even more striking. A large section of the important Lulua people, also long-established as traders, and apparently with relatively powerful chiefs, were said in 1880 to be still virtually without guns: yet already their northern neighbours the Tetera and the Manyema, on the banks of the Lualaba–Congo, in an area barely touched by commerce until ten years before, were falling over each other in their anxiety, by war or labour or exchange, to obtain guns and powder. A year or two more, and these were playing in the Congo basin very much the same predatory role as the Nyamwezi to the east of Lake Tanganyika.[2]

Often, however, those who profited from the new 'turning world' were not the local traditional authorities of the old order, but immigrant adventurers who succeeded in imposing themselves in areas away from their own home districts. East

---

[1] Giraud, *Afrique équatoriale*, chs. xviii–xix; I. Cunnison, *History on the Luapula*, Rhodes–Livingstone Papers, No. 21, London, 1951.

[2] Thomson, *Central African Lakes*, ii. 125. Hutley to L.M.S., 20 Feb. 1880 and 23 Feb. 1880, Central Africa Box III, L.M.S. Archives. For the part played by the Manyema in raiding the northern Congo basin see especially Stanley's later works.

African tribal traditions suggest that this was no new phenomenon, but never can it have happened on such a scale. The history of one notably successful usurper of this kind, Msiri of Katanga, has already been traced; his Nyamwezi countrymen and half-caste Arabs were conspicuous amongst those who established themselves in similar fashion elsewhere. Such a one was 'Matimula', a Swahili from Kilwa, who with the help of a band of professional elephant-hunters set himself up soon after 1870 in the neighbourhood of Karema on the east shore of Lake Tanganyika, selling to King Leopold's International Association a site for their new station out of what he claimed was 'his' land. Another was Nyungu, a member of the Nyamwezi chiefly family of Unyanyembe, who as the result of a profitable intervention in a district south of the main coast road near the present site of Manyoni, ensconced himself there as the local ruler. A third, another half-caste Arab called Songoro, succeeded in imposing his authority over a considerable region on the southern shore of Lake Victoria before he was murdered in a fracas on the island of Ukerewe in 1877.[1]

The subjects of such adventurers as these, and of many others who could be cited, were sometimes coherent tribal units, but more often assorted groups of very mixed origin. To those accustomed to the better articulated, more distinct, social structures of other parts of Africa, and particularly to administrators, whether German or British, the fragmented, variegated picture in central Tanganyika has always been something of a problem. Partly indeed this may be traced to the physical character of the region—to the centuries-old pressure of stronger groups on the fertile perimeter pushing the weaker and smaller into the less hospitable central area: but much more is probably due to the economic development and political turmoil of this period. In the 1930's groups belonging to more than fifty tribes were counted in the Ujiji District, and the Tanganyika Administration's District Books speak eloquently of similar conditions elsewhere.[2] Numerous descriptions might be cited from travellers' and missionaries' accounts of these mixed societies in process of

[1] Tanganyika District Books, Kigoma, Manyoni; Becker, *Vie en Afrique*, i. 259, ii. 78–79, 91–93; Thomson, *Central African Lakes*, ii. 208, 227–8; Southon to L.M.S., 6 July 1880, Central Africa Box III, L.M.S. Archives; Kirk to Derby, 6 Mar. 1878, F.O. Confidential Print 3928.

[2] See, for example, Tanganyika District Books, Kigoma, Tabora, Manyoni, Kilosa.

formation. Slavery brought large numbers of people into Unyamwezi and Ugogo from south and west of Lake Tanganyika; there in turn Nyamwezi traders settled and colonized, while Guha and Manyema from across the lake, likewise seeking their fortunes, spread eastward along the central trade route.[1] Often the displacements of war changed and complicated the tribal picture.

While it was in Unyamwezi and beyond that the coast-men's impact was most conspicuous, its historical importance was more crucial in the north-western, Interlacustrine, region, at all events in Buganda. That, however, is discussed elsewhere in this volume.[2] It was also felt, although with different and less substantial effects, along the trade routes leading inland from Kilwa and its neighbourhood. In the intervening highland region north and west of Lake Nyasa its significance was slighter and more debatable; for here there is some difficulty in distinguishing between the consequences of coastal penetration and those of the Ngoni irruption whose progress was traced in an earlier chapter.[3] In general it may be said that the peoples of this region who succeeded in repelling the Ngoni were also those amongst whom Arab or Nyamwezi influence remained subordinate, while in the smaller and weaker units to the east and south of Lake Tanganyika which suffered severely from Ngoni raiding it became more firmly established. Thus Arab or Nyamwezi adventurers were powerful in Ufipa and Upimbwe to the east, and in Itawa to the south-west, of the lake; and the two former areas eventually fell under the dominance of Mirambo. To the north of Lake Nyasa it was otherwise. Here, among the Bena, Sangu, and Hehe peoples, there were no Arab or Nyamwezi enclaves, and the Arab merchants who visited there never asserted any real political authority. Moreover, there is no doubt that it was the Ngoni who were the chief factor in shaping political development among these peoples. It was in fighting with them, it has been suggested, that the Sangu first became a consolidated kingdom; and their adoption of Ngoni weapons and Ngoni tactics was conspicuous.

Whether it was in fact the cohesion acquired in combating

[1] Becker, *Vie en Afrique*, i. 220–1; E. Hore, *Eleven Years in Central Africa*, 1892, pp. 159–60; Thomson, *Central African Lakes*, ii. 231; Livingstone, *Last Journals*, ii. 198.   [2] See Chapter IX below.   [3] See Chapter VI above.

the Ngoni which was the determining factor in enabling these peoples to keep coastal influence at a minimum is more doubtful. It is true that, while in the early part of the century Zanzibar merchants did travel through the district, by the 1850's they had temporarily at least ceased to do so, and that the reason for this change was, as told to Burton, that the Warori (Sangu) had become so violent and predatory.[1] But it seems at all events arguable that their militant character would have deterred substantial trading settlements even without Ngoni influence; as among the Bemba, the most warlike of the peoples to the south of Lake Tanganyika, where the position of the coastal merchants remained essentially marginal, individual, and non-political, the coast-men simply acting as agents who advanced trade goods against the promise of ivory deliveries—although it was not until well after they were in contact with the Arabs that the Bemba came into direct conflict with the Ngoni, whom they repelled. On the other hand it is perhaps significant that it was in the long run that section of the Bena–Sangu–Hehe complex which had apparently the least contact with the coast that emerged supreme. Except for an interval of some ten or fifteen years of which we have no certain knowledge, coastal traders continued to visit the country; but although in the 1860's at any rate there were established Arab depots among the Bena and the Sangu, among the Hehe we hear only of the visits of occasional Arab caravans. One would infer, therefore, that the Hehe had less opportunity than the other two groups of acquiring coastal weapons, and in fact, in the various descriptions which we have of the Hehe and their encounters, there are few allusions to firearms. When, for instance, right at the end of the century, they captured breech-loading rifles from the Germans, they did not know how to use them. It seems certain therefore that it was principally by virtue of Ngoni techniques rather than as a result of Arab influence that the Hehe in the latter part of this period became the dominant political and military power in this area.

Ngoni influence was also considerable, in one form or another, along the southern trade route, the route leading inland from Kilwa to Lake Nyasa and beyond. To the west of the lake, indeed, there is no doubt of it. It helped to weaken and disrupt,

[1] Burton, *Lake Regions*, i. 303.

as slave-raiding likewise helped to weaken and disrupt, the small and vulnerable societies along its shores; and there is little doubt that the proximity of the Ngoni was one of the chief factors in contributing to the decline of the Bisa, once the foremost trading people between Nyasa and the Katanga copper region. To the east of Lake Nyasa, however, we come again to a region where the extent and nature of the Ngoni impact, in relation to that of influences emanating from the coast, is difficult to determine. Here the Yao were the predominant trading tribe, corresponding approximately to the Bisa in the west and to the Nyamwezi in the north. The Arabs in this region remained essentially the clients of the Yao chiefs, not their rivals or masters. But their penetration of the region long preceded the Ngoni incursions, and Yao tradition makes no mention of the Ngoni as a factor, one way or another, in the Yao rise to power. On the other hand there is an outside force whose impact that tradition does describe—a formidable attack, apparently some time in the middle 1850's, by their Alolo (i.e. Makua) neighbours, before which the Yao scattered because the Makua were armed with guns while they themselves had none.[1] It was this dispersal which proved, paradoxically, to be the starting-point of Yao political hegemony. One reason seems to have been that the Yao themselves quickly acquired firearms. The main demand in this region was for slaves, and the Arab traders very soon found it convenient to supply the Yao with the guns that would secure them. Another reason was perhaps that the market uncertainties of the slave trade were such that it never permanently attracted the more reputable and ambitious of the Arab merchants—it is possible indeed to trace how some of these abandoned the Kilwa trade for the larger prospects of ivory- and copper-dealing to the north. At all events the upshot of the Yao dispersal was not to disintegrate, but to extend, the influence of the Yao. Once equipped with firearms, their chiefs, moving with small kinship groups, asserted themselves over a widening area outside their own homeland; across the Ruvuma in the north, to the south in the Shire highlands, and around the south-western corner of the Lake. During the next twenty years the power of the four or five leading Yao chiefs grew, at the expense both of their less successful Yao

[1] Abdallah, *The Yaos*, pp. 34–58.

rivals and of the coastal traders operating in the region. By the 1880's Islam was making rapid headway amongst them and, although Arab traders of doubtful standing still frequented their courts, by far the greater part of the commerce of the region was evidently again in Yao hands.

The character of the Arab penetration of the East African hinterland was essentially infiltrative and diffuse; it could scarcely be described as colonization. The number of Arabs in the interior cannot have exceeded a few hundreds at the most; that of the Mrima or Swahili of the coast perhaps one or two thousands. And at every point this thinly scattered community was interwoven with the indigenous political and social scene. Its members took part in local quarrels; established themselves as local political figures; took African wives and concubines (Arab wives were a rarity in the interior); and dispensed arms by barter or as wages to their African neighbours and employees. It is characteristic for instance that the Nyamwezi chief Mirambo, even during his phases of most open hostility against the Unyanyembe Arabs, was still able to obtain firearms from them.

Yet within this general truth there are still distinctions to be drawn. In the Interlacustrine kingdoms to the north Arab influence was as it were consular and diplomatic, and in Buganda, where it was chiefly concentrated, was seeking by the end of the 1870's to attain its ends by altering the political balance within the framework of the existing state. Farther south, the position is more difficult to define. There were sizeable Arab settlements, as at Kagei, Tabora, and Ujiji: by 1880 these were in general small enclaves on the fringes of Mirambo's domains within which the Arabs controlled the local situation through puppet chiefs and remained on sparring relationships with the areas beyond. A rather similar state of affairs obtained at Kota Kota on Lake Nyasa, where a miscellaneous settlement of refugees from the lakeside tribes looked to an Arab 'Jumbe', settled at one of the main crossings of the lake, for protection against Ngoni and Yao depredations.[1] But in the first case

[1] See Tanganyika District Books, Kigoma; L.M.S. correspondence, Central Africa Box II, L.M.S. Archives; Livingstone, *Narrative of an Expedition*, pp. 512–13; Elton, *Travels and Researches*, pp. 295–6.

power was exercised only indirectly; in the second its territorial extent was limited and precarious. West of Lake Tanganyika, however, a wholly different situation had grown up; and it was this region which, thanks to its resources of ivory, became the main focus of political developments during the last few years before European influence became paramount. Partly the contrast in Arab–African relationships here was due to the more primitive condition of the peoples of the Upper Congo and to the later date of Arab arrival there. But much more was it the doing of the one outstanding character among the Arabs of the interior—Muhammad bin Hamid, more generally known as Tippu Tip.

Tippu came on his father's side (his mother was a Muscat Arab of Zanzibar) of a Mrima family which had early been concerned in the pioneer penetration of the interior; and being as a young man involved in the disturbances arising out of the Nyamwezi Manua Sera's quarrel with the Unyanyembe Arabs, he seems very early in his trading career to have determined that the most efficient long-term means of securing ivory was to bid unequivocally for political power. He chose as his main field of operations the ivory-rich lands west of Lake Tanganyika. In 1867–8 he attacked and defeated the Bemba chief who held the strategically placed territory called Itawa, just west of the southern end of the lake. Three years later, perhaps in some kind of alliance with Msiri of Katanga, he seems to have killed and replaced by his own nominee the Lunda Kazembe.[1] Leaving to Msiri, however, and his Garaganza followers the Katanga area itself, he thereupon turned north and, establishing a base between the Lomami and Luapula tributaries of the Congo, usurped outright the position of the existing Tetera chief, and thence extended a formidable territorial power. With Stanley's transcontinental expedition he penetrated in 1877 down the Congo, thus making the first breach in the forest barrier which had hitherto checked Arab expansion northward; meanwhile over a wide and increasing area he exercised direct authority—collecting taxes, settling disputes, punishing

---

[1] *Maisha ya Hamed bin Muhammed*, pp. 49–55, 79; Giraud, *Afrique équatoriale*, p. 372; Livingstone, *Last Journals*, ii. 253. The sequence of events, however, is obscure, and according to some accounts it was not the Kazembe himself, but one of his sub-chiefs, who was replaced.

recalcitrant tribes, and appointing regents to act in his absence.[1] It is characteristic of the whole African–Arab relationship in the interior that, across Lake Tanganyika, Tippu's closest affinities were not with his own countrymen, but with the indigenous ruler Mirambo; and that when, after twelve years spent in building up his own 'empire' in the Upper Congo, he sought to transport his accumulated ivory by way of Tabora to the coast, it was not with the Unyanyembe Arabs but with the Nyamwezi chief that he went out of his way to seek alliance. Mirambo's interests—the preservation of a large area of peace and the promotion of trade—marched with his own. The respect was mutual: Mirambo received honourably an embassy headed by Tippu's son, and bade the Arab merchant's caravans pass freely through his own dominions. And he sent assurances to Sultan Barghash at Zanzibar that, if only Tippu Tip could be put in authority at Unyanyembe, this would speedily bring to an end the misunderstanding between himself and the Arab colony there.[2]

But the two 'empires' were to be short-lived. Already from beyond the African continent harsh winds were blowing which were to wither alike Tippu Tip's power to the west of Lake Tanganyika and Mirambo's to the east of it.

During the 1870's there had been a marked quickening of European interest in the central African interior. Much of this interest was evangelistic. Inspired above all by the example of Livingstone, a number of British Protestant missionary societies sent volunteers to establish outposts among the main concentrations of population far inland. Between 1875 and 1879 two Scottish missions took successful root near Lake Nyasa. At Blantyre in the Shire highlands the Church of Scotland set up a station which, after some initial difficulties that brought it into temporary disrepute, settled down to become a prosperous mission. On the lake itself, first at Cape Maclear and later at Bandawe near its northern end, Dr. Laws, a Scots doctor of sterling character and of great practical ability, established the Livingstonia mission of the Free Church. Livingstone's own

---

[1] *Maisha ya Hamed bin Muhammed*, pp. 101, 109, 119; Stanley, *Dark Continent*, ii, chs. iv–vii.

[2] *Maisha ya Hamed bin Muhammed*, pp. 135–7; Becker, *Vie en Afrique*, ii. 168.

parent body, the London Missionary Society, during the same period dispatched pioneering parties to Mirambo's and to the shores of Lake Tanganyika, though these ventures in the long run were to prove less successful than the Nyasa missions, being crippled by the heavy toll in death or disablement of their early members, and long hampered by the length of their lines of communication. The earliest undertaking to be prompted by Livingstone's call, the Universities Mission to Central Africa, met with initial set-backs in its attempt to found a station near the Shire river, but by 1876 it was again pushing inland towards Nyasa along the line of the Kilwa slave route. Meanwhile, following this time the exhortations of the explorer Stanley, the first contingent of the Church Missionary Society in 1877 reached the capital of Buganda; on the way it planted intermediate stations at Mpwapwa in Usagara and at Unyanyembe. Nor were the Roman Catholics idle in the movement towards the interior. In response to the enthusiasm of the fiery French Cardinal Lavigerie, groups of White Fathers founded stations in Buganda, in Unyamwezi, and on both sides of the Tanganyika lake.

All this missionary activity was, in its early stages, essentially non-political—at all events in a nationalist sense. But it was frequently regarded otherwise. As the Zanzibaris at the Buganda court from the first suspected the Christian missionaries as political rivals, so the Mrima Arabs on Lake Tanganyika did their best to frustrate their efforts to set up mission stations there. And the missionary penetration had indeed considerable effect in making known something of the African continent and its peoples and resources to a wider European public. There was by this time a growing commercial and scientific interest, exemplified on the one hand by the efforts of speculators of various nationalities to obtain concessions from the Sultan of Zanzibar, on the other by explorations sponsored by learned bodies, such as the Royal Geographical Society's expedition under the enterprising leadership of the young Joseph Thomson. And more dimly behind these commercial and scientific probings there was already hovering the shadow of interests that were decidedly national and political. There were certainly political ambitions at the back of the German expedition sent out to Lake Tanganyika under the auspices of two German societies which in 1880 combined to form the German African

Society of Berlin; but, above all, the International Association sponsored by King Leopold of the Belgians, whose first expedition had some time before this, in 1877, set forth to found a station on that lake, seemed to the perspicacious to be the presage of a new epoch in East African affairs.[1]

Back at Zanzibar the impact of these new forces, together with the long-standing impediment to the ivory trade represented by the feud between Mirambo and the Arabs of Unyanyembe, was well before 1880 giving concern to Sayyid Barghash—who in 1870 had succeeded his brother Majid as sultan—and to his British mentor the Consul-General Sir John Kirk. Kirk pressed the sultan to consolidate his authority gradually inwards from the coast. Farther inland Zanzibar policy, when that of using the Unyanyembe Arab colony as a cornerstone of political influence had failed, was for a time to support Mirambo.[2] But this support was, at Kirk's instance, abruptly abandoned in 1880, when two members of one of the International Association's expeditions were murdered by a war-party under one of Mirambo's client chiefs. The incident was no more than symptomatic of the general situation in the interior. It was impossible even for a ruler as powerful as Mirambo to guarantee unfailingly the good behaviour of the unruly and predatory forces upon which he depended. Nevertheless it shook virtually beyond repair the confidence placed in him in Zanzibar; and although there was no dramatic decline in Mirambo's fortunes, his dreams of establishing an enduring state underpinned by European support were thenceforward doomed to failure. His 'empire' disintegrated on his death three years later. By 1882 it was above all on Tippu Tip that Barghash was setting his hopes of reinforcing his position in the interior. Tippu's return to the coast, after twelve years' absence, was anxiously awaited.

Up to this point it was still the Unyamwezi region that was in question—Barghash's intention was to appoint Tippu as *wali*, or governor, of Unyanyembe. But within a few hours of the latter's landing in Zanzibar events had taken a different turn. Taking the Arab merchant aside, the local representative of Leopold's International Association, Captain Cambier (apparently

---

[1] See R. Coupland, *The Exploitation of East Africa*, 1939, ch. xiii–xvi; R. Oliver, *The Missionary Factor in East Africa*, 1952, ch. i.
[2] Kirk to Salisbury, 27 July 1878, F.O. Print 3928.

following up some earlier talk between them at Tabora) offered him an alliance. If Tippu would act as its agent in his Upper Congo domain, the Association would supply him with the munitions to defend that domain against all comers and the profits of administration would be shared between them. On hearing of this proposition—which Tippu immediately reported to him—Barghash dramatically changed his own plans. Of what use was it to safeguard the central region if the main ivory lands to which it was but a pathway were to be filched and to send their produce westwards down the Congo? At length he fully grasped the purport of Stanley's crossing of the continent and of Leopold's expeditions since. Two days later he peremptorily forbade all further enlistment in the expedition which the International Association was at that moment engaged in recruiting to ship to the Congo mouth. To Tippu he offered all the resources at his disposal, if only he would return forthwith to his Congo territories and maintain Zanzibar authority there.[1]

Tippu had evidently been weighing up the situation for some time past. A conversation with him is recorded for instance by the Belgian Jerome Becker at Unyanyembe, in which Tippu was frankly balancing what he could discover of the resources of the International Association against what he already knew, from their joint expedition through the Congo forest five years earlier, of the character of Stanley, its most notable representative.[2] And those who most respected his abilities did not doubt that his policies were dictated primarily by his hopes of personal profit. In the event, it was Barghash's bid that he now accepted; and on behalf of the sultan that over the next three years he undertook the consolidation and yet further extension of his Upper Congo territories. Barghash commissioned him to take possession of the whole Congo basin from Lake Tanganyika to Banana, but above all to prevent the westward shipment of ivory down the river.[3] Towards the end of 1883 he set off

[1] *Maisha ya Hamed bin Muhammed*, pp. 139–41; Miles to Granville, 17 Jan. 1883, F.O. Print 4914.
[2] Becker, *Vie en Afrique*, ii. 47.
[3] Kirk to Granville, 23 Oct. 1884, F.O. Print 4914. This letter, and the information gleaned by L.M.S. missionaries on Lake Tanganyika, shows that the general drift of the arrangement between Barghash and Tippu was widely known or suspected. See also Jones to L.M.S., 2 Dec. 1884, Swann to L.M.S., 9 Aug. 1884, Hore to L.M.S., 23 Apr. 1884, Box V, Central Africa, L.M.S. Archives; C. Coquilhat, *Sur le Haut Congo*, Paris, 1888, p. 408.

once more up-country at the head of a large caravan and 1,000 guns. On Lake Tanganyika, which he crossed in the following April, the Zanzibari followers of the French traveller Giraud, after eighteen months of loyal service, deserted to a man upon hearing a rumour that hostilities were imminent between the sultan's forces, under Tippu, and Stanley in the Upper Congo. Tippu first took steps to assert the sultan's authority in the lands immediately south and west of the lake. Then he pushed westwards into Urua, and, somewhat farther to the north, carried fire and the sword across the Lomami and to the banks of the Sankuru river. Finally, going once more down the Congo, he established himself, with a strong force, near Stanley Falls—taking occasion, incidentally, to call to order a fellow Arab trader who had, as he himself had for the time being declined to do, entered into commercial relations on the Congo with the representatives of the International Association. From Stanley Falls some twenty large expeditions were dispatched into the neighbouring regions. Barghash meanwhile from Zanzibar supplied munitions and exhortations.

At the end of three years of rapacious achievement Tippu retraced his steps, once again at the head of a huge ivory caravan, to Zanzibar, where he arrived in September 1886. But even before he left the Congo he knew that at least in Barghash's main design he had been overtaken by events in Europe. At the Berlin Conference at the end of 1884 the seal of approval had been set on Leopold's claim to the whole Congo basin. In the following February, Bismarck's *Schutzbrief* asserted German rights over a large though ill-defined area on the mainland opposite Zanzibar. At that moment a mixed international commission, with German, British, and French representatives, was engaged in reducing the sultan's recognized interests on the mainland to a thin strip of coastline. Tippu found Barghash a disillusioned man. 'Hamid,' he said, 'I must beg your forgiveness. I no longer have any hope of keeping the interior. The Europeans here in Zanzibar are after my possessions. Will it be the hinterland they want? Happy are those who died before now, and know nothing of this. And you, too, you are yet a stranger to these events: but you will see how the land lies.'[1]

[1] *Maisha ya Hamed bin Muhammed*, p. 151. From this source comes also the main

This was the end of the last serious or concerted attempt to assert the authority of Zanzibar over the interior. In fact, it is clear that Tippu himself had made up his mind to come to some kind of terms with the European powers even before he left Stanley Falls.[1] It was not, indeed, the end of conflict between the old order in East Africa and the new. For another ten years sporadic outbreaks of violence, by both Arabs and those African peoples associated with the Arab trading system, continued to challenge the penetration of European influence. But their ultimate failure was not in doubt; and in the light of the changes which in this last phase had come over the economic and political system which the old order represented, it was little to be regretted. Abolitionists might claim with satisfaction that at the coast, while the 'legitimate' commerce in ivory, cloves, and rubber prospered, the traffic in slaves, whether for export or for the coastal plantations, had virtually ceased: yet in the heart of the continent the decade between 1880 and 1890 was perhaps more costly in human life and misery than any that had gone before.

Up to about the middle of the 1870's the political and economic system orientated towards Zanzibar, unscrupulous and unsavoury as it largely was, had still been basically a trading system. In this it had been observed to contrast for instance with the openly predatory methods of the ivory-hunters working southward from the Nile into the Sudan. Zanzibar traders were said to regard with abhorrence the pillage and rapine carried on by these 'Turkish' dealers in the north, who in turn had been abashed on being confronted at the court of Buganda by Zanzibaris who were in the habit of paying in trade goods for what they obtained. But as the use of firearms became widespread, as ivory became scarcer and dearer, as the threat of European competition, both political and economic, became explicit, the situation changed. The place of the cotton, beads, and wire that had formerly gone inland was to an ever-increasing extent taken by guns and powder; and in the interior those

outline of Tippu's campaign; which, however, can in the main be checked by reference to the L.M.S. Archives and to the accounts of International Association informants, especially Coquilhat, *Haut Congo*, pp. 400–14.

[1] Ibid., pp. 408–14. For the relations of the International Association and subsequently the Congo Free State with the Arabs at Stanley Falls, see P. Ceulemans, *La Question arabe et le Congo (1883–1892)*, 1959, especially ch. ii.

who could not purchase these desirable items with ivory purchased them with human beings. Even rulers of the status of Mirambo and Tippu Tip, although they were able at the centre of their spheres of influence to achieve a high measure of peace and security of life and property, did so largely at the cost of allowing a free hand to their subordinates on the perimeter. What this free hand meant in burnt villages, destroyed crops, slaughtered, starved, and kidnapped inhabitants, may be read in half a dozen travelogues of the period. Thus the German traveller Hermann Wissmann records his second journey through the Upper Congo region:

> Where formerly thousands of Benecki, the inhabitants of the strikingly beautiful and prosperous villages, had joyfully welcomed us... where in peace and amity we had been conducted from village to village—we now found a waste, laid bare by murder and fire... the clearings in the bush on both sides of the straight tracks, which three years before had been occupied by the neatly cultivated plots of the Benecki, were now overgrown with grass of a man's height, whilst here and there a burnt pole, a bleached skull and broken pottery were left as reminders of our former friends.

Wissmann also gives a blood-curdling description of one of the camps from which the so-called Arab 'administration' of these districts was carried on—an agglomeration of some 3,000 souls, foul with the stench of corpses, and subsisting on the pillage of the surrounding country.[1] Similar accounts could be multiplied; from the later works of Stanley, from the writings of members of the expedition sent up the Congo to give aid to Emin Pasha in Equatoria, from the experiences of officers in the service of the Congo Free State.

Perhaps the most appalling feature of these accounts is their reiterated emphasis on the traffic in, and merciless exploitation of, human beings. The virtual stoppage of the slave trade on the east coast, combined with the effects of war and famine, made slave labour in the interior cheap and plentiful; and man-dealing there, so far from diminishing, seems to have grown to ghastly proportions. Slaves were now used to undercut the porter-carried ivory of European caravans—it is from this period

[1] H. Wissmann, *My Second Journey through Equatorial Africa, from the Congo to the Zambesi in 1886 and 1887* (trans. M. J. A. Bergmann), 1891, pp. 183-4, 200-4, 209-10.

that there come most of the circumstantial accounts of 'raw' slaves carrying ivory and of women forced to abandon the child at the breast rather than the precious tusk. They were employed to cultivate not only the estates of Arabs, but the plots of Manyema, Nyamwezi, Gogo, and Yao masters grown rich on the profits of the ivory trade, and, nearer the coast, of rubber. They were acquired as wives and concubines; and above all in the Upper Congo, where undoubtedly the ravages of war and ivory-raiding were at their worst, and where cannibalism was in many places already a practice of long standing, captured enemies were frequently eaten by the victorious force. It is revealing to see how even the missionaries, who for long had set their faces steadfastly against political entanglement, and some at least of whom had been ready to welcome the Arabs as allies in the work of introducing civilization and stability, came in these last years to see in effective occupation by European powers the only possible relief from a world of bloodshed and perpetual insecurity.

IX

THE NORTHERN INTERIOR
1840-84

D. A. LOW

JUDGED by the extent of their horizons, two tribes were pre-eminent in the northern interior of East Africa during the middle part of the nineteenth century—in the western regions the Ganda, in the east the Masai; the one constituting a centralized kingdom, the other comprising several autonomous sub-tribes.

From their base upon the north-western shore of Lake Victoria, the Ganda were by the 1870's dispatching royal embassies to Gondokoro, north of the modern Uganda boundary with the Sudan; to Zanzibar, 800 miles away on the south-east; and even to Britain itself. They were, too, extending their more immediate influence southwards down the west side of Lake Victoria to Buzinza at its southern edge, eastwards to the verges of Kavirondo country, whilst probing westwards into the neighbouring kingdoms of Bunyoro, Toro, and Ankole. The immediate range of the Masai was even greater. From the east side of Lake Victoria their orbit struck north of Mount Kenya to the edges of Boran country, curved eastwards to the sea north and south of Mombasa, and then ran southwards to the verge of Gogo and Hehe country south of Mount Kilimanjaro. The magnitude of their arena was unique in East Africa. Yet unlike the Ganda, who had still to reach their zenith, the Masai had probably seen the apogee of their power by the middle years of the century, and had already begun their subsequent decline.

Into this whole region, of which few parts remained unaffected by these two tribes, there came at this time four distinct incursions from without. There was, first of all, a miscellaneous movement of peoples from the north and north-east—preliminary encroachments by the Ethiopians, a decisive advance by the Somalis, raids by the Suk (Pokot), the first thrust of the Turkana

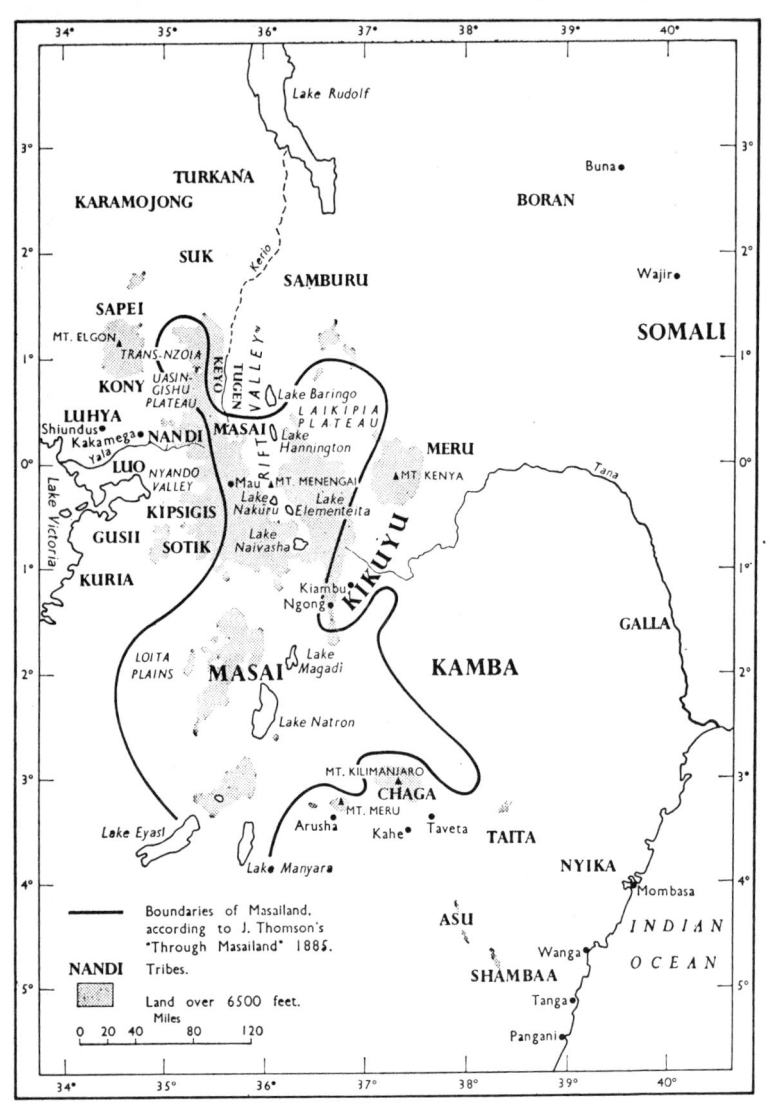

MAP 13. The north-eastern interior, 1840–1884

into the confines of Masailand, the advance of the Karamojong to the Turkwel river region,[1] and the convergence southwards

[1] C. A. Turpin, 'The Occupation of the Turkwel River Area by the Karamojong Tribe', *Uganda J.* 12, 1948, pp. 161–2.

to the shores of Lake Kyoga of the Lango, the Kumam, and the Teso. Secondly, there came traders. These were mostly Arab traders (to use a generic term) who hailed from two main

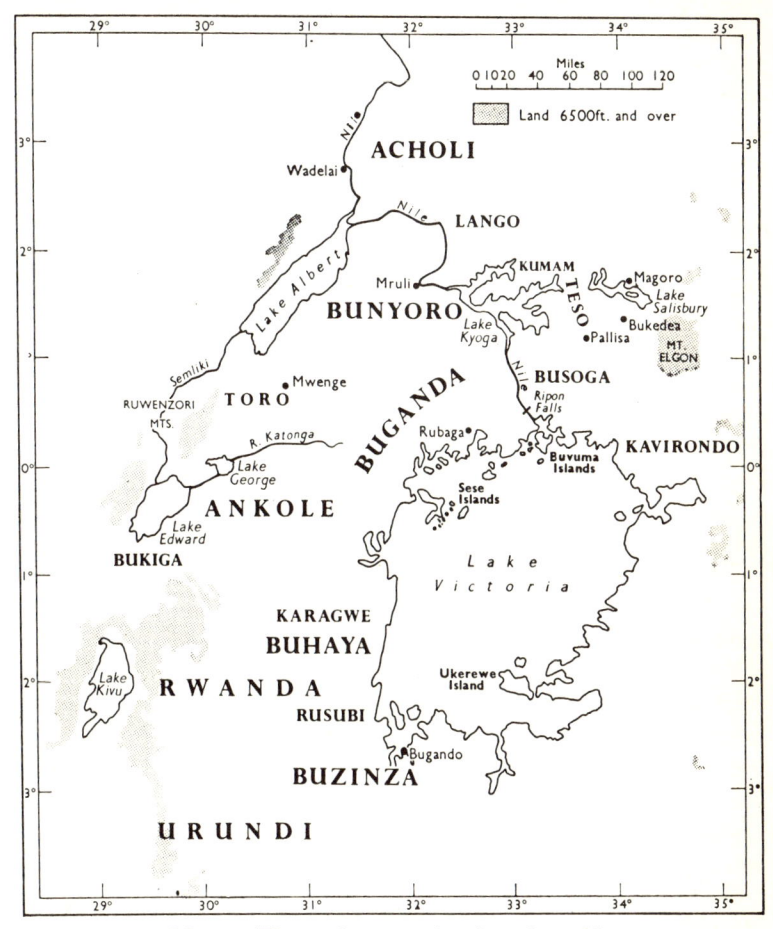

MAP 14. The north-western interior, 1840–1884

starting-points—Khartoum in the north, and the Zanzibar coast in the east. By the 1870's the Khartoumers had reached as far south as Bunyoro and Buganda, while the Zanzibaris had wound round the western side of Lake Victoria to enter these same kingdoms from the south, and had also thrust through Masailand. There were other traders as well, who reached

Samburu country from Brava on the Somali coast,[1] and some too from Ethiopia. Before 1884 there was never any overt clash between any of these different trading interests, yet, since their tentacles had already begun to intertwine, they would, no doubt, have increasingly conflicted, had not later events supervened.

The third invasion might have been the most important of all. In the 1870's the Egyptian government tried to establish an Egyptian empire in the region of the northern lakes of East Africa, and for ten years partially succeeded. They even sought to provide themselves with a base upon the East African coast north of Mombasa. Yet it was probably the fourth, seemingly the most modest intrusion of these years, that had the greatest influence of all. This came from Christian missionaries. At first there was an abortive attempt in the 1850's to establish a Roman Catholic mission at Gondokoro.[2] Simultaneously there were plans for a chain of Protestant mission stations spreading inland from Mombasa.[3] But while these too failed to produce anything more than a single mission station in the Taita hills, both Protestant and Roman Catholic missionaries had by 1884 established themselves in the inland kingdom of Buganda, and had won converts from within its far from decadent tribal society. Each incursion, though largely promoted by external agencies, was at the same time prompted by prevenient occurrences within the northern interior itself.

In the region between Lake Victoria and the sea lay Masailand. Its core was shaped like an hour-glass, slightly tilted to one side, with its base resting on the slopes of Mount Kilimanjaro, and extending westwards past the Arusha on the slopes of Mount Meru to the Loita Hills beyond. It then tapered northwards to a narrow neck around Lakes Naivasha and Nakuru, before broadening out northwards up the Rift Valley to Lake Baringo, north-eastwards on to the Laikipia plateau, and north-westwards to the Mau and the Uasin Gishu, and the trans-Nzoia region beyond.[4]

---

[1] T. Wakefield, 'Routes of Native Caravans from the Coast to the Interior of Eastern Africa', *J.R.G.S.* 40, 1870, p. 322; C. New, *Life, Wanderings and Labours in Eastern Africa*, 1873, p. 460; J. Christie, *Cholera Epidemics in East Africa*, 1876, pp. 212, 238.

[2] C. P. Groves, *The Planting of Christianity in Africa*, vol. ii, London, 1954, pp. 82–85.   [3] Ibid., pp. 109–11.

[4] See end map in J. Thomson, *Through Masai Land*, 1885.

Unless they were forced to take to settled agriculture the Masai had two overriding passions—cattle and warfare. Cattle were both their pride and their sole livelihood. They drank the milk of their animals, quaffed their blood, and ate their flesh. Moreover, it was apparently their belief that it was the decree of the gods above that all the world's cattle belonged to the Masai alone. Two consequences ensued. First, the Masai needed cattle, and not infrequently needed to replenish their herds. Secondly, they had no compunction in laying their hands upon anyone else's cattle within their grasp. To this was added their delight in, and propensity for, warfare. To it they trained all their young manhood, submitting them to a rigorous discipline that did not require, as it seems, an elaborate hierarchy of leaders, since it was steeped in the singleness of purpose of a whole people. Their warfare, however, was never directed to the conquest of territory or subjects. Its sole objectives were the procurement of cattle, and the proving of youth's manhood.[1]

From their base north and south of Lake Naivasha Masai warriors raided to all points of the compass. They would traverse 300 miles and think nothing of it.[2] Catching the unwary, they would surge forward in a rush, themselves protected by a solid phalanx of buffalo-hide shields, with two lines of ostrich plumes waving aloft, sling their rhinoceros-horn clubs, hurl their broad-bladed spears, and vent their spleen upon all within reach. Then, having rustled up the available cattle, they would be gone, as quickly as they had come, leaving naught but a trail of terror behind. They probed westwards to Kuria country on the shores of Lake Victoria; swept north against the Kony and Sapei of Mount Elgon; fought with the Boran and the Rendile; cut across to the Tana; probed south towards Mombasa and even Bagamoyo; curled around Mount Kenya against the Meru and the Tharaka; belaboured Kilimanjaro.[3]

And yet out of this troublous sea of Masai raids there thrust an archipelago of islands which they never overran. Unlike the Zulus, the Masai never fought as a single compact army. They were never, moreover, numbered in anything more than

[1] G. W. B. Huntingford, *The Southern Nilo-Hamites*, 1953, and the authorities there cited.
[2] Christie, *Cholera*, p. 208.
[3] Thomson, *Masai Land*, p. 438; J. L. Krapf, *Travels, Researches and Missionary Labours in Eastern Africa*, 1860, pp. 190, 358, 382.

tens of thousands, and, most important of all, wherever there was high ground and fertile ground (and the two were naturally linked) other peoples held them at bay. For the fertile areas carried populations, not of tens, but of hundreds of thousands—certainly enough to repel most Masai assaults by sheer weight of numbers. Fertile soils, moreover, bore not only people, but forests—forests in which the phalanxes of the Masai found it difficult to operate, and in which their heavy-bladed spears, so ascendant in the open, were no match for bows and arrows tipped with poison. It was the same where the hill-sides were steep. Even the Masai found it difficult to maintain formation as they scrambled up rocky slopes. Against them the murderous fire of arrows from cornices above could be disastrously efficient.[1]

Almost all the Bantu and Nilo-Hamite peoples of this region that survived their attacks enjoyed one, if not all, of these natural advantages. Those, like the Nyika, down by the coast, who did not, were exceedingly vulnerable. They might build elaborate defence works of timber and bush, and stall-feed their cattle, but these expedients generally availed little.[2] The Kikuyu, however, behind their forest fringe; the Taita in their hills; like the Kamba, the Nandi, the Tuken, the Keyo, the Meru, and the others who survived, were all blessed with natural sanctums, from which, unless they were foolish, ventured too far, or neglected their look-outs, they could defy the Masai. There was frequent warfare on the fringes of these fastnesses, but it was rare for the Masai to penetrate very deeply.[3]

In consequence, during most of the nineteenth century there existed a broad equilibrium between the Masai and their neighbours. Apart from the raids of the Suk and the Turkana in the region around Lake Baringo, and the subsequent emergence of the Nandi from their hills, the hold of the Masai upon the plains and open plateaux was never seriously threatened. At the same time, with perhaps two exceptions, they nowhere extended their hold. In the 1850's and 1860's they drove the

---

[1] e.g. H. H. Johnston, *The Kilimanjaro Expedition*, 1886, pp. 396, 538–9; New, *Life, Wanderings*, pp. 469–70; Thomson, *Masai Land*, p. 88; Christie, *Cholera*, pp. 203–4.

[2] Christie, *Cholera*, p. 211; New, *Life, Wanderings*, p. 114; R. F. Burton, *Zanzibar, City, Island and Coast*, 2 vols., 1872, ii. 70–71.

[3] Krapf, *Travels*, p. 362; New, *Life, Wanderings*, pp. 330, 468; Thomson, *Masai Land*, p. 308; Wakefield, *J.R.G.S.* 1870, p. 309.

Nyika and their confrères back nearer the coast, and the Galla north-eastwards across the Tana.[1] But the Masai attempt in the 1830's to extend to the south was checked by the Gogo and the Hehe, and was never seriously repeated.[2]

Instead, the major occurrence for the Masai in the middle years of the century was a long series of civil wars. The reasons for these are still largely unknown. But it is probably necessary to recall at the outset the general circumstances in which they took place. For they could scarcely have occurred upon quite the scale that they did, if the Masai had had to be at all seriously concerned about their outward defences. Nor might they have so turned in upon themselves, if they had been fully occupied in extending their domain. These wars are probably indeed a further reflection of the more general equilibrium which had overtaken the region as a whole. If this is so, then it is not altogether surprising that a people so addicted to the practice of warfare should have turned upon each other for lack of anyone else of substance to fight. It is probably no coincidence that some of these wars were more in the nature of jousts or tournaments. About 1880, for instance, two huge kraals were built in that important corridor between the north and south of the Masai homeland, between Lake Nakuru and Lake Naivasha; in the arena between, the warriors from each side hurled themselves at each other, while their women stood by and cheered them on.[3]

But there were obviously more serious considerations as well. Some of these wars were fought, as we know, over cattle; at a time when cattle were not so easily procured from other peoples (who were careful to hide them away), the simplest method by which a Masai sub-tribe could restock its herds was to attack one of its fellows. Many, too, were fought, no doubt, over grazing and water rights. Several of the critical battles took place near Lake Nakuru; possession of the adjoining corridor gave command of its spectacular chain of lakes, and of its choice of approaches to several areas of good grazing to the northward and southward. Command of the corridor was, moreover, crucial to any attempt to dominate the whole of Masailand.

Here two points may be made. First, it seems clear that from

---

[1] Christie, *Cholera*, p. 211; New, *Life, Wanderings*, pp. 193 sqq.
[2] Thomson, *Masai Land*, p. 414. [3] Ibid., pp. 347-8, 416-17, 438-9.

about the middle of the nineteenth century the Purko, an important Masai sub-tribe from south of the corridor, had embarked on a career of aggrandizement. Secondly, some of these wars had the appearance of being struggles between the adherents of different *laibons*.[1] These key Masai figures were primarily ritual experts, who were mainly dependent for their authority upon their prophetic powers. Anthropologists have been at pains to insist that they enjoyed no legislative or executive authority. Yet by their ritual control over the direction of warfare they exercised a good deal of indirect political power, and constituted each sub-tribe's recognizable 'head'. It seems clear that a major *casus belli* for the Purko and their neighbours (who were followers of the laibon Mbatian) against the Laikipia, a sub-tribe in the north-east, was the support which the Laikipia gave, together with the Samburu, the northernmost sub-tribe of all, to a rival laibon named Koikoiti.[2]

Although the Masai were clearly divided into a number of largely autonomous sub-tribes, who nevertheless grouped themselves about different laibons, these sub-tribes seem at the same time to have fallen into at least two discrete categories—the 'Masai proper' and the 'Kwavi'. Herein, however, lies an enigma. For it is not at all certain how precisely the Kwavi are to be defined.[3] The commonest assertion declared that they were the agricultural as distinct from the pastoral Masai. Certainly the remnants of all those who were called 'Kwavi' were by the end of the nineteenth century to become agriculturalists: but three-quarters of a century earlier many of them were still primarily pastoralists. Yet they seem at that time to have been clustered in two main groups—one around Mount Kilimanjaro, the other north of the Naivasha–Nakuru bottleneck.[4] There is much doubt about the connexion, if any, between them: one theory, for instance, avers that some of the Kilimanjaro group derived from the northerners;[5] another that they gave birth to

---

[1] Cf. H. H. Johnston, *The Uganda Protectorate*, 2 vols., 1902, ii. 834–5.
[2] Here, and elsewhere, Mr. Alan Jacobs and Mr. Frank Holford-Walker have been most generous with their assistance. Both of them have lived with the Masai, the former as an anthropologist, the latter as Moran Officer of the Kenya Government. They are not, of course, responsible for the conclusions reached.
[3] Johnston, *Kilimanjaro*, pp. 312–3, 405–6.
[4] Thomson, *Masai Land*, pp. 414–17.
[5] e.g. F. J. Jackson, *Early Days in East Africa*, London, 1930, pp. 114, 200.

some of the northerners.[1] It is plain, however, that they both encountered the enmity of the interlying group of sub-tribes; and it is certainly convenient, and in accord with much of the evidence, to call the latter 'Masai proper', and the former 'Kwavi', even if their distinguishing characteristics remain somewhat uncertain.[2] For on this reading these wars emerge as, in the main, those in which the 'Masai proper' fought the 'Kwavi'.

Their details are often obscure, and it is all too easy from the paucity of evidence to elide one war into the next, and to distinguish between two that were both really parts of the same one. Nevertheless, one may hazard some account.[3] There is, first of all, a tradition about the early destruction of a sub-tribe called the Ilogolala by a combination of three others—the Matapatu, the Keekonyukie, and the Laitokitok, who lived northwards and westwards of Mount Kilimanjaro. Whether this preceded, succeeded, or coincided with a second war at this period cannot be ascertained; but in a further account a sub-tribe living eastwards of Mount Kilimanjaro, after a repulse in the south from the Gogo, is said to have turned against the Kissongo, a sub-tribe in the west, only to be repulsed once more, and before long to be almost destroyed by a combination of sub-tribes, who allied with the Kissongo against them. There is certainly substantial evidence of internecine conflict around the north and east of Mount Kilimanjaro during the forties and fifties, in which the victors were all undoubtedly 'Masai proper', and the vanquished seemingly 'Kwavi'.[4]

In the meanwhile away to the north-west after an inconclusive encounter, probably in the 1840's, between the Laikipia and the Uasin Gishu—the two northern 'Kwavi' sub-tribes from the plateaux east and west of Lake Baringo—there occurred the so-called Losegallai war. In this the Purko from south of Lake Naivasha, who were undoubtedly 'Masai proper',

---

[1] e.g. Thomson, *Masai Land*, p. 416.
[2] Wakefield, Christie, Thomson, &c., *passim*.
[3] In addition to the evidence collected by Dr. Jacobs and Mr. Holford-Walker, accounts will be found in Thomson, *Masai Land*, pp. 414–17; G. R. Sandford, *An Administrative and Political History of the Masai Reserve*, London, 1919; *Southern Nilo-Hamites*, ch. ii; and other authorities cited by Huntingford, loc. cit.
[4] New, *Life, Wanderings*, pp. 293, 355; Burton, *Zanzibar*, ii. 72; Wakefield, *J.R.G.S.* 1870, pp. 303–6, 315.

allied with the Laikipia against the Losegallai, a 'Kwavi' sub-tribe, who lived between Lake Nakuru and the Mau. The Laikipia, however, apparently attacked the Losegallai on their own—it looks as if an alliance between 'Kwavi' and 'Masai proper' proved difficult to sustain—whereupon the Purko allied with their more immediate neighbours, the Il Damat and the Keekonyukie, and fell on the Losegallai, destroying all but a few remnants. It was probably the ominous threat of this victory to the remaining 'Kwavi' in the north that led the Uasin Gishu to concert an attack upon the Purko, sometime in the early seventies. Having allied with the Siria, a neighbouring sub-tribe from north of Lake Nakuru, they drove the Purko out of the Rift Valley on to the Kaputie plains south and east of Kikuyu. This seems, however, to have provoked a decisive intervention by the laibon Mbatian, who had succeeded his much renowned father, Subet, in 1866.[1] For Mbatian quickly summoned the Kissongo, the Il Damat, and the Loita to assist the Purko against the Uasin Gishu and the Siria. The latter were eventually driven to battle on the slopes of Mount Menengai, overlooking Lake Nakuru, where, it is said, their main body made a last desperate stand on the rim of the crater, till they were literally driven over its sheer sides. The victors then swept on to the Uasin Gishu plateau itself, and scattered the remnants.[2] But upon their withdrawal southwards, the Laikipia from the north-east, who had harassed rather than helped their Uasin Gishu neighbours, occupied Lake Nakuru; whereupon about 1880 the Purko and their neighbours returned to the attack. This time there was no walk-over victory, but it was not long before the Laikipia were routed. On each occasion the laurels went to those who managed to summon neighbouring sub-tribes to their assistance, and it looks as if the critical links were forged by the laibon. Although Mbatian was never even laibon of all the 'Masai proper', none the less, by 1884, he had attained an unrivalled pre-eminence.[3]

These wars had several important consequences. First, if the distinction drawn here between 'Masai proper' and 'Kwavi' is

[1] Wakefield, *J.R.G.S.*, 1870, p. 305; New, *Life, Wanderings*, p. 460.
[2] It is clear from Wakefield, *J.R.G.S.* 1870, p. 323, that this had not happened by 1870; it had happened by 1884, cf. Thomson, *Masai Land*, p. 467.
[3] His sons Lenana and Sendeu fought each other for his inheritance. See Vol. II of this History.

valid, then they saw the elimination as powerful sub-tribes of all those dubbed Kwavi. They certainly witnessed the rout of all those sub-tribes living north of the corridor (except the Samburu, the northernmost of all) and the destruction of at least one sub-tribe near Kilimanjaro. Secondly, 'Kwavi' survivors scattered in all directions. Some Kilimanjaro 'Kwavi' moved north to the Kaputie plains; some settled with the Taveta;[1] some with the Chaga;[2] some with the Arusha;[3] while others moved down south of Kilimanjaro and the Pare Hills to settle on their own.[4] It even seems possibile that it was 'Kwavi' fugitives from Kilimanjaro who, in their desperation, were responsible for the raids upon the coast which some Europeans witnessed after the middle of the century,[5] and which, by being blamed on the Masai, gave the 'Masai proper' an unwarrantably evil reputation. In the north-west remnants of the Losegallai settled in the Nyando Valley down towards Lake Victoria till they were destroyed by the Nandi and the Kipsigis. Some Uasin Gishu took service as mercenaries in Kavirondo; some Laikipia in Kikuyu; while other fugitives joined some earlier Kwavi settlements in the Lake Natron area, and at the two Njemps villages on Lake Baringo.[6] Thirdly, these wars reduced the occupied core of Masailand. Pastoral Masai no longer inhabited the regions eastward of Kilimanjaro; nor were they to be found any longer upon the Uasin Gishu or Laikipia plateaux. Large areas of Masailand were denuded of population.[7] None of this, however, affected the dominion of the remaining Masai over the plains and open plateaux. Of all their neighbours only the Nandi dared to venture forth into what had previously been undisputed Masai territory.

Nevertheless these were not years of unmitigated prosperity for the 'Masai proper', for on several occasions they were severely scourged by disease, both human and animal. To all types of infection Masai were, of course, especially prone. For

[1] Johnston, *Kilimanjaro*, p. 78.
[2] Thomson, *Masai Land*, p. 151.
[3] Wakefield, *J.R.G.S.*, 1870, p. 304.
[4] e.g. the modern Baraguyu of Tanganyika.
[5] Burton, *Zanzibar*, ii. 70 sqq.
[6] Johnston, *Uganda Protectorate*, ii. 796–7, 800–2.
[7] These were among the main areas in which Europeans subsequently settled. See Vol. II of this History. For contemporary evidence see Thomson, *Masai Land*, pp. 408, 467.

the tremendous area which they covered in their raids brought them into contact with a large number of other peoples, any one of which might at any time have an infectious disease to pass on. Moreover, such was the extent of the Masai's movements, that once an infection had been caught by one group of them it was liable to spread to all others with alarming rapidity. Such, we know, was the story in 1869 when a Masai raiding party contracted cholera from the Samburu. It spread like wildfire, and in a month or two was killing off large numbers of Masai all over Masailand.[1] It must have been the same with the extremely serious smallpox outbreak in about 1883-4, and with the pleuro-pneumonia which decimated their herds during the same decade.[2] It was these lethal epidemics, rather than anything that occurred in war, which constituted the most serious inroads upon the strength of the 'Masai proper' at this time, and marked the first stage in their decline from their previous most formidable position.

Amongst the Bantu and Nilo-Hamite neighbours of the Masai, the rise of the Nandi was probably the most notable occurrence of these years. All the Nilo-Hamite peoples on the west side of the Rift Valley were warlike in their ambitions, and although they often stood on the defensive against the Masai they were generally active themselves against the Nilotic and Bantu peoples around the Kavirondo Gulf—the Gusii, the Luo, and the various Luhya groups. Amongst themselves the Kipsigis and the Nandi were very closely linked by tradition and blood, but for military purposes the Kipsigis seem to have allied more frequently with their fellows in the south—the Buret and the Sotik. Their joint raids westwards would put a stop to the endless scuffles amongst the Bantu Gusii[3] and the Nilotic Luo,[4] and it was by no means always that the victory went to the Nilo-Hamites. The Nandi, too, are said to have suffered one serious defeat at the hands of the Luo.[5] Their destiny, however, was soon substantially altered by the collapse

[1] Christie, *Cholera*, pp. 225, sqq.
[2] Thomson, *Masai Land*, p. 564.
[3] P. Mayer, 'The Lineage Principle in Gusii Society', International African Institute, Memorandum XXIV, London, 1949, pp. 3, 10, 12.
[4] A. W. Southall, 'Lineage Formation among the Luo', ibid. XXVI, 1952.
[5] A. C. Hollis, *The Nandi*, Oxford, 1909, p. 50.

of the Uasin Gishu. In the 1850's a Nandi attack upon them had dismally failed. But in the 1870's the tables were turned.[1] In a major action the Nandi first defeated the Uasin Gishu themselves, and then shortly afterwards leapt into the breach formed by the near annihilation of the Uasin Gishu by the 'Masai proper'.

Some time previously the Nandi had taken unto themselves a laibon-like figure (according to Nandi tradition from the Masai—on the principle that someone else's ritual, since it is more mysterious than one's own, is likely to be more efficacious) whom they called their *orkoiyot*. This was a critical development, for it imparted to the Nandi before any of their non-Masai neighbours that *élan* which had previously been the monopoly of the Masai. As their military success under the *orkoiyot*'s ritually inspired guidance expanded, so he became established as the major symbol of their tribal solidarity, and reinvigorated the direction and purposefulness of its energies.[2] Moreover, since with the defeat of the Uasin Gishu the Nandi were no longer thrown on the defensive, they were now able to spare more men for their raiding parties; and with the largely deserted Uasin Gishu plateau on which to pasture their cattle, they were able to enlarge their herds very considerably. By 1884 they were throwing back a Masai raid from the Rift Valley (killing half its participants),[3] and were themselves raiding southwestwards around the shores of the Kavirondo Gulf; westwards through Luhya country to the slopes of Mount Elgon; and northwards against the Karamojong, the Suk, the Keyo, and the Tuken. Of all their neighbours only their kinsfolk the Kipsigis remained immune from their depredations. By 1884 they were by far the most formidable of the tribes between the Rift Valley and Lake Victoria, and were still enraptured by their first flush of success.[4]

Thus it was that the defeat of the Uasin Gishu brought such scant respite to the various Kavirondo peoples. Some of them—particularly the Bugusu in the north-east—used to surround their villages with mud walls ten feet high and two feet thick

[1] This was almost certainly not before the 1870's. See Wakefield, *J.R.G.S.* 1870, p. 309.   [2] Hollis, *Nandi*, pp. xviii, xxii.
[3] Thomson, *Masai Land*, p. 566.
[4] Here, and elsewhere, Mr. A. T. Matson has been most generous with his help; also Mr. John Nottingham.

as some means of defence.[1] But most others remained perilously vulnerable. In any event, both the Luhya and the Luo were very loosely organized politically. Each sub-tribe amongst the former consisted of a number of more or less sovereign clans;[2] whilst the clans of the latter, though localized, were frequently mutually hostile.[3] Only in western Buluhya, amongst the Hanga (Wanga), was there a political authority with any extensive power. Here a substantial little chieftainship existed. The Nabongo of the Hanga was primarily, no doubt, no more than the chosen head of his lineage, but the extent of his authority bore many similarities to that of the neighbouring rulers in Busoga. Such authority in skilful hands gave the Hanga a decided advantage over their neighbours, and in the middle years of the century their chief Shiundu attained a substantial local pre-eminence. At the peak of his power the Hanga dominion extended from the Yala river in the south to southern Bukusu in the north, and from the borders of Samia in the west to Kakamega in the east. Yet by 1884 there were serious difficulties. Even in Shiundu's day the Hanga had been divided; for his brother Sakwa had set himself up in some independence a few miles away, and upon Shiundu's death, some time in the early 1880's, Sakwa took his place as the foremost of the Hanga. Mumia, Shiundu's son and successor, soon found himself deserted by his father's followers; whilst the adjacent peoples naturally took advantage of the weakness which the family rivalry entailed to break loose. It is scarcely surprising that Mumia should now have taken steps to enrol Uasin Gishu fugitives in his service; that he should have encouraged coastal traders to extend their use of his headquarters as their base; and that he should have put out feelers for European help as well. He was anxious to resuscitate a disintegrating organism.[4]

It is generally suggested that the most significant occurrence amongst the tribes of the Mount Kenya region at this time was the expansion of the Kikuyu into the southern tip of Kiambu. But this was probably only one manifestation of a wider process.

[1] G. Wagner, *The Bantu of North Kavirondo*, London, 1949, i. 12–13.
[2] Ibid., pp. 20, 27.
[3] Southall, 'Lineage formation among the Luo', loc. cit.
[4] Ibid., p. 14; Thomson, *Masai Land*, pp. 472 sqq.; E. C. Dawson (ed.), *The Last Journals of Bishop Hannington*, 1888, pp. 209–10; Political Record Book, North Kavirondo District, vol. ii (District Commissioner's Office, Kakamega, Kenya).

Confined by fear of the Masai to a forested area, they and their neighbours had cut back the trees to provide themselves with new ground on which to grow their crops.[1] They were thus provided with a superabundant livelihood—certainly far better than anything they might have won had they spread to the plains. The abundance served, moreover, to diversify their internal trade, so that among the major features of their lives were the regular markets held at selected places every four days or so.[2] It also meant that they suffered less from famine than many of the other peoples of the region,[3] and that the growth of the Kikuyu population was rarely interrupted.

Yet there were now, it seems, under the old conditions, only two main lines of expansion open to them. The first ran northwards to the west side of Mount Kenya. Here they came face to face with the Masai. Only with the defeat of the Laikipia in the 1880's was the major obstacle removed. The second ran southwards into a region formerly occupied by some of the Dorobo, a small, primitive people, renowned for their hunting. By a precise, if complicated, series of arrangements, the Kikuyu procured land from the Dorobo, and made it their own, and by the 1880's had thrust down to within a relatively short distance of the forest edge, even at its southern tip. Kikuyuland was in fact near to bursting its seams. A crisis had not yet arisen, but it was plainly in the offing. The probability was that before long there would be increasing friction with the Masai to the south, and with the Kamba to the east.[4]

It is customary to affirm that the Kikuyu had no chiefs, and this by and large is true. The inhabitants of each ridge into which the country was divided rarely owed allegiance to anyone beyond, and were frequently at war with their neighbours. But on occasions sociological norms can be misleading. For the Kikuyu were throwing up, if not chiefs, then 'prominent individuals' (*athamaki*) of some considerable consequence.[5] In the

[1] L. von Höhnel, *Discovery of Lakes Rudolf and Stephanie*, 1894, i. 302 sqq.
[2] W. S. and K. Routledge, *With a Prehistoric People, the Akikuyu of British East Africa*, London, 1910, pp. 105–6.
[3] Ibid., p. 38.
[4] L. S. B. Leakey, *Mau Mau and the Kikuyu*, London, 1952, esp. pp. 2–8. See also H. E. Lambert, 'The Systems of Land Tenure in the Kikuyu Land Unit', *Communications*, N.S., No. 22, School of African Studies, Cape Town, 1950.
[5] Leakey, *Mau Mau*, p. 35; H. E. Lambert, *Kikuyu Social and Political Institutions*, 1956, pp. 105–6.

middle years of the century one hears bare names, like Mundu Wazeli[1] and Kippingo (presumably 'Kibinge').[2] But thirty years later these become distinct figures like Karuri of Metume, and Wangombe of Gakï. They, and a few others like them, had large numbers of followers and large herds of cattle. It is probably no coincidence that they owed their position mainly to their skill in leading their fellow warriors against the Masai. The cutting back of the forest by the closely packed population within was imposing important changes on the political structure of the Kikuyu. There was even an occasion when Karuri and Wangombe fought each other for the supremacy.[3] That nothing resulted is an indication of the difficulties involved in establishing an overlordship. But attempts were being made.

The political position amongst the Kamba was basically similar. The people of one hill-side fought those of the next. It is possible, however, to exaggerate these internecine conflicts. They were sometimes bloody, but there was frequently more bark than bite.[4] There were likewise amongst the Kamba the same 'prominent individuals', though the requirement of military skill seems here to have been less important than amongst the Kikuyu.

Wealth, a ready flow of language, an imposing personal appearance, and above all the reputation of being a magician and rainmaker, are the surest means by which a Mkamba can attain power and importance, and secure the obedience of his countrymen.[5]

It is worth remembering that the two Districts into which modern Ukamba is divided—Machakos and Kitui—are named after just two such men. Amongst the Taita the importance of 'prominent individuals' was probably less still, though even here they were not unknown. But the Taita seem to have been able to undertake successful raids on a very considerable scale against, for instance, the Shambaa, where not merely chieftainship but a rulership existed.[6] Amongst all these peoples external

[1] Burton, *Zanzibar*, ii. 63.
[2] M. Guillain, *Documents sur l'histoire, la géographie et le commerce de l'Afrique Orientale*, 3 vols., 1856, iii. 295.
[3] C. Cagnolo, *The Akikuyu*, 1933, pp. 101 sqq.
[4] e.g. Krapf, *Travels*, pp. 312–13; G. Lindblom, *The Akamba in British East Africa*, 1920, p. 201.
[5] Krapf, *Travels*, p. 355.
[6] New, *Life, Wanderings*, pp. 325–35.

warfare was one of the most important occasions when local differences were sunk in a greater unity of effort.[1]

Unlike their neighbours farther north, the Chaga had a fairly strong system of chieftainship.[2] Indeed the crux of the story upon Mount Kilimanjaro during these years lay in the successive attempts by various individual chiefs to attain to some supremacy over their neighbours. It was perhaps even before 1840 that a chief named Horombo managed to extend his overlordship from some small beginnings at Chimbii to cover two-thirds of the whole Chaga tribe from Gaseni to Kirua. At the height of his power he spent six years employing thousands of people to build forts for himself and his brother at Keni, with walls often six feet thick and anything up to 400 yards long. Horombo, however, was killed in a furious battle with the Masai, and his empire collapsed. Such men as Rengwa and Mamkinga of Machame, and Masaki of Kilema,[3] sought to step into his shoes, but it was not until the 1860's that any substantive successor appeared. This was young Rindi of Moshi (known to Europeans as Mandara), and within ten years he had become overlord from Tela to Mriti, with the chiefs of Mbokomo and Kibosho his creatures. Yet after a disastrous expedition against Usseri, Rindi found himself a fugitive with only a few followers at Kahe, south-eastwards of, and away from, Kilimanjaro. However, by 1880 he was invited back to his original chieftainship in Moshi, and by 1886 had recovered most of his old paramountcy. But it never covered the whole of Kilimanjaro. The obstacles to achieving overlordship were scarcely less here than amongst the Kikuyu or the Hanga.[4]

It seems probable that preoccupation with these events, together with an understandable satisfaction with their locally abundant material possessions, precluded the Chaga from taking to extensive trading. They were renowned as expert craftsmen in wood and metal, and sold the former to the Masai.[5] (There

---

[1] P. Bostock, *The Taita*, London, 1950, pp. 15, 35.

[2] The reasons for this are set out by Oliver, Chapter VI above. See also Krapf, *Travels*, p. 243; New, *Life, Wanderings*, p. 457; Thomson, *Masai Land*, p. 148.

[3] Krapf, *Travels*, pp. 245, 248-9.

[4] C. F. Dundas, *Kilimanjaro and Its People*, 1924, pp. 66-97; Johnston, *Kilimanjaro*, pp. 95-100; Thomson, *Masai Land*, pp. 213, 227.

[5] New, *Life, Wanderings*, pp. 453 sqq.; Guillain, *Documents*, iii, 284-6; Thomson, *Masai Land*, p. 149; Johnston, *Kilimanjaro*, pp. 439-41.

was even talk of a marriage alliance between Rindi and the daughter of the Laibon Mbatian.[1]) They welcomed traders from the coast, and would sell some of their captives in internecine war as slaves, but despite their strategic position they never seem to have ventured forth on trading expeditions themselves. It was the Kamba who were the main indigenous traders of the northern interior.[2]

A modicum of inter-tribal trade was fairly general over this whole region.[3] There was frequent trafficking, for instance, between the women of the Masai and the women of the Chaga and Kikuyu, who, even whilst their menfolk were fighting, were deemed to be immune from attack; the Masai generally traded hides, and in return procured some of the abundant grain and plantain crops of the hillier areas.[4] For the Kamba, however, unlike any of their neighbours, trade became a major preoccupation. This was partly, perhaps, because they were originally compounded of offshoots from the Taita and Kilimanjaro peoples, and thus had close connexions over a wide area. But the crucial point was almost certainly their greater proneness to famine—greater for them than for any of the other peoples of this region. For whenever famine occurred, they would send for food to Kikuyu country, which was generally not so harshly hit; and then, as famine conditions worsened, whole parties of Kamba would move away from their homeland *en bloc*. A severe famine about 1836 seems to have been of particular importance, for as a result at least one substantial group of Kamba moved to Rabai on the outskirts of Mombasa.[5] They naturally still maintained contact with their fellows in the interior, who in turn were in touch with the Mount Kenya peoples (to whom, for instance, they marketed a special poison that was invaluable for tipping arrows).[6] Such a dispersion placed the Kamba in a unique position, which, if they had not already done so before, they were now quick to exploit.[7] One observer was soon recounting how:

[1] Christie, *Cholera*, p. 230. [2] Lindblom, *Akamba*, p. 12.
[3] Wagner, *Kavirondo Bantu*, ii, 1956, pp. 162–5.
[4] Christie, *Cholera*, p. 203; Thomson, *Masai Land*, p. 308; von Höhnel, *Discovery of Rudolf*, p. 291.
[5] Krapf, *Travels*, pp. 142, 352 sqq.; Lindblom, *Akamba*, pp. 339–40, 351.
[6] Krapf, *Travels*, pp. 315–16. [7] Ibid., pp. 144, 351–2.

the Suahili purvey to the Wakamba cotton fabrics (Americano), blue calico, glass beads, copper and brass wire, ruddle, black pepper, salt, luaha, blue vitriol (zinc) &c., and receive in exchange, chiefly cattle and ivory.[1]

Even by the 1840's weekly Kamba caravans to the coast, carrying anything up to 400 frasilas of ivory, were being reported.[2] Kamba trading parties, moreover, were soon visiting, not only the Mount Kenya region, but the Mau and Gusii country beyond, and Lake Baringo and Samburu to the north.[3] There was even, in the person of Kivoi, a Kamba merchant prince, who was a familiar and respected figure in Arab circles in Mombasa, and who once dreamt of switching his transporting business from the usual route by land to a river route down the Tana.[4] Nineteenth-century African export trade was broadly dependent upon three groups of people—the collectors (or producers) of the original merchandise, the transporters to the coast or base, and the exporters (who dispatched the goods from the continent). Africans were usually, though not always, as we shall see, the initial collectors. Of the tribes of East Africa's northern interior, only the Kamba ever successfully dominated the whole transporting business to the coast on their line of trade.[5]

Nevertheless it was the Zanzibaris who now became the chief exploiters of the trade routes running inland from Mombasa. Their activity here was part of their wider penetration of all parts of East Africa during this period. This was in part prompted by the extensive trade of African peoples, like the Kamba, who revealed the commercial opportunities lying in the interior,[6] but it was mainly determined by the growing world demand for ivory.[7] The whole enterprise was, at the same time, greatly assisted by Sultan Sayyid Said's nurturing of the commercial base at Zanzibar and along the coast[8] (more especially by the encouragement he gave to Indian merchants,

[1] Ibid., p. 353.
[2] Guillain, *Documents*, iii. 211.
[3] Map in Krapf, *Travels*.
[4] Guillain, *Documents*, iii. 24; Krapf, *Travels*, pp. 295, 298; Lindblom, *Akamba*, pp. 298, 149–50, 314.
[5] Krapf, *Travels*, pp. 172, 232, 283; Lindblom, *Akamba*, pp. 12, 311, 352.
[6] Guillain, *Documents*, iii, 289–97, 383.
[7] e.g. Krapf, *Travels*, p. 140. Cf. Chapter VIII, above, pp. 268 sqq.
[8] R. Coupland, *East Africa and its Invaders*, 1938, ch. x.

who were, at a critical moment, ready to inject capital into the trade), and it was immeasurably strengthened by the industrial revolution in the West, which by producing manufactured arms and manufactured goods in large quantities and at a low relative cost made deep trading penetrations both very much safer than before and now potentially profitable.

It was not until about the middle of the nineteenth century,[1] however, that the Zanzibari routes through Masailand were opened up. At the outset there were two main possibilities. An avenue might be cut past Mount Kilimanjaro; another could be hoped for through the Kamba Hills.[2] The approaches from the coast to either could be by the Taita Hills, lying 100 miles inland from Mombasa, or via Usambara and the Pare Hills.[3] The Kilimanjaro route in the 1840's and 1850's ran, very roughly, due west to the southern edge of Lake Victoria;[4] but it never appropriated the trade of the southern and western shores of the lake. This was captured by traders using the Tabora route farther south, who struck north-westwards. Tabora was a much better up-country base than anything upon the Kilimanjaro route; and there seem to have been exceptional difficulties on the stretch between Mount Meru and Lake Victoria's southern shore.[5] There was even less prospect of the coastmen developing the Kamba route, convenient though its stepping stones were. Swahili traders using it reached Kikuyu country by at least 1849,[6] but during the ensuing decade they failed to win the battle for its control.[7] For the time being the Kamba hold upon the Kamba route remained unshaken.

It was not, therefore, until the 1860's that the Zanzibaris thrust out decisively into the hinterland.[8] But by the end of the decade their Masailand routes were firmly established. This break-out had three aspects. First, it sprang from the old Kilimanjaro route, which, frustrated in the west, turned north. Secondly, having failed to win the trade upon the west side of

---

[1] Cf. Guillain, *Documents*, iii. 266; Krapf, *Travels*, pp. 119, 144, 172.
[2] Ibid., p. 119; Burton, *Zanzibar*, ii. 61–68, 116–18.
[3] e.g. Guillain, *Documents*, iii. 279–84.
[4] Krapf, *Travels*, p. 277.   [5] Christie, *Cholera*, pp. 238, 241.
[6] Guillain, *Documents*, iii. 289–97.   [8] Krapf, *Travels*, p. 552.
[7] Cf. the accounts of Guillain, Krapf, and Burton, which deal with the 1840's and 50's, and those of Christie, Wakefield, and New which deal with the 60's and 70's.

Lake Victoria, the traders from Kilimanjaro now appropriated that upon the east side, and especially around the north-east corner. And thirdly, they captured the inland markets of the Kamba. Foiled in their attempt to enter the region westwards of Mount Kenya from the east, the Zanzibaris moved in from the south. Reaching the Aberdare Hills on the western edge of Kikuyu country from Kilimanjaro, they were soon within seventy-five miles of eastern Kikuyu, which their forerunners had tried to enter from the Kamba side. The Kamba thus found themselves outflanked up-country, by men to whom—and the irony is plain—they had boasted of the trading possibilities lying before them. As a result Kamba trade went into decline. By the 1880's the Kamba route was in the coastmen's hands.[1]

Up-country there were at least two areas where the Zanzibaris almost invariably encountered a hostile reception, and which they accordingly avoided. They traded with the Kikuyu on the edges of their country, but rarely moved into its centre. Sensitive and lively, the Kikuyu resented the questionable dealings of some of the traders, and their interference with internecine conflicts.[2] The Nandi, too, refused to countenance them,[3] and can still recall the places where unsuspecting traders were lured to their deaths. The prime objectives of the Zanzibaris were, therefore, Kavirondo, and the regions north of Mount Kenya. They seem, however, to have been checked for a time from going farther north-west by fear of the Suk and Turkana.[4] Nevertheless, trade in Kavirondo had expanded towards Mount Elgon by the seventies;[5] to judge from the size of the Zanzibari stockade at Kitale it soon became substantial.[6] It never, however, crossed the Nile, which was being reached by traders on the Tabora route through Buganda. But about 1880, perhaps as a result of a diminution in the power of the Suk, the Zanzibaris eventually began to move north of Lake Baringo into the regions west of Lake Rudolf,[7] those to the east, as we shall see, being convulsed by the Somalis.

[1] Thomson, *Masai Land*, pp. 572–5.
[2] Cagnolo, op. cit., p. 44; Thomson, *Masai Land*, p. 306; Routledge, *Akikuyu*, pp. 15–16.   [3] Thomson, *Masai Land*, p. 469; Hollis, *Nandi*, p. 3.
[4] Wakefield, *J.R.G.S.*, 1870, p. 323.
[5] See map at front of New, *Life, Wanderings*.
[6] R. Meinertzhagen, *Kenya Diary 1902–1906*, Edinburgh, 1957, pp. 199–201.
[7] Thomson, *Masai Land*, pp. 531, 571.

By this time regular Zanzibari routes through Masailand had been developed. They were primarily dependent upon a series of stepping stones that were characterized by their ability to provide food for substantial caravans. Beyond the Asu and the Taita, Taveta became a key centre. Here a small conglomerate tribe, living upon a river bank, where extensive cultivation was possible, and within a dense forest fringe which provided protection against the Masai, were soon providing food for two to three months' journeyings.[1] From Taveta one route ran westwards to Mount Meru, and then north-west to those Kwavi settlements like Dasikera on the west side of the Rift Valley,[2] where abundant food was again available. From there the caravans moved on to the rich food centres of Kavirondo.[3] Meanwhile an alternative, and increasingly important, route ran from Taveta to Ngong on the southern edge of Kiambu. Here, and at Mianzini farther north, where primitive Dorobo conducted the market, copious provisions from the rich harvests of the Kikuyu could be procured.[4] Thereafter the traders either struck northwards of Mount Kenya,[5] or moved to the two Njemps villages on the shores of Lake Baringo, where Kwavi agriculturists specialized in providing for traders' wants,[6] and from here it was but a short step across the Uasin Gishu plateau to Kavirondo once again. There were variations in the routes taken, and sometimes there was a severe shortage of food, but for the most part these generally very fecund oases helped to keep the caravans moving, and to limit the points of friction along the Masailand routes. The traders were rarely driven to raid for food.

The base for all this trade was not only in Mombasa, but at Vanga, Tanga, Saadani, Pangani, and Bagamoyo farther south.[7]

There are generally three leading men who have a pecuniary interest in the caravan. An Arab of high standing in Zanzibar; an influential coastman, such as the native Governor of a coast town, and one or more Banyan merchants. One or more Arabs often

[1] New, *Life, Wanderings*, pp. 336, 353, 362; Johnston, *Kilimanjaro*, pp. 74–79, 209 sqq.; Thomson, *Masai Land*, p. 117.　[2] Christie, *Cholera*, pp. 226–7.
[3] Wakefield, *J.R.G.S.* 1870, pp. 303–14; Thomson, *Masai Land*, p. 487.
[4] Ibid., pp. 306–7, 448; von Höhnel, *Discovery of Rudolf*, pp. 284 sqq.
[5] Christie, *Cholera*, p. 223; Wakefield, *J.R.G.S.*, 1870, pp. 314–23.
[6] New, *Life, Wanderings*, p. 463; Thomson, *Masai Land*, p. 453.
[7] Burton, *Zanzibar*, ii. 116–18, 145–7.

accompany the caravan, but the real leaders thereof are usually Wasawahili, or Wazalia—domestic slaves born in the family. They are always men who have been long in the trade, and who are thoroughly acquainted with the general features of the country; the people thereof; the language, and their manners and customs. One of the leaders of the caravan from Pangani to Laikepya [*sic*] had made fifteen journeys among the Masai, and the other nine.[1]

Some slaves were bought from the Chaga.[2] Others came down the Kamba route, Masai girls being especially coveted.[3] Some were also obtained from Kavirondo. Here the traders occasionally joined with local people, like the Hanga, in fighting their neighbours.[4] But such intermeddling was nowhere general. Indeed during this period in the whole of this north-eastern region external trade had very little effect upon political developments; there is no evidence, for instance, of a gun traffic. The traders' main interest was always in ivory. In return they gave chiefly cloth, tobacco, wire, and beads.[5] All these items were even sought after by the Masai, who used to employ their dependants, the Dorobo hunters, to collect ivory for the traders, which they exchanged for iron wire to melt down for spears, and to adorn their womenfolk. Some Masai women used to carry 30 pounds of wire strung around their arms, legs, and necks.[6]

Naturally enough with such a warlike people astride it, the Masai route was always a dangerous road to traverse. The traders had to learn to suffer their insults in silence. They never managed to establish an inland base comparable to Tabora farther south; and, while caravans of 500 to 1,000 porters were usual, even these were not immune from assault.[7] Even though the tracks through Masailand were never closed for very long, the Masai remained their unquestioned arbiters; and, sure of their own superiority, always kept the traders at arm's length. Indeed such was the disdain for the traders of at least four of the

[1] Christie, *Cholera*, p. 222.
[2] New, *Life, Wanderings*, p. 473; Krapf, *Travels*, p. 196.
[3] Ibid., pp. 357, 361.   [4] Thomson, *Masai Land*, p. 506.
[5] Burton, *Zanzibar*, ii. 45–46; Guillain, *Documents*, iii, 265–7.
[6] Krapf, *Travels*, p. 365; Christie, *Cholera*, p. 209; Johnston, *Kilimanjaro*, p. 424; Thomson, *Masai Land*, pp. 273, 429, 447.
[7] Burton, *Zanzibar*, ii. 72; Christie, *Cholera*, p. 214; Krapf, *Travels*, p. 363; Thomson, *Masai Land*, pp. 37, 126, 226, 278–9, 455; Johnston, *Kilimanjaro*, pp. 185, 294.

five or six most prominent peoples of this region—the Masai, the Kamba, the Kikuyu, and the Nandi (each for their own reasons)—that they never secured the influence here which they attained elsewhere. These northern routes never became as important for the Zanzibaris as those running inland farther south. Nevertheless, it would be a mistake to dwarf their significance. Although they probably never provided more than perhaps 20 per cent. of the total Zanzibari trade in East Africa, from their relatively difficult and peculiarly restricted sphere, one hears of Mombasa marketing 2,600 frasilas of ivory in the 1840's,[1] and Pangani handling about 1,000 and Tanga 2,000 frasilas in the 1850's (in addition to slaves, camels, cattle, asses, rhino horns, and hippo teeth).[2] Compared with neighbouring areas of comparable size, these were creditable amounts; and in the years that followed the trade was obviously growing. The apparent lack of European interest, which was partly fortuitous, was due in the main to fear of the Masai.[3] This seems to have been deliberately fostered by the northern traders, who were, no doubt, anxious to preserve their hard-won commerce inviolate.[4] Nevertheless, it was these same routes, pioneered by the Zanzibaris, which in due time became the Europeans' highway into the Masailand region.

Change on a tremendous scale was about to occur. Some, however, would aver that but for the subsequent British intrusion, it would have come, not from the east, but from the north, and that before very long Masailand would have been invaded by the Turkana, the Somalis, or even the Ethiopians. These are all conjectures, but it seems clear that as early as the 1880's the Ethiopians were making inroads into Turkana, as part of their campaign to extend their dominions to the south.[5]

As for the Turkana themselves,[6] who were originally a fragment of the Karamojong group of Nilo-Hamite tribes which

[1] For a description of the Mombasa trade, see Guillain, *Documents*, iii. 265–7.
[2] For descriptions of the Pangani and Tanga trade see Burton, *Zanzibar*, ii. 116–18, 145–7.
[3] e.g. Mackay's Journal, 14 Dec. 1876; Mackay to Wright, 10 Jan. 1877, Church Missionary Society Archives CA6/o16.
[4] e.g. Johnston, *Kilimanjaro*, p. 88; Krapf, *Travels*, p. 172.
[5] P. H. Gulliver, 'A Preliminary Survey of the Turkana', *Communications*, N.S., No. 26, School of African Studies, Cape Town, 1951, p. 152.
[6] Ibid., *passim*.

had formerly migrated from the south-eastern Sudan, it seems that it was about 1850 that they began to expand from a base west of Lake Rudolf, eastwards to the lakeshore, westwards to the foot of the Rift Valley escarpment, and south towards the Suk and the Samburu, those two peoples living north-west of Lake Baringo. At an early stage in this process the Suk,[1] it seems, having been driven by the Samburu back from the Kerio river down in the western part of the Rift Valley (a key point in this often barren region), successfully sought help from the Turkana,[2] and with them drove the Samburu from the Kerio. The Suk now became a terror to their neighbours right across the Rift Valley. There was a term, however, to their alliance with the Turkana, for by the 1880's the Turkana had grown more independent. Some Turkana groups were now harassing the Suk;[3] some were thrusting up the Kerio Valley; some had even raided to the edge of Lake Baringo. The Turkana were obviously becoming a formidable people.[4]

Meanwhile, farther east, the Somalis, who had been migrating from the horn of Africa since the twelfth century, were on the move once again. On the coast a Darod Somali group, the Herti, had by the middle of the nineteenth century begun to settle in the rich districts of the Juba river basin. Upstream, towards the middle Juba, more Darod groups, the Muhammad Zubeir, the Telemuggeh, and the Marehan, were advancing southwards too. Sweeping all before them, the Marehan crossed the Juba in force about 1850. Here they met the Tana Galla, who, though they had long since lost contact with the main body of the Galla peoples in Ethiopia, still dominated the whole Tana river area, and spread southwards to the confines of Kamba country, to the Taita Hills and even beyond.[5] At first the Marehan managed to share the available grazing grounds with the flocks and herds of the Galla. But about 1865 they turned on their hosts, smote them hip and thigh, drove the remnants from their pastures, seized most of their cattle,

---

[1] M. H. Beech, *The Suk*, Oxford, 1911, pp. 3-5.
[2] Gulliver, *Turkana*, pp. 145, 154.
[3] Huntingford, *Southern Nilo-Hamites*, p. 78.
[4] Gulliver, *Turkana*, p. 154. Mr. Neville Dyson-Hudson's comments on this passage have been of great assistance; he is not, of course, responsible for the final version.
[5] Krapf, *Travels*, p. 182; Lindblom, *Akamba*, p. 19.

and enserfed most of the survivors. Simultaneously the Masai and the Kamba[1] were bludgeoning the Galla from the south, and by the end of the century there were very few left of this once powerful people.[2] The Marehan and the other Darod groups did not, however, stop there. After 1880 they turned on the Boran, the south central branch of the Galla, who had sought to profit by the Tana Galla's defeat by appropriating some of their grazing, but who now very nearly shared their fate. For the Darod drove them from all the northern parts of the Juba river region, leaving them with only a few footholds at Wajir and Buna. This whole concourse of events was racked, as one might expect, by plot and counterplot, alliance and feud. But the general trends were clear. There occurred the resounding defeat of all the southern Galla; the domination of the remaining peasantry by the nomads from the north, and with it of pagan peoples by Muslims. The invasion of East Africa by the first wave of Somalis had begun.[3]

There can be little doubt that these movements of both the Turkana and the Somalis were primarily impelled by their need for new grazing areas as their own numbers grew. Such movements were on all fours with so many of the migrations that have scored Africa's past. Even so, it seems plain that it was the travels by Somali traders that helped to open the eyes of the Somalis to the new opportunities lying before them; and (for all the paucity of evidence) that it was the expanding network of Ethiopian ivory and gun traffic (southwards into Turkana, for instance) which prompted the imperial ambitions in East Africa of the Ethiopians, that were already in evidence by the 1880's.[4] A similar tide was flowing, wave upon wave, elsewhere. No foreigner had as yet sought to annex Masailand; but prompted by Arab tales of snow mountains in the interior, the early Mombasa missionaries of the Church Missionary Society, Rebmann and Krapf, had 'discovered' Mounts Kilimanjaro and Kenya in 1848 and 1849;[5] the German, Baron von der Decken, made two journeys towards Kilimanjaro in

---

[1] Lindblom, *Akamba*, p. 19; New, *Life, Wanderings*, pp. 193, 199–200, 238.
[2] J. Willoughby, *East Africa and its Big Game*, 1889, pp. 268–70.
[3] An unpublished paper by Sir Richard Turnbull, and the comments of Dr. I. M. Lewis, have been of great assistance with this passage. They are not, of course, responsible for its conclusions. [4] Gulliver, *Turkana*, p. 152.
[5] Krapf, *Travels*, pp. 230–64, 283–99.

1861–2;[1] the Wesleyan missionary, Charles New, tried to climb Kilimanjaro in 1871;[2] and in 1883–4 the explorer Joseph Thomson, excited by the constant accounts of Zanzibaris travelling through Masailand, made the first journey by a European from Mombasa to Lake Victoria and back. A similar cavalcade had meanwhile advanced towards the headwaters of the Nile.

The Zanzibari trade around the western side of Lake Victoria had been opened up as far as Buganda by 1844. Stimulated, no doubt, by the pre-existing Ganda commerce with the south, the Zanzibaris from Tabora had adroitly superimposed themselves upon the north-western routes.[3] The Ganda, however, being fully confident of controlling the traders' activities in their own kingdom, and anxious to oversee the trade going elsewhere, were not disposed to interfere. Buganda indeed became for the traders a major resort. Even by the mid-fifties there were, as a result, coastal traders on the western side of Lake Victoria who had been there for ten years and more; by the seventies there were Zanzibaris present who had not seen the coast for over twenty years.[4] So there were patent differences here from the Masailand routes. The north-western trade was not so tied to moving caravans; trading settlements were more easily established; and different traders could specialize in successive stages along the route. The active co-operation of the interlacustrine kingdoms was usually commercially efficient, and food was no problem. Raids by traders, in fact, proved both inadvisable and unnecessary. Ivory and slaves (still in that order) were once again the main items of export, but among imports cloth was more important than in the north-east, and there was a steadily rising demand for guns. These were wanted not merely by ambitious ivory hunters but by those countries, like Buganda and Bunyoro, which were seeking to enlarge their power, and by those who, in Buzinza, for instance, at the south end of the lake, were heavily involved in internecine struggles.

[1] O. Kersten, *Von der Deckens Reisen in Ost-Afrika*, 1869, vol. ii; Richard Thornton's MS. Diary (Rhodes House, Oxford).
[2] New, *Life, Wanderings*, chs. xx–xxi.
[3] J. M. Gray, 'Ahmed bin Ibrahim—The First Arab to reach Buganda', *Uganda J.*, 11, 1947, pp. 80–82.
[4] R. F. Burton, *The Lake Regions of Central Africa*, 2 vols., 1860, ii. 173 sqq.; H. M. Stanley, *Through the Dark Continent*, 2 vols., 1878, i. 455.

A portion of the trade passed by water across Lake Victoria, and by the 1870's some coastmen were building their own dhows upon it;[1] but because the early canoe traffic was never sufficiently regular or reliable (it was, in any event, firmly monopolized by the Ganda), it was necessary for the Zanzibaris to develop a land route as well. They made various attempts to establish a traders' settlement at the south end of Lake Victoria. Yet even before the death in 1860 of Ruhinda II, Mugabe of Buzinza, the area was too disturbed, and it was worse afterwards.[2] The natural lines of trade by land, however, passed through Buzinza up the longitudinal valleys running parallel to the lakeshore (the people here made sure of a cut of the trade in the shape of a substantial *hongo*, or passage money, which the traders paid in the interests of peaceful movement along a valuable highway).[3] They might thereafter have established a base in Rusubi (the northernmost and most stable of the Buzinza kingdoms) since trade with the important kingdom of Rwanda to the west radiated from here, but the Mukama of Rusubi was none too pleased at the development of a commerce which so benefited the larger kingdoms around him, which were already powerful enough. He therefore made the passage of caravans through his kingdom very expensive, and was not averse to holding them up.[4] So a traders' settlement was founded instead at Kafuro in the friendlier kingdom of Karagwe farther north. From here trade routes extended northwards to Bunyoro and north-eastwards to Buganda; and at least by the early 1860's traders traversing these routes were operating north of Lake Kyoga as far as the modern Uganda boundary with the Sudan.[5] Here they encountered the trade coming down from the north.

It was in 1841 that, setting out from Khartoum, a Turkish frigate captain, named Selim, had found the Nile navigable as far south as Gondokoro, some thirty miles north of the modern

[1] Smith to Wright, 27 Aug. 1877, Church Missionary Society Archives, CA6/022.
[2] P. H. van Thiel, 'Businza unter der Dynastie der Bahinda', *Anthropos*, 6, 1911, pp. 507, 510 sqq.
[3] J. H. Speke, *Journal of the Discovery of the Source of the Nile*, 1863, p. 143. Mr. W. J. Tyler, who has lived in Buzinza as an anthropologist, has generously made his notes available.
[4] Speke, *Journal*, pp. 204, 208.
[5] Ibid. 237, 242; Burton, *Lake Regions*, loc. cit.

Sudan–Uganda boundary. This important disclosure was promptly exploited by official traders in Egyptian government service, and it was not till ten years later that the government monopoly was broken by European consular pressure upon Egypt, and the first private traders moved up the Nile. As the opportunities for extracting ivory from the areas south of the Bahr al-Ghazal appeared extremely promising, their numbers grew rapidly. But the outcome was hell-born. At the outset difficulties arose because the Khartoumers, as they were called (they originally included some Europeans), were unable to find any Africans upon their line of trade who were even tolerably efficient as initial collectors of ivory; the Nilotic tribes of the southern Sudan had not the political or economic organization to match the traders' requirements. The Khartoumers, therefore, brought up alien collectors from the north, whom they ranged in permanent encampments. But they could ill afford to pay these properly. So they began to extend their interests from ivory (which always comprised the main item in their trade) to slaves, as the only available means of augmenting their own profits, and meeting their agents' wages. But what was more, after some initial hesitation, they were swift to intervene in the profusion of internecine and inter-tribal quarrels amongst the Nile's peoples, so as to exploit them to their own advantage. The tribesmen were rarely attracted by the traders' products, like cloth; it was only the prospect of acquiring cattle that moved them, even to enlist as porters; but it was then found that tribesmen and traders could join forces to their mutual advantage in raids upon neighbouring peoples from which the tribesmen procured cattle, and the Khartoumers their ivory and slaves.[1] As it chanced, the openness of the country, and the absence of 'state' organizations, made most of the local peoples peculiarly vulnerable to raiding parties armed with guns; the islamized northerners despised the naked southerners as barbarous pagans, beyond the pale of ordinary humanity; in the absence of food-producing oases hunger, and not only avarice, drove the traders to raid; while the widespread

[1] J. R. Gray, *A History of the Southern Sudan, 1839–1889*, 1961, chs. i–ii. Dr. Gray is not, of course, responsible for the conclusions reached here. See also J. Petherick, *Egypt, The Soudan and Central Africa*, 1861, and *Travels in Central Africa*, 1869; S. W. Baker, *The Albert Nyanza*, 1867, and *Ismailia*, 2 vols., 1874; Speke, *Journal*, chs. xix, xx.

possibility of withdrawing to the comparatively secure highway on the Nile allowed the Khartoumers a greater licence in their dealings with obstructive tribesmen than would have been possible upon most land routes. Unlike the East Coast traffic, the Nile trade, in consequence, abandoned itself to violence. Raids and rapine covered the land. Their memory has not been erased in a hundred years.

The Acholi, living immediately north of Bunyoro and the Victoria Nile, were penetrated by the Maltese trader Debono by the late 1850's, and his agents soon ravaged the land. Before long they were followed by others.[1] In about 1872 Ali Husain, an agent of another northern trader, Abu Saud, went eastwards into Lango country, and obtained large presents of ivory and cattle; but he soon turned upon his hosts, butchered the men, and carried off the women, children, and large herds of cattle to Patiko in Acholi. This was a typical occurrence, except that the Lango, a warlike people, promptly expelled the next intrusion, and do not seem to have been disturbed again.[2]

The Lango were at this time expanding in two waves southwards and westwards to the north-western shores of Lake Kyoga and the Victoria Nile. In the course of their advance they attacked the Acholi upon their right flank, and were soon sending frequent plundering expeditions across the Nile. They also fought the Kumam and the Teso upon the south-east.[3] These two peoples, offshoots of the Karamojong group of Nilo-Hamite tribes, were for a century after 1800 moving out in three directions from their resting-point at Magoro, immediately north of Lake Salisbury. In the van, going westwards, were the Kumam; though closely linked to the Teso, they were by the end of the century adopting the language of their new Lango neighbours.[4] The Teso, meanwhile, thrust northwards up the east side of the Akokoroi river; a second section moved westwards to Soroti, before fanning out towards Amuria on the north, Kaberamaido on the west, and Serere and the northern shore of Lake Kyoga on the south; whilst a third section crossed Lake Salisbury, and advanced southwards to the confines of

---

[1] J. M. Gray, 'Acholi History, 1860–1901', Part I, *Uganda J.* 15, 1951, pp. 121 sqq.
[2] J. H. Driberg, *The Lango*, 1923, pp. 33–34.
[3] Ibid., pp. 31–32.   [4] Ibid., pp. 38–41.

Pallisa and Bukedea (some smaller groups going farther south still).[1] These last two sections—known as the Iseera and the Ngoratok—became the two sections into which the tribe was to be unhappily divided.

These migrations, in part at least, filled up the vacuum left by the Luo upon their march south-eastwards to the northern shore of the Kavirondo Gulf. They led, among other things, to the establishment of a flourishing commerce upon the Lake Kyoga littoral, in which it was the Nyoro who played the leading part. They became anxious, in particular, to obtain ivory, so that they could procure guns and other novelties from the Zanzibaris from the south and the Khartoumers from the north. They were especially fortunate in having not merely iron to retail in exchange, but skilled smiths to fashion it. For their Lake Kyoga neighbours seem to have been ignorant of some of the more important secrets. In the past the Lango and the Teso had used only wooden implements for digging; they were now delighted to procure iron hoes. Nyoro iron went also to the making of Lango and Teso spears; neither people had formerly made their own.[2] Thus Jopaluo traders from Bunyoro sold hoes to the Lango, in exchange not only for hides and goats and produce, but for ivory.[3] Nyoro sold iron hoes and spears to the Teso; and the Lango sold their surplus hoes to the Kumam and Teso for cattle (such was their value that three hoes would fetch a heifer).[4]

At the same time amongst all these peoples, groundnuts and sweet potatoes began to be grown for the first time.[5] Amongst the Teso this development, together with their acquisition of new implements, and the encouragement given by the greater fertility and reliable rainfall of their new homeland, and the consequentially increased crop yields, soon converted these former cattle people from pastoralism to agriculture. Permanent homesteads of some size followed, and the population grew.[6] The foundations were laid for the third largest tribe in twentieth-century Uganda.

South of the Victoria Nile and Lake Kyoga there lay the numerous kingdoms of the Interlacustrine Bantu. In political

[1] J. C. D. Lawrance, *The Iteso*, 1957, pp. 12-15.
[2] Ibid., pp. 13-14, 148; Driberg, *Lango*, pp. 80-81.
[3] Ibid., pp. 30-31, 44.　　　　[4] Lawrance, *Iteso*, p. 148.
[5] Driberg, *Lango*, p. 100.　　　[6] Lawrance, *Iteso*, pp. 13-14, 148.

conformation they were mostly basically similar. They had rulers with royal drums, royal spears, or royal stools, to whom all men bore allegiance. They had territorial sub-chiefs, subject to the ruler alone, as well as royal clans. Their state structure (with its hierarchy of princes and commoners) distinguished them fairly sharply from their Nilotic, Nilo-Hamitic, and Bantu neighbours. Their number depends on the period and the definition. Lists of kingdoms for Busoga, east of the Nile, for instance, vary for this period from eight to over fifty.[1] But it is not difficult to name upwards of forty kingdoms of some substance for the whole area.

They differed in their racial composition. The dynasties to the south of the Katonga river were in origin Hima or Tutsi (the names correspond); to the north Nilotic. They varied too in their social structure. In the north, in Busoga, and in Buganda to the west of the Nile, the social assimilation of classes was substantially complete. To the west in Bunyoro and Toro, the distinctions between Nilote, Hima, and Bantu could, however, still be discerned; while south of the Katonga, in Ankole, in Buhaya, in Buzinza, the division between Hima and Bantu (there were no Nilotics here) was sharp, as indeed in Rwanda and Urundi to the south-west, where there were some pygmoids as well, who formed a third caste. Yet these variations did not affect the most striking differences between these kingdoms— that some of them were large and powerful, whilst others were small and weak. Buganda, for instance, which was to become the most powerful of all, was socially similar to its Busoga neighbours, which were amongst the weakest; while Rwanda, where the ethnic distinctions were sharp, became scarcely less powerful than Buganda, although its structurally similar neighbours in Buhaya and Buzinza were amongst the feeblest of all.[2]

Although the tribal chronicles of these kingdoms abound with brief summaries of wars, they offer little explanation of their cause. This is probably most frequently to be found, in the pasturelands west of Lake Victoria, in the attachment of their Hima-Tutsi aristocracies to cattle. To begin with, the rulers of these kingdoms, if they valued their reputations, needed cattle

---

[1] Cf. Y. K. Lubogo, *A History of Busoga*, 1960, For the origin of the Interlacustrine dynasties see Chapter VI, pp. 180 sqq. above.

[2] A. I. Richards (ed.), *East African Chiefs*, 1960, chs. vii and viii.

to distribute both to faithful and to impoverished followers. It was always tempting to increase one's herds at the expense of a neighbour. Cattle, however, required grazing, and the possession of large herds led as easily to civil embroilment as to foreign attack. Indeed, a superfluity of cattle might lead to debilitating internecine conflict, whereas a shortage of cattle might well encourage a tight offensive war organization. Nevertheless, both shortages and superfluity could as easily eventuate in foreign aggression, either in search of new cattle, or in search of new pastures.[1] Here, it would seem, was the central theme of the career of such a man as Ndagara, who between about 1822 and 1855 ruled Karagwe (at this juncture the foremost of the Buhaya kingdoms). Famous as a warrior, he fought successfully against the neighbouring kingdoms of Kiziba, Kiamtwara, Ihangiro, Rusubi, and even Rwanda. But he never showed any desire to subjugate his enemies. His son, Rumanika, successfully defended himself against two attacks from the direction of Gissaka, and unlike his father drove the Subi from Buhimba; yet even he never sought to conquer the Subi.[2] Ambition for cattle, together with changes in the ecological balance between animals and their pastures, rather than any desire to lord it over others, seems to have been the most potent reason for intertribal warfare in this area.

There were, however, in addition, amongst these Interlacustrine kingdoms, perennial wars of succession, since most of them shared the common disability that most male relatives of the ruler, and almost always all his sons, were equally legitimate successors to his throne. These succession wars had the constitutional merit that they tended to eliminate rival contestants for the throne; and, because the throne was the prize for which princes strove so furiously, they tended to underscore the uniqueness of the rulership—a uniqueness which served to unite the kingdom about it. Nevertheless, the legitimacy of multifarious princely aspirations constituted a grave embarrassment, if not a chronic weakness, to many of these kingdoms. The formerly extensive Bunyoro-Kitara, for instance, had already suffered serious erosion from princely ambitions in the

---

[1] Dr. Derrick Stenning has been of great assistance with this passage.
[2] J. Ford and R. de Z. Hall, 'The History of Karagwe', *Tang. Notes*, 24, 1947, pp. 7–10.

past. It had lost to a princely usurper the fertile province of Toro, as recently as the 1830's.[1] Sometimes there were actual disintegrations. Such indeed was now the fate which in the early 1850's overtook southern Buzinza, where Ruhinda II had ruled for perhaps forty years past. He had long since appointed a number of his sons to fiefs within his kingdom; but no decision had been taken about his successor, and upon his death no one of his sons was able to establish an ascendancy over the remainder. The position was made worse by an increasing number of raids by Nyamwezi and Ngoni from the south. Rwoma, one of Ruhinda's grandsons, held on to Bukara after the death of his father, Kashaija, in 1864, but he was only one of about a dozen petty rulers who were struggling to survive. The truth was that Ruhinda II was the last Mugabe of Buzinza proper, and that upon his death his kingdom broke up.[2]

As in southern Buzinza, these wars were often, moreover, the occasion for intervention by outsiders. Indeed, the temptation for an ambitious prince to call in external forces to his assistance was considerable. In northern Buzinza, for example, in the kingdom of Rusubi, Mankorongo expelled his uncle, Rwesarura, in 1873 with the help of some Ngoni from the south, only to be attacked in 1874 by Rwebogora, a brother of Rwesarura, who had appealed to the Ganda; Mankorongo was unable to feel secure upon this throne until he had survived a costly battle with them.[3] It was much the same in Karagwe to the northward. Here, on Ndagara's death in 1855, Rumanika almost certainly owed his throne to the Arab traders at Kafuro, armed with guns, who gave him their support. Upon his own death in 1878, one of the contesting parties secured help from Rwanda; the successful one from Buganda.[4] Such occurrences prompt the conclusion that a kingdom could never hope to be powerful until it could surmount the crises which accompanied its successions to the throne. For even when one candidate emerged triumphant, the cost to his kingdom was frequently high, both in confusion at home, and in obligations contracted abroad. An overall triumph might not, however, result, and a ruler

[1] J. W. Nyakatura, *Abakama ba Bunyoro Kitara*, 1947, pp. 119–22.
[2] van Thiel, *Anthropos*, 1911, loc. cit.
[3] Ibid., pp. 508–9. Also Father Betbeder's notes (1938) on van Thiel, a copy of which Mr. Tyler kindly made available.
[4] Ford and Hall, *Tang. Notes*, 1947, pp. 9, 12.

might succeed to the throne only by recognizing, at least tacitly, a new kingdom, which an importunate brother had carved out for himself from his father's domain; this was frequently the story at this time in Busoga.[1] It might even be, as in Gissaka, as we shall see, that a succession war entailed the complete destruction of a kingdom. Whatever the circumstances succession conflicts were almost always peculiarly critical, since they provided golden opportunities for intervention by astute and powerful neighbours, and especially now that guns were replacing spears in their armoury.

The greatest of these kingdoms down in the south was the large and important Tutsi kingdom of Rwanda. Its saga in these years exhibited most of the themes which were most typical of them. To begin with, its ruler, Mutara II, exploited a disputed succession in Gissaka, sometime in the 1840's, to conquer and absorb this small neighbouring kingdom to the eastward. His successor, Kigeri IV, was often faced with plots against his life, no doubt because his inheritance was worth contending for. But he successfully surmounted them all, by means of a complete ruthlessness towards all suspects, thereby adding to his own strength, and to the cohesion of his kingdom. Meanwhile, in a belligerent career, he first gained control of the largest island in Lake Kivu—thus consolidating his rear—and then tried to invade the important kingdom of Urundi to the southward. But the attempt failed, and was never repeated. He does not appear to have struck eastwards, probably because, after some initial hesitation, he was anxious to maintain his trading connexions in that direction through Rusubi. Instead his main campaigns were launched northwards. Here he thrust deeper and deeper. By the 1880's, he was harrying the Ankole, storming the Kiga, and striking out towards Lake Edward. The secret of his success lay in his reorganization of the military potential of his tribe (a reform executed at this time by three of the most powerful of these kingdoms). He bought guns from Arab traders; and arranged that all able-bodied Rwanda should be liable to fight, but that they should not all fight together. They were to serve instead in successive batches. (It is said that every Rwanda who lived whilst he reigned saw at least one of his many campaigns.) In this way Kigeri was never without a

[1] Lubogo, *Busoga, passim*; L. A. Fallers, *Bantu Bureaucracy*, 1956, pp. 134-5.

fighting force at his disposal, and continuous warfare was made possible. Nevertheless, although he undoubtedly ensured that Rwanda remained one of the two or three most formidable kingdoms of the region, he never added new provinces to his father's domain, and he never broke the deadlock with Urundi. Cattle, not conquest, it would seem, comprised his chief preoccupation.[1]

Meantime in the north-west, Buganda's star was already well in the ascendant by the middle of the nineteenth century. It is important to bear in mind that in all essentials it belonged to the same cultural constellation. Yet one at least of the reasons for its ascendancy lay in the peculiarities of its institutions governing succession. There was no royal clan; a ruler's sons took the clan of their mothers. Nor were princes given territorial jurisdiction. Only sons of the drum—the sons of a former ruler—could aspire to the throne, and by the nineteenth century all unsuccessful ones were customarily exterminated. The choice of ruler, moreover, was fairly effectively exercised by the two most important commoners—the Katikiro, or head of the administrative hierarchy, and the Mugema, the foremost clan-head in the country. It was not dependent, as in Ankole, for instance, upon an aspirant's ability to survive a protracted war with his brothers. There were still acute tensions when a ruler died, and these were not necessarily obliterated immediately upon his successor's accession. But for over half a century before 1884 Buganda knew no open civil conflict.

Yet this was not the only secret of Buganda's success. It had besides a material basis. For Buganda enjoys an average rainfall of at least thirty inches; what is better, it is reliable and well distributed. For the tropics her soils are unusually well supplied with plant nutrients, so that agriculture was always rewarding. The basis of life was neither cattle nor cereals, but permanent groves of plantains, yielding food all the year round, with greater certainty and at a lower cost in labour than almost any other plant. Some of Buganda's neighbours enjoyed similar advantages, but none to the same degree. For most of the peoples of East Africa, food production was a major occupation; not so for the Ganda; they could leave it almost entirely to their women.

[1] A. Pagès, *Un Royaume hamite au centre de l'Afrique*, 1933, bk. ii, chs. iii and iv.

The result was a surplus of time and energy, which could be devoted to more rewarding activities. So the Ganda dressed in barkcloth; built (by East African standards) relatively substantial buildings; fashioned weapons with care and artistry; and constructed elaborate canoes. They supported, moreover, a complex state organization and a large governing bureaucracy. The former frequently demanded considerable military levies, collected taxation, and enforced corvées, such as those which maintained the roads and causeways that radiated so dramatically from the capital; while the latter, never demeaning themselves to manual labour—not even to the herding of cattle —demanded regular gifts of labour, service, and produce.[1]

By the nineteenth century this superstructure was dominated to a remarkable degree by the personal authority of the kabaka. Even a mid-nineteenth century ruler—Suna reigned till 1856; Mutesa till 1884—was not wholly inviolable. It was not until six years after Mutesa's accession that he began to feel secure; he remained a murderous despot (largely in his own defence) for some years thereafter; and as late as 1879 there were threats to his life.[2] But with two possible exceptions (his own mother and his official sister) the kabaka's relatives had by this time been firmly subordinated; the independence of the priests of the traditional religion had been curbed; and the clans had lost much of their control over land, and most of their territorial jurisdictions, to the kabaka's personal appointees, who now constituted an authoritarian hierarchy of administrative chiefs, subject to his sole control; and lest any of them should act independently, he had his own personal agents posted in their areas, who had few public duties, and reported directly to him. Moreover, Kabakas Suna and Mutesa I in reforming the military organization of the kingdom in the middle part of the century spread throughout the country military specialists under the authority of the Mujasi, or general, who acted as a

[1] See C. C. Wrigley, 'Buganda: an outline economic history', *Economic History Review*, 2nd ser., 10, 1957, pp. 69–74. The most useful accounts of nineteenth-century Buganda are: Sir Apolo Kagwa, *Basekabaka be Buganda*, *Mpiso za Baganda*, and *Ebika bya Buganda*, 1927, 1906 and 1908; J. Roscoe, *The Baganda*, 1911; R. P. Ashe, *Two Kings of Uganda*, 1889; C. T. Wilson and R. W. Felkin, *Uganda and the Egyptian Sudan*, 1882; J. W. Harrison, *A. M. Mackay*, 1890; Speke, *Journal*; J. W. Grant, *A Walk across Africa*, 1864.

[2] Litchfield to Wright, 3 Jan. 1880, Church Missionary Society Archives, CA6/015.

further check upon independent action; while with the advent of guns, heralding a revolution in warfare and the *locus* of power, the kabaka was careful to keep their distribution under his own personal control. At the same time officials were now being so constantly switched from post to post, that none of them could develop a strong personal control over any particular area. Their careers were now completely dependent upon the favour of a monarch who could depose, or even put to death, on the slightest whim; and did. This elaborately centralized autocracy gave Buganda a decided advantage over its weaker neighbours.

Its hub was the kabaka's court. Life there demanded a good head, a knowledge of etiquette, the ability to fawn upon superiors, and a gift for intrigue. Round about there existed the largest urban area in the East African interior, with a population numbered in tens of thousands. Support for the system was forthcoming because advancement within it was open to anyone with the necessary wits to invoke a sufficiency of helping hands; and because its rewards were substantial—in terms of personal honour and fortune for the successful, and in largesse for the generality. Here, indeed, was now an essentially acquisitive society, whose over-riding concern was always to increase the material prosperity of its kabaka, his chiefs and their people; and for this during the nineteenth century it turned its eyes primarily abroad.

Its exotic requirements were obtained from Zanzibar. In about 1852, for some unrecorded misdemeanour, the coast-men who had been trading in Buganda for about eight years past were suddenly expelled.[1] As a result their influence might have been as negligible here as it was amongst the Nandi and Kikuyu. But about ten years later they were allowed to return,[2] and so rapidly did their trade with Buganda then grow that by 1869 it warranted a special embassy to the kabaka from the Sultan of Zanzibar.[3] No doubt the Ganda had recalled the help which some Zanzibaris had accorded Kabaka Suna (in raiding the Soga).[4] They certainly wanted guns. But they were keenly anxious too to secure those other novelties which the traders

[1] J. H. Speke, *What led to the Discovery of the Source of the Nile*, 1864, p. 259.
[2] Speke, *Journal*, p. 243.
[3] Kirk to President, Royal Geographical Society, *Proc. R.G.S.*, 16, 1871–2, p. 186.
[4] Gray, *Uganda J.*, 1947, p. 81.

had to offer, like cotton cloth, which their chiefs now wore in place of the traditional barkcloth. Zanzibar trade-goods, however, satisfied only a part of the cupidity which had now overtaken the Ganda (trade-goods had, in any event, to be paid for). All sorts of other supplies were wanted as well. Yet Buganda never seems (apart from a little ivory) to have produced very much for export. It certainly never conducted a trade in any way comparable to Bunyoro's iron trade around the shores of Lake Kyoga. Instead, most of its foreign requirements were procured, not by commercial, but by military and political means; not by trade, but by raids.

An essential constituent of Buganda's increased prosperity was, therefore, a commitment to war, for here lay the source of its superabundance. Suna's and Mutesa's reforms of the military system, with the creation of the Mujasi and his professional subordinates, were matched at this time by a contemporaneous expansion of the very substantial canoe-fleet, under the Gabunga. By the seventies Mutesa had begun to collect guns; in 1872 he was said to have 1,000 of them, which was certainly more than any of his neighbours.[1] He then reconstituted a portion of his regimental system, so as to form a musket-armed striking force, mostly of young courtiers; and by these means kept Buganda abreast of military developments amongst its neighbours. Nevertheless, it was still the tribal levies, armed with spears, that provided the rank and file of Buganda's warbands. These raided into Busoga to the east, attacked Bunyoro, Toro, and Ankole on the west, and the Buhaya and Buzinza kingdoms farther south. The canoes often assisted, and themselves raided the Buvuma islands, the Kavirondo coast, and Ukerewe island at the south end of Lake Victoria. From these expeditions were brought back not just cattle, but men, women and children, produce, ivory, and hoes, which were then distributed in the capital and throughout the provinces, or, in the case of some of the captives, and the ivory, bartered to the traders. Sixty such raids by land and water are listed for the twenty-seven years of Mutesa's reign.[2] As a result, his own household contained many thousands of women, and

[1] Baker, *Ismailia*, ii. 98.
[2] Wrigley, *Economic History Review*, 1957, p. 72, footnote 5. See Kagwa, *Basekabaka*, pp. 51–136, *passim*.

great stores of 'danegeld' and loot. In one such raid against Buzinza in 1884, an army, assisted by a canoe fleet under the Gabunga, was said to have carried away 5,000 head of cattle, and several hundreds of prisoners.[1]

The external political relations of the Ganda in the middle part of the nineteenth century were intrinsically complementary. By contrast with the eighteenth, and the early part of the nineteenth century, they never made any but one very minor attempt at this time to extend their territory. While subjecting people to their domination, they were apparently content to leave them their independent existence. They were quick to exploit civil dissensions, either (as, for instance, in Ankole) by direct invasion, or (as more frequently in Busoga), by forcing the contestants to bid for Ganda support.[2] But upon such occasions it was the kabaka's primary purpose to procure a decisive influence over the succession to the throne, so as to have neighbours who would co-operate in his service, whenever they were ordered,[3] and pay him tribute in kind.

Such demands, it should be noted, like the raids with which they were pressed, were essentially *ad hoc*, and it is generally misleading to submit them to some European legal formula, which is almost sure to imply a continuing and defined relationship that did not exist. There is no evidence, for instance, that the Ganda ever ruled any of their neighbours. Although politically Ganda power was a dominating consideration for all of them, even this varied in its intensity. In the southern and most of the central kingdoms of Busoga, in Koki, in several of the Haya kingdoms, and in parts of Buzinza, the domination of Buganda grew real and effective. But it had obvious limits elsewhere. We have already noted that Mankorongo of Rusubi defied a Ganda onslaught against him on behalf of one of his rivals. In Ankole, in the years following the death of Mugabe Mutambukwa in 1878, the Ganda failed on at least two major occasions to overthrow his eventual successor, Ntare, in the interests of their own candidate, Mukwenda. Despite Ganda raids Ntare continued as Mugabe of Ankole until 1895, and his gravest difficulties were caused, not by raids by the Ganda, but

---

[1] van Thiel, *Anthropos*, 1911, p. 512.   [2] Lubogo, *Busoga, passim*.
[3] The most famous instance was Luba of Bunya's murder of Bishop Hannington on Kabaka Mwanga's orders in 1885; see Harrison, *Mackay*, p. 265.

by raids from Rwanda.[1] In Toro to the north, Buganda also failed, some time in the early seventies, to maintain its candidate upon the usurper's throne,[2] as about the same time it failed to make a protégé of the ruler of Bunyoro.[3]

Bunyoro, indeed, during these years constituted Buganda's most serious concern. For despite the impotence which had overtaken this age-old kingdom in the early part of the nineteenth century, by the middle years it was showing signs of a notable recovery. Kamurasi (the mukama who died in 1869), had owed his throne to Lango intervention, and had been severely attacked by the Ganda; at one stage he had even been forced to flee to an island on Lake Albert for refuge. But he re-established the mukama's authority, and hurled the Ganda back. Bunyoro already enjoyed the major share of the trade upon the Lake Kyoga littoral. She was soon trading too, not just with the Zanzibaris, but (unlike Buganda) with the Khartoumers as well. Kamurasi's successor Kabarega, indeed, enlisted Khartoumers in his service, collected guns, and created new military regiments (*abarusura*).[4] Such energy upon the part of their foremost neighbours was anathema to the Ganda. It seemed to be the one serious threat to the enlargement of their prosperity. By the 1870's a determination to humble Bunyoro had come to dominate Ganda policy.

Upon the Upper Nile the decade of the 1870's was distinguished by the attempt of the Egyptians to extend their empire to the great lakes. This was the logical conclusion to the establishment of Khartoum and the Egyptian empire in the northern Sudan forty years before, and was part of a simultaneous expansion towards Darfur on the west and the Somali coast on the east. In a sense it involved the restoration of the monopoly over the Nile trade which the government had relinquished in 1851. Following the trebling of the tax on the

---

[1] Kagwa, *Basekabaka*, 1927 edition, pp. 322 sqq.; A. G. Katate and L. Kamugungunu, *Abagabe b'Ankole*, 1955, i. 109 sqq.
[2] Kagwa, *Basekabaka*, pp. 304 sqq.
[3] R. M. Fisher, *Twilight Tales of the Black Baganda*, 1912, pp. 160–1; Nyakatura, *Abakama*, pp. 142–8.
[4] The most useful accounts of nineteenth-century Bunyoro are: Fisher, *Twilight Tales*, pp. 145 sqq.; Kagwa, *Basekabaka*, pp. 297–8; 'K.W.', 'Abakama ba Bunyoro-Kitara', Part III, *Uganda J.* 5, 1937, pp. 53–67; Nyakatura, *Abakama*, pp. 117 sqq.

trade in 1863, most of the Europeans had withdrawn. Middle Eastern private traders thereafter reigned supreme, some of them enjoying princely monopolistic rights. But in the middle 1860's European explorers unmasked their malefactions, and disclosed the rich trading prospects in the kingdoms verging on the lakes.[1] David Livingstone's trans-African journeys, and the accounts of Krapf and Rebmann of great lakes and snow mountains in the interior, had by 1850 turned the attention of those Europeans interested in African exploration from West Africa (where the main mysteries, Timbuktu, and the course of the Niger, had been laid bare) to East Africa. Following the paths tramped out by Zanzibari traders, the first Europeans, Burton and Speke, reached Lake Tanganyika in 1858. The same year Speke saw Lake Victoria, and with Grant on a second expedition was able to vindicate his claim that it was the source of the White Nile, when he set eyes on the Ripon Falls on 28 July 1862. Meanwhile the Germans Werne and Schweinfurth, and the Welshman Petherick, had been pushing up the Nile in the wake of the Khartoumers. In 1863 Speke and Grant met their successors, Samuel Baker and his wife, at Gondokoro on their way southward to 'discover' Lake Albert in the following year.[2] The only potentate to react positively, however, to the explorers' revelations was the Khedive Ismail of Egypt, and it was not long before he had determined upon a two-pronged intervention, by which he would kill the slave-trade as an affront to his own conscience, and as a means of winning favour in Europe; and, at the same time, enrich his own coffers by an essay in economic imperialism. In 1869 he appointed Lake Albert's 'discoverer', now Sir Samuel Baker, to lead an Egyptian government expedition up the Nile to the lakes.[3]

Baker was delighted to be returning to his old explorer's haunts upon such an ambitious venture. The khedive allowed him to equip a lavish expedition, and on 8 February 1870 he led it south from Khartoum. But it promptly stuck in the Sudd —the marsh barrier that had frustrated so many travellers in the past—and it was not until 15 April 1871 that the vanguard

---

[1] Gray, *Southern Sudan*, chs. i–iii, *passim*.
[2] See R. Coupland, *The Exploitation of East Africa*, 1939, ch. vi.
[3] Baker, *Ismailia, passim*; G. Douin, *Histoire du règne du Khédive Ismail*, tome iii, 1ᵉ partie, 1936, ch. ix; 2ᵉ partie, 1938, chs. i–ii.

reached Gondokoro amongst the Bari. Baker's initial task was to appropriate the trade and settlements of the monopolist Abu Saud. After much procrastination and some open hostility, Abu Saud withdrew with his accumulated stocks of ivory, and the Khartoumers who remained in the region accepted service under Baker. The crucial step was the establishment of a government monopoly over ivory, which took effect on 9 April 1872. Meanwhile the alternatives confronting Baker in the wider task of establishing Egyptian rule over the Africans in the area became exemplified in his relations with the Bari on the one hand and the Acholi on the other. It was not long before Gondokoro was running short of food, and with little prospect of growing much more, Baker was soon, like all his trading predecessors, ravaging the countryside for supplies. He quickly earned the hostility of the Bari, who have ever since execrated his name as no better than a trader's. But the Acholi farther south, who quickly proffered their assistance against the incipient hostility of Abu Saud's settlements in their country, hailed his curb upon the traders' nefariousness, welcomed Baker's own presence amongst them, and have zestfully treasured his memory ever since.[1]

By 1872 there was very little left, however, of the four years for which Baker's services had originally been commissioned. So on 7 April 1872 he crossed the Victoria Nile to enter Bunyoro, the first Interlacustrine kingdom lying immediately in his path. He was thus the first agent of a foreign empire-building power to enter the lakes area. The portent was not immediately appreciated, even by the most astute of its rulers. For naturally enough, as the Bari and the Acholi had already made clear, Africans saw the approach of outsiders primarily from within their own terms of reference.

In Bunyoro at this time, Kabarega had been ruler for about three years. His most immediate concern was to prevent his kinsman, Rionga, who aspired to his throne, from procuring help (as Kabarega himself had) from the Khartoumers. He therefore proposed that Baker should help him to expel Rionga from his refuge north of the Nile. To this Baker agreed. But he in turn had a request to make. He wished Kabarega to

[1] e.g. A. C. Beaton, 'A Chapter in Bari History', *Sudan Notes*, 17, 1934, p. 190; Gray, *Uganda J.*, 1951, p. 127.

accept the protection of the khedive. Aroused by the warnings of some scheming Khartoumers, Kabarega refused. No harm would have resulted, but for a number of tactless actions by Baker and his agents which now ensued, which provoked a swarm of Nyoro onslaughts that buzzed about their ears. They were forced to withdraw north of the Nile, and Baker was lucky to escape with his life. But Kabarega suffered too, for Baker immediately transferred Egyptian support to his rival, Rionga.[1] Such apostasy—as it seemed to Kabarega—determined Bunyoro's hostility towards Baker and his successor in a number of encounters that followed. This hostility was fully reciprocated, and until the end of the century, Bunyoro, in European circles, was invariably reckoned to be implacably hostile. For this there was little warrant, but, later on, the consequences for Bunyoro were to be disastrous.

Meantime Buganda had taken note of the Egyptian advance too. It can safely be said that the kabaka's primary concern throughout the next five years was to forge an alliance with Egypt against Kabarega's alarmingly resurgent Bunyoro. Even whilst Baker remained in Bunyoro, Mutesa dispatched envoys to meet him, almost certainly in order to frustrate any potential alliance between Bunyoro and the Egyptians;[2] and when Baker retreated, and supported Rionga, Mutesa promptly offered him an army to fight Kabarega.[3]

Baker, however, was soon away northwards, telling the khedive (most erroneously) that his commission had been largely fulfilled. His successor as governor of the Equatoria Province was Charles Gordon, of Chinese, and later of Khartoum, fame. In outlook Gordon was considerably more pacific, not to say scrupulous, than Baker. Horrified by conditions upon the upper Nile, he first concentrated upon putting an end to raids against riverain peoples. But his main objectives were always the lakes and their kingdoms. He took no steps to establish effective Egyptian control over the northern peoples of his province. He concerned himself instead with pushing steamers and stations up to the source of the Nile.[4]

[1] Baker, *Ismailia* loc. cit.; Douin, *Khédive Ismail*; Fisher, *Twilight Tales*, pp. 161–6; Nyakatura, *Abakama*, pp. 148–52.
[2] Baker, *Ismailia*, ii. 280–3.   [3] Ibid., pp. 461 sqq.
[4] By far the best of the numerous accounts of Gordon's work in Equatoria is to be found in B. M. Allen, *Gordon and the Sudan*, 1931. See also Douin, *Khédive*

On his arrival at Gondokoro in 1874, Gordon was met by some envoys from Buganda, who were almost certainly there to forward Mutesa's plans for a joint alliance against Bunyoro.[1] This Gordon never grasped. Instead he looked upon their advent as a providential opportunity for promoting the peaceful establishment of Egyptian control over one of the key kingdoms in the lakes area. He sent them back to Buganda, therefore, with an envoy of his own, the American, Chaillé-Long. At Mutesa's court Long was royally received, and it was probably Mutesa's cordiality, as much as Long's own inventiveness, that made Long declare on his return to Gondokoro that Mutesa had accepted Egyptian protection. To maintain this vital initiative (as it not unnaturally seemed) Gordon dispatched a second subordinate, Linant de Bellefonds, on a further embassy to Buganda. But Mutesa's prime interest in both these envoys lay, not in any question of protection by Egypt, but in the hope that they might open the way to his eagerly hoped for alliance against Bunyoro. He specifically declared his desire for this to both of them. But whilst Long declared he must refer the question to Gordon, Linant was actively discouraging.[2]

So it is scarcely surprising that Linant's reception by Mutesa was less cordial than Long's, and the position was made very much worse by Linant's own tactlessness, which was all too reminiscent of Baker in Bunyoro. Mutesa, however, having seen the grave consequences that Kabarega's precipitate hostility had entailed, held his hand, and the situation was relieved by the arrival from the south end of Lake Victoria, at the very moment when Linant was in Buganda, of another European, Stanley the explorer. For unlike Long and Linant, Stanley showed himself ready to support the kabaka's ambitions, by fighting, for instance, for Mutesa against the islanders of Buvuma. Here was important evidence that the khedive might not be the only potentate who could enlist Europeans in his

Ismail, tome iii, 3ᵉ partie, chs. i–iv; G. Birkbeck Hill, *Colonel Gordon in Central Africa*, 1881.

[1] Gordon to Baker, 17 Apr. 1874, T. D. Murray and A. S. White, *Sir Samuel Baker, a Memoir*, London, 1895, p. 220.

[2] Egyptian General Staff, *Provinces of the Equator, Summary of Letters and Reports of H.E. the Governor General*, pt. i, year 1874. Cairo, 1877; C. Chaillé-Long, *Central Africa*, 1876; L. de Bellefonds, 'Voyage de service fait entre . . . Fatiko et . . . Uganda', *Bulletin Trimestriel de la Société Khédiviale de Géographie du Caire*, ser. i, 1876–7, pp. 1–104.

service. Certainly when Stanley suggested that he should send for more Europeans to come to Buganda, Mutesa embraced the idea with alacrity.[1]

Linant's report upon his largely fruitless mission led Gordon to the important conclusion that an advance through Buganda would after all be as likely to provoke hostilities as an advance through Bunyoro. He accordingly decided to bypass Buganda, and proceed straight up the Nile's banks.[2] Realizing that the Egyptians might be on the eve of an advance, the kabaka too suddenly became apprehensive, and in January 1876 sent Gordon a personal appeal. This time he asked that Gordon should not attack Bunyoro. (This unusual request was doubtless prompted by Mutesa's desire to play for time, till his own Europeans, for whom Stanley had written, had arrived. For if Gordon were to overrun Bunyoro on his own, the frontiers of Buganda would lie immediately open before him.) Yet if Gordon were adamant, Mutesa asked him in lieu to attack Bunyoro only in alliance with Buganda. (In this way not only would Mutesa's long standing ambition to demolish Kabarega be fulfilled, but there would be ample opportunity for Buganda to win Egyptian goodwill as the surest guarantee against an Egyptian advance upon Buganda itself.[3]) Once again Gordon failed to appreciate Mutesa's intentions.

Gordon instead now organized a force to excise Mruli from Bunyoro, and there installed the Egyptian protégé, Rionga. He then sent his trusted subordinate, Nur Aga, to procure permission from Mutesa for the establishment of Egyptian garrisons up the Nile to its source. Mutesa imagined that Nur Aga was the vanguard of an Egyptian expedition directed against him; so he quickly lured him to his capital, Rubaga. Upon hearing of this, Gordon assumed that contrary to his own more recent forebodings Mutesa had submitted; but he was quickly disillusioned. Far from Egypt annexing Buganda, Mutesa had annexed Gordon's garrison. At this juncture Gordon's threatening messages might have precipitated disaster. Indeed it was probably only the cool head of Mutesa, and the careful advice of the Zanzibaris at his court (who had no doubt heard of the

[1] Stanley, *Dark Continent*, i, chs. ix–xvi.
[2] Gordon to Stanton, 21, 24 Sept. 1875, 26 Feb. 1876, *Sudan Notes*, 10, 1927, pp. 26–30, 36–37.
[3] Hill, *Gordon*, pp. 159–60.

recent peaceful withdrawal of the khedive's expedition from the east coast of Africa) that saved Nur Aga and his subordinates from massacre. Realizing, on second thoughts, the delicacy of the position, Gordon, in a moment of inspiration, entrusted the highly critical task of withdrawing Nur Aga to the mild-mannered Dr. Emin, who effected it without incident.[1] Thus it chanced that Buganda never clashed with the Egyptians, and that unlike Bunyoro she was not at this point dubbed hostile to Europeans. The whole story, however, is replete with those misunderstandings on both sides which were typical of imperial advance here and elsewhere.[2]

By the time the immediate crisis was over, Gordon had decided to take his long-desired leave. He quickly agreed to recognize the independence of Buganda, and it was never again threatened by the Egyptians. For on Gordon's return in 1877 as governor-general of the united Sudan, he very soon realized that Egypt could no longer afford a determined advance to the lakes. By 1880 the Egyptian garrisons which straddled Bunyoro from Magungo to Mruli had been withdrawn. After 1881 all that remained of the enterprise begun by Baker was the rump of the Equatoria Province under the governorship of Dr. Emin, with its headquarters at first at Lado and then at Wadelai to the north of Lake Albert. Emin occasionally visited the Acholi, and might with greater resources and greater initiative have manipulated their friendly advances so as to establish effective Egyptian authority over their country.[3] But with bankruptcy, revolt, and European intervention in Egypt, an active policy in the Sudan was impossible. By 1884 the success of the Mahdists' revolt in the north presaged the beginning of the end.

As it happened, the Egyptian intervention had made no immediate difference to the balance of power between Buganda and Bunyoro. Yet it had served to sharpen the rivalry between them. Conflict with the Egyptians had, of course, killed Kabarega's Nile trade. In view of Buganda's hostility, it was, therefore, doubly important for him to maintain Bunyoro's commerce

[1] Gray, *Uganda J.*, 1947, pp. 88-91.
[2] J. M. Gray, 'Mutesa of Buganda', *Uganda J.* 1, 1934, pp. 22-49, is still useful.
[3] G. Schweitzer, *Emin Pasha, his Life and Work*, vol. i, London, 1898; Mrs. R. W. Felkin (ed.), *Emin Pasha in Central Africa*, 1888.

with the Zanzibaris to the southward. Standing in the path, however, lay the breakaway kingdom of Toro, and when, about 1876, Nyaika the son of Kaboyo (the original usurper) raided the Mwenge area, which was still at this time part of Bunyoro, Kabarega in retaliation, upon the Egyptian withdrawal, launched a formidable expedition against him, which overran Toro, killed numbers of its people, and captured hundreds of cattle. Two years later Nyaika of Toro died, and his sons warred for the succession. Immediately Kabarega seized the opportunity to crush Toro's independence, and dispatched a second expedition, which captured the successful claimant, Mukabirere, with most of his sons. When another prince set himself up as mukama in his place, Kabarega dispatched a third expedition; this failed to track down the new claimant, but forced him into hiding. At the same time a series of direct attacks upon Bunyoro by Ganda armies were all repulsed. For the time being Bunyoro's dramatic recovery continued unchecked.[1]

In the meantime, however, Buganda was making a rather more systematic endeavour to control the trade routes to the south upon which victory ultimately depended. The assertion of its control over these routes was the purport of the raid which the kabaka planned against Ukerewe island at the south end of Lake Victoria in 1878,[2] as it was, no doubt, of the two expeditions that he dispatched to Karagwe after Rumanika's death in that year, which successfully bolstered his eventual successor. In the early 1880's, moreover, in two expeditions against petty principalities in Buzinza, Mutesa helped his old enemy Mankorongo of Rusubi to unseat Kazuna of Kimwani, and defeated Rwoma of Bukara. About the same time some Ganda agents established control over the Zanzibari emporium in Karagwe, whilst others took the quite unusual step of actually occupying Butundwe and parts of Bugando in southern Buzinza.[3] Thanks to its military reforms, and more especially to its acquisition of guns, the influence of Buganda down the west side of Lake Victoria was, by the 1880's, supreme in a way that it had never quite been before. Buganda's most promising triumph,

[1] Fisher, *Twilight Tales*, pp. 166–70; Kagwa, *Basekabaka*, pp. 305–7.
[2] Wilson to Wright, 8 Feb. 1878, Church Missionary Society Archives (hereinafter C.M.S.) CA6/025.
[3] Ford and Hall, *Tang. Notes*, 1947, pp. 12, 16; van Thiel, *Anthropos*, 1911, pp. 509, 512. Mr. Tyler's notes.

however, occurred at this time in Toro. Here despite his succession of victories, Kabarega had failed to consolidate his hold, and when Namuyonjo, a Toro prince, appealed to the Ganda for assistance, they successfully installed him upon the throne of his fathers; with the result that a Ganda protégé now held the rulership of this key western kingdom, and stood astride Bunyoro's sole commercial highway to the outside world. It was potentially a most powerful position for Buganda to have suddenly attained.[1]

Yet, as it happened, the matrix in which Buganda's future was being determined was compounded not of these external achievements, but, nearer at home, in the teaching of missionaries at the kabaka's court. When Stanley had suggested that Europeans should come to Buganda, he had meant Christian missionaries. Kabaka Mutesa had already exhibited considerable interest in the rudiments of Christianity that Stanley had expounded to him, and had given every sign of genuinely desiring to hear more. It is unlikely, however, that he viewed the advent of missionaries in quite the same terms as Stanley, let alone as the missionaries themselves. Mutesa almost certainly, as we have seen, was interested in procuring European support for his ambitions. That Stanley's Europeans would be teachers of religion did not deter him in the least. After all the Zanzibaris had taught religion, and so had Stanley, but that had not prevented them from forwarding Buganda's political interests, or from trading in cloth and guns. Gordon too, the Egyptian opponent, was known to be a religious man, but that did not prevent him from serving the khedive. There was in any event little remarkable to the majority of Africans in the suggestion that a man's temporal fortunes were intimately linked with his relationship to the unseen. All teachers of religion who came to Buganda in the nineteenth century specifically affirmed as much. It remained a consideration of major importance. For there were no foreigners—no 'godless' European traders for instance—who came to this region until at least 1890 to gainsay the idea.

[1] Kagwa, *Basekabaka*, p. 307. They had held this position once before, and had then lost it (ibid., pp. 304–5). They were to lose it once again, but (almost certainly) only because of the debilitating effects of their own revolutions in 1888.

So it is scarcely surprising that when the first members of the Church Missionary Society, who had answered Stanley's call, reached the south end of Lake Victoria in January 1877, Mutesa should have eagerly summoned them forward.[1] Nur Aga's garrison had recently withdrawn. There was, however, no evidence as yet that it was not *recueillir pour mieux sauter*. Upon the missionaries' arrival Mutesa inquired after their bibles and guns.[2] It was something of a shock for him to learn that they were solely concerned with religion, and it was not very difficult for the Zanzibaris present, who feared this intrusion into their preserve, to suggest that there must be some deep-seated plot— that these Europeans, for instance, were really Egyptian spies.[3] Such a suggestion soon seemed confirmed by the arrival early in 1879 of a second C.M.S. party, who had come up the Nile with the blessing of Gordon.[4] Mutesa, however, once again held his helm steady. He probably felt that whatever the risks it would be wise to hold on to these European missionaries. They could easily be watched, and they might yet prove useful.

Had this been all, there was no particular reason why the missionaries should have had any recognizable impact upon Ganda society. But it was not all. Buganda, as it chanced, despite its external successes, and its autocratic ruler, was a society subject to a good deal of flux. Since the start of the century there had been a succession of reforms. There were now very few loyalties, except those to the kabaka, which remained inviolable. There was, in particular, by the time of Mutesa much discontent with the traditional religious cults. Men were freely averring that the priests were rascally extortioners.[5] They had certainly lost their power. It was long since a kabaka had required a religious role to boost his personal authority. The kabaka, moreover, performed no ritual ceremonies which were in any way essential to tribal law and custom, so he was free to toy with any new cult that came to his country, since he would not thereby be infringing any traditional taboo. He may, indeed, have wanted to foster a new religion, so as to deliver the *coup*

---

[1] Two letters from Mutesa, one dated 10 Apr. 1877, the other undated. C.M.S., CA6/022.
[2] Wilson to Wright, 6 July 1877, ibid. CA6/025.
[3] Mackay to Wright, 14 July 1879, ibid. CA6/016.
[4] Wilson and Felkin, *Uganda and the Sudan*.
[5] Wilson to Wright, 19 Apr. 1878. C.M.S., CA6/025.

de grâce to the traditional priests, who were now his last remaining rivals. New beliefs were certainly entertained seriously.

Of the advancing world religions, the first to intrude was Islam, brought up from the coast by Zanzibari traders. Very little is known of the means by which it became established; but there can be no doubt of the outcome. By the 1860's Islam had won Ganda converts; mosques were being built; the Islamic calendar had been adopted, and for ten years after 1867 Kabaka Mutesa observed Ramadan.[1] Amongst all these Interlacustrine kingdoms such a result only occurred in Buganda. It was accordingly a remarkable coincidence that amongst all these kingdoms it was only to Buganda, too, that Christian missionaries came. There were suggestions in the 1860's that they should go to Bunyoro; but they proved abortive.[2] The C.M.S. at first proposed to advance to Buganda only after establishing stations at the south end of Lake Victoria and in Karagwe,[3] and only then should it prove safe to do so. But, because of Mutesa's urgent entreaties, Karagwe was initially by-passed, and because of the turmoil there following Rumanika's death, and the shortage of missionaries, it was never in the end entered at all. Anglican missionaries became established in the uniquely advantageous circumstances of Buganda alone.

The main religious question on their arrival was simply: was Christianity a better religion than Islam? In 1879 the answer became confused by the arrival of Roman Catholic White Fathers, who claimed to preach Christianity too, but of a much purer cast.[4] Between the European missionaries themselves, and between them and the adherents of Islam, there now took place colloquies at court, reminiscent of sixteenth-century Europe in their manner, bitterness, and barrenness of result.[5] Amid such confusion it was hardly surprising that there should before long have come a counter-attack from those who still clung to the traditional cults. Faced with this challenge, the missionaries' judgement deserted them. They omitted, for instance (unlike the

---

[1] Kagwa, *Basekabaka*, chs. xiv–xv.
[2] e.g. Venn to Rebmann, 26 Feb. 1864, C.M.S. CA5/LI.
[3] 'Instructions delivered by the Committee of the C.M.S. to the Members of the Mission Party proceeding to the Victoria Nijanza', C.M.S. pamphlet 1876.
[4] A. Nicq, *Le Père Siméon Lourdel*, 1932, is still the best account of the early years of the White Fathers' Mission.
[5] e.g. Harrison, *Mackay*, pp. 120 sqq.

explorer Speke), to pay court to the queen mother and royal sister—after the kabaka, perhaps the most powerful persons in the country. It was their pressure which now forced Kabaka Mutesa to abandon his flirtation with Christianity.[1] After his request for baptism had been rejected by both of the missions, since it did not appear to be accompanied by a desire to face its full implications, it was never renewed. Mutesa never adhered to Christianity in the way that he had to Islam. The missionaries, however, in fits and starts were allowed to continue their work, and the fact that Islam had already turned the soil eased their task greatly.

The setback at court was probably a blessing in disguise. Argumentation there would probably have led nowhere. The missionaries were now forced instead to concentrate upon those who would come as individuals to their mission stations. The foremost C.M.S. missionary, Alexander Mackay, had already made the crucial decision to work within the terms allowed to him, and not to press unduly for privileged conditions.[2] His teaching, and that of those in both missions who worked like him, was soon much in demand. Christianity enjoyed, of course, the (at this stage) invaluable association with western technology, whose wonders were apparent to all. But the original desire of those who came to be taught was probably to learn to read and write so that they might enjoy an advantage over their fellows, which, in due course, would bring them to the notice of some senior chief, or even the kabaka, when he required an amanuensis. For they mostly consisted of ambitious youths about the kabaka's court. The Buganda court was a lively place; in the intervals between raids there was much opportunity for disputation, and it was natural that some of its young men should seek to put their time to profitable use. The process was made easier because the missionaries became closely assimilated to the life which surrounded them. Buganda was an open society. Zanzibaris, for instance, had become chiefs. No missionary accepted a chieftainship, but it soon appeared that their mission stations were not unlike the establishments which senior Ganda chiefs maintained near the capital, and that their relationship to their followers was closely akin to that between other prominent

---

[1] e.g. Litchfield to Wright, 7 Jan. 1880, C.M.S., CA6/015.
[2] e.g. Mackay to Wright, 17 Apr. 1879, C.M.S., CA6/016.

chiefs and the numerous young men sent to serve under them. The missionaries thus became closely linked with a select band of followers. What was more, there were some who visited them and became attracted by their teaching, who belonged to the households of other chiefs; and it sometimes happened that such individuals became proselytizers on their own; so that Christian communities began to spring up not only within the missionaries' own households, but within those of some other prominent Ganda chiefs too.

As understanding deepened, it appears to have been the person of Christ, and such standards as the demand for humility —a revolutionary concept for a thrusting society—that made the greatest impact. But firm adherence did not come quickly. Many (to begin with all) would desert when conditions became hostile. Yet increasingly they came back once the crisis was past, because, for some at least, the dilemmas were real. One could not revert to the discredited traditional cults. Life was still not so secular that one could easily stand still. There might be no choice but to go forward. Courage was in any event a requirement for life at the court. It was gradually enlisted on the side of conversion. The first baptisms took place in 1882. Two years later they numbered over a hundred for each of the missions.[1]

In other parts of East Africa converts were won too, but always from amongst the detribalized—from among absconders from tribal society, refugees from tribal wars, and freed slaves. But the converts in Buganda belonged to the lower echelons of the most powerful hierarchy within the most flourishing polity in the whole of East Africa. The phenomenon was unique. It was, moreover, to be uniquely important. Not merely here, but for a greater distance around, Christianity was to prevail over both Islam and the traditional cults; and Buganda itself, because it alone amongst the peoples of the Interlacustrine region provided so fecund a field for the first Europeans who settled in the region, now contained some uniquely auspicious ingredients that before the end of the century were to give it both an unrivalled pre-eminence amongst its fellow kingdoms,

[1] The most recent accounts of this early period are: R. Oliver, *The Missionary Factor in East Africa*, 1952, pp. 39–41, 67–69, 73–78; D. A. Low, 'Religion and Society in Buganda, 1875–1900', 1957, pp. 1–7; J. V. Taylor, *The Growth of the Church in Buganda*, London, 1958, pp. 19–46; H. P. Gale, *Uganda and the Mill Hill Fathers*, London, 1959, pp. 1–37.

the largest single territorial expansion in its history, and an especially advantageous relationship with its subsequent European overlords.

Hence, by 1884, the northern interior presented an altered but, in many respects, not very different face from that displayed in 1840. Man's search for his livelihood; the assaults of his enemies; the rites attending his passage through life; the vagaries of the weather; the ubiquity of the unseen; these remained as before. Only upon the crests was there any rise and fall in the tides of men. But what of these? The Somali thrust towards Mount Kenya was still in its embryo. The Egyptian advance had proved abortive. Buganda and Rwanda remained the giants among the Interlacustrine kingdoms, and the Masai were still unchallenged in the east. True, there had come traders. But their impact about Masailand had been limited, and they were now more interested in the ivory farther north. As for the Nile trade, that had been cauterized by the intervention of the Egyptian government; Khartoum traders were no longer to be encountered.

Yet there were enough breakers on the waters to give warning of rocks and currents lying beneath. In the north-east the Kamba had declined from their previous eminence; the Masai were not the power they had been; the 'Kwavi' had disintegrated. The Kikuyu, however, were burgeoning forth, and Nandi were flaunting their fledgling triumphs. There was considerably more undertow in the north-west. The Zanzibaris were heavily committed in Buganda, to whose strength they had contributed greatly, and from whose energies they drew considerable profit. The gap hereabouts between small and large kingdoms had undoubtedly widened, and although none of them had freed itself wholly from internal infirmities, there were several substantial polities. Amongst them Buganda held a primary, and, by 1884, a most powerful position; but Bunyoro, if now suddenly threatened, had engineered an uncommonly vigorous recovery, while Rwanda had never been more formidable. At the same time there were everywhere signs of an incoming tide breaking the bar. Some of its flecks of white spray catch the eye—Baker speeding for his life from Kabarega's warriors; Gordon deciding not to quarrel with Mutesa;

Thomson hot afoot through Masailand; Mackay the centre of an eager group of Ganda 'readers'; Emin bewildered amidst the relics of his province—for these portended the greatest transformations. The last two represented the only two toeholds possessed by Europeans in the northern interior. The events attending their inception had warned just a few of its peoples of the possibilities in store in any further encounters. But the rest remained unapprised.

X

THE WIDER BACKGROUND TO PARTITION AND COLONIAL OCCUPATION

JOHN FLINT

BEFORE the 1830's European influence, where it existed at all in tropical Africa, was slight, indirect, and confined to regions near the sea coasts. Even where Africans had long contact with Europe, as in West Africa, direct relations with Europeans tended to be monopolized by coastal peoples exploiting their geographical position for commercial and political advantage. By the 1840's, however, there had already emerged in Europe powerful forces which would eventually put an end to the isolation of African cultures in the interior of the continent.

In the later decades of the eighteenth century Europe and the European settlements in North America began to experience profound upheavals in their political structures, economic systems, and ways of living together in society. Popular revolutions in America and France began a ferment of political and social speculation as their aristocratic governments crumbled before the onslaught of powerful new emotions —individualism, egalitarianism, democracy, and nationalism. The industrial revolution in Britain, slowly spreading to Western Europe and America in the nineteenth century, intensified this ferment, shifting the centres of wealth and power, transforming political and economic theory, and demanding governmental policies which recognized the new competitive system of production.

These were some of the forces which, directly and indirectly, drove Europeans into Africa in the second half of the nineteenth century. The rapid increase in Europe's population and manufacturing capacity made possible by the spread of the factory system intensified the demand for tropical raw materials:

THE BACKGROUND TO PARTITION

cotton and dye-woods for the textile industries, gums and waxes, vegetable oils for the manufacture of soap and for lubricating the machines. The industrial revolution also helped to remove barriers which prevented the penetration of Africa. The steamship transformed the great rivers and lakes into potential highways, and medical advances gave some protection to Europeans against tropical diseases.

As important as these material forces was the new attitude to Africa brought about both by industrial changes and by the political revolutions in France and America. In earlier times Africa had been regarded chiefly as an inexhaustible source of slave labour for the plantations in America and the West Indies. The inventiveness and scientific curiosity which had produced the spinning jenny and the steam-engine also made scientifically minded Europeans feel that their ignorance of African geography was 'a reproach upon the present age'.[1] Such men began to organize and finance expeditions whose original motive was simple curiosity, but which soon became involved in commercial and religious ambitions.

Religion was the most powerful force which drove Europeans into Africa. The evangelical missionary movement had its origins in Europe, and was born of the conditions created by the industrial revolution in Britain. It was a reaction against the godless and depraved conditions of the new urban working classes, uprooted violently from their traditional rural life. The Methodists and evangelical Anglicans attempted with considerable success to reclaim them for religion. By accepting the new social system of the towns, and propagating the middle-class doctrines of thrift, self-help, and *laissez-faire*, the revivalists gained powerful support from industrial manufacturers and traders.

It was a natural development to extend this missionary attitude into Africa and other tropical areas. Political and economic theorists were destroying the image of Africa as a slave labour reserve. Slavery affronted the principles of *laissez-faire* economics. Men should be free to sell their labour to the highest bidder; slavery was unnatural and wasteful.

[1] Words which were used at the foundation meeting of the African Association in 1790, *Proceedings of the Association for Promoting the Discovery of the Interior Parts of Africa*, London, 1790.

The evangelicals brought a violent emotional fervour to reinforce the cold intellectualism of the economists. To them slavery was an affront against God, a blasphemy and an abomination, a visible expression of Satan's will. Men were prepared to die fighting it.

Before 1838 the missionary and humanitarian forces were occupied almost entirely with the problem of slavery in the West Indies. In 1807 the slave trade was made illegal for British subjects, in 1833 the institution of slavery was abolished in British colonies, and five years later the last vestiges of slavery were abolished in the British West Indies. Almost at once the humanitarians turned their attention to West Africa.

Despite its abolition by Britain the Atlantic slave trade continued, and even increased, after 1807. The naval patrols were powerless to control the huge coastline from which traders of many nationalities operated. It was this situation which prompted the humanitarians to enunciate a new policy for the 'regeneration' of Africa—the policy of 'legitimate commerce'. If the slave trade could not be stamped out by force, then it must be eradicated by the intense competition of a superior economic system, the *laissez-faire* system. If African peoples and rulers could be given a living demonstration of legitimate commerce by traders penetrating great rivers like the Niger in their steamships, slavery would die a natural death. The missionary and the trader thus became allies in a God-given task. It was a characteristic new attitude to 'native peoples', a fusion of Manchester free-trade principles with evangelical piety, which could scarcely fail to appeal to Victorian England. Even the organizational structure of Protestant missionary activity symbolized the alliance between religion and lay society. The Anglican organization, the Church Missionary Society, was financed, controlled, and administered by influential laymen, and not by the ecclesiastical authorities.[1]

Though the Church Missionary Society had pioneered missionary work in East Africa in the 1840's, it was David Livingstone's travels and writings which revealed to a horrified public the existence of an East African slave trade which rivalled the cruelties and brutalities of the Atlantic trade. His programme of action was very similar to that which had

[1] For details see R. Oliver, *The Missionary Factor in East Africa*, 1952, pp. 4–5.

been urged in West Africa; the interior must be penetrated along the waterways and lakes by missionaries, upright and godly settlers, and by traders, to give Africans an insight into a superior commercial and social system, and into Christian morality and religion.[1] The result was an intense Protestant missionary effort during the 1860's and 1870's to penetrate inland and establish stations near the great lakes.[2] Protestant activity stimulated the Roman Catholic Church to enter the field. The Holy Ghost Fathers founded a station for freed slaves at Bagamoyo in 1868, but the great Catholic impetus came from the foundation by Cardinal Lavigerie in the same year of the *Société de Notre Dame d'Afrique*, or White Fathers. At first the new order worked only in North Africa, but in 1878 Pope Leo XIII sanctioned Lavigerie's plan for the occupation of inland stations in East Africa which began almost at once with the foundation of stations on Lakes Victoria and Tanganyika. All this activity, Protestant and Catholic, meant that by 1885 there were about 300 Europeans living in East Africa, nearly all of them missionaries.[3]

At the same time missionaries and explorers were trying to interest British merchants and shippers in the commercial prospects of East Africa. Though prospects for trade were in fact poor, there was some response, especially from Scotsmen influenced by the appeal of Livingstone and the activity of the Scottish missions which followed him. In 1872 William Mackinnon, a self-made Glasgow shipowner and a member of the Free Church of Scotland, began running the ships of his British India Steam Navigation Company to Zanzibar. After a meeting of Glasgow business men in 1876 he decided to begin the construction of two roads to Lake Nyasa and Lake Victoria from the coast. In 1874 Glasgow merchants had subscribed £10,000 to place a steamer on Lake Nyasa, and in 1878 James Stevenson of Glasgow formed a trading company to put more steamers on the lake and bring out local trade.[4] Other European nations were building up a trade based on Zanzibar. Hamburg

[1] Ibid., pp. 9-15.
[2] Ibid., ch. i, *passim*, and *Cambridge History of the British Empire*, iii, 1959, pp. 88-92.
[3] Oliver, *Missionary Factor*, p. 49.
[4] M. J. de Kiewiet, 'History of the Imperial British East Africa Company, 1876-1895' (unpublished Ph.D. thesis, London, 1955), pp. 2, 17-20, 25; R. Coupland, *The Exploitation of East Africa, 1856-1890*, 1939, pp. 302-4.

merchants had begun in the 1840's by exporting cowrie shells to West Africa, where they were used as currency, but by the 1870's the Germans were dealing in all local products and held a fifth of the trade. By this time Zanzibar trade, much of it a transit trade with the mainland, was worth over £2 million annually, and had doubled in ten years.[1]

It would be a profound mistake to imagine that this expansion of European missionary and commercial activity in East Africa was to lead inevitably to the establishment of colonial rule. Politically the British, with their Indian and South African territories, and their powerful navy controlling the Indian Ocean, were in the strongest position. But the British had no desire at all to found new colonies in East Africa. Prevailing economic theories insisted that colonies were a bad investment, yielding less in trade returns than they cost in administrative expenditure. The problems of abolishing slavery (illegal in British colonies since 1834), or of persuading Parliament to vote supplies, were enough of themselves to deter any prudent British politician. Even if these objections had been overcome, there would have been no need for the establishment of colonial rule. The British government's task was to create the conditions necessary for legitimate commerce and Christianity to expand and drive out the slave trade. Politically, this meant that the British Foreign Office, acting through its consular officials, sought to strengthen local states which seemed likely to maintain the peace and order necessary for commercial expansion.[2] In East Africa the Sultanate of Zanzibar was an obvious vehicle for such a policy; the only difficulty was that it was a slave-trading state. In 1841 Britain appointed a consul in Zanzibar and began the long process of forcing the sultan to cut down the extent of the slave trade and to seek compensation in expanding his 'legitimate' trade and his political control over the mainland. The principle of Zanzibar independence,

[1] Coupland, *Exploitation*, pp. 319–22, and H. Brunschwig, *L'Expansion allemande outre-mer du XVième siècle à nos jours*, 1957, pp. 78–79.
[2] Compare the way in which the British deliberately fostered the growth of strong African states in West Africa, assisting the political ambitions of the city-states of the Niger delta, and in Nupe actually negotiating an arrangement whereby British traders were under the Emir's 'protection'. See K. O. Dike, *Trade and Politics in the Niger Delta*, Oxford, 1956, for the Oil Rivers, and J. E. Flint, *Sir George Goldie and the Making of Nigeria*, London, 1960, pp. 23–26, for policy in Nupe.

enshrined in the Anglo-French Declaration of 1862, was an important part of this policy.[1] After the appointment as consul of John Kirk, who had been with Livingstone on the Zambezi, British pressure on Zanzibar increased to such a pitch that in 1873, under threat of force, Sultan Barghash had to prohibit the sea-borne slave trade completely. Zanzibar was now ready, as the British saw it, to go forward as a respectable, enlightened state, and to expand its authority over the East African mainland, maintaining law and order in favour of British explorers, missionaries, and traders.

In the light of later events such a policy seems naïve and fantastic. Yet British politicians, officials, traders, and missionaries of the time did not doubt that it was realistic or feasible. They were able to think so because of an extraordinary combination of circumstances in Britain and Europe. First and foremost the policy depended on the superior development of British manufacturing processes. In general British exports could outsell those of any of her rivals without the need for the tariff protection of a colonial régime. In normal times such a superiority might have provoked less fortunate rivals to compensate themselves by seeking colonies from which British competition could be excluded by tariffs, but in Africa this did not occur until the 1880's. Since 1815 Europe and the Americans had been absorbed by problems of nationalism and democracy, and half-hypnotized by the persuasive theories of the British economists. France was convulsed by revolutions in 1830, 1848, and 1870, and did not finally establish a solid basis for the Third Republic until the mid-1870's. Germany and Italy did not achieve national unity until 1870. Austria had enough to do to keep the shaky Hapsburg monarchy in being, and had to share power with Hungary after 1867. Russia, attempting both to maintain Tsarism and to absorb western techniques, had a vast field for colonization in her Asian territories. The United States of America was absorbed in the conquest of Indian and Mexican lands on the western frontiers, and weakened by internal disunity which almost destroyed the Union in the Civil War of 1861 to 1865. Britain was thus granted a unique opportunity to pursue the expansion of her

[1] For full text see E. Hertslet, *The Map of Africa by Treaty*, 3rd. edn., 3 vols., 1909, ii. 718, no. 222.

already supreme economy, and to extend her cultural, religious, and political influence, unmolested, unobstructed, and unrivalled by any other power. But this situation could not last for ever. When the European nations at last achieved stable régimes and began to survey the world and dream of 'places in the sun' the British would be forced to rethink their African policies.

The first challenge to the idea of using Zanzibar to control East Africa came not from Europe, but from Egypt. This was ironical, for behind the Egyptian challenge lay many of the same attitudes which had prompted the policy in Zanzibar. Since the opening of the Suez Canal in 1869 Egypt had been of particularly vital importance to British interests in India and the East. Yet there was no idea of establishing direct control over Egypt. Though Disraeli was prepared to buy the Khedive Ismail's shares in the Canal Company in 1875, and to acquire Cyprus in 1878, he refused to follow German suggestions that Britain should take Egypt as a colony or protectorate.[1] The right course seemed to be to allow the khedive to modernize his country, and create an efficient army which could secure the Suez Canal from interference. Private capital was poured into Egypt, and European officials were given key posts in the administration. A vital part of Ismail's policy was the need to expand up the Nile to secure complete control of the flow of water upon which Egyptian agriculture was absolutely dependent. Ismail wanted Europeans, and particularly Englishmen, to undertake this task, and to attract them he not only put forward strategic arguments, but fitted his plans into the framework of the legitimate commerce theory, arguing that a trade in Sudanese ivory would drive out the slave trade. In 1869 Sir Samuel Baker, the discoverer of a second source of the Nile in Lake Albert, was appointed governor-general of 'The Equatorial Nile Basin', and by 1872 he had pushed Egyptian authority south to Lado. In 1873 Baker was succeeded by Gordon, and three years later Gordon was in Bunyoro sending messengers to Kabaka Mutesa of Buganda. He had already come to the conclusion that a permanent control of the Nile sources was not possible along the overland route to Cairo. In January 1875 the Khedive Ismail accepted Gordon's plan to seize territory on the east coast and establish a series of military

---

[1] G. E. Buckle, *Life of Beaconsfield*, vi, London, 1920, p. 353.

posts linking the coast with Uganda. In November over 500 troops in four warships were landed at Brava commanded by a British officer in the khedive's service. They disarmed the Zanzibar troops, hauled down the sultan's flag, and left a garrison. They went on to occupy Kismayu and Lamu, and laid claim to the East African coast in the name of Egypt.[1]

Here was a nice problem of judgement for the British Foreign Office: was Zanzibar or Egypt to undertake the opening up of the lakes region and the east coast? Kirk pressed the claims of Zanzibar, Gordon those of Egypt. The humanitarian interests in Britain supported Zanzibar; they were not convinced that the Egyptians had a genuine anti-slavery impulse.[2] The Foreign Office, remembering the humiliation which the sultan had suffered in 1873, and fearing that further humiliation might permanently embitter his attitude to Britain, decided for Zanzibar. Pressure from Britain induced the khedive to order a withdrawal from the coast in January 1876. Already the Egyptian expedition had found that it was too weak to move inland and join Gordon.

Britain had backed the right horse. Ismail still had ambitions in East Africa, and continued to demand a seaport on the east coast; in May 1876 Egypt officially 'annexed' the territories around Lakes Victoria and Albert.[3] But it was all a sham. Ismail's lavish projects were already creating misgivings. Between 1879 and 1882 they would produce a complete breakdown of the experiment of Egyptian independence, with profound results for East Africa.

The Egyptian 'invasion' opened Sultan Barghash's eyes to the difficulties of his position. Since 1873 he had been suspicious of the British, and had even offered the Germans a protectorate which Bismarck rejected.[4] With the Egyptian withdrawal his attitude changed completely. He declared that he 'had never before understood the force of the disinterested friendship of Her Majesty's Government for his country' and was 'greatly delighted'.[5] Now painfully aware of the weakness of Zanzibar,

[1] Coupland, *Exploitation*, pp. 271–80.
[2] J. M. Gray, 'Sir John Kirk and Mutesa', *Uganda J.* 15, 1951, p. 8; Oliver, *Missionary Factor*, pp. 87–89.   [3] Coupland, *Exploitation*, p. 293.
[4] de Kiewiet, 'British East Africa Company', p. 10.
[5] Kirk's description of what the sultan said, quoted by Coupland, *Exploitation*, p. 293.

Barghash began to look upon Britain as a source of capital and military and technical skills. In 1875 the sultan paid a visit to England and witnessed the might of her industrial civilization. In 1877, after the Egyptian affair, Barghash decided, with the advice of Kirk who was acting on Foreign Office suggestions, to build up an efficient army. A British naval officer, Lloyd Mathews, became 'General' of the Zanzibar Army. Ammunition and modern guns were presented by the British government, and soon there was a force of a thousand men which could be used in extending the sultan's authority over the mainland. Barghash himself began to modernize the government, a police force was raised, a water-supply for Zanzibar town established, and the streets were lit.[1]

This zeal for development fitted in very well with the plan which William Mackinnon and his associates were making to bring 'legitimate commerce' to the East African interior. In 1876 the sultan had approved Mackinnon's road-building plans; later in the same year Mackinnon was considering schemes for establishing colonies of Africans at strategic points along the new roads, and for leasing a coastal port. By January 1877 Mackinnon and his friends were discussing the formation of a company to rule and develop the mainland under a lease from the sultan, modelled on 'the East India Company of old'.[2] In March 1877 Mackinnon's agent Gerald Waller obtained the sultan's agreement in principle to a scheme whereby the company would rule the mainland as the sultan's vassal. Meanwhile Mackinnon consulted the British government, and was told that it had no objections to the scheme, though 'formal sanction' could not be given. This presumably meant that the government would take no responsibility for the area by declaring a protectorate or granting a royal charter to the company, but it satisfied Mackinnon. In April 1878 Waller returned to Zanzibar, accompanied by an Arabist, G. P. Badger, to negotiate the final concession. Though the sultan still favoured the principle of the scheme, the detailed negotiations went badly. Kirk complained that Badger was 'making an ass of himself' and offending the sultan by displays of bad temper.

[1] Coupland, *Exploitation*, pp. 237 sqq.
[2] H. Waller to Mackinnon 12 Jan. 1877, Mackinnon Papers, quoted by de Kiewiet, 'British East Africa Company', p. 23.

The sultan seemed to be raising conditions, demanding that the company begin administration almost at once. In May the talks were broken off.[1]

'There is a secret history of the failure of the former Mackinnon scheme which I will not commit to paper', wrote the Foreign Office official Villiers Lister some years later.[2] The secret history is now known. Badger's conduct was not entirely due, as Kirk imagined, to his eccentric character. He had been prompted to wreck the scheme by Lord Salisbury, who had become Foreign Secretary just before the negotiators left for Zanzibar, but after Mackinnon had been told that there was no objection to his plans. Salisbury privately hinted to Badger that he should use his position as interpreter to give the sultan the impression that the scheme was against his interests.[3] Salisbury's motive is not known. Perhaps he suspected trickery by Mackinnon, and really wished to protect Barghash; more probably he feared that Mackinnon's company would involve the British government in the affairs of the mainland, and felt there were no sufficient British interests to warrant interference.

Salisbury's action in 1878 was but the first of a series of deliberately missed opportunities. The Gladstone government in power from 1880 to 1885 continued to believe that Britain must avoid direct commitments in Zanzibar, and that British interests could be best upheld by maintaining and extending the power of the sultan. Yet it was becoming apparent that other powers were beginning to show interest in Africa. Nationalism in Germany, fulfilled in its internal demands by the creation of a unified *Reich* after the Franco-Prussian war of 1870, now began to find expression in economic nationalism and the demand for national expansion overseas. Theorists like Frederic Fabri,[4] Hübbe-Schleiden,[5] and Ernst von Weber,[6] revived the earlier

---

[1] The above account is based on Coupland, *Exploitation*, pp. 306–15, and de Kiewiet, 'British East Africa Company', pp. 32–40.

[2] Memo. by T. V. Lister, 25 Apr. 1885. F.O. 84/1737.

[3] This intrigue was unearthed by Miss de Kiewiet, 'British East Africa Company', pp. 43–44. The documents showing Salisbury's part in the affair are Badger to Salisbury, 3 July 1878. F.O. 84/1528, and Sultan Barghash to Salisbury, 18 May 1878. F.O. 84/1527.

[4] F. Fabri, *Bedarf Deutschland der Kolonien?*, Gotha, 1879.

[5] W. Hübbe-Schleiden, *Deutsche Kolonisation*, Hamburg, 1881.

[6] E. von Weber, *Vier Jahre in Afrika, 1871–1875*, Leipzig, 1878.

anti-Free Trade doctrines of the economist List,[1] and demanded colonial territories for the settlement of surplus German population, and for markets and raw materials for the rapidly expanding German industries. German culture was extolled, and Germany's mission to civilize 'barbarous' races proclaimed.[2] The international Conference in Brussels, called by King Leopold of the Belgians in 1876 ostensibly to discuss methods of peaceful and co-operative action to develop legitimate commerce and attack the slave trade, broke up into a number of jealous national sub-committees. Leopold proceeded with his plan to create a private empire under his personal control in the basin of the Congo. At the same time a 'colonial group' began to emerge in France and to gain influence. In January 1878 the monarchist President MacMahon resigned, and the Republic at last achieved a stable form. French politicians now began to look outside France; plans for a railway from Algeria to Senegal were seriously considered, treaty-making expeditions began to wander along the upper Niger, whilst French traders and a consul appeared on the lower Niger in competition with the British, threatening that the region would soon be annexed to France. Jules Ferry, an Alsatian with an intimate knowledge of French industry, and a firm believer in the need for protected colonial markets as an outlet for industrial exports, was at this time Minister of Education. Alarmed by the activities of Leopold's agents in the Congo, Ferry provided the Italian adventurer de Brazza with funds for an 'educational and scientific mission' to the districts north of the Congo mouth. The result was a series of treaties which established the French colony of Gaboon. In a short time French tariffs were driving established British traders away. The British government began to worry about the position of British traders on the Congo and Niger rivers, and to think about ways in which they could be protected from foreign encroachments, if possible without entailing new responsibilities or expenditure of government money.

Meanwhile events in North Africa were forcing Gladstone and his colleagues towards intervention in Egypt. The attempts

---

[1] G. F. List, *Das nationale System der politischen Oekonomie*, Stuttgart, 1841.
[2] Brunschwig, *L'Expansion allemande*, pp. 94–96; M. E. Townsend, *Origins of Modern German Colonialism*, 1921, pp. 26–31.

of the khedives to create an independent Egypt using European personnel and capital had not met with success. Inefficiency and corruption, and usurious interest rates demanded by European financiers, had produced chaos and bankruptcy by the middle of the 1870's. The German Chancellor Bismarck had several times urged Britain to take control of Egypt, but British politicians had rightly suspected that his motive was to embroil them in difficulties and weaken Britain's international position. Disraeli, the Conservative Prime Minister in 1878, was convinced that Egypt would be merely 'an expensive encumbrance'.[1] Nevertheless neither Britain nor France, both of whose subjects held large investments in Egypt, could afford to stand aloof, if Egypt's government collapsed. Early in 1879 the Khedive Ismail had been forced to establish a ministry including one French and one British official, so as to give security to investors' interests. When Ismail a few months later tried to rid himself of these ministers, and deal directly with the bondholders, Britain and France intervened, replaced the khedive by his son Tewfik, and established a joint control over Egypt. This sparked off a nationalist movement among the Egyptian army officers, led by Arabi Pasha, who by the middle of 1882 were in control of Egypt.

Though Gladstone's government was now prepared to intervene, it continued to shun any idea of establishing any permanent or legal form of British control. The object of intervention would be to 'restore the khedive's authority'. Gladstone wanted the intervention to be international, in the name of Turkey, but the French at first insisted that it be limited to Anglo-French forces. By a combination of maladroit diplomacy and political changes in France the outcome was that the British fleet alone bombarded Alexandria on 3 July 1882. Troops were landed, and in September defeated the nationalist forces at Tel-el-Kebir. Britain now controlled Egypt. Yet British politicians, whether Liberal or Conservative, refused to admit that the control was more than temporary. Solemnly and repeatedly Gladstone promised that Britain would withdraw as soon as the khedive's power had been restored, and he was assuming that the work might take months, or a year or two at the most. This was a profound miscalculation. Before

[1] Quoted by H. Hinsley, *C.H.B.E.* iii, 1959, p. 108.

Egyptian finances could be made to support a stable régime, and pay interest to the bondholders, the entire structure of the administration, and even the society itself needed to be reorganized. This was a work for decades, as the consul-general, Sir Evelyn Baring, soon realized. In reality the British occupation was as 'permanent' as any other colonial acquisition, and Egypt's interests, including the need to control the waters of the Nile which alone made life in Egypt possible, became British interests. But by persisting in the belief that the occupation was soon to come to an end, the British estranged France and impeded and confused their own attitude to Egyptian problems, sometimes (as with the dispatch of Gordon to the Sudan in 1884) with disastrous results. They also permitted international control of Egypt's finances to continue, which meant that Britain had to curry favour with European powers, especially Germany, in order to secure agreement to financial reforms in Egypt. This often entailed the sacrifice of British interests elsewhere in Africa.

The British occupation of Egypt helped to set off the complicated series of events which led to the entry of Germany as a colonial power in Africa and revolutionized the situation in East Africa. The German Chancellor Bismarck wished to exploit Britain's dependence on the German support needed to secure international consent to financial reforms in Egypt. This did not at first mean that Bismarck wished Britain to smile upon the establishment of a German colonial empire at her expense. He was no believer in colonies, which he regarded as useless and expensive liabilities. He poured scorn on the colonial enthusiasts, and refused their demands. 'I want no colonies', he said in 1871 when it was suggested that Germany should seize colonies from defeated France, 'They are good only for providing offices. For us colonial enterprise would be just like the silks and sables in Polish noble families, who for the rest have no shirts.'[1] In 1872 he rejected a petition for annexation from Fiji; in 1874 he refused a request for German protection from the Sultan of Zanzibar, who was smarting under the British threats which had forced the abolition of the sea-borne slave trade. He welcomed and encouraged proposals

[1] H. von Poschinger (ed.), *Fürst Bismarck als Volkswirt*, 1889, i. 63.

by which other powers would set up régimes capable of giving security for German traders, hoping that the United States would establish a Samoan protectorate in 1875, and urging France, Italy, and Britain to carve up North Africa in 1878. In 1880 German merchants were firmly told that they could get no support for their plans for colonization in New Guinea. As late as 1882 the brothers Denhardt could get no support for a request from Simba of Witu for German protection.

After the occupation of Egypt Bismarck hoped that the British government would show its gratitude for German co-operation by facilitating the activities of German merchants overseas. He hoped to be able to stave off the demands of German merchants by persuading Britain to undertake the burden of protection in certain areas. The test came in South West Africa. A German merchant, F. A. E. Lüderitz, had succeeded in obtaining claims to monopolies and land rights at Angra Pequena, and demanded German protection for his settlements. In February 1883 Bismarck asked Britain to declare a protectorate over the region, but received no definite reply. In August the German government told Lüderitz that German protection might be granted, in so far as it did not conflict with British claims. In September the British were again pressed to state what their claims were. Left to themselves, the British government were prepared to admit that there were no British claims to the area, but the intervention of Scanlen, the Prime Minister of Cape Colony, altered the tone of the British reply. On 21 November, pressed once more by the German ambassador, the Foreign Office stated that although they claimed no sovereignty over Lüderitz' concession area, 'any claim to sovereignty by a Foreign Power' between Angola and the Cape frontier 'would infringe their legitimate rights'.[1]

This in itself might have been enough to provoke Bismarck into establishing German protection in South West Africa. When no reply was received to a demand for further elucidation of the British claim made in December 1883, a German protectorate over Angra Pequena was announced in April 1884. But Bismarck had decided to do more than this. He had determined

[1] Quoted in W. O. Aydelotte, *Bismarck and British Colonial Policy*, 1937, p. 36. This account of the Angra Pequena affair is largely based on this work, by far the best analysis of the problem.

to found colonies elsewhere and to do all in his power to embarrass the British position in Africa. The British government must be shown that German demands could not be ignored. Other considerations, too, played their part in convincing Bismarck that Germany must begin a colonial career, though he still remained unconvinced of the value of colonies. If the colonial programme were directed against Britain, this might soften the implacable anti-German feelings of France, embittered by her defeat in 1870 and the loss of Alsace and Lorraine. Franco-German co-operation on the basis of mutual colonial antagonism to England might even result in greater stability in Europe. Bismarck also calculated that a German colonial policy would enable him to recreate a solid majority in the Reichstag.[1]

Britain now faced a most uncomfortable period of German opposition. In July 1884 Germany annexed the Cameroons, forestalling the British consul by a few days. In the previous month French and German pressure had forced Britain to abandon the treaty she had concluded with Portugal by which she had hoped to secure British mercantile interests on the Congo by recognizing Portuguese control. In August the German representative joined with the French in blocking British plans for financial reform put before the London conference on Egyptian affairs. In September the British were forced to recognize German claims to South West Africa from Angola to the Cape frontier. Meanwhile France and Germany had come together to challenge the British position on the Niger and that of her ally Portugal on the Congo. A Conference of the Powers was summoned to Berlin in November to draw up rules for international control of both areas. Britain was only able to preserve her position on the Niger through the breakdown of the shortlived and unstable Franco-German *entente*,[2] but the

---

[1] Erich Eyck, in *Bismarck and the German Empire*, 1950, pp. 273 sqq., argues that German party politics were the main motive behind Bismarck's decision, and that diplomacy merely provided the opportunity. This view seems to give too little weight to the Angra Pequena affair, which convinced Bismarck that the British were unaware of the diplomatic facts of life. His later refusals to acquire more colonies, and the continuing neglect of existing German colonies, would indicate that the German colonial enthusiasts had little real power in politics.

[2] And by the success of the Niger traders, led by George Goldie Taubman, later Sir George Goldie, in purchasing the French Niger firms at the last minute. For details see Flint, *Goldie*, pp. 67–68. For a full account of the diplomacy of the Berlin Conference see S. E. Crowe, *The Berlin West African Conference, 1884–1885*, 1942.

Congo was placed under King Leopold's Congo Independent State and Portuguese claims were rejected.

The German humiliation of Britain in Africa during 1884 might perhaps have produced the opposite result to that which Bismarck intended. Instead of appreciating how dependent Britain was upon German favour, the Liberal government might have been provoked to take swift action in other areas to forestall possible German moves and to retaliate. Some of the younger and more ardent spirits in the government were so inclined, particularly the radicals Chamberlain and Dilke.[1] There was an opportunity for retaliation in East Africa where Harry Johnston, at that time a young naturalist commissioned by the Royal Geographical Society, had secured treaties with local rulers around Mount Kilimanjaro. In July 1884 Johnston wrote to Lord Edmund Fitzmaurice urging that Kilimanjaro be colonized by English settlers as a British protectorate.[2]

The letter arrived just as Germany seemed about to make a move in East Africa. On 1 October 1884 Gerhard Rohlfs, a noted explorer well known for his advocacy of German colonial expansion, was appointed German consul-general in Zanzibar.[3] In November the German press carried reports that Germany intended to establish a protectorate over Zanzibar.[4] The German government denied these rumours but the British were uneasy. Kirk, acting on instructions, secured a declaration from Sultan Barghash in December that he would cede no sovereign rights without British consent.[5]

Johnston's proposal thus fell upon sympathetic ears, and for the first time the Foreign Office lent its support to a proposal for direct British action in East Africa. But Sir John Kirk was not so enthusiastic; he felt that Africans would resist alienation of their land, and saw in the scheme a dangerous departure from the traditional policy of controlling East Africa through Zanzibar and one which could only alienate the sultan. Yet Kirk agreed that the German threat was real and that something had to be done; there were mysterious German travellers inland, and a German man-of-war on the coast. He therefore

[1] J. L. Garvin, *Life of Joseph Chamberlain*, i, 1932, p. 497.
[2] R. Oliver, *Sir Harry Johnston and the Scramble for Africa*, 1957, pp. 66–67.
[3] Coupland, *Exploitation*, p. 398.
[4] de Kiewiet, 'British East Africa Company', p. 62.
[5] The declaration is printed in Coupland, *Exploitation*, p. 388.

urged that the traditional policy be intensified and strengthened; the area should be developed by British persons, but under the sultan's flag.[1]

The forward spirits in the cabinet, including Chamberlain and Sir Charles Dilke, now made what they could of this proposal, canvassing the support of the Foreign Secretary Lord Granville, the Colonial Secretary Lord Derby, and Lord Kimberley at the India Office. On 5 December 1884 Kirk was instructed to urge the sultan to send a military expedition to Kilimanjaro to negotiate treaties with the local chiefs recognizing Zanzibari authority, and to establish posts *en route*. Kirk was to go with the party, and if the chiefs refused to accept Zanzibari authority he was to make treaties on behalf of Great Britain.[2]

All these decisions had been taken without the active participation of the Prime Minister. On 14 December Gladstone attacked the plan at a meeting of the cabinet. Later, writing to Dilke, he described himself as 'puzzled and perplexed at finding a group of the soberest men among us to have concocted a scheme such as that touching the mountain country behind Zanzibar with an unrememberable name'.[3]

Whilst Gladstone irritably restored his authority over his erring colleagues, their fears were becoming realities in East Africa. On 14 November Carl Peters had arrived in Zanzibar with three associates. Peters's group was the most extreme within the German colonial movement. The main body of colonial advocates were organized in the *Kolonialverein*, founded in 1882 with the aim of identifying the colonial demand with German nationalism, to which end they had begun to issue a newspaper, the *Kolonialzeitung*, in 1884. It was a respectable and sober body, attempting to influence the German government through its influential members like Prince Hohenlohe-Langenburg.[4] Peters and the extremists found such methods altogether too slow and uninspiring, and resolved to act independently to force the government's hand. In March 1884, with thirty others, he founded the *Gesellschaft für Deutsche Kolonisation*. This

---
[1] Coupland, *Exploitation*, pp. 386–7.
[2] Granville to Kirk, 5 Dec. 1884. F.O. 84/1676.
[3] S. Gwynn and G. M. Tuckwell, *Life of Sir Charles Dilke*, 1917, ii. 83–84.
[4] M. E. Townsend, *The Rise and Fall of Germany's Colonial Empire*, 1930, pp. 81–83.

was in fact a commercial company designed to establish a colony; its manifesto stressed the urgent need for direct action before other powers swallowed the whole of Africa. The project was financed by the sale of shares in the *Gesellschaft* which soon raised 175,000 marks. Various projects were submitted to the German Foreign Office for approval, but Peters was merely told not to meddle in overseas policy. He therefore decided to keep secret his scheme for a German East African colony. He and his companions embarked from Europe, in disguise. On arrival at Zanzibar they were told by the German consul that they could expect no official support, and that they would travel on the mainland at their own peril.[1] Nevertheless Peters set off undaunted. By 17 December he was back in Zanzibar with treaties in his pocket which purported to place Usagara, Ungulu, Uzigua, and Ukami under German protection. At about the same time the Denhardt brothers were negotiating in Witu with 'Sultan' Simba.

Returning to Berlin in February 1885 Peters succeeded in maturing his plans with such precision that it is difficult to believe that Bismarck had been unaware of them. On 26 February the Kaiser signed the *Schutzbrief* placing the territories claimed by Peters under German protection, and on 3 March, the day after the delegates to the Berlin Conference on the Niger and Congo had departed, the *Schutzbrief* was published. It was in reality what the British would have called a 'Royal Charter' entrusting the administration of German East Africa to the *Gesellschaft für Deutsche Kolonisation*.[2]

The establishment of the German protectorate brought to an abrupt end the British policy of controlling East Africa through Zanzibar. The Germans had successfully defied the sultan's rights, despite the fact that these were supposedly guaranteed by the Anglo-French agreement of 1862. Moreover it was clear that further German inroads on Zanzibar claims were to come. If the German chartered company wished to pay its way it would have to acquire a part of the sea coast in order to

[1] Brunschwig, *L'Expansion allemande*, pp. 115–20; Townsend, *Rise and Fall*, pp. 132–3; C. Peters, *Die Gründung von Deutsch-Ostafrika*, 1906, pp. 63 sqq.

[2] Townsend, *Rise and Fall*, p. 133, argues that Peters got his way by threatening to sell the treaties to King Leopold if the *Schutzbrief* were not granted, but she does not reveal her source.

escape Zanzibar customs duties and be able to levy its own. However loudly the Germans might claim that the inland regions had never been a part of the sultan's dominions, they could hardly pretend the same for the coast. There would have to be a frontal attack on the sultan's position.

Britain could have strangled the new German colony at birth by preventing the acquisition of a coastal foothold. But a resolute defence of Zanzibar would probably have needed the rapid declaration of a British protectorate over the whole of the sultan's dominions, the dispatch of numerous treaty-making expeditions to the interior, and the stationing of enough warships in the Indian Ocean to overawe local German naval strength. Such belligerency was not to be expected from Gladstone and his colleagues. The need for German diplomatic support was now more urgent than it had been at the time of the Angra Pequena affair. The Egyptian situation had become a nightmare—the Sudan had risen under the Mahdi and the popular British hero General Gordon had been killed at Khartoum in January 1885. Russia, taking advantage of British preoccupations in Egypt and the Sudan, began to advance into Afghanistan, and Gladstone's government, unable to face public opinion in England with yet another humiliation, tried to stand firm and threaten the Russians with war at a time when England had not a single ally in Europe. When the Russians occupied Pendjeh at the end of March war seemed imminent.[1]

The Sultan of Zanzibar was but a pawn on this wide chessboard, to be sacrificed if necessary. His position was not even that of a pawn which is so strategically placed that its player will defend it as a minor piece. Whilst British policy in Egypt was still fluid and undecided the approach to the Nile headwaters from the East African coast could not be an important British interest. There remained only British trading and missionary interests. In West Africa, especially on the Niger, French threats had provoked British protectorates in defence of such interests. But the French were protectionist and Roman Catholic; German colonies were free-trading and Protestant. If Germany wished to assume the burden of administering

[1] *C.H.B.E.*, iii, 122–6. The crisis died down in May when the two powers began negotiations for a settlement by arbitration.

regions which were open to British traders and missionaries no true free-trader could possibly object. 'If Germany becomes a colonising Power', exclaimed Gladstone, 'all I can say is "God speed her".'[1] When Kirk asked the Foreign Office to give more support to the sultan, who had reacted strongly to the German moves and sent Mathews with troops to Kilimanjaro,[2] he was told by Lord Granville that 'Her Majesty's Government are favourable to German enterprise in districts not occupied by any civilised power'.[3] German inroads on the sultan's dominions thus proceeded without impediment from Britain. In April Clemens Denhardt obtained a treaty from Simba of Witu, who was anxious to escape from Zanzibar control. Shortly afterwards Dr. Jühlke obtained treaties near Kilimanjaro in the region which Johnston had wished to colonize, and where Mathews's troops had recently been. The British made no counter-demonstration when, at the end of May, it became known that a German naval squadron would sail for Zanzibar 'to bring the Sultan to a more correct bearing'.[4] When these warships arrived in August Barghash was forced to agree to a series of concessions, withdrawing his protests at German colonization in the interior and at Witu, abandoning his own claims to these places, accepting a new customs régime on the coast which would allow passage of goods to Usagara, and virtually handing over Dar es Salaam to German control.[5] In all this Kirk had to play a most humiliating role, using his influence with Barghash to secure surrender to Germany.

Meanwhile the British had begun to try to salvage what they could from the wreck of British influence in East Africa. These attempts were made within the limits of traditional policy and the diplomatic situation; action must not involve the government in expense or direct responsibility, nor offend the Germans. The obvious way to do this was to try to resurrect the scheme for a private British concession which could operate

[1] *Hansard*, 2nd ser. ccxcv, col. 979. He was speaking on 12 Mar. 1885 in a debate on the Foreign Office vote.   [2] Coupland, *Exploitation*, pp. 409–11.
[3] Granville to Kirk, 20 May 1885. F.O. 84/1722.
[4] Malet to Granville, Af no. 191, 30 May 1885. F.O. 84/1714.
[5] The customs arrangements were complicated, and in some ways were favourable to Barghash in the strictly financial sense. In theory they also respected Zanzibar sovereignty over the coast. As they were soon abandoned for a more open German control they have not been analysed in detail here. For full details see Coupland, *Exploitation*, pp. 429–47.

within an area agreed to by Germany. The initiative came from Consul Frederick Holmwood, Kirk's chief assistant in Zanzibar. In April 1885, whilst on leave in England, he prepared a memorandum designed to arouse the enthusiasm of James Hutton and the Manchester merchants, and to reinvigorate William Mackinnon's earlier interest. In glowing terms he described the prospects of economic development in the regions between Mombasa and the Great Lakes, alleging the existence of an immediately profitable trade, especially in ivory, infinite possibilities for new export crops, and the prospect of a near-paradise in Kenya for European settlers with abundant African labour. He urged the immediate formation of a British East Africa Association to control the region's development, and perhaps to obtain a charter from the British government. It might well be the beginning of a new East India Company.[1]

Mackinnon and Hutton took up these ideas, and began to enlist influential support, including that of Lord Aberdare, a retired Liberal politician who was a friend of Granville and had once been Gladstone's Home Secretary. The chartered company idea was in the air. In 1881 a charter had been given to the British North Borneo Company which allowed it to rule territory under concessions from the Sultans of Sulu and Brunei. Lord Aberdare was the chairman, and Hutton a director, of the National African Company, which was at present actually negotiating with the Foreign Office the details of a charter to rule the Niger region under concessions from local rulers.[2] On 22 April Lord Aberdare placed Holmwood's scheme before the Foreign Office, and asked for government support.[3]

Lord Granville was so anxious to conciliate the Germans that he instructed his ambassador in Berlin to disclose the scheme in full, and virtually to seek German permission for it.[4] Replying most politely, the Germans asked that the plan be deferred

---

[1] Holmwood to James Hutton, 10 Apr. 1885, enclosed in Aberdare to Granville, 22 Apr. 1885. F.O. 84/1737.

[2] For details of the important part played by both Hutton and Lord Aberdare in the acquisition of the Niger charter see Flint, *Goldie*, pp. 45–46, pp. 50–51, pp. 56–57.

[3] Aberdare to Granville, 22 Apr. 1885, with enclosures. F.O. 84/1737.

[4] Granville to Malet, Af no. 171, 25 May 1885. F.O. 84/1711.

until Germany had defined her sphere more closely, so that the British scheme would not 'interfere' with German operations.[1] Bismarck was now disposed to be more conciliatory, the crisis in Afghanistan had passed, the short-lived German attempt to create an anti-British accord with France in Africa had failed, and the British had mended their ways since their unfortunate behaviour over South West Africa. He therefore accepted in principle a British proposal that a joint commission should determine exactly the Sultan of Zanzibar's boundaries.[2]

This commission, on which France obtained representation, began its work in October. The sultan was not represented, and as a quasi-judicial inquiry the whole thing was farcical. The Germans continued to make treaties in the disputed areas whilst the commision was actually in session. The estimates of the extent of the sultan's authority varied from the forty miles of the British commissioner Col. H. H. Kitchener, to less than ten miles on the part of the German commissioner. 'Unanimity' was finally secured when all three commissioners placed their signatures to a list of places which all had agreed were in the sultan's dominions, so that the German minimum was in effect accepted.[3]

Whilst the members of the commission had been so busily wasting their energies, Mackinnon and his associates had made some progress. In June 1885 Lord Salisbury replaced Gladstone as Prime Minister, and though his attitude to the problem did not differ much from that of the Liberals, his diplomatic position was slightly stronger, and he was prepared to act more firmly towards Germany. Wishing to obtain at least the bases for possible counter-claims, he prompted Harry Johnston to make over his treaties in Kilimanjaro to Mackinnon.[4] In March 1886, by which time the Liberals were again briefly in power with Lord Rosebery as Foreign Secretary, Hutton and Mackinnon obtained official approval for an expedition to Kilimanjaro which succeeded in September in confirming some of Johnston's treaties.[5] The idea was gaining ground that Britain should attempt to preserve a position on the mainland which could be developed by Mackinnon and his friends.

[1] Hatzfeldt to Munster, 3 June 1885. F.O. 84/1718.
[2] Malet to Granville, Af no. 194, 3 June 1885. F.O. 84/1714.
[3] P. Magnus, *Kitchener: Portrait of an Imperialist*, 1958, p. 70.
[4] de Kiewiet, 'British East Africa Company', pp. 69-70.
[5] Ibid., p. 71; Coupland, *Exploitation*, pp. 468-70.

By the autumn of 1886 Bismarck wished to bring the Zanzibar affair to a close. He was still unconvinced of the value of colonies, and wished to limit the expansive designs of Peters and the Denhardt brothers. As usual, the demand for a settlement was accompanied by threats; if Germany did not get her way she would support France on Egyptian questions. In October negotiations were begun in London, and within a fortnight they were concluded.

Coolly disregarding any views which Sultan Barghash might have on the matter, the two powers proceeded bilaterally to define that unfortunate monarch's possessions. The islands of Zanzibar, Pemba, Mafia, and Lamu, the coast from Tungi Bay to Kipini to a depth of ten miles, the towns of Kismayu, Brava, Merca, and Mogadishu with a radius of ten miles around each, and Warsheikh with a radius of five miles, were all declared to be the sultan's. Britain agreed to use her influence with Barghash to obtain for the German East Africa Company a lease of customs dues at Dar es Salaam and Pangani, to promote a friendly settlement of rival claims in Kilimanjaro, and to persuade the sultan to adhere to the Berlin Act. The coastline of Witu was defined as German. Germany promised to adhere to the Anglo-French Declaration of 1862 (i.e. to maintain the integrity of the sultan's dominions!). Another clause revealed that Britain and Germany had mapped out the lines of future partition of the mainland. The territory between the Ruvuma and Tana rivers was divided along a line from the Umba river to Lake Victoria and the northern portion declared to be a British, the southern a German, 'sphere of influence'. Each agreed not to interfere in the other's sphere, and within its own sphere 'not to make any acquisitions except protectorates'. France was later persuaded to adhere to these arrangements in return for Anglo-German recognition of a French protectorate over the Comoro Islands, where the Sultan of Zanzibar may have possessed some rights.[1]

Britain thus preserved something of the sultan's dominions, and created an inland sphere which Germany agreed to leave alone. The way seemed clear for Mackinnon to proceed with his plans to develop the inland region through a concessionary company. Yet the extent to which Britain had altered her

[1] Hertslet, *Map of Africa*, iii. 882-6; Coupland, *Exploitation*, pp. 472-6.

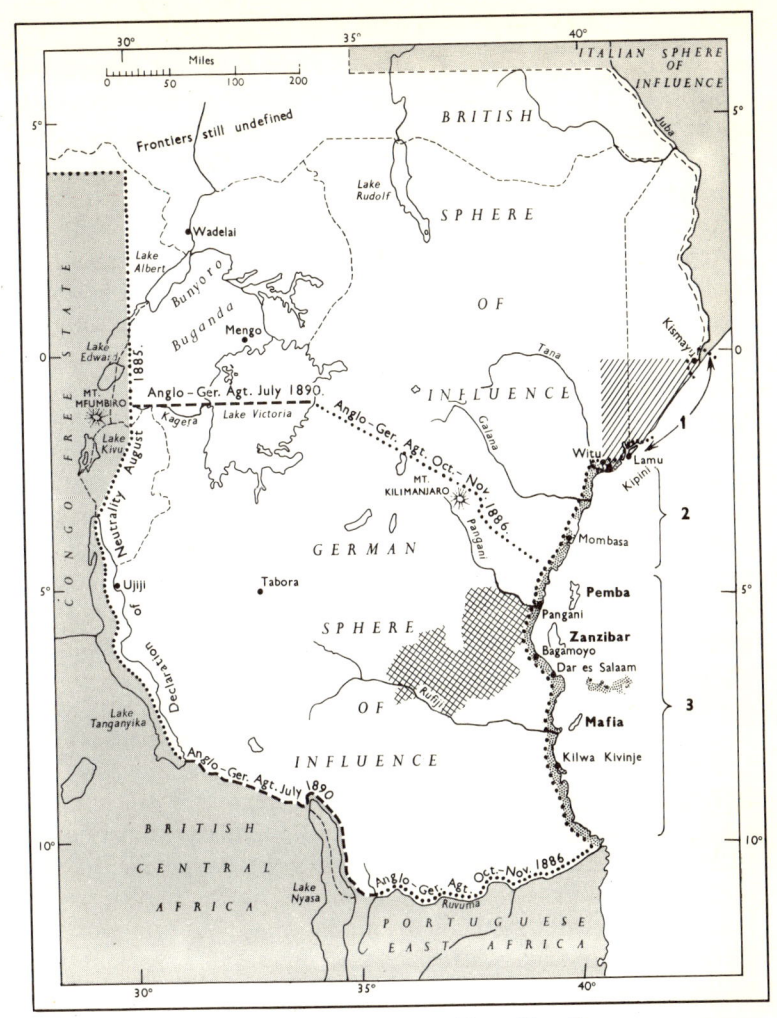

MAP. 15. Partition of East Africa, 1884–1891

.... Agreements of 1885/1886. – – – – Agreements of 1890/1891.

Approximate area claimed by Germany as Witu Protectorate October 1889. German claim renounced, July 1890.

Approximate area claimed by Germany under protectorate of February 1885.

Sultan of Zanzibar's coastal dominions as acknowledged by Anglo-German-French Delimitation Commission, 1886.

1 Zanzibar northern Ports leased to I.B.E.A. Co. 1889.
2 Northern coastline of Zanzibar leased to I.B.E.A. Co. 1887.
3 Southern coastline of Zanzibar leased to German E.A. Co. April 1888. Sold to Germany December 1890.

attitude as a result of the 1886 agreement must not be exaggerated. The traditional attitudes were still very pronounced, even in the agreement itself. The preservation of the sultan's authority, or what was left of it, remained the official policy. The delimitation implied that Britain intended to establish 'protectorates' in her 'sphere of influence', but the idea of the protectorate at this time was far removed from outright annexation. It was essentially a device for preventing other powers from securing control. Usually local rulers ceded only their right to have external relations with other states, and their internal administration remained in being. British officials and politicians were unable to find special economic or strategic reasons for more positive intervention in East Africa. During his visit to the coast Kitchener had come to the conclusion that Mombasa was of vital strategic importance for the maintenance of British seapower in the Indian Ocean, and had urged that it should be acquired by lease from the sultan, to counteract the German lease of Dar es Salaam. Though the Foreign Office lent some support to the proposal, it was frustrated by the opposition of the Admiralty and the Treasury.[1]

The British government also refused to become involved directly in the attempts to relieve Emin Pasha. During 1886 it became known that this remarkable German, holding the Egyptian governorship of Equatoria Province, was still holding out against the Mahdi, with the aid of Sudanese troops, in an area which controlled the upper reaches of the Nile. In December letters were received from Emin asking for British help.[2] The Anglo-German agreement of 1886 made no provision for the area west of Lake Victoria, thus Uganda and the southern Sudan lay open to annexation by either power. The danger of a German expedition to relieve a renowned German subject was obvious. If it succeeded the result might be German control of the upper Nile, and the encirclement of the British 'sphere'. An agitation was begun in England for a British relief expedition; *The Times* even referred to Emin as a 'second Gordon'.[3]

Lord Salisbury set his face against any such involvement. He

---

[1] de Kiewiet, 'British East Africa Company', p. 72; Coupland, *Exploitation*, pp. 470–1.
[2] Emin to Felkin, 7 July 1886, in *Scottish Geographical Magazine*, 1886, pp. 715–16.
[3] *The Times*, 15 Dec. 1886.

could see no real British interests on the upper Nile, and Egypt's interests were not yet Britain's. The occupation of Egypt might very soon come to an end.[1] He was also afraid that a British expedition might suffer the same fate as Gordon's mission to Khartoum. Public reaction might then turn against the Unionists as it had against Gladstone's Ministry in 1885. He was quite prepared to allow the Germans the dubious privilege of rescuing Emin.[2] The initiative was left to private interests, led by Mackinnon, who raised a fund of £10,000 to which the Egyptian government added an equal amount. An expedition led by H. M. Stanley reached Emin in 1888, and finally withdrew him and his forces from Equatoria in May 1889, abandoning the region to the Mahdists.[3]

The British government's indifference and reluctance to become involved directly in East Africa presented a real problem to Mackinnon and his associates. Mackinnon, remembering the failure of the earlier scheme, wanted some tangible evidence of government support. The obvious technique was to obtain a royal charter like those granted to the British North Borneo and Royal Niger companies. To work a concession without a charter would have meant that the company, in the event of disputes with the Germans, would have been able to rely only on the diplomatic resources of its suzerain, the sultan. How ineffective these resources were had already been more than amply demonstrated.

In February 1887 Sultan Barghash, fearing further German aggression and prompted perhaps by Acting Consul Holmwood,[4] telegraphed to Mackinnon asking him to negotiate a concession and form an association which could obtain the British government's approval. In March, informing the government of the steps he had already taken to this end, Mackinnon insisted

[1] Lady G. Cecil, *Life of Robert, Marquis of Salisbury*, 4 vols., 1921-32, iii, pp. 231-6.   [2] Minute by Salisbury, Oct. 1886. F.O. 84/1775.
[3] By the time the relief expedition was organized Mackinnon had received the charter for the Imperial British East Africa Company, and he hoped to enlist Emin as the company's governor of the Lakes region. When Emin told his followers that he had accepted this offer they rebelled and imprisoned him whilst Stanley was away bringing relief to a rear-column. When Stanley returned he practically forced Emin to leave, fearing a German expedition which had already set out. (See H. M. Stanley, *In Darkest Africa*, 2 vols., 1890).
[4] de Kiewiet, 'British East Africa Company', p. 74; P. L. McDermott, *British East Africa or I.B.E.A*, p. 9.

that he would require 'more than ordinary support'.[1] In May, when his agents were in Zanzibar, he elaborated this to mean that he would require the same guarantees as those obtained by the Borneo and Niger companies. Lord Salisbury refused to go so far. As long as Mackinnon's plan cost nothing to the taxpayer, and infringed no other rights, it deserved all proper assistance, but Mackinnon could not be given 'more than any other British subject would receive'. The British government must not become entangled in the administration of East Africa.[2]

Despite this coolness Mackinnon persisted with his plans. Amongst his associates there were those who could tell him that the Borneo and Niger companies had first needed to establish claims to effective rule before they could overcome the reluctance to grant a charter.[3] In May 1887 Mackinnon's newly formed British East Africa Association obtained a concession from Sultan Barghash granting full judicial and political authority, including the right to levy customs duties, over his mainland possessions from Kipini to the Umba river, for fifty years. In return the Association was to pay the sultan not less than the present amount which he received in customs duties from the region.[4] By the end of the year the Association was claiming sovereign rights to a depth of 200 miles from the coast by treaties made with the inland peoples.[5] At the same time Mackinnon was trying, through Stanley, to enlist Emin Pasha as the Association's governor of the Lakes region.[6] The Association was already in dispute with the German Witu Company formed in 1887, and early in 1888 began a long dispute with Italy over the sultan's northern ports, Kismayu, Merca, Mogadishu, and Warsheikh. The Association was thus

[1] Memo. by Sir Percy Anderson on conversation with Mackinnon, 21 Mar. 1887. F.O. 84/1860.
[2] Minute by Salisbury, 18 May 1887. F.O. 84/1863, used by de Kiewiet, *British East Africa Company*, p. 75.
[3] Sir Thomas Brassey, an original subscriber to the East Africa Company and its first Vice-President, had been a director of the North Borneo Company. Lord Aberdare was chairman of the Royal Niger Company, and James Hutton was a director of the same company.
[4] McDermott, *I.B.E.A.*, p. 9. [5] Ibid., p. 10.
[6] See above, p. 377, note 1. The subscribers to the British Emin Pasha relief expedition were virtually identical with those who were later to be shareholders in the Imperial British East Africa Company. Stanley, *In Darkest Africa*, i. 35 gives a full list of the subscribers.

laying claim to be the government of the British sphere of influence in East Africa by virtue of the sovereign rights conceded to it by the sultan and the African rulers.

In April 1888 the members of the Association made an agreement to form a company which might obtain 'a Charter or Charters incorporating the Company as a British corporate body and under British protection or otherwise'. The company would act both as a trading and development agency, and as an administration. £250,000 was subscribed; Mackinnon was the largest shareholder with £25,000 and other members of his family together subscribed the same amount. Philanthropic interest was very pronounced; Sir Thomas Fowell Buxton of the Anti-Slavery Society subscribed £10,000; Burdett-Coutts, whose wife, of the great banking family, was famous for charitable works, subscribed £10,000; and Alexander L. Bruce, who was connected by marriage with David Livingstone's family and had a strong interest in Nyasaland, subscribed £5,000. Past and present members of the British consular staff in Zanzibar were well represented. Sir John Kirk, now no longer consul-general, Francis de Winton the acting consul, and Frederick Holmwood who had done so much to promote the scheme, each subscribed a modest £1,000. The Sultan of Zanzibar was not entirely forgotten; the agreement laid down that he was to receive, without subscribing capital, one founder's share entitling him to 2 per cent. of all profits after eight per cent. dividend had been paid on the ordinary capital, and his share was to be worth one-hundredth of the proceeds of any voluntary sale of the company.[1]

Clearly the promoters of the new company had great hopes that a royal charter would be granted, and their expectations were soon justified, for on 3 September 1888 the British government, almost without discussion, granted the charter. Lord Salisbury now agreed to what in March 1887 he had declared impossible.

The reason for this abrupt change of front lay in developments in the Egyptian situation. When Mackinnon had asked for a charter in March 1887 Salisbury had at that time determined to make a serious attempt to withdraw from Egypt. He sent Sir Henry Wolff to Constantinople to negotiate a British

[1] McDermott, *I.B.E.A.*, appendix II, pp. 280-1, also p. 278.

withdrawal within five years, with guarantees for re-entry in case of invasion or anarchy. Wolff's mission came within sight of success; an agreement was actually signed, but not ratified. The condition that Britain should be given re-entry rights raised such opposition from France and Russia that it came to nothing.[1] The failure of this attempt convinced Salisbury that withdrawal with safeguards was impossible. He would therefore have to make the best of the occupation. The Tripartite Agreement of December 1887, between Britain, Italy, and Austria, brought Britain closer to the German system of alliances, and Salisbury could therefore hope for less German obstruction in Egypt. By the spring of 1888 Egyptian finances were in balance, and the need for international agreement on reform therefore less important. Salisbury now accepted the fact that the British occupation would be a long one.[2]

Once this had sunk home Egypt's interests became British interests. It now became a fixed purpose of Salisbury, and of later Liberal imperialists like Lord Rosebery, to prevent any power with the technical skill to interfere with the flow of water from obtaining a foothold on the Nile. The reconquest of the Sudan would one day have to be undertaken, preferably when Egypt could pay for it.[3] European powers would have to be prevented from controlling the sources of the Nile in Uganda; preferably this must be done without help from the Exchequer, and without too many awkward questions in Parliament.

Mackinnon's plans fitted in perfectly with this new strategic thinking. His company promised, at no cost to the taxpayer, to control a part of the East African coast from which the sources of the Nile could be most easily reached. Mackinnon's interest in Emin Pasha showed his genuine desire to obtain control of the Lakes region, and it was known that he had plans for a railway from the coast to Uganda. The Vice-President, Sir Thomas Brassey, was one of the greatest of British railway-builders overseas. The grant of the royal charter incorporating the Imperial British East Africa Company was an announcement to the powers of Europe that the company was henceforth

[1] Cecil, *Salisbury*, iv. 36–52.
[2] Ibid., pp. 134 sqq.
[3] Ibid., pp. 139–40. For a full discussion of the strategic significance of the Nile waters see W. E. Langer, *The Diplomacy of Imperialism*, 1951, pp. 102–8.

THE BACKGROUND TO PARTITION

not merely the agent of the Sultan of Zanzibar, but an arm of British Imperial policy.[1]

Once the charter had been granted to the British company a period of intense rivalry set in, despite the fact that the British and German governments had agreed in June 1887 to discourage annexations in the Lakes region. But nothing had been said about protectorates. Both the British and the German companies felt threatened by the possibility of encirclement by the other; the British feared a pincer movement linking the German Witu and East Africa companies in Uganda; and the Germans feared that Mackinnon's company might join with that of Cecil Rhodes in establishing an all-British route 'from the Cape to Cairo' by controlling the area between Lakes Tanganyika and Victoria.[2]

Thus, though Salisbury was interested only in keeping the Germans away from Uganda and the sources of the Nile, and though Bismarck had little sympathy for the German companies' expansionist plans, the fact that the local administrations were in private hands generated intense Anglo-German friction. On the coast Mackinnon attempted to squeeze out the German Witu Company, countered their attempts to seize Lamu island, and even carried on negotiations with Italy over the sultan's northern ports to prevent the Germans obtaining them, despite rebukes from the British Foreign Office.[3] His efforts met with some success; in August 1889, by arbitration, Lamu was awarded to the Sultan of Zanzibar, who promptly ceded it to the British company. The same concession gave the northern ports to the company, which agreed to transfer them to Italy in November. In December 1889, after the breakdown of negotiations by which Mackinnon would have purchased the German Witu Company, Manda and Pate islands were occupied by his agents.[4]

[1] For a full text of the charter see McDermott, *I.B.E.A.*, appendix III, pp. 282 sqq.

[2] For the Cape to Cairo idea, its origins and ramifications, see Oliver, *Sir Harry Johnston*, p. 143 and L. Weinthal, *The Story of the Cape to Cairo Railway and River Route*, London, 1923. *The Times*, 22 Aug. 1888, contained an article by H. H. Johnston, then British Commissioner in Nyasaland, advocating the Cape to Cairo strategy.

[3] de Kiewiet, 'British East Africa Company', pp. 141–59. By its charter the company was forbidden to conduct negotiations with foreign powers without the consent of the Secretary of State. [4] Ibid., pp. 163–76; Cecil, *Salisbury*, iv. 217–18.

Rivalry in the interior was even more serious. Initially the focus for this rivalry was the presence of Emin Pasha in Equatoria. The Germans were not content to leave the 'rescue' of their compatriot to Stanley and the British. In the autumn of 1888 they had formed a German Emin Pasha Relief Committee, and money was subscribed to send Carl Peters to Equatoria. Despite the fact that the German coastal area was in open revolt, and that the German company's powers had been superseded by the appointment of Wissmann as Imperial Commissioner in the spring of 1889,[1] Peters persisted with his plans. Though repudiated by the German government and by Wissmann he set off with 300 men in June 1889. The British company replied by sending an expedition under Frederick Jackson to open a route to Wadelai on the upper Nile and support Stanley.

After both expeditions had left the coast Stanley returned to it in December 1889, having 'rescued' Emin and abandoned Equatoria. Events determined that Buganda should provide a new focus to replace Equatoria; in October 1889 the Muslim faction had deposed Kabaka Mwanga, and the Anglican missionaries appealed to the British company to establish its rule in the kingdom. But Peters arrived first, and in February 1890 he persuaded Mwanga, who had in the meantime regained control, to sign a treaty establishing 'friendly relations with the German Emperor'. When Jackson arrived in April Peters had gone, but faced by opposition from the Catholic party the Englishman could not persuade Mwanga to sign a treaty.[2]

Little news of what was happening in Uganda reached Europe until the summer of 1890. In the meantime the diplomats were moving towards a settlement over the heads of the treaty-makers. In the autumn of 1889 the German government suggested negotiations on outstanding disputes,[3] and since the end of the year Salisbury had been working for a settlement by arbitration. Negotiations began in earnest after the fall of Bismarck from power in March 1890. The ambitions of the British government and those of the British company were by no means identical; Mackinnon was by now determined to make a territorial link with Cecil Rhodes's newly chartered

---

[1] See pp. 386–7 below.   [2] Oliver, *Missionary Factor*, pp. 128–39.
[3] *German Diplomatic Documents 1871–1914* (ed. E. T. S. Dugdale), 4 vols., 1928–31, ii. 25–30.

company in the south, and in May he signed a 'treaty' with King Leopold. In return for recognition of his right to certain territories around Lake Albert, Leopold promised to hand over, when convenient to Mackinnon, a strip of territory five miles wide which would link Lake Albert Edward with Lake Tanganyika.[1] Lord Salisbury was much more concerned with the alarming news coming from Uganda, than with the 'all-red route'. In March 1890 it had been rumoured that the Imperial Commissioner, Wissmann, was about to leave for Uganda, and on·the last day of that month the Germans announced that Emin was to lead a large caravan to Buganda. The Foreign Office pressed the British company to make some counter-move,[2] and in April it was announced that Frederick Lugard would lead an armed expedition to establish the company's authority in Buganda. In the first week of May news was received of Peters's treaty with the kabaka. Salisbury was so alarmed that he was prepared not only to abandon the idea of the strip to Lake Tanganyika, but to make a sensational and hitherto unprecedented offer to cede territory in Europe to keep the Germans from Uganda.

On 13 May, meeting the German negotiator Count Hatzfeldt, Salisbury played this trump card. He began with a list of formidable demands; Germany must recognize Uganda as within the British sphere, she must abandon Witu, she must accept a British protectorate over Zanzibar and Pemba islands. In return Britain would drop the claim to a strip of territory to Lake Tanganyika, use her influence to persuade the sultan to sell the German coastal leases outright, and Britain would cede the island of Heligoland to Germany.[3]

An African reader might find it strange, even ludicrous, that

---

[1] Langer, *Diplomacy of Imperialism*, pp. 118 sqq.; H. M. Stanley, *Autobiography*, 1909, pp. 412–13; Cecil, *Salisbury*, iv. 285–6; de Kiewiet, 'British East Africa Company', pp. 198–9; and Agreement between I.B.E.A. Co. and Congo Free State of 24 May 1890 and Salisbury to Leopold, 21 May 1890 in F.O. 84/1082.

[2] This has long been a matter of some dispute. The company, when pressing the British government for compensation for the loss of its charter, argued that it had occupied Uganda in response to appeals from the government. As the costs of the Uganda occupation largely caused its bankruptcy, the company claimed that the government was morally liable. See I.B.E.A. Co. to F.O. 23 Nov. 1894 in *Africa* No. 4, 1895, C. 7646. Detailed researches have tended to confirm the company's claim, e.g. de Kiewiet, 'British East Africa Company', pp. 201–2.

[3] Cecil, *Salisbury*, iv. 288–90.

the German government should have seized eagerly the chance of acquiring a small, sandy, and almost barren little island in the North Sea in exchange for thousands of square miles of East African territory. But there were special reasons to tempt the Germans in Heligoland. In 1887 construction work had begun on the Kiel Canal, linking the Baltic and North Seas, nationalists were pressing for a programme of naval expansion, and the new Kaiser supported them. The strategists had long argued that the possession of Heligoland was essential if the German navy was to be able to exploit the usefulness of the Kiel Canal in time of war.[1] Thus the Germans were soon biting at Salisbury's bait. Beside Heligoland, Count Hatzfeldt was told, 'our East African interests merely come forward as matters for concession'.[2] Salisbury had some difficulty with his cabinet, and with the Queen, who was uneasy that British subjects in Heligoland should be thus 'bartered away',[3] but these scruples were overcome, and the agreement was signed on 1 July. By its provisions the British company obtained a solid block of territory, the thorn in its side at Witu was removed, and above all Uganda and the Nile sources had been secured.[4] If the Cape to Cairo scheme was ever to come to fruition there still remained Mackinnon's 'treaty' with Leopold.[5]

A further British attempt to prevent Italy from reaching the Nile from the east coast also affected the frontiers of the British East African sphere. Though the British company and the Italian government had agreed to establish a joint administration of Kismayu, and the company had ceded the four other northern ports to Italy in November 1889, the inland frontier remained undefined. Mackinnon would have liked to arrange a boundary line, but Lord Salisbury was more concerned with the implications of Italian activity farther north. In 1889 the Italians had concluded the Treaty of Ucciali with

[1] Cecil, *Salisbury*, iv. 291; *Die grosse Politik*, viii, nos. 1676–82, pp. 17–18.
[2] *German Diplomatic Documents*, ii. 37.
[3] *Letters of Queen Victoria* (ed. G. E. Buckle), 3rd ser. i, London, 1930, p. 613.
[4] For full text see Hertslet, *Map of Africa*, iii, no. 270, pp. 899–906.
[5] In May, 1894, the British government made an agreement with Leopold, which would have regularized the Mackinnon–Leopold 'treaty', and allowed Britain to construct a telegraph and a railway between Lakes Tanganyika and Albert Edward, through a lease, 25 kilometres wide. Intense opposition from Germany forced the two parties to withdraw this provision from their agreement. See Hertslet, *Map of Africa*, ii, nos. 163 and 164, pp. 578–84.

Emperor Menelik of Ethiopia, which, so the Italians claimed, gave Italy a protectorate over Ethiopia. If this claim could be substantiated it would have given the Italians control of the sources of the Blue Nile. Another Italian claim to incorporate Kassala within the Italian colony of Eritrea would have given Italy control of the River Atbara, a tributary of the Nile. In the autumn of 1890 Baring, the consul-general in Egypt, visited Italy to negotiate directly with Signor Crispi, the Italian Prime Minister. Salisbury told Baring to 'insist on the command of all the affluents of the Nile, so far as Egypt formerly possessed them,'[1] but Crispi argued that these rights had been abandoned by Egypt to the Mahdists.

In February 1891 Crispi's government fell, largely as a result of his expensive colonial adventures. His successor, the Marchese di Rudini, was willing to come to terms. Agreements were signed in March and April in which Britain recognized the extravagant claim of Italy to a protectorate over Ethiopia, but the boundaries of the supposed protectorate were so drawn that Italy was kept away from the main Nile. Kassala was declared to be Egyptian (though Italy could occupy it temporarily for military purposes), and Italy bound herself not to build any works on the Atbara which might alter the Nile water-level.[2] By thus defining Italy's claims the agreement established a frontier between the British company's sphere and the Italian possessions to the north. British East Africa thereafter was undefined only in the north-west, where its territories touched those of the Mahdist State,[3] which it was assumed would one day revert to British-controlled Egypt. The 'scramble' for East African territory was virtually over. The struggle for the Nile continued, but henceforth the British had to guard only the western approaches to the Nile basin.

[1] Salisbury to Baring, 31 Aug. 1890, quoted in Cecil, *Salisbury*, iv. 330–1.
[2] For full text see Hertslet, *Map of Africa*, iii, no. 288, p. 948, and no. 289, pp. 949–50; see also M. Shibeika, *British Policy in the Sudan, 1882–1902*, London, 1952, pp. 319–27; Langer, *Diplomacy of Imperialism*, pp. 110–12, and Cecil, *Salisbury*, iv. 324–34.
[3] Though in fact the frontier with Ethiopia was undefined, for the Italians failed to subdue that state, which regained its independent stature after the defeat of Italy at Adowa in March 1896. Thereafter it could not be expected that the Ethiopian government should recognize a frontier drawn up by two alien powers. In Dec. 1906 Britain and Ethiopia signed an agreement demarcating the border, see Hertslet, *Map of Africa*, ii, no. 103, pp. 445–6.

Both the British and German governments entered upon the control of their East African territories with no intention of spending state funds on colonial administration. Both hoped to avoid responsibility by entrusting territorial administration to chartered companies, but in doing so they created formidable problems. The chartered company was a commercial concern with a limited amount of capital. On the Niger the chartered company had been financially successful because it paid for its administration out of an already large trade, which could be brought out cheaply and quickly along huge navigable rivers. Cecil Rhodes's company could draw on his immense private fortune, and could tempt speculators with land and mineral rights. But in East Africa there was little hope of mineral wealth, little trade within reach of the coast which could be brought out economically, and no large navigable rivers reaching far inland.

The German company was the first to find itself in difficulties. In February 1885 the *Gesellschaft für Deutsche Kolonisation* delegated its authority in East Africa to a directorate headed by Peters, which in turn became the German East Africa Company. Peters's desire for expansion into the interior, the establishment of stations, and the rivalry with the British East Africa Association, soon made the acquisition of new capital essential. German banking houses lent some support, reorganizing the company in 1886, but Peters's expensive policies were continued. There was little to offset unremunerative expenditure on expansion; experiments with plantations only lost money, and profits from trading were difficult to secure in face of competition from other German firms in Zanzibar.

When the Arabs and other coastal peoples led by Abushiri rose in revolt in August 1888,[1] the German company lacked the resources to meet this challenge, and its finances and administration collapsed. In asking the German government for military aid the company undermined the very basis of its existence, for if forced to spend funds in East Africa the German government had no longer any reason for entrusting the administration to a chartered company.

Bismarck, by arguing that the suppression of Abushiri's revolt was a crusade against the slave trade, was able to secure

[1] See Chapter XII below.

credits from the Reichstag for a military expedition under Wissmann as Imperial Commissioner. Soon after his arrival in East Africa Wissmann assumed control of the civil and military administration, leaving only the collection of customs duties in the hands of the company. In November 1890 the company formally ceded all its political and administrative rights and functions to the imperial government, receiving in return handsome financial compensation and monopolistic rights as a private company in German East Africa.

The British company was at first more successful than its rival. It possessed rather more capital, which it kept more or less intact so long as its activities were confined to the coast. But rivalry for the interior created a desperate financial situation. After Lugard's expedition in December 1890, the annual cost of occupying Uganda was about £40,000 or £50,000, from a total capital of only £250,000. The company hoped that the construction of a railway to the Lakes, with a government guarantee for investors, would enable trade to flourish, and revenue to expand, and Lord Salisbury showed sympathy towards the idea, making it possible for a survey to be started.[1]

The railway scheme was a long-term solution, and gave no immediate help to the company, which by July 1891 was practically bankrupt. The directors therefore announced that unless the government would immediately guarantee interest on capital for a railway, its administration would have to be withdrawn from Uganda. Salisbury was in no position to go as far as this. In the event private interests stepped into the breach. A subscription of £40,000 was raised, mainly from English and Scottish Protestants who were afraid that if Lugard were withdrawn the Catholic party would bring about a French protectorate. The company thereupon promised to continue its occupation of Uganda until the end of 1892.

The situation became even more complicated after the return of the Liberals to power in July 1892. The Liberals were divided in their attitude to the Egyptian occupation. The Foreign Secretary Lord Rosebery, supported by the 'Liberal Imperialists' shared Salisbury's view that withdrawal was

[1] The survey was to cost £25,000, paid for out of government funds. For a more detailed account of Salisbury's attitude to the railway see Cecil, *Salisbury*, iv. 309–14.

impossible, and control of the Nile a British interest. But more traditional and orthodox Liberals, Gladstone the Prime Minister and Sir William Harcourt the Chancellor of the Exchequer, supported by radicals like John Morley and Sir Charles Dilke, wished to prepare for evacuation of Egypt.[1] And if there was no advantage in retaining Egypt, there could be none in retaining Uganda.

As the company was due to withdraw in December 1892, Rosebery was forced to bring the issue forward in cabinet. After bitter disputes, a compromise was patched up whereby the government would pay the company to administer Uganda for the first quarter of 1893, but withdrawal would then take place.[2] This breathing space allowed the 'retentionists' to mount a public campaign, Lugard returned to England in October 1892 and began a full programme of speeches, letters and articles in the press, and lobbying. Protests and memorials deluged the Foreign Office, the majority from religious organizations.[3] Not since the death of Gordon had the country been so stirred by an African issue. The Cabinet considered tempting offers which might have relieved them of the problem;[4] and its members continued to quarrel amongst themselves. In an attempt to postpone the question it was decided in November to send a commissioner to Uganda to report on the situation. In securing the appointment of Gerald Portal for the post Rosebery virtually won the day, for he knew that Portal favoured retention, and took pains to let Portal know that this view would be supported by himself.[5]

Whilst the Parliamentarians, unaware that anything had been decided, continued acrimoniously to discuss Uganda, Portal was effectively committing the government in East Africa. A true imperialist, wanting a direct British administration, he was contemptuous of the company's administration.

---

[1] A. J. P. Taylor, 'Prelude to Fashoda', *English Historical Review*, 65, 1950, pp. 255 sqq.

[2] The cabinet almost broke up on this issue. Harcourt accused Rosebery of 'Jingoism with a vengeance', and Rosebery twice threatened to resign. See A. G. Gardiner, *Life of Sir William Vernon Harcourt*, ii, London, 1923, pp. 192 sqq.; Lord Crewe, *Lord Rosebery*, ii, 1931, p. 406 and p. 422; and M. Perham, *Lugard: the Years of Adventure 1858–1898*, 1956, pp. 409–10.   [3] Perham, *Lugard*, pp. 425–6.

[4] Leopold of the Belgians offered to administer Uganda under British authority; Rhodes offered to do so if he could obtain a subsidy of only £25,000 a year. See Perham, *Lugard*, pp. 417–19.   [5] Perham, *Lugard*, pp. 427–31.

THE BACKGROUND TO PARTITION 389

Arrived in Uganda, he enlisted Sudanese troops and signed on some of the company's agents as government officials. When the day came for the company's withdrawal on 1 April 1893 Portal hoisted the Union Jack.[1]

Portal's report was not published until April 1894.[2] In the interval the political situation had altered radically. Gladstone had quarrelled with the majority of his colleagues who favoured increased naval expenditure, and in March, pleading impaired faculties and his eighty-four years, he resigned.[3] Rosebery was chosen to succeed him, and on 10 April Portal's report was at last published. It strongly advocated the retention of Uganda and the building of the railway.[4] Two days later it was announced that 'Her Majesty's Government have determined to establish a regular administration, and for that purpose to declare Uganda to be under a British protectorate'.[5] Measures to put this decision into effect passed through Parliament in June. The Imperial British East Africa Company surrendered its charter and had to be content with £250,000 in compensation. Only £50,000 of this sum was found by the British taxpayer; the rest was 'borrowed' from Zanzibar at 3 per cent. interest. If Uganda was to be a British protectorate then communication with the coast also needed to be under British control. A sub-commissioner, responsible to the Zanzibar consul-general, was therefore appointed to administer what is now Kenya. Thus the whole of the British 'sphere of influence' came at last under direct governmental control.

Though British administration had come to East Africa in name, it would still be years before it was present in actuality—the Uganda railway did not reach Lake Victoria until 1901. The slowness with which Britain exploited her position is perhaps a reflection of the attitudes and motives which had prompted her intervention. Not a single British politician of stature in either party had shown any interest in East Africa

[1] Ibid., p. 448; Langer, *Diplomacy of Imperialism*, p. 123.
[2] The delay is somewhat mysterious, particularly in view of the death of Portal in Jan. 1894 as a result of disease contracted in Uganda. Rosebery may have deliberately held up publication pending the government reshuffle as a result of Gladstone's retirement. See Perham, *Lugard*, pp. 448 sqq.
[3] J. Morley, *Life of Gladstone*, London, 1908, ii. 561–7.
[4] *Africa* No. 2 (1894) C. 7303, lix. 361, *Reports relating to Uganda by Sir Gerald Portal*.
[5] *Hansard*, 4th ser., vol. xxiii, col. 223, 12 Apr. 1894.

for itself.[1] The arguments in favour of acquiring East African territory always rested on strategic considerations arising from interests elsewhere, in the Indian Ocean but above all in Egypt and the Nile Valley. In pursuit of these ends an East African Empire had been won which of itself seemed economically valueless. It is not therefore surprising that British governments were in no hurry to cover the country with a network of administrative stations, or to pour capital into its development.

[1] Though there were, of course, a number of minor political figures, including many back-benchers in Parliament, who voiced the aspirations of the missionary party, vitally interested in the christianization of Uganda. The Imperial British East Africa Company also had its champions in Parliament. But none of these could be counted among those who made important decisions of policy.

XI

THE BRITISH SPHERE[1]
1884-94

MARIE DE KIEWIET HEMPHILL

WHEN Carl Peters returned to Zanzibar in December 1884, with his ten treaties of 'eternal friendship', a new period in the history of East Africa was begun. It is true that the Anglo-German partition did not immediately bring about any profound or widespread changes in the internal politics of the area now known as Kenya and Uganda. It took Great Britain nearly a decade after the publication of the German *Schutzbrief* to make up her mind to establish her Uganda and East Africa Protectorates. But while British and German ambassadors and foreign secretaries were still quarrelling over the correct placement of boundary lines on maps, and members of the House of Commons were rising to declare that they wanted nothing to do with any new overseas empire, the first efforts were being made to bring European administration to East Africa.

The European scramble and partition seemed to burst upon East Africa with little clear warning. The transition to European rule came much more slowly. In some areas, the impact of European ideas preceded by many years the arrival of the first European administrator; in others, no white man appeared for a long time to come. In the late 1880's and early 1890's newly arrived administrative officers relied heavily upon the experience of the travellers and the missionaries who were already familiar and influential figures in East Africa. At the coast, the administration hired the employees and adopted the policies of the long-established British consulate at Zanzibar. Lake-bound caravans followed the routes and stopped at the camps already marked out by the explorer

[1] The author wishes to acknowledge with gratitude the assistance she has received from Dr. L. A. Fallers, Mr. A. T. Matson, and Mr. H. B. Thomas.

Joseph Thomson and the missionary Bishop Hannington. It was a fact of the utmost importance that in Buganda Christian

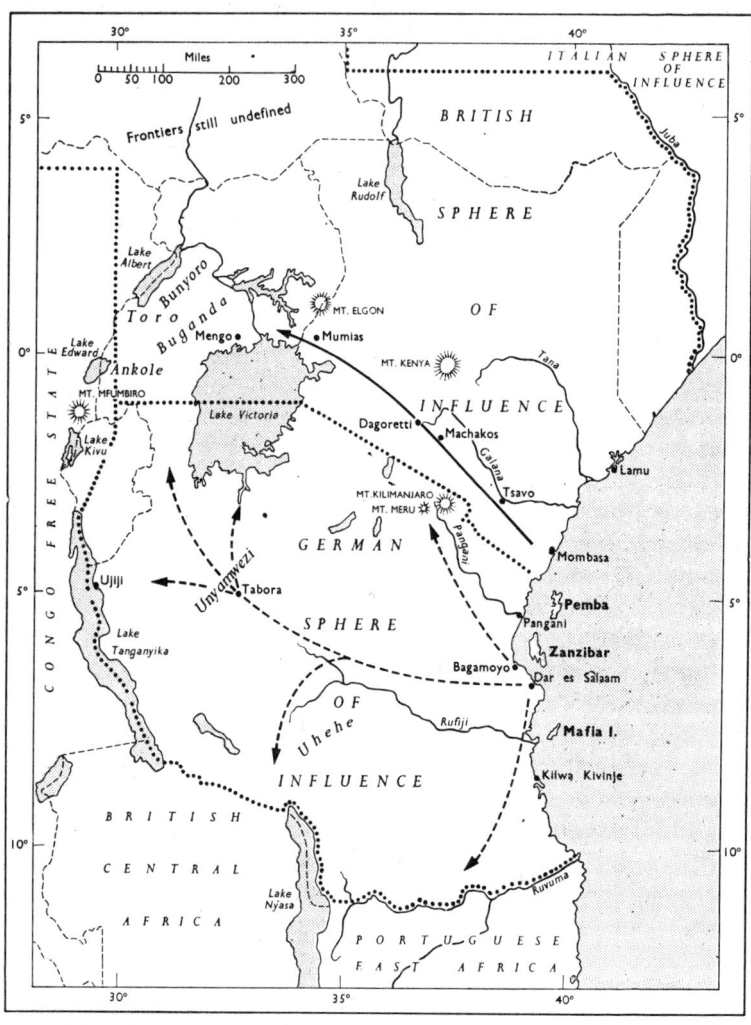

MAP 16. German and British advance into the interior, 1884–1893

religion and moral teaching preceded rather than accompanied the arrival of European government.

The areas in which the early administrators made a direct and decided impact were small and, viewed from Europe, the work accomplished did not always seem of much significance. In the ten years after 1884, the great majority of the European administrative agents were stationed either at the coast or in Uganda. At first, most of the tribes in Kenya were affected little or not at all by the coming of the Europeans. In some areas of what later became the Uganda Protectorate the African societies were influenced only indirectly, through the momentous events taking place in Buganda and Bunyoro.

If the first attempts to organize an administration were hesitant and partial, still they were full of the greatest significance for the future history of East Africa. The inexperienced young Englishman who was sent to Machakos, or to Kampala or Fort Portal, was seeking to establish a totally new relationship with the African society in which he lived. When he was clumsy or cruel or successful, when he negotiated and implemented a treaty or settled a dispute or joined in a war, he established a series of patterns and precedents which, for better or for worse, gave direction to the subsequent history of European and African relations. The ideas and the work of a few missionaries and soldiers and administrators in the Buganda of the early 1890's formed and guided the nature of the whole future Uganda Protectorate. The failure to establish an effective relationship with any of the major tribes in Kenya had results just as important for the future East Africa Protectorate and Kenya Colony. To an Englishman at home, the establishment of a few forts in Bunyoro or the enlistment of a Kamba police force might look like a very insignificant undertaking, but for future generations of Africans, these early successes and failures provided the basis upon which to build a new, revolutionary organization of society and government.

In British East Africa, as in Nigeria and south-central Africa, the early work of administration was handed over to a chartered company. The Imperial British East Africa Company, founded by a group of philanthropists, businessmen, and professional empire-builders, held a royal charter from 1888 until the Foreign Office took over full control in 1895. The I.B.E.A. Company was by all odds the least distinguished of the African chartered companies. Poorly conceived, badly managed, and grossly undercapitalized, the company was destined from the start to a short existence.

It has often been remarked in the history of imperialism that underinvestment in a decent administration is sometimes worse than the most thoroughly arbitrary rule, just as undercapitalization of a business or a colony is sometimes worse than the most thorough exploitation. Britain's unwillingness to accept full responsibility for her sphere of influence in East Africa gives to the history of the late 1880's and early 1890's a character of fluidity and confusion, a lack of definite shape or form or direction. The use of the chartered company was an expedient. The declaration of the sphere of influence was an evasion. The sphere of influence was a peculiar sort of possession, a good deal more than unclaimed territory, a good deal less than a colony or a protectorate. The chartered company, though perhaps better than nothing, was never more than the poorest of substitutes for a colonial government. The early administrators in East Africa, for all their goodwill, were themselves victims of a remote situation created by the complexities of international politics and British finance. In Uganda, in Kikuyu, and at the coast, Europeans were sometimes forced to resort to unsavoury expedients, to act with decision within situations which they poorly understood, and, above all, to economize drastically where economy was unwise or dangerous.

Britain's traditional concern with East Africa was paramount influence rather than rule. Until 1890 the object of her policy was to protect the territory already acknowledged as within her sphere and to push forward claims to areas still in dispute. Not only the government, but also the company, the missions, and the African societies concerned became deeply involved in Anglo-German rivalry. The company carried much of the burden of upholding British interests in Uganda and in Witu and the off-shore islands at the northern limit of the British sphere. The missionaries played a leading role in the quarrel over the control of Uganda. At the coast, Witu became a centre of intrigue and discontent, and in Buganda, local politics and religion became thoroughly and unhappily entangled in the European disputes.

By its charter the I.B.E.A. Company, like other chartered companies of this period, was given wide powers to govern under concessions or treaties obtained from local rulers. The authority thus delegated by the British Foreign Office was wielded by an oddly assorted group of fifteen directors. Of these, six were businessmen, three were prominent humanitarians and philanthropists, and five were retired or semi-retired army officers or civil servants who had spent most of their working lives in India or Africa. This mixture of backgrounds gave the East Africa Company the multiple character of a philanthropist dedicated to fighting the slave trade and bringing civilization to Africa, of an imperialist seeking to add new lands to the British empire, and of a businessman interested in commerce and bound to protect his investment. These three interests were joined and given cohesiveness by another, which has been called the appeal of unexplored Africa, the appeal which made African travel books best-sellers, and which brought fame and the adulation of Victorian England to Livingstone and Stanley.

William Mackinnon, the company's chairman, was an ageing Scotsman who, starting out in life as a grocer's assistant, became the founder of a large steamship line and one of the wealthiest men in Britain. In Mackinnon's character deep and rigid piety was mixed with exceptional financial ability. Mackinnon, with several of the other directors, was one of the last of a group of men to believe wholeheartedly in the concept of 'legitimate commerce', which purposed to fight the slave trade by developing other, competitive forms of trade. This concept, in which humane and commercial objectives were theoretically combined to the mutual benefit of both, was a typical product of Victorian humanitarian thinking. The character of the late nineteenth-century imperialist was in part made up of a moral certainty that the annexation of African territory would bring as much benefit to Africa as it would to Britain. In the twentieth century some of this confidence has been lost, but so too have some of the conditions which inspired it. To Mackinnon and to many others of the late Victorian period, east and central Africa, cruelly decimated by the Arab slave trade, presented an almost total picture of human misery.[1]

Mackinnon was not, however, one of the most impressive

[1] See the excellent summary by Margery Perham in *Lugard, The Years of Adventure, 1858–1898*, 1956, pp. 712–13.

examples of the late Victorian imperialist. As a religious man, he was genuinely concerned with the sufferings of Africa, and as a man of business, he hoped to open up Africa to European commerce. Yet his way of going about these objectives was curiously impractical. His ideas were large and poorly thought out. He was changeable and unreliable, and he had an impulsive way of dealing with daily affairs. A much better type of imperialist was George Mackenzie, the calm chief of the company's central administration in Mombasa, or Gerald Portal, who reorganized the government of Zanzibar under the British protectorate and who, in 1893, as British Commissioner, reported to the government on the situation in Uganda. Frederick Jackson, who was many years later to become Governor of Uganda, led the first of the company's caravans into the far interior. More remarkable than any of these men was Frederick Lugard, one of the greatest architects of empire Britain has ever produced. Lugard was a young man with little experience of Africa when he was hired by the company as administrator for Uganda. His work there, although it was of a controversial nature, was the most substantial and distinguished to be achieved in British East Africa in the years before the declaration of the protectorates.

The East Africa Company was officially incorporated by royal charter in April 1888. A year before, in May 1887, a concession had been obtained from the Sultan of Zanzibar leasing to the company the narrow strip of coastline between Vanga and Kismayu which, with the similar German coastline to the south, was all that was left of Zanzibar's mainland possessions. The sultan, Barghash ibn Said, was willing, if not eager, to sign the concession agreement in favour of the company. There was every reason why this should be so. The Anglo-German partition had broken up his empire on the mainland, and in the spring of 1887 he had reason to fear further annexations by Germany and by a new contender, Portugal. In these circumstances Barghash, understandably on the 'verge of distraction',[1] turned to the power which had protected him in the past. On 17 May Carl Peters arrived in Zanzibar with twenty Germans; the British East Africa Company's concession was signed a week later, on 24 May.

[1] Holmwood to Salisbury, 14 Mar. 1887. F.O. 84/1852.

The African response to the European annexations varied widely. Most of the tribes cannot have been aware of its real meaning; the stack of treaty forms collected in the Foreign Office files in London represented, to the chiefs and elders who agreeably inscribed them with their X-marks, considerably less than the familiar ceremony of blood-brotherhood. Europeans might be anything from objects of curiosity to dreaded conquerors. To some Africans they were, perhaps, at first no more than a new source of decorative beads or cotton cloth; to others, more sophisticated, they were allies or enemies in the struggle for power and dominance over other tribes.

Africans interpreted the European advance in terms of experience familiar to them. The form of their social and political organization, the nature and prosperity of their economic life, the past history of their relations with other tribes, and their experiences of the alien influence of Arabs, Christian missionaries, and European travellers, all influenced the African attitude to the establishment of European administration. The inhabitants of the centralized, prosperous kingdoms in the Lake Victoria basin were better able to understand and to adapt their institutions to European rule than the pastoral and nomadic peoples of the plateau or the loosely organized, decentralized tribes of the highlands and the lowlands near the coast. By and large, tribes dependent upon trade had much to gain from the European; tribes dependent upon raids and extensive grazing lands had much to lose from the *pax britannica*. Those already used to dealings with Arab and Swahili caravans found it immeasurably easier to make the initial adjustment to the coming of the European than those who were proud, isolated, and traditionally hostile to the caravan traffic. On the other hand, in centres of strong Muhammadan influence, Africans joined with Arabs in opposition to Christian missionaries as well as to British administrators. European travellers prepared or, in some cases, impeded the way for administrators. Missionaries, themselves the bearers of one of the most potent forces for social revolution, served as guides and advisers to both Europeans and Africans in the establishment of the new colonial relationship.

The initial reaction to the European annexations was less pronounced in the British sphere than it was to the south, where the principal Arab slave routes and settlements were located, and where the Germans wasted neither time nor tact in planting their standard on East African soil. At the coast, at any rate, the agents of the British company, starting administration in 1888, managed to avoid serious disturbances. This did not mean that the traders and plantation owners of Mombasa were any more willing than the traders of Bagamoyo and Ujiji to see their monopoly of commerce and influence pass to the European. In fact, one prominent Arab described the feeling on the British coastline 'as to the selling of themselves and of their land'[1] in much the same terms as were to be used by a later generation of East Africans. In Mombasa riots broke out in June directed both against the sultan's concession to the British company and against the near-by mission stations.

At the coast, as in Buganda and elsewhere, it was the missionaries who received much of the force of the first hostile reaction to the European annexations. In spite of expansion in some areas, this was a difficult period for the missions. Isolated from the outside world, as many of them were, deeply committed to the African societies among which they worked, and absolutely opposed to slavery and the slave trade, they were in natural opposition to all those who stood to lose by the introduction of European civilization and the increase in European power.

In most ways the position of the Mombasa missions after 1884 was far less precarious than that of the Buganda missions. Their proximity to the coast and to the British consul at Zanzibar gave them a measure of security their brothers in Buganda did not know. At the same time, the character of their work, largely carried out among freed and runaway slaves, put them into conflict with the slave owners and slave traders of Mombasa. Since domestic slavery was still legal in the Zanzibar dominions, the owners could, and did on more than one occasion, demand the return of their runaway slaves.

It is significant that the arrival in Mombasa of George Mackenzie, the company's administrator, was seized upon by the slave owners as the occasion for a fresh demand for the

---

[1] Euan-Smith to Salisbury, 19 Nov. 1888. F.O. 84/1910, quoting a conversation with Nasr bin Sulaiman.

return of the slaves harboured at the mission stations. Although Mackenzie was able to satisfy both the owners and the missionaries by paying off the owners and freeing the slaves, in doing so he symbolically admitted his inability to challenge Arab power. The failure to challenge or to provide a substitute for Arab power was to be the main characteristic of the East Africa Company's administration at the coast.

The company was publicly supported by the sultans of Zanzibar, but Khalifa, who succeeded Barghash in March 1888, made little pretence about his dislike for all Europeans. Euan-Smith, who followed John Kirk in 1887 as consul-general, believed that Khalifa actively encouraged Arab opposition to the European companies, but in fact he can have had little influence upon events. Khalifa's reign marked the end of the old sultanate; the Anglo-German delimitation and the two concession agreements had stripped the office of all real power on the mainland, and in Zanzibar itself the arbitrary authority of the sultan was denied by Arabs and British alike. Although the office regained some of its dignity under Khalifa's successor, Ali, it was never to regain its power.

The kingdom of Buganda was better prepared than any other part of the East African interior to adjust itself to the European annexations. For a people so isolated from the outside world, the Ganda had achieved a considerable degree of political and social sophistication, and their economy, by any African standard, was exceedingly prosperous. Their response to Christianity had been remarkable. Already there were among the Ganda a group of literate, fully aware men who were able both to understand and to respond effectively to the challenge of European political control.

The death of Mutesa and the succession of Mwanga in October 1884 brought a change in the attitude to the European missions. Mwanga was only eighteen when he became kabaka; he was inexperienced and possessed of what the C.M.S. missionary, Alexander Mackay, aptly described as a 'fitful and fickle' character.[1] The Christian converts were becoming increasingly liberated from tribal authority, and Mwanga, therefore, looked

[1] J. M. Gray, 'The Year of the Three Kings of Buganda, Mwanga—Kiwewa—Kalema, 1888–1889', *Uganda J.* 14, 1950, p. 15.

upon them and upon their foreign teachers as a threat to his position of privilege and power. He feared the growth of European influence, and when alarmed by the Arab traders' accounts of Joseph Thomson's travels through Masailand, he directed his first act against the young men and boys of his court, among whom both missions had made their first converts. Three young followers of the C.M.S. missionaries were put to death in January 1885, and Mwanga made a plain statement of his dislike of all missionaries. All through the year 1885, the missionaries' position deteriorated steadily, but the real crisis did not come until toward the end of the year. By this time, news of the German annexations had reached Buganda. Upon hearing reports of the approach of the Anglican bishop, James Hannington, who was travelling to Mengo along the eastern route by which it was traditionally believed the conquerors of Buganda would come, Mwanga ordered his murder. Hannington was killed in October 1885 by Luba in Busoga. The lives of the missionaries at the capital were for a time in danger, but they were spared to witness, in the following May, a massacre of their Christian pupils which has already gone down in history as one of the most impressive martyrdoms of modern times. The immediate pretext, if not the whole cause, for the tragedy was the refusal of the young Christian pages at the court to yield to Mwanga's unnatural desires. When about thirty of them refused to recant their new beliefs, they were collectively burnt alive at Namungongo. Like the martyrdoms of the early Christians in Europe, this and the persecutions which followed it served to increase, rather than decrease, the appeal of the new religion.[1]

In the years after Mwanga's succession to power a new ruling class composed of a group of highly intelligent young Christian converts grew up at the capital. This group, which was eventually to provide the government of the protectorate with the nucleus of a new administration, gradually became divided into mutually antagonistic Protestant and Catholic factions. The missionaries themselves contributed substantially to the political rivalry among their converts, even though

[1] See Roland Oliver, *The Missionary Factor in East Africa*, 1952, which has been used for the above account, and D. A. Low's unpublished thesis, 'The British and Uganda, 1862–1900' (D.Phil., Oxford, 1957).

the religious issue itself soon became almost lost in the struggle for power. Under Buganda's hierarchical form of society, the leaders could claim the support of large numbers of men whose loyalties had little to do with religion. By 1888 it was estimated that each party commanded a thousand fighting men.

A third, smaller party, composed of the resident Arab traders and their friends and religious followers among the Ganda, was in a position of predominance at the court until September 1888, when Mwanga, having been caught in a scheme to rid the country of all three parties, was summarily deposed. Kiwewa, Mwanga's eldest brother, was declared kabaka, and the offices of state were split among the Roman Catholic, Protestant, and Muslim chiefs. This cooperative arrangement lasted only about a month, before the Muslims, in a rapid and cleverly managed manœuvre, ousted the Christian chiefs from the capital and seized power for themselves.

The Muslim attempt to seize power in Buganda was only one part of the Arab reaction to the European occupation simultaneously going on in many parts of East Africa. It has not been definitely established that these widespread occurrences were co-ordinated in any way, or, as Mackay believed, directed from Zanzibar. Sultan Barghash's representative, Sulaiman bin Zeher, who arrived in Buganda early in 1887, may or may not have been acting upon orders from the sultan when he aided and encouraged the Muslim party through the events of 1887-8. In Buganda, as in such far-flung places as Bagamoyo, Ujiji, or far north on the Benadir coast where the Somalis fought against occupation by the Italian 'unbelievers', the Arab–Muslim uprising was directed against the European religion as much as against European power. The Muslim Ganda followed up their victory by driving the two groups of Christian missionaries to the south of the lake, and by attempting to force conversions to their religion. Kiwewa himself, upon refusing to be circumcised, was deposed in favour of his more tractable younger brother, Kalema.

Mwanga took refuge at the south end of the lake, first with the Arabs and then with the French missionaries; later, with the aid of the trader Charles Stokes, he moved north again to the island of Bulingugwe. In September 1889 Mwanga's party was joined by the main body of Christian Ganda refugees who had collected in Ankole. In early October they marched into the capital and, with much bloodshed, ousted Kalema and the Muslims. After an uneasy month in which the Muslims consolidated their forces and obtained the support of Kabarega, Mukama of Bunyoro, Mwanga and the Christians were once more defeated, but were able, in turn, to drive out the Muslims in February 1890. This time Mwanga's victory was nearly complete.

The lasting significance of the successive crises of 1887–90 lay in the gradual transfer of power within Buganda from the hands of the kabaka to the hands of the young Christian chiefs. Apolo Kagwa, an Anglican convert, who later became the real ruler of Buganda under the regency established in 1897, was already in a position of leadership in 1888, and he became the katikiro, or prime minister, upon Mwanga's reinstatement as kabaka in 1890. The rise of the new oligarchy which was so ably and successfully to lead Buganda through the transition to European rule was accomplished amidst several years of internal dissension and instability. After the Christian victory of 1890, the offices of state and the great chieftainships were shared out more or less equally between the Roman Catholic and Protestant parties. The religious and political differences which split the ruling class made this an unstable arrangement. During 1890 the issue of European rivalry over the possession of Uganda widened the rift still further. It was believed that the establishment of British rule would bring both political power and religious prestige to the Anglican party, which came to be known as the Wa-Ingleza, and would correspondingly diminish the power and influence of the Catholic or Wa-Fransa party.

Mwanga himself, too unsure of his own position to act as a peace-maker or an arbiter, was a somewhat dilatory partisan of the Fransa party. During the critical months in early 1890 he acted under the advice of the White Fathers to throw his influence on the side of Germany, to sign the treaty offered by Peters, and to refuse the treaty offered by Jackson. The upshot was that the Anglo-German Agreement of July 1890, which settled the international dispute in favour of Britain, only served to widen the rift between the two parties in Buganda. On the eve of the arrival in Buganda of Frederick

Lugard, the British company's administrator, the Ingleza party was preparing to celebrate its victory, while the Fransa party, led by an extremely angry Mwanga, was forming itself into the party of opposition.

The Christian revolution in Buganda took place against the background of a struggle between Buganda and Bunyoro for supremacy in the Lake Victoria basin. In the years between the death of Mutesa in 1884 and the first deposition of Mwanga in 1888, warfare along the frontier between the two countries was almost continuous. The Ganda, masters of a centralized, stable kingdom, were engaged in establishing themselves as the dominant power in the east. With the young Mwanga's succession, they became more aggressive and sent into the field large armies which pushed Kabarega's armies deep into Bunyoro to the north and north-west. Frequent successful Ganda attacks across the River Kafu upon Mruli, upon Kabarega's capital at Juaya, and as far afield as Kibiro on Lake Albert are recorded for the years 1884–7. Kabarega lived in continual fear of Ganda invasions; it is recorded that during the years 1885–8 he was himself forced to retreat and to move his capital as many as five or six times.

The pastoral kingdom of Bunyoro was meanwhile reasserting her power over an extensive, diffuse empire in the south-west. Kabarega's mercenaries were continuing to expand the areas tributary to Bunyoro through the modern district of Toro, and were intensifying their raids for cattle across the River Semliki, into Busongora and as far south as the salt lake at Katwe and the borders of Ankole. On the north, relieved of the earlier pressure of the Egyptian garrisons on the Victoria Nile, Kabarega was attempting to establish his control over chiefs formerly protected by Egyptian authority, and was beginning to exert an influence among the tribes and clans of Acholiland.

The rise of the Mahdi in the Sudan and the closing of the Nile route to the north indirectly affected the course of events within Bunyoro. Already before the death of Gordon at Khartoum in 1885, the Egyptian garrisons in the Equatoria Province had become isolated between the advancing dervishes on the north and the hostility of Kabarega on the south. The scholarly, gentle Emin Pasha, Governor of Equatoria, gradually losing control over his men, began to concentrate his troops in the region of Wadelai. The Acholi garrisons were weakened by disloyalty and by desertion. Discipline among the Sudanese troops became virtually ineffective, and the Acholi suffered untold indignities at their hands.[1] Acholi rebellions against the increasingly severe oppressions of the Egyptian garrisons took place at Dufile early in 1885 and at Patiko early in 1886. In the latter year the Acholi on the east bank of the Nile organized resistance to the Egyptians and refused to supply the garrisons with grain.

These unstable conditions greatly aggravated the pattern of Acholi inter-tribal and clan warfare. The Sudanese soldiers themselves joined in one of the most serious and protracted of the clan feuds, which had as its chief result the establishment of an enduring antagonism between the Padibe and the Payera. In 1886 the Labongo induced Rwot Ogwok and the Egyptian garrison at Padibe to join them in an attack upon the Payera. In the ensuing battle Rwot Ochama of Payera was killed; the hitherto friendly Payera were permanently alienated and driven to find support from Kabarega. This was soon followed by an attack by the Padibe and the Egyptians upon the Payera and the Labongo. In April 1887 the Payera joined the Nyoro in an expedition against Kachope Kamurasi, successor of the Jopaluo chief Mpina, who had remained loyal to Egyptian authority.

The withdrawal and decline of the Egyptian garrisons opened the way for Kabarega to make a thrust for power in the north. At the same time Emin Pasha, against whom Kabarega had maintained a persistent hostility, made an attempt to obtain Kabarega's friendship in order to open a line of communications and a possible route for withdrawal through Bunyoro and Buganda. In early 1886, the Russian explorer Junker, who had been travelling in Emin's province, succeeded after some difficulty and a long delay in reaching Kabarega's capital from the north. In June of the same year another explorer, the Italian Gaetano Casati, arrived in Bunyoro

---

[1] The history of events in Acholiland has been taken from articles by Sir John Milner Gray in the *Uganda Journal*; 'Rwot Ochama of Payera', 12, 1948, pp. 121–8; 'Acholi History, 1860–1901' I, 15, 1951, pp. 121–43; 'Acholi History, 1860–1901' II, 16, 1952, pp. 32–50. See also R. M. Bere, 'Awich, A Biographical Note and a Chapter of Acholi History', *Uganda J.* 10, 1946, pp. 76–78.

with a commission from Emin to negotiate with Kabarega for a free passage for trade goods and mails from Buganda through Bunyoro to Wadelai on the Upper Nile.

Casati's negotiations with Kabarega precipitated a crisis in the internal politics of Bunyoro. On the one hand, Kabarega, already involved in an advance upon the areas vacated by the Egyptian garrisons, feared reprisals and a renewed Egyptian advance upon Bunyoro. On the other hand, he looked upon an alliance with Emin, who possessed guns and a store of ivory, as a possible source of strength in his struggle with Buganda. The land-holding Nyoro chiefs, represented by Kategora, the mukama's chief minister, pressed for friendship with Emin, while the military party, composed of a heterogeneous band of mercenaries which had grown rapidly with the expansion of the Bunyoro empire, opposed it. The mercenaries, who numbered about one thousand in 1886, were in a position of strength. They held a virtual monopoly of Bunyoro's arms, and were on close terms with the resident Arab traders who controlled the importation of arms and ammunition through Ankole from the German sphere to the south.

In November 1886 Kategora was poisoned. Simultaneously, Kabarega announced that he would no longer recognize the authority of the chiefs. The military party and the traders thus won a position of supremacy at the mukama's court, and the chiefs were progressively alienated from their customary role in the government of Bunyoro. Casati's fate remained uncertain through the next months, while the chiefs tried unsuccessfully to overthrow their rivals and Kabarega and the mercenaries engaged in an abortive attempt to invade Acholi-land. In June, hostilities between the two parties were temporarily halted by a large-scale Ganda invasion across the River Kafu. With the withdrawal of the Ganda armies in July Casati's position became increasingly dangerous. By the end of the year Kabarega was openly hostile, and in January 1888 Casati was made a prisoner, but managed to escape with the assistance of the Bunyoro chief at Kibiro, and to return by Emin's lake steamer to Wadelai.

Kabarega and the military party now launched an intensive raid upon Kavalli's and the country at the south-western extremity of the Albert lake. Simultaneously, the Acholi on the north, encouraged, as Emin believed, by emissaries from Kabarega, were growing intensely hostile to Egyptian authority. Emin, forced to give up all idea of an alliance with Kabarega, and strengthened by the arrival of Stanley with the Emin Pasha relief expedition, ordered the confiscation of Nyoro boats on the lake and at the end of May sent a large expedition to destroy the salt markets at Kibiro. Emin's soldiers sacked and burnt villages through the countryside near Kibiro, but this did not in the least diminish Kabarega's raids which, in the next months, were growing bolder.

By this time the rapidly disintegrating state of affairs in the Equatoria Province had been brought to a crisis by Stanley's arrival at Lake Albert. While Stanley returned through the Congo to bring up the remnants of his rear column, the withdrawal of the Egyptian garrisons remaining in Acholiland was accomplished amidst punitive expeditions by the Sudanese soldiers and reprisals by the Acholi. At Patiko the garrison laid waste the surrounding countryside and were in turn attacked and driven to Pabo by Awich of Payera, son and successor to Rwot Ochama. The Pabo garrison's retreat to Dufile was marked by further hostilities, and when Selim Bey, commander of the Dufile station, withdrew to Wadelai he left behind him a wake of burned Acholi villages.

When Stanley returned to the lake in January 1889 he found that the August before a large body of Emin's troops, led by Fadl al-Mula, had rebelled against their governor upon hearing of Stanley's proposal to withdraw the Equatoria garrisons. The singularly resolute Stanley thereupon delivered what amounted to an ultimatum to withdraw and, in May, he set out for the east coast, leading a reluctant Emin with those soldiers and followers who chose to accompany him. About one thousand armed Sudanese, with several thousand civilians, were left in the Equatoria Province. These soldiers divided into two groups; one, led by Fadl al-Mula, stayed in the region of Wadelai, and the other, led by Selim Bey, retreated to Kavalli's on Lake Albert.

The recent history of events in Buganda and Bunyoro and the differing patterns of society and politics within the two kingdoms had a considerable effect upon the establishment of British rule in Uganda. The highly centralized kingdom of

Buganda, with a social and political life closely bound to agriculture and to a complex system of land tenure, was able successfully to adapt its own institutions to the British administration. The pastoral Bunyoro empire was less centralized and less stable than Buganda and, in the twenty years which preceded 1890, these tendencies were accentuated by Kabarega's policy of expansion and conquest and by the ascendancy of the military party and the alienation of the landed chiefs.

Buganda's entrance into the modern world was made with the help of the Christian missions, and it was accompanied by the rise to power of a group of revolutionary, literate, and imaginative Christian chiefs. The struggle for power between the Christian parties in Buganda only temporarily obscured the fact that the leaders of both the Fransa and the Ingleza were prepared to accept the coming of European rule. Bunyoro, in the same period, was occupied in building up her military strength, establishing close ties with the Arab traders in arms and ammunition, defending herself against Buganda on one side and the Egyptians on the other, and in extending her empire in Toro. While the most powerful elements in Ganda society identified their interests with those of the European administration, the most powerful elements in Nyoro society did the opposite. Kabarega's hostility to Britain was the product of his long-standing fear of conquest, by the Egyptians, by the Ganda and, finally, by Buganda's British allies.

For some years after the Anglo-German partition, British interests in East Africa were focused upon the kingdom of Buganda. The 'great object' of the East Africa Company was 'how to reach the Lake and its populous surroundings'.[1] By 1890 the establishment of an administration in Buganda had become not only the chief goal of the company, but also the chief justification for its stations at the coast and along the route into the interior. The company's overriding emphasis upon Buganda determined the way money was spent in Mombasa and the way economies were made in Kikuyu. It is hardly an exaggeration to say that Buganda was the frame of reference through which the early administrators viewed all other areas of the British sphere.

[1] Pelly to Mackinnon, 5 Sept. 1889, Mackinnon Papers.

This meant that the wider area later included in the Uganda Protectorate was important largely in terms of its relation to Buganda and to the security of the British administration there. Busoga, Ankole, Toro, and Bunyoro became problems or assets, dependencies or allies or enemies, according to how they were seen through the eyes of the Ganda. In 1894 Busoga was still held to be a tributary to Buganda; Kasagama of Toro served to shield Buganda's western frontier; and Kabarega of Bunyoro was an irreconcilable enemy. Similarly, the area which was to become the East Africa Protectorate and subsequently Kenya was of secondary importance to the new administrations. The stretch of territory from the coast to the lake was, quite simply, something to be got through, as quickly and as easily as possible, on the way to Kampala.

The future productivity of the highlands, populated by white settlers, financed by foreign capital, and heavily subsidized by imperial grants-in-aid, could not be anticipated by the British investor of the 1890's. At the end of the nineteenth century, East Africa to the east of the lake was thought to be 'comparatively valueless';[1] the Foreign Office bluntly called it a 'sterile region'.[2] From an administrative as well as an economic point of view, the area was less manageable and less attractive than Uganda. There was no centre of population in the east to compare with the lake kingdoms in the west. The pastoral and nomadic Nilo-Hamitic peoples, scattered thinly over a wide area, were difficult to contact. The Kikuyu might have been made the nucleus for the development of an administrative network but, like the other Kenya tribes, they had no paramount chief or king, no centralization of authority like that of the Ganda. The Swahili from the coast had for some years before 1890 been spreading their influence through the British sphere, but their Muhammadan religion and their traditional connexion with the slave trade made them, at least at first, unacceptable as agents of European civilization.

Stations were built by the company at Machakos in Kamba country and at Fort Smith in Kikuyu but, for the most part, they were looked upon simply as convenient places for lake-bound caravans to halt and restock with provisions. At the

---

[1] P. L. McDermott, *British East Africa or I.B.E.A.*, 1895, pp. 103-4.
[2] Perham, *Lugard*, p. 175.

coast the company stationed agents at Kismayu, Lamu, Witu, Malindi, Takaungu, and Vanga, but they did little but supervise the collection of customs. The central administration at Mombasa was larger—according to one estimate, it employed as many as twenty Europeans—and more active but, again, most of its work was connected either with customs or with preparing caravans for the interior.

One observer remarked in 1889 that the British coastline remained relatively tranquil simply because no changes were made by the company; while the remark was not literally accurate, it came close to describing the general state of affairs. Poor management was responsible for a great deal of inefficiency and sheer waste. The disorganization of the Mombasa administration was commented upon by every traveller who visited the town. In the London office there was a serious lack of co-ordination among the directors; plans were confused, made and changed overnight; orders and counter-orders were telegraphed to Mombasa; and if a comprehensive plan of operations was ever drawn up, it was certainly never followed. A great deal of this was justly attributed to the character of Mackinnon. Salisbury had never had much confidence in him, for 'he has none of the qualities for pushing an enterprise which depends on decision and smartness';[1] Kitchener's advice in 1888 was to 'get rid' of him;[2] by 1890 even his fellow directors were beginning to grow exasperated with his impractical ideas and his poor tactical sense.

Yet not even Rhodes or Goldie could have made a financial success of the East Africa Company. A number of factors outside the company's control worked against it from the start. Of these the most serious was the nature of East Africa itself, its limited range of exportable produce, and its lack of waterways navigable from the coast into the distant interior. In addition to this, the preoccupations of Anglo-German rivalry accounted for a great deal of lost time and spent resources. Undercapitalization drastically limited the company's activities; East Africa in the 1880's was not an attractive field for investment. The company had few means of raising revenue

[1] Salisbury to Goschen, 10 Apr. 1890, quoted in Lady G. Cecil, *The Life of Robert, Marquis of Salisbury*, 4 vols., 1921–32, iv. 281.
[2] R. Coupland, *The Exploitation of East Africa, 1856–1890*, 1939, p. 469.

outside of trade in ivory. Although powers of taxation were granted by the charter, it was not yet feasible to tax in the interior, and the provisions of the Zanzibar commercial treaties virtually prohibited taxation at the coast. Under the terms of their concession from the Sultan of Zanzibar, the company had to pay back in rental almost all of their revenue from customs.

Most of the East Africa Company's European staff was recruited outside East Africa, in Britain or in British regiments overseas. In West Africa during this period, established European trading agents with knowledge and experience of local conditions were hired by the Niger Company as the nucleus of their administration; in East Africa, where there was no comparable group of traders, the administration was largely made up of young men of no particular experience or qualifications who were just making a start in life. The company's 'raw Scots lads' were the butt of many a caustic joke. East Africa in the early 1890's was a training ground for the protectorate administrations; many of the company's employees were absorbed into the British service, and some of them rose to distinction there.

The greatest problem for any European enterprise in East Africa, whether chartered company or colonial government, was the lack of an economical means of transport. It took about six weeks for a caravan to march from Mombasa to Kikuyu, and three months for the journey from the coast to Buganda. Human portage cost in the region of £250 a ton; at this figure, only ivory could be a profitable source of commerce from the interior. The company made several unsuccessful attempts to improve its method of transportation and to find a substitute for its Zanzibari porters. Pack animals—donkeys and camels—and carts pulled by Cape oxen were imported and tried out along the Mombasa route. A steamer was chartered for use up the Rivers Tana and Juba. Roads were levelled off inland from Malindi and from Kibwezi toward Tsavo. Finally, a few miles of a 24-inch tramway were shipped out to Mombasa in 1890, and the first rails of the so-called Central African Railway were laid to the accompaniment of fireworks, ceremony, and celebration. Like the pack animals, which died, and the roads, which washed away with the seasonal rains, the railway was a 'sorry fiasco',[1] never used except for what one sarcastic witness called 'occasional picnic parties' from Mombasa.[2] The company needed only two years' experience of East Africa before its directors began, with increasing desperation, to press the government for a railway subsidy.

In Zanzibar sweeping administrative changes were made under the British protectorate declared in 1890. The transition to the protectorate was achieved smoothly, partly because it was in some ways only a formal recognition of a relationship which already existed, and partly because Ali bin Said, who succeeded Khalifa as sultan in February 1890, was the most tractable and co-operative ruler who had yet occupied that office. Ali cannot have understood the full meaning of the protectorate; he was ignorant of European affairs and the explanation given to him by the consul-general, Euan-Smith, was partial and misleading. In any case, Ali had no choice in the matter, as he himself realized, and at least on the surface he accepted the protectorate with 'joy and gratitude.'[3]

A few months after the protectorate was declared, Gerald Portal arrived in Zanzibar as consul-general and commissioner for the British sphere of influence. Portal was an able, energetic man, impatient of inefficiency and disorganization. During his years with Lord Cromer at the British agency in Cairo he had received an excellent training in administrative reorganization which he put to good use in Zanzibar. His first act was to make the administration solvent by cutting the sultan's personal share of the public revenue to one-third. The remaining two-thirds were spent upon new building and upon improvements in the administration of the harbour and the town. New port regulations were put into effect. Badly needed judicial reforms were begun. Prohibitions were placed upon the importation of arms, ammunition, and liquors. The value of Zanzibar's trade and its ability to compete with the ports in German East Africa were vastly increased by declaring it a free port. By the end of 1892, when Portal left for Uganda,

[1] R. P. Ashe, *Chronicles of Uganda*, 1894, p. 7.
[2] G. H. Portal, *The British Mission to Uganda in 1893*, ed. Rennell Rodd, 1894, p. 28.
[3] Euan-Smith to Salisbury, 19 June 1890. F.O. 84/2062 (most secret), quoted in L. W. Hollingsworth, *Zanzibar under the Foreign Office, 1890–1913*, 1953, p. 43.

the creation of the new Zanzibar administration was well under way. Portal's reforms worked for the general good of the town and its inhabitants, but this was, perhaps, poor compensation for Zanzibar's inevitable decline following the Anglo-German partition. The sultan himself, with some of the associates and advisers who had profited under his régime, were losers under the new financial and administrative arrangements. The sultan had already lost his power on the mainland; under the protectorate he lost much of his power in Zanzibar itself.

One of the by-products of the decline in the authority of the Sultan of Zanzibar was a temporary increase in the influence of the Sultan of Witu. For a short period of time the balance of power in the northern portion of the Zanzibar dominions shifted, and the coast Arabs looked for leadership not to Zanzibar, but to Witu. During the long and tedious Anglo-German dispute over the possession of Witu, the sultan had built up a strong stockade and consolidated his following of runaway slaves and outlaws enlisted from all over the coast. By 1890, when the international dispute was settled in favour of the British and a protectorate was declared over Witu, the sultan was in a strong position to oppose the British company.

Serious Arab opposition to the Zanzibar anti-slavery decree of August 1890 was directed from Witu. The next year, following further difficulties, a large garrison was established by the company in Witu. Still the sultan and his followers continued actively to oppose British authority. The company, unable to continue to pay the cost of the Witu garrison, withdrew their forces in July 1893. Finally, in August of the same year, the Sultan of Zanzibar's forces captured and destroyed the Witu stockade; the Sultan of Witu fled, and was later captured and executed.

Witu was one of four bases from which the company planned to open up routes into the interior. With this end in view, the Tana was explored in 1889 and 1891, but the unrest in Witu kept the route from being developed. On the Juba, where the hostility of the Somalis resulted, in 1893, in the murder of a European and the withdrawal of the company's garrison, the story was much the same. Two other routes—from Malindi and from Mombasa—were successfully used by company caravans.

The Malindi route, up the Sabaki river to Tsavo, was explored by Lugard before he left for Uganda, but it was passed by in favour of the more direct route from Mombasa through the Taru desert to Tsavo. From Tsavo the company's caravans marched to Machakos and on to Fort Smith in the highlands. Between Kikuyu and the lake the country was poorly known, sparsely populated, and lacking in food-supplies. The general direction taken for this, the most difficult part of the journey, varied little from the route first used by Joseph Thomson in 1883–4 from Kikuyu to Lake Naivasha, north along the meridional rift towards Lake Baringo, and then west to Mumia's in upper Kavirondo. In 1889 Jackson and Gedge succeeded with difficulty in finding a way westwards from Lake Naivasha through the Mau forest. Two years later Captain Eric Smith reached the Sotik country by a route south of Jackson's. Meanwhile, in February 1891, James Martin had discovered the route through Nandi country by way of Eldama Ravine, which was recommended by Captain J. R. L. Macdonald's railway survey party of 1892–3 as the best route for the projected Uganda railway.

Most of the Kenya tribes were affected only indirectly or not at all by the European caravans which passed through the country along their narrow track to the lake. A few Africans, at isolated points along the caravan route, entered into an intensive relationship with the European, a relationship which had as its chief purpose not the establishment of an administration, but the supply of caravans with food and hospitality. The treaties of protection signed by the company in 1888–90 with sections of the Taveta, Taita, Nyika, Pokomo, Galla, Kamba, and Kikuyu, were looked upon primarily as deeds of possession or as guarantees of co-operation or safe conduct for caravans. The sites for the stations at Fort Smith and Machakos were selected for their location in rich, grain-producing country; the smaller station on the River Tsavo was used as a storage depot; the station at Mumia's in Kavirondo, strategically located at the River Nzoia crossing, was a depot, a halting place, and a centre from which to restock for the march to Kikuyu.

This essentially opportunistic attitude towards Kenya affected the African societies concerned in widely different ways. In upper Kavirondo the chief Mumia welcomed European caravans; he gave them shelter and a plentiful supply of food; he supplied guides, arranged for the care of sick porters, and encouraged caravan leaders to use his village for halts which sometimes lasted more than a month. As a result of Mumia's shrewd alliance with the British, his once small village grew rapidly in size and prosperity and became the trading, and, later, the administrative centre for Kavirondo. Mumia himself was to wield an influence under the Nyanza administration which was far in excess of the original importance of his family or tribe.[1]

Near the coast the Taita and the Nyika, used to frequent dealings with Swahili traders, supplied the company with porters and mail-carriers for the short march from Mombasa to the Tsavo and on to Machakos. The company's relations with these tribes were fairly uneventful except for an incident in 1892, which resulted in a punitive expedition against the Taita of Mwanda and Mgangi. The Pokomo, well off the main caravan route, were left to their own devices after the first rush for treaties and, with the defeat of Witu and the decline of the Galla, they were relieved from the pressures of their nearest neighbours.

In Machakos, where the company's first inland station was established in 1889, the Kamba were enjoying a period of unusual prosperity. Their attacks against the Galla on the east and north-east and against the Masai on the south-west were growing yearly more successful, and they were continuing to spread through Ulu and into Kikumbuliu. Their trade with the coast Swahili was flourishing, their herds had increased, and, until the locust famine of 1894, they suffered no natural disasters to compare with the epidemics which attacked the Masai in the 1880's. After initial difficulties caused by an inept station superintendent, Machakos station prospered. John Ainsworth, an able administrator who later became Chief Native Commissioner in the East Africa Protectorate,[2] was appointed superintendent of the Machakos station in 1892. Ainsworth helped the Kamba to organize a series of posts

---

[1] I am heavily indebted to A. T. Matson, of Kapsabet, Kenya, who has most generously supplied me with a memorandum upon which the account of the history of western Kenya has been based.

[2] F. H. Goldsmith, *John Ainsworth, Pioneer Kenya Administrator, 1864–1946*, 1955.

through Machakos and into the modern Kitui district to watch for Masai war parties and to check slave caravans. Young men from Machakos acted as mail-carriers to Mombasa and Fort Smith, formed a volunteer force to protect the station, and served as labourers on the cart road being built from Kibwezi towards Tsavo. By 1893 Ainsworth was writing home for more trade goods, not only for the familiar white cotton cloth but for commodities as imaginative and as foreign to East African tribal life as umbrellas.

Three events of great symbolic importance for the future of East Africa took place in Ukambani in the early 1890's. In 1891 the first mission station in the Kenya interior was established at Kibwezi. In 1892 an Indian trader named Adamjee Alibhoy settled at Machakos.[1] In 1894 East Africa's first white settler, an Englishman named James Watt, arrived with his family to set up farming near the station.[2]

While in Machakos the Kamba were helping to lay the groundwork for a European administration, and in upper Kavirondo Mumia was rapidly increasing his power, in the highlands the Kikuyu were suffering from a prolonged and bitter struggle with European caravans and station agents. At the time of the European annexations the Kikuyu were expanding rapidly in a southerly direction through the area which today forms the district of Kiambu. They were growing more aggressive in their occasional raids upon both the Kamba and the Masai, and they were greatly feared by Jumbe Kimameta and the other Arab and Swahili caravan leaders who stopped at Ngong, on the border of Masailand and Kikuyu, to restock with food. Kikuyu trade with the Masai and the Kamba may have been increasing, but trade was still a minor factor in the economic life of the inhabitants of the agriculturally rich highlands.

The initial adjustment of the Kikuyu to the European was made difficult by two opposing circumstances. On the one hand, the prosperous Kikuyu, protected from their neighbours by a thick belt of forest, were traditionally isolated, proud, and aggressively hostile to outside interference. On the other

---

[1] C. Ehrlich, 'The Economy of Buganda, 1893–1903', *Uganda J.* 20, 1956, p. 21.
[2] See also A. T. Matson's article, 'The Freelanders: Cautionary Tale of a Social Idea', *Kenya Weekly News*, 7 Aug. 1959, which describes another early, unsuccessful attempt at colonization.

hand, the very fertility of the highlands, and relative poverty of the plateau between Ngong and the lake, meant that no caravan, European or Swahili, could cross through the British sphere without restocking with Kikuyu produce. The Kikuyu experience of European caravans was from the outset unpleasant. In the course of the 1880's, two large, well-armed caravans, one led by a German explorer, G. A. Fischer, and the other by a Hungarian traveller, Count Samuel Teleki, forced their way across Kikuyu country. The hostile reaction of the Kikuyu to Fischer and Teleki earned them an 'evil reputation'[1] among European travellers, and other caravans led by Bishop Hannington, Joseph Thomson, and Frederick Jackson were careful to skirt the highlands on their way to the lake.

The company's first Kikuyu station was built by Lugard in 1890 at Dagoretti, outside the area of cultivation in the present day Kiambu district. Like other European travellers, Lugard remarked with pleasure upon the 'bracing' air in the highlands, the cold nights, the flowers and shrubs which reminded him of home. 'Kikuyu was like a garden of Eden', wrote Bishop Tucker some years later, 'the climate was almost English.'[2] The productivity of the highlands led more than one caravan leader to ponder upon the feasibility of white settlement. After 1892 the company agent in Kikuyu tended a flourishing kitchen garden; Captain Macdonald, when he stopped there in 1892, enjoyed a repast of 'mutton chops and vegetables that would not have disgraced an English table'.[3]

For a time George Wilson, the first company superintendent at Dagoretti, 'succeeded admirably'[4] with the Kikuyu, but within a few months a nearby chief, Wyaki, became hostile, attacked the station, and forced Wilson to withdraw to the coast. In 1891 the station, renamed Fort Smith, was rebuilt upon a new site within the cultivated area, close to Wyaki's village. A year later the company, which was by now in serious financial difficulties, decreed that Fort Smith and Machakos must be made self-supporting. This decision, the disastrous

---

[1] F. J. Jackson, *Early Days in East Africa*, 1930, p. 171.
[2] A. R. Tucker, *Eighteen Years in Uganda and East Africa*, 2 vols., 1908, i. 207 and 208.
[3] J. R. L. Macdonald, *Soldiering and Surveying in British East Africa, 1891–1894*, 1897, p. 56.
[4] F. D. Lugard, *The Rise of Our East African Empire*, 2 vols., 1893, ii. 535.

product of undercapitalized chartered company government, had its most serious results in Kikuyu. During the course of the year 1892 European agents at Fort Smith occupied themselves with a series of punitive expeditions and raids for grain. By this time the well-built brick station house had grown into a fortress, surrounded by a ditch, protected by thick coils of barbed wire, and entered by way of a drawbridge. In 1893 the station was 'practically in a state of siege'.[1] The next year a group of several thousand Masai who had settled near the station agreed to give the superintendent protection in return for permission to raid the Kikuyu crops for food. For their conduct of affairs at the Kikuyu station, which they maintained until the establishment of the Protectorate administration in 1895, the company received severe criticism from Gerald Portal. 'By refusing to pay for things,' he wrote, 'by raiding, looting, swashbuckling, and shooting natives, the Company have turned the whole country against the white man.'[2]

While Europeans were making their first impact upon the Kikuyu and the Kamba, large-scale tribal readjustments were being made throughout western Kenya as a result of the decline and dispersal of the Masai.[3] The decline of Masai power, which nearly coincided with and greatly facilitated the European penetration, followed gradually upon a series of disasters. The rinderpest epidemic of the 1880's carried away large numbers of Masai cattle, and disease and famine weakened the tribes over a wide area. Inter-section rivalries, as those between Purko and Laikipia in 1889 and 1890, and the struggle for leadership between Lenana and Sendeu, sons of the Laibon Mbatian who died about the year 1890, also combined to destroy Masai dominance and to give other tribes the opportunity to expand and increase their power. Masai raids continued as far afield as the Galla country, where at Golbanti the missionary Houghton and his wife were murdered by a raiding party in 1886, and to the north against the Suk (Pokot), into Kitosh and Samburu, and west into Kavirondo. Frequent raids in the east were noted by company agents, who were impressed by the evident ability

---

[1] Portal to Rosebery, 31 Jan. 1893. F.O. 2/60.
[2] Portal to Lady Alice Portal, 3 Feb. 1893, Portal Papers.
[3] The following account has been largely taken from the memorandum by A. T. Matson, cited above.

of the Kamba, the Kikuyu, and the Taita to defend themselves. The Nandi, the Suk, the Kamba, and to a lesser extent the Kikuyu, were growing yearly more successful in organizing their own counter-attacks and raids for cattle, and this, in turn, further contributed to the weakening of Masai fighting power.

In the west the Nandi gained more than any other tribe from the Masai decline. Their earlier successes against the Uasin Gishu and Rift Valley Masai had strengthened and given cohesiveness to the tribe, and with the dispersal of the Masai to the north and east new areas were opened for raids and expansion. The Nandi raided both close to home and far into the north and north-west; they were feared by Arab and Swahili caravan leaders as well as by their African neighbours, and European caravans, searching for a route to the lake, avoided Nandi country. When the East Africa Protectorate was declared in 1895, the Nandi were firmly established as the dominant power in the west. A few years earlier, in 1890, an event occurred which was to have a considerable effect upon the future struggle between the Nandi and the British administration. In 1890 the senior laibon, Kimnyole, who was believed to be responsible for misfortunes which had beset the Nandi, was killed. Before his death, Kimnyole prophesied the European conquest and future disasters for the tribe. The accuracy of these prophesies greatly increased the prestige and power of the institution of the laibon, and Koitatel, Kimnyole's son, was consequently able to organize and to unify Nandi resistance to the establishment of British rule.

The balance among the tribes in western Kenya was gradually altered by the dispersal of the Masai and the rise of the Nandi. Other factors, such as the cattle epidemics, the famine of 1887, the drought of 1890, the famine of 1892, and the locust epidemic of 1894 also affected the relative strength of the tribes. On the whole, these natural disasters weakened the tribes on the plains more than the northern hill tribes. New and traditional patterns of inter-tribal warfare were accentuated by the settlement of colonies of Uasin Gishu Masai among many of the tribes. Within the area of Kavirondo, the Nandi as well as the Kipsigis, the Arab slavers, and Ganda raided the Nyanza tribes, who were split by continual inter-tribal and clan warfare. The Keyo were being forced off the plateau grazing grounds, but the

Tuken were reoccupying low-lying areas to the south and east of the Kamasia hills. The Luo, raided by the Kipsigis and the Nandi, had been driven to the north and south of the Nyando Valley. The Kitosh were being pushed to the south-east by the Teso; at this time they possessed large herds of cattle and were frequently raided by the Nandi and by the slavers. The Sapei were moving to the west, and the Kony to the north. The Kipsigis, like the Nandi, were engaged in expansion and intensive inter-tribal warfare, to the south of the Amala river against the Masai, to the north and west against the Luo, and to the north-east against the Masai settlements in the Nyando Valley.

While the Nandi were pressing to the north and north-west, the Somali, the Galla, and the Suk were pushing the tribes in a general movement southwards. The Suk, who were becoming a strong cattle-owning people, largely escaped the cattle epidemics and were expanding their grazing areas. The Turkana raided into Samburu and Rendile and west against the Suk. The Karamojong made raids upon the tribes in the area of Mount Elgon, and against the Kitosh and the Nandi, but their movement southwards was checked in the mid-1890's by the epidemics, and they were forced to turn to hunting and to trading with the Swahili caravans.

During the 1880's and 1890's Arab and Swahili caravan traffic in the British sphere was increasing as a result of the Masai decline, the hostilities in German East Africa, and the Arabs' continual search for new sources of ivory and slaves. Most of the traders, like Jumbe Kimameta, were from Pangani, but evidence of an increase in the number of caravans, particularly slave caravans, was also noticed at several of the British ports. In the east, traders marched from Witu up the Tana to Mount Kenya. At centres all through Machakos and Kitui the Kamba served as middlemen for Swahili dealers in ivory purchased from the Dorobo, and in the slave captives of inter-tribal warfare. In the west the caravans followed routes from Taveta across Masailand to Lake Naivasha, through Sotik and Kipsigis, and up the lake shore to Mumia's. Some, notably Kimameta, traded north into Suk and Turkana country and as far afield as Samburu and Lake Rudolf.

As the caravan traffic increased the traders began to make their influence felt upon tribal and inter-tribal politics. Colonies and centres of Muhammadan influence comparable in kind though not in size or importance with Tabora in German East Africa were established by the 1890's at Kitoto's, at Sapei, and at Mumia's. Mumia allied himself with the traders as he allied himself with the British and they joined him in raids on other tribes. Attacks in Kitosh and upon Njemps in 1889 and 1893 by the traders 'Fat' Sudi and Abdullah were not dissimilar to the massacres observed by Livingstone and by Stanley in the upper Congo. On the eve of the declaration of the British protectorates the slave-trade in the British sphere had reached substantial proportions, and the Arab and Swahili caravans were rapidly becoming a significant influence upon the course of events in Kenya.

In December 1890 Lugard, the first British administrator in Buganda, reached Mengo with a force of Zanzibari porters, fifty Sudanese and Somalis, a poor supply of arms and ammunition, and an old Maxim gun. He was given a mixed reception. Both groups of missionaries, with their converts, looked upon his arrival as an event likely to be favourable to the Ingleza party and unfavourable to the Fransa party. Mwanga and the Fransa, who were the stronger of the two parties, therefore at first opposed Lugard while the weaker Ingleza supported him. Lugard, for his part, made a noble effort to deal fairly with both parties, but his own military weakness made him dependent upon the goodwill of the Protestants. Even when his forces were increased at the end of January with the arrival of Captain W. H. Williams, his military strength still compared extremely poorly with the Fransa's 2,500 fighting men or with the Ingleza's similar, though smaller, number.

On 26 December a treaty giving the company 'suzerainty' over Buganda was signed, after stormy negotiations, by the kabaka and the chiefs of both factions. This accomplished, Lugard spent the next months in an attempt to settle some of the outstanding differences between the Ingleza and the Fransa. One of the chief sources of trouble between the two Christian parties arose from the land settlement by which they had divided the kingdom after their victory over the Muslims in 1890. Under this unwritten agreement the offices of state and the territorial chieftainships were shared out equally among the leading Ingleza and Fransa chiefs. It was understood that if a chief of any rank or position in the feudal hierarchy changed his religious or party allegiance, he must also forfeit his office and his land. This arrangement proved highly unstable. When Lugard arrived on the scene the Catholics, supported by the White Fathers, were pressing for a removal of the obstacles to a change of party. Since this would immediately result in a strong movement to join the dominant Fransa party, the weaker Protestants held fast to the existing arrangement, and Lugard, whose military position depended upon the preservation of a balance between the parties, sided with them.

In May Lugard led a large Christian army to the borders of Bunyoro to win an easy victory over the exiled Muslim party. The expedition against the Muslims temporarily united the Christians, but it also removed the persistent threat of a Muslim attack which, in the past months, had kept the Christians from fighting each other. It was in the midst of this potentially explosive situation that Lugard left Williams in charge at Kampala, with a small force, while he marched on a seven months' tour in the country to the west of Buganda.

Lugard set out in June from Buddu, and thence through Ankole, where he made a treaty with the representative of Ntare, the Mugabe of Ankole. Ntare, whose kingdom was being pressed upon the north by Nyoro raids, was at this time friendly to Buganda and had given hospitality to the Christian refugees of the civil wars of 1888–90. By the terms of Lugard's treaty the protection of the company was extended to Ankole, but for several years to come this was to be no more than a formality. The real purpose of the treaty lay in Ntare's undertaking to check the traffic in arms and ammunition which passed through Ankole from the German sphere to the defeated Ganda Muslims and to Kabarega of Bunyoro.

After establishing a post at the Salt Lake at Katwe, Lugard marched north through Toro to Kavalli's. There he enlisted in the service of the company Selim Bey and his Sudanese, who numbered several hundred armed men and many thousand followers. With the assistance of the Sudanese Lugard was able to initiate the military campaign which in the next few years resulted in the containment and conquest of the Bunyoro empire.

The British attitude to Bunyoro was based partly upon Kabarega's known hostility towards the former Egyptian garrisons in the southern Sudan. Lugard's view of Bunyoro was equally influenced by his view of Buganda. He adopted almost without question Buganda's traditional hostility to her chief rival; the Ganda were friends and the Nyoro were, therefore, enemies. In 1891 Kabarega was doubly an enemy because of his support of the Ganda Muslims.

For military and strategic reasons Lugard felt it was necessary to protect his position in Buganda by checking the arms traffic from Karagwe through Ankole and Toro and by cutting off the expansion of the Bunyoro empire in Toro. Daudi Kasagama, a prince in exile of the former royal house of Toro, who had accompanied Lugard from Buganda, was restored as Mukama of Toro under the company's protection. Kasagama undertook to prevent the importation of arms and ammunition, and Lugard, in turn, established a line of forts along the border of Toro, garrisoned by the Sudanese, to protect Kasagama from invasion by Bunyoro.

When Lugard returned to Buganda at the end of December 1891 he found a letter waiting for him from the company ordering him to evacuate Buganda at once. This letter, which was followed in January by another, postponing evacuation until the end of 1892, influenced Lugard's conduct of the events of the next few weeks. His main purpose was no longer to prevent war at all costs, but was instead to force Mwanga and the leaders of both parties to accept and to acknowledge publicly the authority of the company.

Relations between the Christian parties had not improved while Lugard was away, and the quarrel over the land question had become increasingly serious. Toward the end of January matters came to a crisis over a relatively trivial incident of theft, which led to the murder of a member of the Ingleza party by Mongoloba, a member of the Fransa party. Interpreted in terms of Buganda law, the murder was committed in self-defence, and Mwanga, when appealed to for judgement, therefore acquitted Mongoloba. Lugard refused to accept the verdict of acquittal and pressed for a conviction. Mwanga, encouraged by the Fransa chiefs, held fast to his judgement.

The incident was interpreted by all parties as an issue of

authority between Lugard and the forces of the company on the one hand, and Mwanga and the Fransa chiefs on the other. The capital became unsettled; war seemed imminent. On the night of 22 January Lugard issued forty guns and a keg of powder to the Protestants, and two days later, when the Fransa began to collect for battle, he gave out the remainder—something above 450—of the guns stored in Kampala fort. It is not clear which party fired the first shot in the battle which followed. Lugard's Maxim gun was decisive, and after a brief battle Mwanga and the Fransa chiefs fled to the island of Bulingugwe.

Although upon the day after the battle of Mengo Lugard offered the kabaka an honourable reinstatement, Mwanga, influenced by Monsignor Hirth, the French bishop, did not finally return to the capital until the end of March. By this time the victorious Protestants, in control of the capital, were able to demand the largest share in a new land settlement. By the agreement between the Christian parties of 5 April the old system of dividing offices and estates was abolished, and the principle of geographical partition was accepted. The Catholics as a political party were restricted to the province of Buddu and, by an agreement reached in May, the Muslim party was given the three small counties of Gomba, Busuju, and Butambala as a buffer between the Catholics on the south and the Protestants on the north, east, and west. Under a treaty ratified on 11 April between the company and the kabaka and chiefs, the suzerainty of the company was confirmed, and Mwanga and the leaders of both parties agreed to fly the company's flag.

The company's administration in Buganda left a mixed legacy to the government of the British protectorate. The war between the Christian parties, and the defeat of Mwanga and the Fransa opposition, made possible, for the first time, the secure establishment of British power. The treaties of April 1892 gave future administrators both the authority and the juridical basis upon which to negotiate new treaties and new land settlements. Lugard and Williams left their successors with three major problems: the revision of the inequitable land settlement within Buganda, the pacification of the Ganda Muslims, and, most difficult, the decision upon the question of Britain's future relationship to Toro and Bunyoro.

Gerald Portal, the British Commissioner, arrived in Kampala on 17 March 1893. On 1 April the Union Jack replaced the company's flag upon Kampala fort and, a few weeks later, on 29 May, a new treaty of protection in the name of Her Majesty's Government was signed, without serious opposition, by Mwanga and the chiefs. Negotiations for an agreement between the Christian parties were more difficult. The immediate result of Lugard's restriction of the Catholics to Buddu had been to empty the capital of all but a handful of Catholics, and to put the Protestants in control of Mwanga and the central administration. In the interval between Lugard's departure for Europe in May 1892 and Portal's arrival the next year, Williams had tried unsuccessfully to persuade Monsignor Hirth and Alfred Tucker, Anglican Bishop of Uganda, to reach a compromise acceptable to both parties. It was left to Portal to settle the land question for the time being with an agreement signed on 22 April which gave the Catholics a larger portion of territory, a narrow road between Buddu and Mengo, and a more substantial share in the offices of state. A Catholic *mujasi*, chief of the army, and a *gabunga*, chief of canoes, were appointed as duplicates to Protestants holding the same offices, and Stanislas Mugwanya, who had become *kimbugwe* in the Christian revolution of 1890, was made Catholic *katikiro*, theoretically equal to Apolo Kagwa, the Protestant *katikiro*.

Gerald Portal left Kampala at the end of May 1893 to return to England to report upon the question of the retention or abandonment of Buganda. Pending a definite government decision, Captain Macdonald was left in charge of the provisional British administration, with orders to cut back his commitments and his expenditure as drastically as possible. It was Portal's intention to restrict the British administration to the kingdom of Buganda; accordingly, before his departure he sent Major Roderic Owen to withdraw the Sudanese garrisons in Toro. The events of the next months, however, led the strategically minded Macdonald and Owen to abandon the idea of withdrawal and to reaffirm Lugard's former policy of protecting the British position in Uganda by checking the power of Bunyoro.

The first crisis began in early June, when the Ganda Muslims, who had been demanding a larger share in the Buganda land settlement, entered into discussions for an alliance with their co-religionists, the Sudanese, who had been brought into Uganda by Lugard. Macdonald feared a combined Islamic movement against the Christian Ganda and the European occupation. Therefore when Selim Bey, at the new headquarters at Entebbe, gave indications of disloyalty, he disarmed the Sudanese at Kampala and gave orders to Apolo Kagwa for a Protestant mobilization. On 18 June the Protestants easily defeated the Muslims in a battle in the neighbourhood of Rubaga.

The Muslims retreated towards Toro where Owen, meanwhile, was deeply involved in protecting Kasagama against Bunyoro. Upon his arrival a few months earlier he had found the Sudanese garrisons in a state of disorganization and disloyalty. The Nyoro were continuing their attacks upon Toro, and the Sudanese, forced to support themselves by raiding, were laying waste several hundred square miles of country in southern Bunyoro. Owen followed Portal's orders only to the extent of withdrawing the garrisons from the two westernmost of the forts established by Lugard. He then marched to the Salt Lake, where he strengthened Fort George as a deterrent to the Manyema who were raiding into Busongora across the border from the Congo. Back in Toro, Owen, by now impressed with the extent of the Arab arms traffic to Bunyoro from German East Africa and with the consequent importance of maintaining the Sudanese garrisons in Toro, built a new post, later named Fort Portal, near Kasagama's capital.

At this juncture the Ganda Muslims arrived in Toro, where they entered a series of negotiations with Owen. Then, moving west past the forts, they set out in the direction of the Semliki, and were attacked and defeated by a force of Christian Ganda sent by Macdonald from Kampala. The Muslims returned to Buganda, and after further local fighting the rebels dispersed and the loyal Muslims were given the single province of Butambala. Selim Bey was arrested, tried for mutiny, and deported from Uganda.

The next phase of the conflict in the west did not take place until after the arrival in November of the new British Commissioner, Colonel H. E. Colvile. Macdonald, who erroneously believed that Kabarega owed 'allegiance' to his 'suzerain,

Mwanga'[1] and who by this time fully agreed with Owen that the security of Buganda depended upon the defeat of Bunyoro, told Colvile of his plan to launch an attack deep into Kabarega's territory. Colvile, as it happened, had brought with him instructions to send an expedition to the Nile basin to protect British interests against a Belgian advance from the Congo State. This objective neatly provided further justification for an attack upon Bunyoro. At about the time of Colvile's arrival, moreover, Kabarega himself launched an attack upon Toro, and succeeded in driving Kasagama out of his capital into the mountains.

The Nyoro were marching into Busongora and simultaneously preparing an attack upon northern Busoga when, in January, a large Ganda army, led by Colvile and Macdonald, entered Bunyoro. In the prolonged campaign which followed the Ganda army was victorious. After an unsuccessful Nyoro counter-offensive later in the year, Kabarega was decisively defeated and forced to flee across the Nile to the east.

The military campaign of 1894 had tragic results for Bunyoro. Wide areas of country were devastated by the war and by the famine and disease which followed. A chain of forts established along a line from Kibiro through Hoima to the River Kafu cut off western from eastern Bunyoro and opened a line of communication between Buganda and Lake Albert. Bunyoro lands were lost to the west and to the south. Kasagama was installed at the head of a Toro Confederacy, enlarged to cover the whole stretch of country between Lakes Albert and Albert Edward. Buganda's borders were vastly extended to the north; the Catholics were given the western portion of this new territory, and the Protestants the eastern portion.

The aggrandizement of Buganda at the expense of Bunyoro, and the alienation of Bunyoro's western provinces, was to be a source of enduring bitterness throughout the history of the Uganda Protectorate. There was little historical justification for this decisive outcome of the long struggle for supremacy between the two dominant kingdoms in the Lake Victoria basin. It was, rather, the product of a complex series of circumstances, already narrated, which began in the 1860's and 1870's with Baker's visit to Kabarega's capital and Stanley's visit to Mutesa's

[1] Macdonald, *Soldiering and Surveying*, p. 296.

capital, and which resulted in the evangelization and annexation of a friendly Buganda and the alienation and conquest of a hostile Bunyoro.

There was another, less tangible reason for the conquest of Bunyoro. Lugard, Macdonald, Owen, and Colvile were all military men. They were accustomed to leading active lives, and they had been schooled in a tradition of frontier warfare. Just as the creation of British India had involved the progressive elimination of troublesome frontiers, so the creation of the British empire in Africa was to involve the establishment of new, larger political units. To the British soldier, a hostile frontier such as Buganda's frontier with Bunyoro presented a challenge and a provocation to conquest.

The British administration in Buganda, meanwhile, was being carried on smoothly, without any notable changes by the Commissioner. The terms of the land settlement were confirmed by Colvile in 1894, and the principle of religious liberty, without a penalty in loss of land for a change in religious allegiance, was proclaimed. The Catholic party, with an increased share of territory and a consequently increased stake in the new régime, had become reconciled to British rule. Mwanga, who had been the chief loser in the tumultuous events of the past few years, remained temperamental and unreliable but largely ineffective. He had declared himself a Protestant after the Catholic defeat of 1892, and in the course of the next two years he twice made an unsuccessful attempt to change his religious alliance and to obtain the support of the Catholic party against Colvile and the British administration. The Commissioner saw little of Mwanga; the Christian oligarchy was firmly established in control of the government of Buganda, and the kabaka had become little more than a constitutional monarch.

Already, in 1894, it was becoming evident that the Ganda themselves were to play the leading role in the future history of the protectorate as the disseminators of Christianity and the artisans of a new social order. Within Buganda, fundamental social and political changes had taken place within a very short time as a result of missionary activity, the civil wars of 1888–92, and the new land agreements of 1892 and 1893. The division of offices and estates negotiated by Lugard and Portal consolidated and broadened the position of the Christian ruling class and reaffirmed Christianity as the religion of legitimacy. The physical partition of Buganda into Protestant and Catholic provinces resulted in the dislocation and migration of large numbers of peasants. The solidity of the old social order was shaken,[1] and this served to prepare the way for the introduction of European ideals on society, government, and religion. It was the religious re-orientation which was most marked in the first few years following the civil wars; Christianity, which already flourished near the capital, spread widely into the provinces and grew in strength and number of adherents. A Native Church Council had been set up by the C.M.S. as early as 1885. In 1893 the first Anglican deacons, including Henry Wright Duta and Nikodemo Sebwato, were ordained. In the same year Anglican converts were sent out all over the country to set up reading houses, or 'synagogi'. In mid-1894 it was reported that there were 'nearly one hundred teachers out in the country, all supported by the Native Church',[2] The Roman Catholic missionaries reported a similar success and noted with some amazement the avidity of the Ganda for instructing and spreading their new religion among themselves.[3]

Outside the borders of Buganda, the Ganda were spreading their influence abroad in neighbouring countries. The civil wars and the wars of conquest carried Ganda refugees and armies far afield, and there were always colonies and centres of Ganda influence left behind when the armies withdrew. In Ankole, colonies of Ganda had been established since 1888. In Bunyoro, evangelists and traders followed in the wake of the victorious Ganda armies. The delay in declaring British protectorates over Busoga, Ankole, Toro, and Bunyoro was partly responsible for the early spawning of the influence of the Ganda. Although the original protectorate included only the kingdom of Buganda, British treaties were made and agents or garrisons established far distant from the British headquarters at Kampala. Administrative officers were sent to Busoga in 1893 and to Toro and Mumia's in Kavirondo in 1894. Treaties of protection in the name of the British Government were made with Kasagama of Toro, with the representative of Ntare of Ankole, and with Busoga chiefs.

---

[1] Oliver, *Missionary Factor*, pp. 183–4.
[2] Tucker, *Uganda and East Africa*, p. 307, quoting a letter from Roscoe.
[3] Oliver, *Missionary Factor*, pp. 186–7.

In Busoga, where the chiefs of the several decentralized states paid tribute to the Kabaka of Buganda through the Sekibobo, chief of neighbouring Kyagwe, the impact of events in Buganda was felt most strongly.[1] A treaty was made by Jackson with chief Wakoli as early as 1890, and Wakoli's village became the headquarters in Busoga for caravan traffic to Buganda. Caravan leaders intervened on more than one occasion in Busoga inter-state warfare, and in 1892, when Busoga became involved in the Buganda civil war, Williams led a Ganda military expedition into the country. The first permanent administrative station was established in Busoga in 1893. The absence of a paramount chief or king, and the persistence of inter-state disunity and warfare created administrative problems and made it difficult for the British agent to control Mwanga's tax-collectors, who continued to carry on their activities independently of the British Busoga administration.

In Ankole too the Ganda were beginning to make their influence felt.[2] During the Ganda civil wars the defeated Christians had collected in Ankole, and, after the departure of the army, a number of armed men were left behind. Some of these Ganda refugees settled in permanent colonies in Ankole, and others were formed into a band of mercenaries led by Nuwa Mbaguta, who later became Nganzi of Ankole. In 1894 a new treaty of protection was signed at the Mugabe Ntare's request by Mbaguta and a British representative. Ntare's superstitious fear of meeting a white man delayed the establishment of a British post in Ankole until after his death in 1895.

Of the tribal groups later absorbed into the Uganda Pro-tectorate, only the Kiga in the extreme south-west, and the Acholi, the Lango, the Teso and the Karamojong all to the north of Lake Kyoga, remained uninfluenced by the arrival of British administrators in Buganda. The British treaties[3] made in the north in 1894, with Chief Ali of Wadelai and with Kavalli and other chiefs west of Lake Albert, were designed to protect the British sphere against Belgian agents from the Congo State rather than to prepare the way for the establishment of British

[1] L. A. Fallers, *Bantu Bureaucracy*, 1956.
[2] F. Lukyn-Williams, 'Nuwa Mbaguta, Nganzi of Ankole', *Uganda J.* 10, 1946, pp. 124–35.
[3] H. B. Thomas, 'More Early Treaties in Uganda, 1891–96', *Uganda J.* 13, 1949, pp. 171–6.

rule. A group of Emin's former Sudanese troops, who remained in the region of Wadelai, continued sporadic raids east of the Nile, but otherwise the Acholi were left to themselves for nearly a decade after 1889. Clan and tribal conflict between the Padibe and the Payera and others continued, and its proportions and severity were increased by the Acholi possession of arms and ammunition obtained by trade with Bunyoro, with the Khartoum traders, and by trade or capture from Emin's Sudanese. Over a period of years the Payera were substantial victors. Awich, Rwot of Payera, consolidated his clan and conquered new territory until, by the late 1890's, the Payera lands extended from Kitgum to Pakwach and the Murchison Falls.[1]

While British administrators were making their first impact upon East Africa, in Britain the fate of the British sphere was still being debated in and out of Parliament. The Uganda Protectorate was at last formally declared in June 1894. The proclamation of the East Africa Protectorate followed the next year, in July 1895. These events passed almost unnoticed in East Africa, where the British administration was entering its seventh year.

In a few areas of East Africa, total and revolutionary changes had already been made in response to the European annexations. In most areas Africans continued to lead their traditional lives, influenced less by Europeans than by the weather, by sickness and famine, and by the shifting patterns of inter-tribal politics and warfare. To the west of Lake Victoria the European impact was profound; to the east of the lake it was slight, and scattered through a few isolated depots at the coast and along the caravan route into the interior. A map of British influence drawn in 1894 would show a thin red line stretching through the area of the East Africa Protectorate, and a much heavier colouring from a centre in Buganda, spreading out along an arc into southern Busoga, Ankole, Toro, and western Bunyoro.

The highly centralized kingdom of Buganda, already in process of revolutionary growth and change, was able to respond most effectively to the establishment of British rule. After initial opposition by Mwanga and the Catholic party, both groups of chiefs allied themselves with the British administration and, in

[1] Bere, *Uganda J.*, 1946; Gray, *Uganda J.*, 1952.

turn, their position of supremacy within the government of Buganda was confirmed. A system of indirect rule through the kabaka and chiefs was established by Lugard and reaffirmed by Portal and Colvile. The Ganda emerged from the first years of colonial rule in control of their own institutions, with enhanced prestige and opportunities to extend their influence abroad.

The ruler of Buganda's chief rival, Bunyoro, was meanwhile engaged in an attempt to extend the borders of his large, decentralized empire. A number of factors, notably the traditional hostility between Buganda and Bunyoro, and Kabarega's fear of the spread of Egyptian and European influence, placed Bunyoro in the disastrous position of opposition to the Ganda–British alliance. As a result, by the mid-1890's Bunyoro had suffered the loss of her recently regained empire in Toro, the conquest of her western provinces, and the alienation of territory to the south along her border with Buganda. The further conquest and pacification of Bunyoro remained a grave problem for future British administrators.

Although the Uganda Protectorate declared in 1894 included only the kingdom of Buganda, British influence had already been informally extended over a much wider area through conquest, the signature of treaties, and the establishment of stations. The expansion of the administration in Busoga, Ankole, Toro, Bunyoro and, later, Acholiland had still to be accomplished.

While the form of the wider area of the Uganda Protectorate was beginning to take shape, the future of the East Africa Protectorate remained unclear. The East Africa Company had established there no centre of administration to compare with the administration in Buganda. There had been no tribal response to the European annexations among the scattered, loosely organized peoples east of the lake to compare with the response of the Ganda. European contacts with Africans were made only at small outposts, at Machakos and in Kikuyu country, along the caravan route to the lake. Tribal readjustments to the decline of the Masai continued over a wide area, quite independently of the European administration. In the west, the rise of the aggressive, dominant Nandi created a centre of potential opposition to British rule.

In the eyes of British statesmen and administrators the East Africa Protectorate was still viewed primarily as a highway to

Uganda. Within a few years this almost untouched country was to become the scene of great activity, as the rails of the Uganda railway were laid from the Indian Ocean to Lake Victoria. The problem of paying for the railway, and the difficulties of governing the immense, geographically varied, politically heterogeneous and economically poor area of the East Africa Protectorate, were already, in 1895, giving rise to proposals for white settlement.

XII

THE GERMAN SPHERE
1884–98

G. S. P. FREEMAN-GRENVILLE

THIS chapter brings to a close the narrative of the pre-European history of what has now become Tanganyika. The various agreements concluded among the European powers between 1885 and 1891 for the first time set precise limits to the territories with which this history is concerned. But in the case of German East Africa they were limits which cut sharply across the lines of commercial and political intercourse whose development during the mid-nineteenth century was traced in an earlier chapter. In the west the powerful Arabs and their satellite chiefs of the Upper Congo, gradually cut off from their ivory markets in Zanzibar, were confronted directly with Leopold of Belgium's Congo Independent State; and although for a time Tippu Tip, the most prescient and statesmanlike of the Arab traders, used all his influence to preserve a *modus vivendi* between the two, the eventual upshot was war and Arab defeat. In the vicinity of Lake Nyasa the main Arab and Yao slavers, their lines of communication similarly cut, found themselves in the British or Portuguese spheres of influence, and reacted with varying degrees of hostility to the curtailment of their activities which this entailed: only in 1895 was Mlozi, the most formidable of the Arab slave traders in the region, defeated and hanged. And whereas in the north there existed, at all events in Buganda and the other Interlacustrine states, political structures which were solid and large enough to form the nucleus of government in the new era, in the German sphere of influence, shorn of the areas which had mainly nourished the slave and ivory trades, there were few political units that were either extensive or firmly established. The two largest, the Hehe kingdom in the southern highlands and the Nyamwezi empire of Mirambo, had both grown out of the special conditions prevailing in the past fifty years, and Mirambo's dominions had largely disintegrated on his death in 1884. Only the Hehe and associated tribes, under their able chief Mkwawa, preserved a substantial reputation for cohesion and military power.

The background of the German colonial movement, and Germany's role in the phase of partition, are described in more detail elsewhere. German interest in East Africa had first shown itself in the middle of the century, when German business houses were established in Zanzibar, and German missionaries and explorers—the former, however, under English missionary auspices—had been among the first of those to venture into the East African interior. By 1871 German trade had come to account for no less than one-quarter of the exports of Zanzibar. In 1873 a Society for the Exploration of Equatorial Africa was founded, followed in 1876 by a German national branch of Leopold II's International African Association. In 1880 these combined as the German African Society of Berlin, and in that and the following year sent out exploratory expeditions. In 1884 a further society, the *Gesellschaft für Deutsche Kolonisation* (the Society for German Colonization) was set up to take more instant measures; it was widely feared that other nations would have seized all the areas available for exploitation before Germany could secure a share. But for a long time the movement received no encouragement from the German government. When in 1870 Sayyid Barghash, through the German consul, had applied for German protection, Bismarck had not even troubled to reply. In relation to East Africa, it was not until 1885 that this attitude was abruptly reversed.

Chronologically the establishment of German East Africa falls into easily distinguished periods. Between 1885 and 1887 little more was done than to establish a series of fictitious treaties with allegedly sovereign chiefs. When, between 1888 and 1890, the Germans began to extend their influence over the coastal region, they were faced by an uncoordinated but desperate resistance, led, it is true, by an Arab, but supported by the mass of the peoples of the northern coast and the tribes of the immediate hinterland. It was not until after this rising had been subdued that any serious attention could be paid to the effective occupation of the interior, and this was only concluded in 1898 with the death of the noble Mkwawa, the chief of the Hehe, who committed suicide rather than see his people subject to a foreigner. The subjection of the Hehe was crucial, for although sporadic risings continued until the great blaze of the Maji-Maji rebellion of 1905–7, the Hehe themselves remained quiescent, acknowledging thereby that German rule was an established fact.

The first treaty-making expeditions were organized before the German government's support had been secured and indeed in the face of official disavowal. In September 1884 Carl Peters, a founder member of the *Gesellschaft für Deutsche Kolonisation*, set out from Trieste with Count Otto Pfeil, Karl Jühlke, and August Otto. For secrecy they were disguised as mechanics and travelled steerage under false names. None the less, when they reached Zanzibar on 4 November, they were met by a telegram from Bismarck warning them not to expect government protection in their enterprise. On 10 November they crossed to the copal port of Saadani, and marched into Uzigua up the Wami valley. By 17 December Peters had returned to Zanzibar with twelve alleged treaties with the chiefs of Uzigua, Usagara, Ukami, and Ungulu.[1] All these treaties followed a similar pattern, for which one example may serve. At the petty village of Msowero in Usagara the 'Sultan' Mangongo offered 'all his territory with all its civil and public appurtenances to Doctor Carl Peters as the representative of the Society for German Colonisation for exclusive and universal utilisation for German colonisation'. He declared that in no wise and at no time had he been dependent upon the Sultan of Zanzibar, and that the treaty was to be 'for all time'. Peters, for the Society, merely promised 'special attention' to the grantors in colonizing their territories.

Even had the African signatories been literate, the concept of the treaty was virtually unknown amongst Africans of the interior, and even more was the language employed. None of these persons were 'sultans' or sovereigns in the sense proposed by Peters: even had they been so, African customary law never treats land as a property and its possession is ordinarily

[1] For Peters's treaties, see A. Chéradame, *La Colonisation et les colonies allemandes*, Paris, 1905, p. 85.

corporate rather than vested in the rulers; and consequently it could not be ceded by them. It was moreover at least questionable whether the 'sultans' were not in some sense subject to Zanzibar, for there are numerous cases of matters being referred to Zanzibar for settlement from these areas. Such treaty-making was in short a gigantic and deliberate fraud.

By 7 February, treaties in hand, Peters reached Berlin. By 27 February he obtained from the Kaiser a *Schutzbrief* (charter of protection) accepting the suzerainty of the territories, placing them under his Imperial protection, and granting to the German Colonization Society, so long as it remained German, jurisdiction over the natives as well as German and other nationals in the treaty areas.[1] Peters thus received *carte blanche*: as compared with the royal charters accorded to British chartered companies, the document was remarkable in that it did not forbid or restrict the import of arms, ammunition, or alcohol, or the setting up of monopolies; it did not enjoin respect for native law, custom, or religion, nor even the abolition of slavery. On this basis, later in the year, the society was reconstituted as the *Deutsch Ost-Afrika Gesellschaft* (German East Africa Company), with a broader basis of capital support.

It was an omission of singular tactlessness that up to this moment these proceedings had never been made known to the Sultan of Zanzibar. On 25 April he was apprised by the German consul that Germany claimed the whole area lying across the main caravan route to Tabora and the Lakes. Now at Mamboya in Usagara Zanzibar had a military post, and it could not be pretended that there was no effective occupation. Sayyid Barghash therefore protested strongly to Berlin, at the same time dispatching General Mathews to occupy the Kilimanjaro area. He passed from place to place, raising the sultan's flag. He was shortly followed by Peters's comrade Jühlke. By mid-June the latter had concluded eight further treaties, in the Bondei country and the plains below the Usambaras, and up to and including Moshi, where the faithless Chaga chief Mandara had so recently acknowledged the sultan's flag. Save perhaps Mandara, not one of the African parties to those

---

[1] For a discussion of the sudden change in Bismarck's policy, see Chapter X above, pp. 364–6.

treaties was in reality a sovereign chief. But on 7 August the paper treaties were reinforced by the arrival of five German warships off Zanzibar. On 11 August the sultan received the answer that the Kaiser desired friendly relations and to negotiate a treaty, but that negotiations could not begin until the protest was withdrawn. Barghash had had previous experience of naval bombardment, and he could see from the palace that the decks were cleared for action. He acceded, and on 19 August likewise conceded Dar es Salaam to the Germans as a port, with control of the customs, on condition that it should not be fortified or garrisoned by them.

In 1862 Great Britain and France had signed a Declaration pledging themselves to maintain the independence of the Sultan of Zanzibar, and it was now agreed by these powers and Germany to appoint a tripartite commission to delimit the extent of the sultan's dominions. It began work in January 1886 and, in spite of the protests of the British representative, Colonel H. H. Kitchener, the Foreign Office and the Quai d'Orsay gave instructions, for reasons of European policy, that its finding be unanimous. These instructions paved the way for a German interpretation. In Kitchener's view the sultan's authority extended throughout the coastal belt up to forty miles inland, and along the main caravan routes and to within twenty-five miles of Tabora and Ujiji, as well as in the Kilimanjaro region recently visited by General Mathews. Nevertheless on 9 June 1886 a 'unanimous' agreement was reached; and, on 1 November 1886 an Anglo-German Agreement was signed, restricting the sultan's authority to a ten-mile coastal belt, and, so far as the present Tanganyika is concerned, ignoring the caravan routes and the claims near Kilimanjaro. It delimited a German sphere of influence from the Ruvuma to the Umba river (the present border with Kenya), and inland thence to Lake Jipe and thence again to Lake Victoria. Britain further undertook to support the German East Africa Company in negotiations for a lease of the customs at Dar es Salaam and Pangani, and even to promote a friendly settlement between Germany and the sultan in regard to the Kilimanjaro area. Sick, and indeed shortly to die, Sultan Barghash could not but concur. He prophesied that the German régime would be of no long duration, bitterly telling the British consul that, had Britain acceded to his earlier request for protection, he would not have lived to see his dominions thus carved up.

Thus, on 25 May 1887, Hauptmann Leue arrived by warship at Dar es Salaam with seven German assistants and a bodyguard of twelve Arabs provided by the sultan. He demanded as a residence the house of the Arab liwali, and was refused with chill firmness. According to tradition Leue then held a ceremony of *ubani*, the formal offering of a gift in return for the use of land, by which he purchased Dar es Salaam from the jumbe, while the liwali was withdrawn to Zanzibar.[1] He then took prompt steps to secure the allegiance of neighbouring villages; and was shortly followed by a party of Lutheran missionaries, who settled near the present Dar es Salaam Secretariat, and a party of Benedictines who established themselves not far from the present cathedral and also twelve miles inland at Pugu. The fate of both missions was shortly to be in the balance.

On 28 April 1888 a further agreement was made between the Germans and Zanzibar, by which the new Sultan Sayyid Khalifa conceded the administration of all his territories between the Umba and the Ruvuma, provided that it was carried out in his name, under his flag and subject to his sovereign rights. Sulaiman ibn Nasr al-Lemki, later Liwali of Dar es Salaam, was sent to all the ports to proclaim this new agreement. Almost immediately parties of Germans arrived to take over the ports and the customs throughout the whole length of the coast. In the words of the Tanga poet Hemedi bin Abdullah bin Said al-Buhriy:

'Kilwa na Dari-Salama	'At Kilwa and Dar es Salaam
kuna Wazungu nakama	There was a plague of Europeans;
mtu hapati kusema	There was no free speech:
nti wamezizuiya.'	They had throttled the country.'[2]

Indeed, apart from al-Lemki's tour, there had been no preparation for the German arrival, and the Germans took no trouble to make themselves welcome. Particular offence was given by their requests for houses. At Tanga some dogs, the property of Germans, entered the Friday mosque during Ramadan.

[1] C. Velten, *Prosa und Poesie der Suaheli*, 1907, pp. 292–3.
[2] Hemedi bin Abdallah bin Said al-Buhriy, *Utenzi wa Vita vya Wadachi kutamalaki Mrima* (ed. and trans. J. W. T. Allen), 1960, p. 32.

The Liwali and people protested to the sultan, but he was powerless and could only recommend peaceable and courteous conduct towards the Germans. There was a general fear among the Arabs of German interference with the slave, ivory, and rubber trades, which was shared by the Indians who financed them. Another principal grievance was the compulsory registration of land. Among the chiefs of the coast and the interior there was the fear of the loss of *hongo*, the toll paid by the caravans, which was a main source of revenue.

In this atmosphere, Zelewski arrived in August 1888 at Pangani, where he felled the mast on which the sultan's flag had flown, and set up that of the company in its stead: this closely resembled the German flag. This was clearly contrary to the agreement of 28 April, and gave immediate offence. There were similar incidents, which evoked similar anger, at Bagamoyo, Tanga, and Kilwa. At Pangani local hostility was so intense that the gunboat *Möwe* was obliged to withdraw. It was then that Abushiri ibn Salim al-Harthi, a member of an Arab family which for more than a century had intrigued against the Busaidi in Zanzibar, united public indignation with determination to resist by force. The spark lit at Pangani travelled up and down the coast like a bush fire, and into the interior to include Usambara, Usagara, and even the warlike Hehe near Iringa. General Mathews reported that 'the feeling is against all Europeans', but feeling was rather against both Germans and subsequently the British, for throughout, according to Swahili tradition, there were friendly relations between Abushiri and the French mission of the Holy Ghost Fathers at Bagamoyo. At Magila, near Tanga, it was only with some difficulty that Bishop Smythies and a number of missionaries of the Universities' Mission to Central Africa were saved through Abushiri's personal intervention. In Dar es Salaam the Lutheran missionary had to flee with his family; while on 13 January 1889 two German Benedictine brothers and a nun were murdered, two brothers escaping; three other brother and a nun were fortunate to be taken prisoner. They were later ransomed by the French Holy Ghost Fathers. On 17 January Abushiri's supporters mounted a night attack on the fort which the Germans had hurriedly constructed in Dar es Salaam; and on 25 January Abushiri was decisively defeated.

At Mikindani and Lindi the Germans were forced to evacuate, while at Kilwa Kivinje several thousand Yao tribesmen prevented a force from the *Möwe* from landing across the shallow beach to relieve the two German officials in the fort.

In such circumstances it was clear that the German company could not continue. There was the loss of German prestige to consider, and Bismarck was anxious lest the sultan be drawn into the matter. He found as a pretext for government intervention the putting down of the slave trade, which would certainly have increased had the revolt succeeded, and he obtained an agreement with Britain by which her ships joined those of Germany in a blockade of the whole coast. A single vessel from Italy gave the blockade an international character, while Portugal agreed to close the Mozambique coast to trade in arms and ammunition. At the same time Bismarck dispatched the distinguished explorer Hermann von Wissmann to take control as Imperial Commissioner, with a hastily recruited army of 600 Sudanese from Egypt; 50 Somalis, and 350 Zulus from Mozambique. He set up his headquarters at Bagamoyo, six miles from Abushiri's stockaded encampment at Sanganzeru, near Dunda, and at once repudiated the armistice that the Germans had agreed with Abushiri shortly before his arrival. Abushiri believed he had been tricked by an attempt to gain time. He was already desperate for men, arms, money, and supplies: in his despair he sent to Tanga for the astrologer-poet Hemedi bin Abdallah bin Said al-Buhriy. A search of the prophetic books presaged only disaster.[1] Wissmann attacked in May 1889, routing Abushiri's forces; but in spite of all Abushiri was not dismayed; he withdrew farther inland, recruiting some 5,000 to 6,000 Maviti, a warlike conglomeration of the Yao and other tribes which had been disturbed by the Ngoni invasions. In October 1889 he again marched on Bagamoyo, but was utterly routed by a force under Gravenreuth. He fled again but this time found no support. He was discovered at Kwa Mkoro, half naked and starving, and taken to Pangani, where he was hanged on 15 December 1889. After his collapse Wissmann easily regained possession of Kilwa, Lindi, and Mikindani, and within a year of his arrival restored some quietude to a restless and sullen coast.

[1] Hemedi al-Buhriy, *Vita vya Wadachi*, pp. 54 sqq.

It remains to consider the results of these events in terms of politics before proceeding to the establishment of German power inland. During the course of Abushiri's resistance the British and German governments had been seeking solutions to their many differences in various parts of the world. Agreement was reached on 1 July 1890, and Britain undertook to influence the sultan to cede to Germany all his remaining coastal possessions between the Umba and the Ruvuma. Three months earlier, on 1 April 1890, the German government had taken over direct control of the colony from the German East Africa Company. By the end of the year it had paid Sayyid Ali bin Said (who had now succeeded his brother Sayyid Khalifa as sultan) 4 million gold marks (£200,000) in return for the final cession of his mainland dominions lying between the German possessions and the sea. If the price was low, it could be said that Abushiri's rebellion had been most costly to Germany, that she still had to establish an administration and introduce services into her new possession, that her actions so far had aroused sullen and hostile suspicion, and that what she had bought the right to exploit and administer, even the coast and the caravan routes and settlements, had only been loosely administered by Zanzibar. The company had been a costly failure, both in money and in prestige. If Peters had secured treaties, the company had failed to maintain law and order or to show any profit. In Dar es Salaam it had done no more than erect a petty plant for the washing and sorting of copal. It had not succeeded in developing plantations or any kind of trade. It had provoked an armed rebellion. It had, moreover, done great damage to the peaceful commerce of the Arab and Indian traders who had multiplied greatly since the beginning of the century. Germany had thus reached a position from which she could not retract, and the costly suppression of Abushiri's rebellion emphasized that a more co-ordinated and coherent means of colonization was needed. The appointment of Wissmann as Imperial Commissioner meant a change of policy at the centre, in that Bismarck now accepted direct responsibility for the future of the colony. It is clear that the company could have made no progress until Abushiri had been disposed of, and it would also seem clear that the direct intervention of the German government could alone have effected the occupation of the coast and the interior.

The occupation of the interior proceeded more by exploration than by any planned expansion from a centre, and the comparative lack of communications and the extreme political fragmentation of the territory made it most difficult to bring under control. The differing political traditions of the tribes have been described in previous chapters, varying from organized kingship to separate and often petty clan leaders.[1]

There had already been some penetration of the interior before the Abushiri rebellion. When Peters returned to Germany in February 1885, he had left behind Count Pfeil who, by June, had travelled through Usagara, Uhehe, Mahenge, and Ubena as far west as Lake Nyasa and south to the Ruvuma, securing a whole series of treaties. In August 1885 Lieutenant Rochus Schmidt secured no less than twenty-one treaties in as many days in Uzigua, Ukwere, Ukami, and the Rufiji area. Farther inland, however, the tribes continued to do much as they liked. In Tabora in 1885 Siki, Mirambo's successor, had an efficiently organized army. By 1886 Giesecke, a representative of a Hamburg firm, had set himself up there as a trader, and in September the Russian explorer Junker passed through and learnt that Siki had murdered him and seized all his goods. In 1889 Siki forced the withdrawal of the White Fathers Mission from Kipalapala, apparently as a reprisal for the German occupation of the coast. It was plain that this area demanded priority of attention.

In 1890 von Soden was appointed Governor, and Wissmann, who had reasonably hoped to be given the post, was sent to take possession of the extreme south-west of the territory. A large part of his expedition passed up the Zambezi through Nyasaland, carrying even a gun-boat in sections, which was launched on Lake Nyasa. He built two remote and permanent fortified *bomas* at Langenburg and Bismarckburg from which the country could be held. The reduction of Tabora and the country beyond was reserved to Emin Pasha, who had joined the German service after his successful 'relief' by H. M. Stanley and had enjoyed a period of rest at Bagamoyo.[2] Emin Pasha's

[1] See Chapters VIII and IX above.
[2] For an account of the international developments leading up to the 'relief' or 'rescue' of Emin Pasha from the Sudan Equatoria Province, see Chapter X, pp. 376-7, and Chapter XI, pp. 403-6, above. Emin fell off the balcony after a celebration dinner at a house still to be seen at Bagamoyo. Local tradition that he attempted suicide cannot be substantiated.

expedition of over a thousand persons was accompanied by Wilhelm Langheld and by Franz Stuhlmann, who was later to achieve distinction not only as an administrator but also as an agriculturist, botanist, zoologist, and ethnographer. Emin passed through Morogoro to Mpwapwa, which at this time was the farthest German station inland. In Ugogo he met with some opposition, which he disciplined with the plundering and burning of villages, and was able to report that 'now quiet reigns everywhere'. He then reached Tabora, where he reinstalled the Zanzibari Arab liwali as the liwali of the German government, and forced Siki to deliver up the murdered Giesecke's property. He failed, however, to bring Siki's forces to battle or to destroy them. Siki continued to be a thorn in the German flesh for several more years. In 1892 his son attacked a German column, for which his headquarters, though not his army, were destroyed. It was then learnt that he intended to assemble a great army to destroy all the Europeans in the country, and he almost succeeded in wiping out a column under von Prince. On 9 January 1893 a strengthened column attacked his fort, and after two days' fighting stormed it. Siki blew himself up in the magazine, an act which saved the Germans further trouble and which reopened the main caravan route in this area for the first time in six years.

Emin then proceeded to Lake Victoria and, reaching Bukumbi mission on the shores of Speke Sound, took the opportunity to disperse Arab slave traders and to contract a number of further treaties with neighbouring chiefs. He next pressed on by canoe across the lake, and reached Bukoba on 30 October, being warmly welcomed by the Haya, who complained that they were troubled by raiders who had crossed the lake from Uganda. In Bukoba he was joined by Stuhlmann, who had had some trouble in Usambiro through which he passed along the south side of the lake. By February 1891 Emin was in treaty relations with many of the Bukoba chiefs. Then, without orders, he set out through Karagwe for Uganda. Although he had been apprised of the definition of the Anglo-German boundary in September, it would seem that he was determined to extend the German possessions still farther. On 5 April 1891 he received a letter of recall which he disregarded, pushing on into the Congo. Here, late in October 1892, he was murdered.

Langheld had been left behind at Tabora to deal with raiding Ngoni. He then came up and made himself active on both sides of the lake, and in 1892 set up a new station at Mwanza under Sergeant-Major Hoffmann. There they were later joined by Stuhlmann, and it is no mean tribute to their work that they had no trouble with the Sukuma, but, on the contrary, were welcomed as deliverers from the Arabs and the slave trade.

In sharp contrast to these men was the story of Peters. He had spent 1886–7 in Zanzibar, in charge of the organization of the company. He then returned to Germany, but came back again in 1888, nominally as the head of the German Emin Pasha Relief Expedition, but in reality to see if he could not seize Uganda and parts of the Congo for Germany. He did not meet Emin Pasha until 1890, when the latter had already been relieved by H. M. Stanley and was already on his upward journey to the lake at Mpwapwa. A good example of his, and of much German, behaviour during this epoch is to be found in Peters's *New Light on Dark Africa*.

Having left Uganda Peters first stopped at Bukumbi, where the White Fathers had set up their earliest mission in East Africa in 1878. Here he stayed and engaged in more treaty-making with the Sukuma chiefs, amongst whom one, a woman, he never so much as saw. She was unable to receive him, so she stated, because she was drunk. Nevertheless her bare word of agreement to a treaty was good enough for Peters. From Bukumbi he went on to Mpwapwa, where near by a Gogo 'grinned impudently' at him as he ate breakfast. For this he was given 'a lesson with the hippopotamus-hide whip'. It then turned out that the offender was the son of the local sultan, come on a friendly mission to pay his father's respects. Not unnaturally the Gogo gathered for war. The sultan made two overtures for peace, to which Peters replied: 'The Sultan shall have peace. It shall be eternal peace. I will show the Wagogo what the Germans are. . . . Plunder the village, set fire to the houses, and smash everything to pieces that will not burn!' Nor was this all. He sent to a neighbouring Nyamwezi caravan: 'Come and help us. If we capture the herds of the Wagogo you shall have a share in the booty.' He then advanced on other villages. 'Everywhere the same spectacle was repeated.

After a short resistance the Gogo fled in all directions; burning brands were thrown into the houses, and the axes did their work in hewing to pieces what could not be burnt. By half-past four, twelve villages were thus burnt down.'[1]

Peters had indeed shown the Gogo what the Germans were. The Chaga had yet to learn. After the Emin Pasha relief expedition he first returned to Germany, where he was warmly fêted, and sent back in 1891 as temporary Imperial Commissioner in the Kilimanjaro District. Here, after some fighting, Wissmann had already persuaded Mandara, the principal chief, to accept German rule; but shortly afterwards Mandara died, and was succeeded by his son Meli, who was less inclined to the Germans. The Gogo precedents were followed, even women and children were flogged, villages burnt and plundered; and murder was done even to the point where an African youth was hanged, allegedly for stealing cigarettes, in truth—according to one of the women who fled to a neighbouring chief—for interference with Peters's African concubines. She too was caught, flogged, and finally hanged. It was inevitable that conduct of this sort should become known. In 1893 Peters was recalled to Germany, and, after three separate investigations which took four years to complete, he was deprived of his commission for 'misuse of official power'. But it seems that official as well as public memory was short. In 1906 he was reinstated, and later a statue was erected in his honour in Dar es Salaam as pioneer and architect of the new colony.

On Kilimanjaro Peters was succeeded by von Bülow, whose cousin was later to become Chancellor of Germany and who was a warm admirer of Peters. In 1892 a marauding German soldier was murdered, and Bülow thereon determined to attack Meli. He was utterly routed, and a further expedition had to be sent under Captain Johannes. Johannes brought peace—Bishop Tucker praised him as a just and sympathetic administrator—but he was also responsible for a large-scale alienation of land to German settlers which later led to a grievous land hunger on the lower slopes of the mountain.

In these ways the west, the centre, and the north of the territory were gained and consolidated. There were several minor expeditions against the Gogo, who were not unnaturally

[1] C. Peters, *New Light on Dark Africa*, 1891, pp. 512, 522, 527, 529.

resentful at the loss of the *hongo* for which they had so mercilessly pestered Burton, Speke, and other travellers. A number of other expeditions secured the south. In 1894 there was a fresh Arab revolt in the Kilwa hinterland under Hasan bin Omari, while in 1895 a Yao, Machemba, led a revolt further south against the newly imposed hut tax. Both of these were quelled without serious difficulty. The main focus of opposition to Germany in the south was Mkwawa, the Hehe leader whose headquarters were near Iringa. After his defeat of the Ngoni in 1882, he had attacked the Nyamwezi, the Gogo, and the Sagara, and had made himself ruler from what is now the Central Line railway to more than a hundred miles south of Iringa. Throughout this area, including the road to Tabora, he raided the caravan trade, and had relations with the ports of Lindi and Sudi, far to the east of his domains. In 1888 he allied himself with Abushiri, but afforded him little help. He continued his attacks on caravans and, in 1891, when Zelewski was sent against him with a column, almost destroyed it. A doubled force was sent against him in 1894 led by the governor, von Schele, in person. von Schele destroyed Mkwawa's fort at Kalenga, but Mkwawa himself escaped to maintain his opposition. So long as Mkwawa lived, it was clear that there would be no parley. The Germans accordingly set up a new station at Iringa under Tom von Prince. He waged a constant warfare against Mkwawa, who displayed great tactical skill and personal courage. At length Mkwawa was worn down and his forces dispersed. On 19 July 1898 he was found by Sergeant Merkl, who fired into his body. It was already cold. Some hours before Mkwawa, seeing the end, had taken his own life.

With this event Germany could at length be said to be in effective control. It was no wonder, as Magdalene von Prince wrote, that there was 'frenzied jubilation in Iringa'. In 1905 and 1906 the Germans had to face a massive revolt in the central and southern coastal areas in the Maji-Maji rebellion; so effectively had the work of von Prince and others been done in the interior that the Hehe hardly stirred nor indeed was the farther hinterland affected.

It is a matter for astonishment with what small forces and what men this conquest was achieved. In 1887 the personnel of the German East Africa Company had numbered a mere fifty, of whom a third were in Zanzibar, a third on the coast, and only a third inland. Not one of these persons seemed suitable for the task in hand: such independent observers as the trader O'Swald, Admiral Knorr, and Arendt, the German consul-general, describe the company personnel as misfits, throw-outs, gangsters, freebooters, swashbucklers, and bullies, wholly unfitted for the tasks either of colonization or of the betterment of the peoples under their control.[1] During 1885–7 the company set up ten stations on the mainland, all of which Pfeil was to describe as 'a costly failure'. No wonder. Many of these people remained when the German government took direct control, and the new administrators came exclusively from the army, not a few from the ranks. When Wissmann was ordered to quell Abushiri, 1500 German officers volunteered, but he was only permitted to select twenty-five. If among them there were a small number of men like Tom von Prince, who confessed to nausea when he thought of the prospect of the slaughter of a less well armed enemy, the scientifically minded Wissmann, or the scholarly Stuhlmann, there were others like Bülow, a character similar to Peters, and Zelewski, whom the Swahili named *Nyundo*, 'the hammer'. Many of these officers and non-commissioned officers simply sought to escape the boredom of drills and barracks. Their attitude was summed up by Eduard von Liebert, who became governor in 1896: 'It was impossible in Africa to get on without cruelty.' It would be tedious to list their brutalities, the liberal application of the *kiboko*, the whip of hippopotamus-hide, and the frequency with which the death sentence was lightly awarded.

With the appointment of von Soden as governor in 1890 some semblance of order commenced in central government. In 1891 the first financial secretary was appointed, and shortly afterwards departments of justice, health, public works, and agriculture were set up, the last under Stuhlmann, who was largely responsible for the elegant planning of Dar es Salaam. By 1896 the *Kolonialblatt*[2] reported a German strength in Dar es Salaam of 37, together with 118 coloured persons, mostly

[1] F. F. Müller, *Deutschland—Zanzibar—Ostafrika*, 1959, pp. 223, 225–6, 227: 'Den Ton in der Kolonie geben die brutalen Konquistadoren und die arroganten Herrenmenschen an.'   [2] Ibid., pp. 765–7.

Sudanese or Zulu soldiery. Administratively the colony was divided into fourteen stations, at which there were the following strengths, divided into German and coloured personnel: Moshi: 5: 100; Kilwa Kivinje: 6: 106; Marangu: 1: 21; Iringa: 12: 257; Kilimatinde: 7: 84; Mpwapwa: 4: 60; Perondo: 1: not given; Pangani: 6: 87; Lindi: 7: 121; Ujiji: 8: 133; Tabora: 7: 119; Bukoba: 8: 122; Mwanza: 5: 72; Kilosa: 5: 48; others (mostly on expeditions): 7: 112; total: Germans: 126; coloured: 1,560. Considering the vast area of 370,000 square miles, and the fact that the Germans and their Sudanese, Somali, and Zulu troops in most places suffered incessant ill-health from malaria and other diseases, these were small numbers indeed. Each station was commanded by a semi-military commissioner known as Bezirksamtmann, to whom were directly responsible liwalis and akidas, with village headmen, jumbes, as the lowest stratum, or, in certain areas, hereditary chiefs. Thus in Bukoba, among the Nyamwezi and the Sukuma, on Kilimanjaro and in the Usambaras and a number of other areas, the hereditary rulers were recognized as far as their loyalty made it practicable to do so; in the areas where hereditary rulers had not been the custom and where a tribe was no more than a headless mass of clans the Germans had perforce to adapt the system which Zanzibar had employed in the coastal region before their advent. This was hardly a matter of choice, for the Swahili were the only literates available for employment in any trusted capacity. This local government was responsible for the maintenance of law and order with the assistance of the small bodies of troops under the Bezirksamtmann, and later for the provision of labour for public works, and, after 1896, for the collection of the hut tax. The Swahili were thus spread throughout the country, not only as local government officials and interpreters, but also as foremen, artisans, and personal servants. The tribes of the interior spoke not less than some one hundred different languages, some widely spoken like those of the Sukuma and the Nyamwezi, others only by very small groups. The spread of the Swahili was a potent factor in making their language widely spoken as that of general communication and in giving the country the beginnings of unity, and not less in the conversion of a number of areas to Islam. Before this period the Swahili language and Islam had scarcely penetrated the interior.

Not only were the German personnel and their auxiliaries small in number, but their financial resources were extremely limited. In 1891–2 the revenue of the colony was only 1,458,000 marks, or slightly less than £73,000. Expenditure was 3,409,000 marks, so that more than half had to be met by a grant-in-aid from Germany. By 1895–6 the expenditure had risen to 32,327,000 marks (£1,616,350), of which no less than 26,111,400 marks (£1,305,570) represented the grant. The revenue had thus reached 6,215,600 marks (£310,780), or more than four times the figure of five years previously. These figures illustrate a remarkable development of commerce. In the earliest stage the Germans could do no more than seize control of and impose customs on existing trade, much of which had been in slaves and ivory. As to the first, they were nervous of the consequences of wholesale emancipation, and no more was done in 1890 than to forbid the trade and to make the process of emancipation easier. At the same time the ivory trade had declined sharply, partly because the greatest source of supply was cut off in the Congo, partly on account of the indiscriminate shooting of elephant that had taken place in the colony in preceding years. The early 1890's witnessed a number of experiments in the planting of coffee, cotton, tobacco, tea, vanilla, and sisal, of which the first bulbils were obtained from Florida. The last especially flourished, and soon sisal plantations had spread up and down the coast, and inland from Tanga and Dar es Salaam. From the experiments of this time the chief sources of the colony's future wealth derived, and there was a considerable development also in the native rubber and copra. Such development was, of necessity, a slow business since the plantations required not only capital but men of experience in tropical agriculture. Nevertheless by 1893 prospects seemed good enough for a private company to venture to undertake to build a railway from Tanga inland. It failed for lack of funds, and the matter was only revived under government auspices in 1902. As compared with the vast area of the country as a whole, these enterprises were small indeed, and even by the end of the German period a mere 3,515 square miles of land had been alienated to German settlement, less than 1 per cent. of that available. A decree of 1895 had made all land Crown Land, but the alienations were most numerous on the lower slopes of Kilimanjaro and in the region of Tanga.

Development was hampered, moreover, at the beginning by the slowness and inefficiency of communications. The earliest German colonists had perforce hugged the coast and the caravan routes, on which last from time immemorial human porters had been the only means of transport. A number of mules and donkeys had been imported in 1897, but their use was largely frustrated by the tsetse fly. The construction of bridges and roads began in 1894: in the north direct roads started from Tanga to Kilimanjaro and from Pangani through Mgera to Tabora: a fan of roads linked Saadani, Bagamoyo, and Dar es Salaam with Mpwapwa, and thence to Tabora and on to Ujiji and to Mwanza. This great central road branched south at Kilosa for Mahenge and Iringa, and on to Songea and to Lake Nyasa, which could also be reached by a direct route from the coast roughly parallel to the Ruvuma from Lindi to Songea. There were a number of subsidiary roads.

Seeing that all this had taken place since the Abushiri rebellion, it was a remarkable achievement. It is true that it had been necessary to introduce an arbitrarily imposed *corvée*, but this was by no means abnormal considering the difficulties of introducing a cash economy. After the collapse of Abushiri many Indian traders who had fled returned to the colony and played an important part in developing trade. The Germans gave them every encouragement. When Caprivi succeeded Bismarck, he told the Reichstag: 'We want them; they have connections with the interior of Africa and we should not be in a position to replace them. We shall strive to make use of their strength.'[1] If the Indians complained that taxation was higher in German East Africa than in neighbouring British colonies, they were certainly making a better living than they would have done in India. As to German settlers, they were receiving every possible encouragement.

One section of the population deserves especial mention, since they, the missionaries, more than any at this period, represented the civilizing mission of the west. The arrival of the French Holy Ghost Fathers and White Fathers, and the

[1] No. 15, F.O. 84/2128, quoted in L. W. Hollingsworth, *The Asians of East Africa*, London, 1960, p. 62.

Anglican Universities' Mission to Central Africa and Church Missionary Society had antedated the scramble for Africa. The year 1886 saw the foundation of the Lutheran Berlin Evangelical Missionary Society for East Africa: it sent its first pastors to Dar es Salaam in 1889. The importance of preparation for the mission field was emphasized by the creation of a chair of Swahili in Berlin in 1887, of which the Lutheran missionary C. G. Büttner was the first occupant. In 1889 the Benedictine Congregation of St. Ottilien was assigned the whole area lying south of a line drawn directly west of Dar es Salaam, the northern sector being assigned to the Holy Ghost Fathers as far as the boundaries of the White Fathers near the Lakes. This Benedictine Congregation was largely financed by the Freiherr von Gravenreuth, a leading member of the German East Africa Company. In 1891 the Protestants were reinforced by the Moravians, who agreed with the Lutherans to a separate sector on the line Tabora–Lake Rukwa–Tukuyu. The White Fathers were already established near Tabora, but there was no question of setting up separate Catholic and Protestant areas. In 1893 in the Kilimanjaro area, however, it seems that the Lutherans had come to regard the Anglican Church Missionary Society as a threat to German national prestige, and succeeded in elbowing them out of Moshi. Likewise in the Usambaras they placed great difficulties in the path of the Holy Ghost Fathers, and these were only overcome by the establishment at Gare of a Trappist monastery.[1]

Apart from these incidents there was little friction; and not only were the seeds of Christianity planted but, especially at Bagamoyo and in the Benedictine missions—which shortly stretched from Lindi on the coast to Peramiho within a hundred miles of Lake Nyasa—both industrial and agricultural skills as well as the rudiments of reading and writing were soon taught. In the Zanzibari period there had only been some casual Koran schools on the coast—of which the only building survives at Bweni Kuu, opposite Pangani: and in this period the German government neither inherited nor attempted to set up an educational system. But soon there was no mission without some sort of

[1] This mission has now been transferred to the Institute of Charity (Rosminian Fathers), and I am grateful to the present Superior, the Very Rev. Fr. F. Kennedy, for this information.

school, the better offering four, or even six, years of education. Many missions, moreover, offered medical facilities, especially where women's orders had been introduced. The first of these were the Filles de Marie from Mauritius. The missions likewise achieved much in the settlement of freed slaves, many of whom they purchased so that they might obtain freedom. Some of these, bereft of their masters, would have had nothing had not the missions provided food, roof, and employment. The Benedictine Mission at Kurasini, on the outskirts of Dar es Salaam, was especially constructed with this end in view. The White Fathers in Bukoba and the Holy Ghost Fathers on Kilimanjaro introduced the cultivation of coffee in those regions. Irrespective of denomination, the Christian missions provided a counter to the spread of Islam and of the Swahili culture, even if shortly they all came to use the Swahili language. As intermediaries between the German colonists and the African they provided a link of a type that the African could comprehend; and the fact that they evangelized to a great extent through the medium of education enabled the Africans to see some of the practical advantages of German administration. To these early missionaries Africa owes an untellable debt.

The poem of Hemedi bin Abdallah bin Said al-Buhriy, which was quoted in the early part of this chapter, and which was completed in 1891, is one of the very few written documents recovered so far which illustrate indigenous opinion at this period. Such opinion was demonstrated rather by acts, by the resistance of Abushiri, of Mkwawa, of Siki and their followers, a resistance which was not solely that of self-interest. From the time of Peters's arrival to the defeat of Mkwawa there is a stretch of nearly fourteen years. It is not possible to trace the change of opinion from resistance to passive acceptance, and from that to the beginnings of co-operation. But such changes were steadily taking place. It is too easy to exaggerate the brutalities of Peters and others. In Germany itself there were grave misgivings which were to find expression in the radical changes of attitude and policy that followed the Maji-Maji rebellion. It must be remembered that it was not only in German colonies that the majority found alien rulers scarcely tolerable and revolted when circumstances allowed. Only the few to whom the arrival of the colonists brought new avenues of employment or other immediate advantages could find the Germans welcome. Such feelings are strongly evidenced in a traditional history of Lindi written down at the behest of Wissmann's interpreter Velten.[1] The acceptance of alien rule meant the acceptance of much that was immediately unpleasing. Nevertheless it was the essential prelude to much eventual good that only the discipline of tutelage could bring.

[1] Velten, *Prosa und Poesie*, pp. 265 sqq., especially pp. 277–8.

EPILOGUE

ROLAND OLIVER

From the outside viewpoint, especially from that of the colonial powers, the definite assumption of full political and financial responsibility by the British and German governments for their possessions in East Africa must inevitably seem the turning-point between one period and another. A young colonial administration might in practice be a factor of very small importance in the country it was supposed to rule. It might have no contact at all with three-quarters of the area within its frontiers, and exercise only the shadowiest kind of suzerainty over the remaining quarter. It was nevertheless, from the metropolitan point of view, an entity which had to be supplied with officials and a grant-in-aid; and even if the officials numbered a dozen and the grant-in-aid barely reached five figures, there would still be a host of matters, political and legal, financial and military, requiring reference to Europe and the attention of an increasing circle of metropolitan bureaucrats and legislators. On however small a scale, the colonial period had begun.

Viewed from the inside, however, from the standpoint of the East African peoples themselves, the formal beginning of the colonial period was not a fact of much significance. Colonial rule did not become effective at the stroke of a pen. It spread its influence slowly outwards from a few growing-points, negotiating here, conquering there, very conscious of its military weakness and its financial dependence, never, if it could be avoided, taking more than a single bite at a time. At the initial growing-points, at Zanzibar or Kampala, at Bagamoyo or Mombasa, para-colonial influences had been active for some time before the formal start of the colonial period. On the other hand, even towards the end of that period, there were peripheral areas, like Karamoja and Turkana and the Northern Frontier District of Kenya, where colonial rule was still a very marginal influence on the lives of the people who lived there. Viewed from the inside, colonial rule was not so much a static condition as a moving frontier of western influences, very difficult to distinguish from the other western influences which were not projected into Africa by European governments.

There is indeed a very real sense in which the period of colonial rule was only the last and most intensive stage in the opening-up of Africa to itself and to the outside world. If most of Africa had been, from mesolithic times onwards, a backward continent, it was not because the people who lived there were less intelligent or otherwise less gifted than others, but because their natural environment was more difficult. Its biggest difficulty, and in East Africa bigger than almost anywhere else, was the difficulty of internal communication, the absence alike of navigable rivers and of beasts of burden. Although the opportunities for sheer material subsistence were not by any means unfavourable, the restriction of commerce to what could be carried on the human head had resulted in a terrible narrowness of mental horizons, a real poverty in the currency of human ideas and inventions. It is true that the isolation had never been total. The knowledge of iron-working had spread to the remotest corners of the land. Ideas of kingship and social organization had travelled with migrant groups and evolved along a hundred separate lines of local development. The cultivation of Indonesian and, later, of South American food-crops had spread from tribe to tribe into the very heart of the continent.

Yet the isolation, if not absolute, was very severe. The proof lies in the sharp contrast between the East Africans of the narrow coastal belt and those of the wide interior. Thanks to the Indian Ocean seaways, the Swahili had access to a universal, monotheistic religion. When they prayed, they turned towards a holy city a thousand miles away and in another continent. They knew of the brotherhood of Islam, if not yet of the brotherhood of man. Even their illiterate masses knew of the existence of writing and its use for transmitting a deposit of learning and culture through the generations. They knew by sight the men of other lands. They saw their ships. Even the humblest had their loins girt with imported cottons. Their rich men sat on Persian carpets and supped off celadon.

All this was hidden from the men of the interior until, in the early nineteenth century, Islam made its break-through, only just in time to be followed by Christendom. Despite the untold sufferings of individuals caught in the web of slavery, many East African peoples made great strides during the middle years of the century in learning from the outside contacts that came to them. If one considers the Buganda of Suna and Mutesa I, one is bound to ask how much farther the process of response to outside challenge might not have gone, even without the ultimate challenge of colonial intervention. Certainly it could have gone much farther than in the event it had the chance to do. Yet the mid-nineteenth century opening-up of East Africa was, in its economic aspect at least, not a sound and natural development but a *tour de force*. In 1890, as in 1800, every article moved into or out of the East African interior was carried by human porterage, at six or seven shillings a ton-mile. With the stoppage of the slave trade, the only viable export of the interior was ivory, by now a rapidly waning asset. Only cheap mechanical transport could overcome the defect, and that required heavy capital outlay which in its turn posed inescapable political problems. Given another fifty years of political independence, the spiritual and educational revolution in Buganda might conceivably have gone on, though probably not nearly as fast as it was to do under colonial rule; but there could never have been a railway from Lake Victoria to the sea, or the cotton and coffee industries on which the foundations of later economic development were laid. For the integration of East Africa with the general progress of mankind in the world outside, a drastic simplification of the old political diversity was an inescapable necessity. It was a problem which, judging by historical precedent, only a period of colonial tutelage could solve.

SELECT BIBLIOGRAPHY

CHAPTER I

The East African Environment [1]

BUXTON, P. A.: *The Natural History of Tsetse Flies: an account of the genus Glossina* (Diptera), London, 1955.
COLE, S.: *The Prehistory of East Africa*, Harmondsworth, 1954 (Chapter 2).
COOKE, H. B. S.: 'Observations relating to Quaternary Environments in East and Southern Africa', *Geological Society of South Africa*, Annexure to vol. 60, 1957.
DIXEY, F.: *The East African Rift System* (Colonial Geology and Mineral Resources Supplement Series), London, 1956.
EAST AFRICA ROYAL COMMISSION, 1953–55. *Report*, Cmd. 9475, H.M.S.O., London, 1955 (Chapters 2, 20, and 21).
EDWARDS, D. C.: 'The Ecological Regions of Kenya: their classification in relation to agricultural development', *Empire Journal of Experimental Agriculture*, 24, 1956, pp. 89–108.
GILLMAN, C.: 'A Vegetation-types Map of Tanganyika Territory', *Geographical Review*, 39, 1949, pp. 7–37.
KENDREW, W. G.: *The Climates of the Continents*, 5th edn., Oxford, 1961 (Chapter 11).
MATHESON, J. K., and BOVILL, E. W. (eds.): *East African Agriculture*, London, 1950 (Chapter 1, Section 2).
MILNE, G. et al.: *A Provisional Soil Map of East Africa (Kenya, Uganda, Tanganyika and Zanzibar) with Explanatory Memoir*, London, 1936.
MOFFETT, J. P. (ed.): *Tanganyika: a review of its resources and their development*, Government of Tanganyika, 1955 (Chapters 1 and 2).
NILSSON, E.: 'Ancient Changes of Climate in British East Africa and Abyssinia: a study of ancient lakes and glaciers', *Geografiska Annaler*, 22, 1940, pp. 1–79.
—— 'Pleistocene Climatic Changes in East Africa', *Proceedings of the First Pan-African Congress on Prehistory, Nairobi, 1947*, Oxford, 1952, pp. 45–55.
SANSOM, H. W.: *The Climate of East Africa, based on Thornthwaite's Classification* [East African Meteorological Department, Memoirs, Vol. 3, no. 2], Nairobi, 1954.
SCHNELL, R.: *Plantes alimentaires et vie agricole de l'Afrique noire: essai de phytogéographie alimentaire*, Paris, 1957 (Chapters 3 and 4).
SNOWDEN, J. D.: *The Grass Communities and Mountain Vegetation of Uganda*, London, 1953.
TOTHILL, J. D. (ed.): *Agriculture in Uganda*, Oxford, 1940 (Section 2).
WAYLAND, E. J.: 'Rifts, Rivers, Rains and Early Man in Uganda', *Journal of the Royal Anthropological Institute*, 64, 1934, pp. 333–52.

[1] RUSSELL, E. W.: *The Natural Resources of East Africa*, Nairobi, 1962, is an additional reference. It contains 7 maps on the scale of 1 : 4,000,000.

SELECT BIBLIOGRAPHY

NOTE ON MAPS

(a) Topographical

Central and East Africa	1:4,000,000	Bartholomew, Edinburgh, n.d.
Africa	1:2,000,000	Kenya and Uganda, Series 2201; Tanganyika. G.S.G.S. 2871.
Africa	1:1,000,000	Uganda; Marsabit; Lake Victoria; Mombasa–Nairobi; Tabora; Dar es Salaam; Lake Nyasa; Lindi. G.S.G.S. 2465, and later series.
East Africa	1:500,000	G.S.G.S. 4355 and E.A.F. Series, 1940– .
Kenya	1:1,000,000	In two special sheets. S. K. 41, 1961.
Uganda	1:1,000,000	1963, Series 1301.
Tanganyika	1:2,000,000	5th edn., 1963.
Zanzibar	1:63,360	In two sheets. G.S.G.S. Y 744, 2nd edn., 1959.
Pemba Island	1:63,360	In two sheets. G.S.G.S. 2639, 1942.

Topographical series, mainly on the scales of 1:250,000 and 1:50,000, published by the D.O.S. and the East African Surveys, are available for each of the mainland countries; but these do not provide a complete coverage. There are large-scale maps of the main towns of East Africa.

(b) Other maps

Geological map of East Africa	1:2,000,000	1954
Mean annual rainfall map of East Africa	1:2,000,000	In two sheets, 1959.
East Africa Royal Commission 1953–5, 1955:		
On the scale of	1:3,000,000	Map 1, East Africa. General, showing alienated lands. Maps 2 and 3, East Africa. Population, tsetse-fly, and rainfall (prospects of obtaining 20 inches and 30 inches respectively).
On the scale of	1:2,000,000	Map 4, Kenya boundaries, land units, population, tsetse-fly, and rainfall (prospect of obtaining 20 inches).
Atlas of Tanganyika	1:3,000,000	3rd edn., 1956.
Atlas of Kenya	1:3,000,000	1962 edn.
Atlas of Uganda	1:1,500,000	1st edn., 1962.

CHAPTER II

The Stone Age of East Africa

AFRICAN PREHISTORY—GENERAL

ALIMEN, H.: *The Prehistory of Africa* (trans. A. H. Brodrick), London, 1957.
ARKELL, A. J.: *Early Khartoum*, London, 1949.
—— *Shaheinab*, London, 1953.
BALOUT, L. (ed.): *Actes IIe Congrès Panafricain de Préhistoire, Alger, 1952*, Paris, 1955.
CATON-THOMPSON, G., and GARDNER, E. W.: *The Desert Fayum*, 2 vols., London, 1934.
CLARK, J. D.: *The Stone Age Cultures of Northern Rhodesia*, Claremont, Cape, 1950.
—— *The Prehistoric Cultures of the Horn of Africa*, Cambridge, 1954.
—— (ed.): *Proceedings of the 3rd Pan-African Congress on Prehistory, Livingstone, 1955*, London, 1957.
—— *The Prehistory of Southern Africa*, Harmondsworth, 1959.
LEAKEY, L. S. B.: *Stone Age Africa*, Oxford, 1936.
—— *Proceedings of the 1st Pan-African Congress on Prehistory, Nairobi, 1947*, Oxford, 1952.
—— *Adam's Ancestors*, 4th edn., rewritten, London, 1953.

PREHISTORY OF EAST AFRICA

BERNARD, E. A.: 'Les Climats d'insolation des latitudes tropicales au quaternaire', *Bulletin de l'Académie Royale des Sciences Coloniales*, Brussels, 1959, pp. 344–64.
BISHOP, W. W.: 'Miocene Mammalia from the Napak Volcanics, Karamoja, Uganda', *Nature*, 182, 1958, pp. 1480–2.
CLARK, J. D.: 'The Newly Discovered Nachikufu Culture of Northern Rhodesia', *South African Archaeological Bulletin*, 5, 19, 1950, pp. 86–98.
—— 'A Provisional Correlation of Prehistoric Cultures North and South of the Sahara', *South African Archaeological Bulletin*, 9, 33, 1954, pp. 3–17.
—— 'The Natural Fracture of Pebbles from the Batoka Gorge, Northern Rhodesia, and its Bearing on the Kafuan Industries of Africa', *Proceedings of the Prehistoric Society*, new series, 24, 1958, pp. 64–77.
COLE, S.: *The Prehistory of East Africa*, Harmondsworth, 1954.
—— 'The Oldest Tool-maker', *New Scientist*, 6, 1959, pp. 678–80.
—— 'How fast did Man evolve?', *New Scientist*, 11, 1961, pp. 208–10.
COOKE, H. B. S.: 'Observations relating to Quaternary Environments in East and Southern Africa', *Geological Society of South Africa*, annexure to vol. 60, 1957.
FLINT, R. F.: 'Pleistocene Climates in Eastern & Southern Africa', *Bulletin of the Geological Society of America*, 70, 1959, pp. 343–74.
—— 'On the Basis of Pleistocene Correlation in East Africa', *Geological Magazine*, 96, 1959, pp. 265–84.

FOSBROOKE, A. et al.: 'Tanganyika Rock Paintings', *Tanganyika Notes and Records*, 29, 1950, pp. 1–61.
HOWELL, F. C.: 'A Preliminary Note on a Prehistoric Donga (Maclennan's Donga) in Central Tanganyika', *South African Archaeological Bulletin*, 10, 38, 1955, pp. 43–52.
KENT, P. E.: 'The Pleistocene Beds of Kanam and Kanjera, Kavirondo, Kenya', *Geological Magazine*, 70, 1942, pp. 117–32.
KOENIGSWALD, G. H. R. VON, GENTNER, W., and LIPPOLT, H. J.: 'Age of the Basalt Flow at Olduvai, East Africa', *Nature*, 192, 1961, pp. 720–1.
LANNING, E. C.: *Uganda's Past*, East African Literature Bureau, Nairobi, 1957.
LE GROS CLARK, W. E., and LEAKEY, L. S. B.: *The Miocene Hominoidea of East Africa* (British Museum, Natural History), London, 1951.
LEAKEY, L. S. B.: *The Stone Age Cultures of Kenya Colony*, Cambridge, 1931.
—— *The Stone Age Races of Kenya*, London, 1935.
—— 'A New Fossil Skull from Eyasi, East Africa', *Nature*, 138, 1936, pp. 1082–4.
—— 'Skull of *Proconsul* from Rusinga Island', *Nature*, 162, 1948, p. 688.
—— *Olduvai Gorge*, Cambridge, 1951.
—— 'Results of Recent Research in Kenya', *Actes IIIe Congrès International de Sciences Préhistoriques et Protohistoriques, Zurich 1950*, 1953, p. 169.
—— 'Recent Discoveries at Olduvai Gorge, Tanganyika', *Nature*, 181, 1958, pp. 1099–1103.
—— 'A New Fossil Skull from Olduvai', *Nature*, 184 (1), 1959, pp. 491–3.
—— 'Recent Discoveries at Olduvai Gorge', *Nature*, 188, 1960, pp. 1050–1.
—— 'New Finds at Olduvai Gorge', *Nature*, 189, 1961, pp. 649–50.
—— EVERNDEN, J. F., and CURTIS, G. H.: 'The Age of Bed I, Olduvai Gorge, Tanganyika', *Nature*, 191, 1961, p. 478.
—— and OWEN, W. E.: 'A Contribution to the Study of the Tumbian Culture in East Africa', *Coryndon Museum Occasional Papers No. 1*, Nairobi, 1945.
—— and REEVE, W. H.: 'Report on a Visit to the Site of the Eyasi Skull', *Journal of the East Africa Natural History Society*, 19, 1 and 2, 1946, pp. 40–50.
LEAKEY, M. D.: 'Notes on the Ground and Polished Axes of East Africa', *Journal of the East Africa Natural History Society*, 17, 3 and 4, Nairobi, 1943, pp. 182–95.
—— 'Report on the Excavations at Hyrax Hill, Nakuru, Kenya Colony, 1937–1938' (with contributions by L. S. B. Leakey and P. M. Game), *Transactions of the Royal Society of South Africa*, 30, 1945, pp. 271–409.
—— and LEAKEY, L. S. B.: *Excavations at the Njoro River Cave: Stone Age cremated burials in Kenya Colony*, Oxford, 1950.
LOWE, C. H. VAN RIET: 'The Pleistocene Geology and Prehistory of Uganda, Pt. II', *Geological Survey of Uganda*, Mem. No. 6, Uganda, 1952.
OAKLEY, K. P.: 'Dating the Stages of Hominoid Evolution', *The Leech*, 28, 1958, pp. 112–15.
O'BRIEN, T. P.: *The Prehistory of Uganda Protectorate*, Cambridge, 1939.
POSNANSKY, M.: 'A Hope Fountain Site at Olorgesailie, Kenya Colony', *South African Archaeological Bulletin*, 14, 55, 1959, pp. 83–89.

SMOLLA, G.: 'Prähistorische Keramik aus Ostafrika', *Tribus*, 6, 1956, pp. 35–64.
WAYLAND, E. J.: 'Rifts, Rivers, Rains and Early Man in Uganda', *Journal of the Royal Anthropological Institute*, 64, 1934, pp. 333–52.
—— and BURKITT, M. C.: 'The Magosian Culture of Uganda', *Journal of the Royal Anthropological Institute*, 62, 1932, pp. 369–90.
WEINER, J. S.: 'The East African Fossil Men', *New Scientist*, 10, 1961, pp. 534–56.

CHAPTER III

See under CHAPTERS VIII and IX below

CHAPTER IV

The East African Coast until the Coming of the Portuguese

PRIMARY SOURCES

(a) Note on the Geographers

The *Periplus of the Erythraean Sea* has been edited by H. Frisk, Goteborg, 1927, and translated by W. H. Schoff, New York, 1912. A critical edition and translation by A. G. Mathew is in preparation for the Hakluyt Society.

The *Geography* of Ptolemy and the *Geographi Graeci Minores* were edited by C. Müller, Paris, 1883/1901 and 1855 respectively.

The *Christian Topography of Cosmas Indicopleustes* has been edited by E. O. Winstedt, Cambridge, 1909, and translated by J. W. McCrindle, Hakluyt Society, London, 1897.

There is great need for satisfactory texts and translations of the Islamic geographers. Al-Masudi has been edited and translated by A. C. Barbier de Meynard and P. de Courteille, 9 vols., Paris, 1861 sqq.

Al-Idrisi has been translated by P. Jaubert, Paris, 1836, but this version should be treated with caution. It is preferable to use G. Ferrand, *Relations de voyages et textes géographiques*, 2 vols., Paris, 1913.

Dimashqi has been edited by A. F. Mehren, Petersbourg, 1866, and translated by him, Copenhagen, 1874.

Ibn Battuta was edited and translated by C. Defrémery and B. R. Sanguinetti, 4 vols., Paris, 1862 sqq. An English translation of his *Travels* by H. A. R. Gibb is in process of publication by the Hakluyt Society.

(b) Chronicles

The *Arabic Chronicle of Kilwa* (B.M. MS. O.R. 2666): edited by S. A. Strong, in *Journal of the Royal Asiatic Society*, 1895, pp. 385–430. Throughout this chapter the translation of Dr. Freeman-Grenville has been used.

The Portuguese variant of the Kilwa Chronicle is contained in the *Decadas da Asia* by J. de Barros, ed. A. Baião, Coimbra, 1930.

Chronicle of Pate, edited and translated by A. Werner, in *Journal of the African Society*, 14, 1914, pp. 148–61 and 15, 1915, pp. 278–97.

Chronicle of Lamu, edited and translated by W. Hichens, in *Bantu Studies*, 12, i, Johannesburg, 1938, pp. 1–33.

Kitab al-Zanuj, edited and translated (*Il Libro degli Zengi*) by E. Cerulli, in *Somalia, scritti vari editi ed inediti*, Rome, 1957, pp. 231–357.

SECONDARY SOURCES

Among books that should be specially mentioned are:

AXELSON, E.: *South-East Africa, 1488–1530*, London, 1940.

DESCHAMPS, H.: *Histoire de Madagascar*, Paris, 1960.

FREEMAN-GRENVILLE, G. S. P.: *The Medieval History of the Tanganyika Coast*, Oxford, 1962.

—— *The East African Coast: select documents*, Oxford, 1962.

KIRKMAN, J. S.: *The Arab City of Gedi*, Oxford, 1954.

—— *Gedi: the palace*, The Hague, 1962.

PRINS, A. H. J.: *The Swahili-speaking Peoples of Zanzibar and the East African Coast, Arabs, Shirazi and Swahili*, Ethnographic Survey of Africa, London, 1961.

The following articles should be consulted:

ALLEN, J. W. T.: 'Rhapta', *Tanganyika Notes and Records*, No. 27, 1949, pp. 52–59.

CHITTICK, H. N.: *Annual Reports* of Department of Antiquities, Dar es Salaam, 1957–1962.

—— 'Notes on Kilwa', *Tanganyika Notes and Records*, No. 53, 1959, pp. 179–203.

FLURY, S.: 'The Kufic Inscriptions of the Kizimkazi Mosque, Zanzibar', *Journal of the Royal Asiatic Society*, 1922, pp. 257–64.

FREEMAN-GRENVILLE, G. S. P.: 'A New Hoard and some Unpublished Variants of the Coinage of the Sultans of Kilwa', *Numismatic Chronicle*, N.S. 14, 1954, pp. 220–4.

—— 'Coinage in East Africa before Portuguese Times', *Numismatic Chronicle*, N.S. 17, 1957, pp. 151–75.

—— 'Medieval Evidences for Swahili', *Journal of the East African Swahili Committee*, 29, i, 1959, pp. 1–14.

—— 'East African Coin Finds and their Historical Significance', *Journal of African History*, 1, 1960, pp. 31–43.

GRAY, J. M.: 'A History of Kilwa', *Tanganyika Notes and Records*, No. 31, 1951, pp. 1–24, and No. 32, 1952, pp. 11–37.

—— 'The Wadebuli and the Wadiba', *Tanganyika Notes and Records*, No. 36, 1954, pp. 22–42.

HUNTINGFORD, G. W. B.: 'The Azanian Civilization of Kenya', *Antiquity*, 7, 1933, pp. 153–65.

KIRKMAN, J. S.: 'The Excavations at Kilepwa', *The Antiquaries Journal*, 32, 1952, pp. 168–84.

—— 'Historical Archaeology in Kenya, 1948–1956', *The Antiquaries Journal*, 37, 1957, pp. 16–28.

KIRKMAN, J. S.: 'Excavations at Ras Mkumbuu on the Island of Pemba', *Tanganyika Notes and Records*, No. 53, 1959, pp. 161–78.

—— 'The Tomb of the Dated Inscription at Gedi', *Royal Anthropological Institute Occasional Papers*, No. 14, 1960.

MATHEW, A. G.: 'Recent Discoveries in East African Archaeology', *Antiquity*, 27, 1953, pp. 212–18.

—— 'Chinese Porcelain in East Africa and on the Coast of South Arabia', *Oriental Art*, new series, vol. ii, 2, 1956, pp. 50–55.

—— 'Songo Mnara', *Tanganyika Notes and Records*, No. 53, 1959, pp. 154–60.

MATTINGLY, H.: 'Coins from a Site-find in British East Africa', *Numismatic Chronicle*, 12, 1932, p. 175.

MONNERET DE VILLARD, U.: 'Note sulle influenze asiatiche nell'Africa Orientale', *Rivista degli studi orientali*, 17, 1938, pp. 303–49.

WALKER, J.: 'The History and Coinage of the Sultans of Kilwa', *Numismatic Chronicle*, 16, 1936, pp. 43–81.

—— 'Some New Coins from Kilwa', *Numismatic Chronicle*, 19, 1939, pp. 223–7.

CHAPTER V

The Coast—1498–1840

The extreme unevenness of the information available makes treatment of this period most difficult. The first thirty years are known in exhaustive and even controversial detail: for the rest there are numerous and unsatisfactory gaps. The *History of Pate* continues unbroken: the Arabic and Portuguese versions of the *History of Kilwa* end in 1502. For the eighteenth and nineteenth centuries there are local histories of Mombasa and Lamu, and, c. 1900, C. Velten recorded the traditions of a number of towns on the Tanganyika coast. For the northern sector the *Kitab al-Zanuj*, (apud E. Cerulli, *Somalia*), a late nineteenth-century redaction of earlier manuscripts, adds fresh information, and there are unpublished manuscripts from Pemba and Pate. For the coast as a whole there is invaluable material, albeit uneven in quality, in the *District Books* compiled under the direction of the Tanganyika Administration from 1923 onwards. The *Habari za Wakilindi* portrays the eighteenth- and nineteenth-century Usambara kingdom. In this period, too, Swahili poetry began to be written down. E. Axelson, *South-East Africa, 1488–1530*, gives the most complete account of Portuguese manuscripts so far. Fifteen thousand further manuscripts from Goa, Lisbon, and the Vatican, await publication by the Central African Archives: Dr. Axelson has kindly informed me that they add details to our knowledge without altering its main sense. As yet there is no systematic publication of inscriptions. In numismatics, apart from published works, remarkable new material from Mogadishu has been examined by the writer.

The following is a select list of manuscripts and of printed works which is not intended to compass all the works relied upon for this chapter.

1. MANUSCRIPTS

Alley, W.: Diary, 1667-8, transcribed from an original of whereabouts unknown, transcription in the possession of Sir John Gray.

Sharpeigh, A.: India Office Marine Records VII (N4-F).

Dallons, P.: Mauritius Archives, Vol. G.A. 11, no. 119.

Morice: *Projet d'un établissement à la côte orientale de l'Afrique, questions de M. Cossigny, réponses de M. Morice*—Rhodes House MS. Afr. r. 6.

De Curt: Reports, 29 Jan. 1791, University of Chicago, Rare Books Library.

Akida Diwani la Simba: Karatassi ya Orza ukoo wazazi wa Kipumbwi ya Mkwaja wa Bimbini (n.d. ? *c.* 1880) (Shajara ya Mkwaja). In the writer's possession.

Ndagoni MS. 1: Beit al-Amani Museum, Zanzibar, 27 Shaaban 1267 (A.D. 1851).

Ndagoni MS. 2: Zanzibar, as above, various documents from Ndagoni, Pemba.

Jambangome MS.: Zanzibar, as above, *c.* 1606–60.

Habari za Wakilindi: School of Oriental and African Studies, London.

2. PRINTED WORKS

Albuquerque, A.: *Cartas de . . . seguidas de documentos que as elúcidam*, 7 vols., Lisbon, 1884–5.

Axelson, E.: *South-East Africa, 1488–1530*, London, 1940.

—— *The Portuguese in South-East Africa, 1600–1700*, Johannesburg, 1960.

Barros, J. de: *Decadas da Asia*: first decade, ed. A. Baião, Coimbra, 1930; 2nd to 4th decade, Lisbon, 7 vols., 1777–8.

Bocarro, A.: *Decadas da historia da India*, Lisbon, 1876. [Extracts printed and translated in Theal, below, iii, 254–435.]

Burton, R. F.: *Zanzibar, City, Island and Coast*, 2 vols., London, 1872.

Büttner, C. G.: *Anthologie aus der Suaheli-Literatur*, 2 vols., Berlin, 1894.

Castanheda, F. L. de: *Historia de descobrimento e conquista da India pelos portuguêses*, 4 vols., Coimbra, 1924–33. [Originally published Lisbon, 1833.]

Cerulli, E.: *Somalia, scritti vari editi ed inediti*, vol. i, Rome, 1957.

Coupland, R.: *East Africa and its Invaders, from the Earliest Times to the Death of Seyyid Said in 1856*, Oxford, 1938.

Duarte Barbosa: *The Book of Duarte Barbosa*, vol. i (ed. M. L. Dames), Hakluyt Society, London, 1918.

Freeman-Grenville, G. S. P.: 'Coinage in East Africa before Portuguese Times', *Numismatic Chronicle*, N.S. 17, 1957, pp. 151–75.

—— 'Swahili Literature and the History and Archaeology of the East African Coast, *Journal of the East African Swahili Committee*, 28, ii, 1958, pp. 7–25.

—— 'Some Recent Archaeological Work on the Tanganyika Coast', *Man*, 58, 1958, 155, pp. 106–12.

SELECT BIBLIOGRAPHY

FREEMAN-GRENVILLE, G. S. P.: 'East African Coin Finds and their Historical Significance, *Journal of African History*, 1, 1960, pp. 31–43.
—— *The Medieval History of the Tanganyika Coast*, Oxford, 1962.
GOES, D. DE: *Cronica do Felicissimo Rei Dom Emanuel*, Lisbon, 1566. [Extracts printed and translated in Theal, below, iii, 1–142.]
GRAY, J. M.: Rezende's Description of East Africa in 1634', *Tanganyika Notes and Records*, 23, 1947, pp. 2–28.
—— 'A Journey by land from Tete to Kilwa in 1616', *Tanganyika Notes and Records*, 25, 1948, pp. 37–47.
—— Portuguese Records relating to the Wasegeju, *Tanganyika Notes and Records*, 29, 1950, pp. 85–97.
—— 'Sir John Henderson and the Princess of Zanzibar', *Tanganyika Notes and Records*, No. 40, 1955, pp. 15–19.
—— *The British in Mombasa, 1824–1826*, Kenya History Society, London, 1957.
GROTANELLI, V. L.: *Pescatori dell'Oceano Indiano*, Rome, 1955.
GUILLAIN, M.: *Documents sur l'histoire, la géographie et le commerce de l'Afrique Orientale*, 3 vols., Paris, 1856.
HEEPE, M.: 'Suaheli-Chronik von Pate', *Mitteilungen des Seminars für Orientalische Sprachen zu Berlin, Erste Abteilung, Ostasiatische Studien*, 31, iii, Berlin, 1928, pp. 145–92.
HICHENS, W. (ed.): 'Shaibu Faraji bin Hamed al-Bakariy al-Lamuy, "Chronicle of Lamu"', *Bantu Studies*, 12, i, Johannesburg, 1938, pp. 1–33.
—— (ed.): Sh. Sayyid Abdallah bin Ali bin Nasir bin Sh. Abubakar bin Salim, *al-Inkishafi* (The Soul's Awakening), London, 1939.
HOLLIS, A. C.: 'Notes on the History of Vumba, East Africa', *Journal of the Anthropological Institute*, 30, 1900, pp. 273–97.
INGRAMS, W. H.: *Zanzibar, its History and its People*, London, 1931.
JACKSON, M. V.: *European Powers and South-east Africa*, London, 1942.
KIRKMAN, J. S.: *The Arab City of Gedi*, Oxford, 1954.
—— 'Historical Archaeology in Kenya, 1948–1956', *The Antiquaries Journal*, 37, 1957, pp. 16–28.
KRAPF, J. L.: *Travels, Researches and Missionary Labours in Eastern Africa*, London, 1860.
LANCASTER, J.: *The Voyages of Sir James Lancaster, Kt., to the East Indies* (ed. C. R. Markham), Hakluyt Society, London, 1877.
MATHEW, A. G.: 'The Culture of the East African Coast in the 17th and 18th Centuries in the Light of Recent Archaeological Discoveries', *Man*, 56, 1956, 61.
—— 'Chinese Porcelain in East Africa and on the Coast of South Arabia', *Oriental Art*, new series, vol. ii, 2, 1956, pp. 50–55.
MBARAK ALI HINAWY: *al-Akida and Fort Jesus, Mombasa*, London, 1950.
OSORIO, J.: *Da Vida e Feitos de El-Rei D. Manuel*, 2 vols., Oporto, 1944.
OWEN, W. F.: *Narrative of Voyages to Explore the Shores of Africa, Arabia and Madagascar*, 2 vols., London, 1833.
PEARCE, F. B.: *Zanzibar, the Island Metropolis of East Africa*, London, 1920.
PRINS, A. H. J.: *The Coastal Tribes of the North-Eastern Bantu*, Ethnographic Survey, London, 1952.

PRIOR, J.: *Voyage along the Eastern Coast of Africa . . . to St. Helena*, London, 1819.
RAVENSTEIN, E. G. (ed. and trans.): *A Journal of the First Voyage of Vasco da Gama*, Hakluyt Society, London, 1898.
RIGBY, C. P.: *Report on the Zanzibar Dominions*, Bombay, 1861, printed as appendix II in Mrs. C. E. B. Russell, *General Rigby, Zanzibar and the Slave Trade*, London, 1935.
SANTOS, J. DOS: *Etiopia Oriental*, Evora, 1609. [Extracts printed and translated in Theal, below, vii, 1–370.]
SOUSA, M. DE FARIA Y: *Asia portuguèsa*, 6 vols., Oporto, 1945–7. [Originally published Lisbon, 1666–74.]
STANLEY, H. E. J. (ed.): *The Three Voyages of Vasco da Gama, from the 'Lendas da India'* of Gaspar Correa, Hakluyt Society, London, 1869.
STIGAND, C. H.: *The Land of Zinj*, London, 1913.
STRANDES, J.: *Die Portugiesenzeit von Deutsch- und Englisch-Ostafrika*, Berlin, 1899. English translation—trans. J. F. Wallwork and ed. J. S. Kirkman, 'The Portuguese in East Africa', *Transactions of the Kenya History Society*, vol. 2, Nairobi, 1961.
STRONG, S. A. (ed.): 'The History of Kilwa', *Journal of the Royal Asiatic Society*, 1895, pp. 385–430.
THEAL, G. M. (ed.): *Records of South-Eastern Africa*, 9 vols., London, 1898–1903.
VELTEN, C.: *Märchen und Erzählungen der Suaheli*, Berlin, 1898.
—— *Desturi za Wasuaheli*, Göttingen, 1903.
—— *Prosa und Poesie der Suaheli*, Berlin, 1907.
VOELTZKOW, A.: *Witu-Inseln und Zanzibar-Archipel*, Stuttgart, 1923.
WELCH, S. R.: *Portuguese and Dutch in South Africa, 1651–1806*, Cape Town, 1951.
WERNER, A. (ed.): 'A Swahili History of Pate', *Journal of the African Society*, 14, 1914, pp. 148–61 and 15, 1915, pp. 278–97.

CHAPTER VI

See under CHAPTERS VIII and IX below

CHAPTER VII

Zanzibar and the Coastal Belt, 1840–1884

1. UNPUBLISHED MATERIALS

At the Public Record Office, London: *Foreign Office Records, Series F.O. 84*, (*Slave Trade*). See especially 'Zanzibar' volumes.
In Zanzibar: *Secretariat Archives*.
The *Kirk Papers*, which were temporarily deposited on loan in the library of Rhodes House, Oxford, have since been returned to the Kirk family.

2. PUBLISHED MATERIALS

Primary Sources
British and Foreign State Papers, London 1839–1886.
BURTON, R. F.: *Zanzibar, City, Island and Coast*, 2 vols., London, 1872.

SELECT BIBLIOGRAPHY

GUILLAIN, M.: *Documents sur l'histoire, la géographie et le commerce de l'Afrique Orientale*, 3 vols., Paris, 1856.
HAMERTON, A.: *Report on the Affairs of the Imam of Muscat*, Bombay, 1844.
HERTSLET, E.: *The Map of Africa by Treaty*, 3rd edn., 3 vols., London, 1909.
Parliamentary Papers, 1873, LXI. *Correspondence respecting Sir Bartle Frere's Mission to the East Coast of Africa*.
Proceedings of the Commission on the Disputes between the Rulers of Muscat and Zanzibar, Bombay, 1861.
RIGBY, C. P.: *Report on the Zanzibar Dominions*, Bombay, 1861. (Reprinted as Appendix II in Russell, below.)
SALIL IBN RAZIK: *History of the Imams and Seyyids of Oman* (trans. G. P. Badger), Hakluyt Society, London, 1871.
SALME, PRINCESS (EMILY REUTE): *Memoirs of an Arabian Princess*, London, 1888.
Selections from the Records of the Bombay Government, No. XXIV, new series, Bombay, 1856.
VELTEN, C.: *Prosa and Poesie der Suaheli*, Berlin, 1907.

Secondary Sources

BRADY, C. T. (jnr.): *Commerce and Conquest in East Africa*, Salem, 1950.
COUPLAND, R.: *East Africa and its Invaders, from the Earliest Times to the Death of Seyyid Said in 1856*, Oxford, 1938.
—— *The Exploitation of East Africa, 1856–1890*, London, 1939.
CRASTER, J. E. E.: *Pemba, the Spice Island of Zanzibar*, London, 1913.
CROFTON, R. H.: *The Old Consulate at Zanzibar*, London, 1935.
INGRAMS, W. H.: *Zanzibar, its History and its People*, London, 1931.
LYNE, R. N.: *Zanzibar in Contemporary Times*, London, 1905.
—— *An Apostle of Empire, being the Life of Sir Lloyd William Mathews, K.C.M.G.*, London, 1936.
OLIVER, R.: *The Missionary Factor in East Africa*, London, 1952.
PEARCE, F. B.: *Zanzibar, the Island Metropolis of Eastern Africa*, London, 1920.
RUSSELL, MRS. C. E. B.: *General Rigby, Zanzibar and the Slave Trade*, London, 1935.
STIGAND, C. H.: *The Land of Zinj*, London, 1913.
WERNER, A.: 'The Wahadimu of Zanzibar', *Journal of the African Society*, 15, 1916, pp. 356–60.
WILSON, C. H.: *History of the Universities' Mission to Central Africa*, London, 1936.
Zanzibar, an Account of its People, Industry and History, Local Committee of the British Empire Exhibition, Zanzibar, 1924.

CHAPTERS III, VI, VIII, AND IX

The Interior up to the Period of Partition

These four chapters are all to a large extent based upon a common literature, of which the main categories are works of discovery and travel, of ethnography and social anthropology, and works which record, with varying degrees of accompanying interpretation, the teachings of traditional history. For such sources the classic categories of 'primary' and 'secondary' are hardly appropriate. This common literature is arranged, therefore, in general and regional sections. There follow sections indicating the sources more particularly relevant to specific chapters.

1. COMMON PRIMARY AND SECONDARY SOURCES FOR CHAPTERS III, VI, VIII, AND IX

(a) General

BAKER, S. W.: *The Albert Nyanza*, 2 vols., London, 1867.
—— *Ismailia; a narrative of the expedition to Central Africa for the suppression of the slave trade*, 2 vols., London, 1874.
BAUMANN, H., and WESTERMANN, D.: *Les Peuples et les civilisations de l'Afrique* (traduction française par L. Homburger), Paris, 1948.
BECKER, J.: *La Vie en Afrique*, 2 vols., Paris, 1887.
BURTON, R. F.: *The Lake Regions of Central Africa*, 2 vols., London, 1860. New edn., 2 vols., London, 1961.
CAMERON, V. L.: *Across Africa*, 2 vols., London, 1877.
CASATI, G.: *Ten years in Equatoria, and the Return with Emin Pasha* (trans. J. R. Clay, assisted by I. W. S. Landor), 2 vols., London, 1891.
CHAILLÉ-LONG, C.: *Central Africa: naked truths of naked people*, London, 1876.
CHRISTIE, J.: *Cholera Epidemics in East Africa*, London, 1876. Important for Masailand trade routes.
DAWSON, E. C. (ed.): *The Last Journals of Bishop Hannington*, London, 1888.
ELTON, J. F.: *Travels and Researches among the Lakes and Mountains of Eastern and Central Africa*, London, 1879.
ERHARDT, J.: 'Reports respecting Central Africa as collected in Mambara and on the East Coast', *Proceedings of the Royal Geographical Society*, 1855, pp. 8–10.
GIRAUD, V.: *Les Lacs de l'Afrique Equatoriale; voyage d'exploration executé de 1883 à 1885*, Paris, 1890.
GRANT, J. A.: *A Walk across Africa*, London, 1864.
GUILLAIN, M.: *Documents sur l'histoire, la géographie et le commerce de l'Afrique Orientale*, 3 vols., Paris, 1856.
HÖHNEL, L. VON: *The Discovery of Lakes Rudolf and Stephanie* (trans. N. Bell), 2 vols., London, 1894.
HORE, E. C.: *Tanganyika. Eleven years in Central Africa*, London, 1892.
JOHNSTON, H. H.: *The Kilimanjaro Expedition*, London, 1886.
—— *The Uganda Protectorate*, 2 vols., London, 1902.
JUNKER, W.: *Travels in Africa . . . 1875–1886* (trans. A. H. Keane), 3 vols., London, 1890–2.
KERSTEN, O.: *von der Deckens Reisen in Ost-Afrika*, Leipzig, 1869.
KRAPF, J. L.: *Travels, Researches, and Missionary Labours . . . in Eastern Africa*, London, 1860.
LINANT DE BELLEFONDS, E.: 'Voyage de service fait entre le poste militaire de Fatiko et la capitale de M'tesa roi d'Uganda', *Bulletin trimestriel de la Société Khédiviale de Géographie du Caire*, sér. I, 1876–1877, pp. 1–104.

SELECT BIBLIOGRAPHY 469

LIVINGSTONE, D.: *Last Journals* (ed. Horace Waller), 2 vols., London, 1874.
—— *The Zambezi Expedition of David Livingstone* (ed. J. P. R. Wallis), 2 vols., London, 1956.
—— and LIVINGSTONE, C.: *Narrative of an Expedition to the Zambezi and its Tributaries*, London, 1865.
MACQUEEN, J.: 'The Visit of Lief ben Saeid to the Great African Lake', *Journal of the Royal Geographical Society*, 15, 1845, pp. 371–6.
NEW, C.: *Life, Wanderings and Labours in Eastern Africa*, London, 1873.
PETHERICK, J.: *Egypt, the Soudan and Central Africa*, Edinburgh, 1861.
—— and K. H.: *Travels in Central Africa and Explorations of the Western Nile Tributaries*, 2 vols., London, 1869.
SCHWEINFURTH, G., et al. (eds.): *Emin Pasha in Central Africa*, London, 1888. Being a collection of his letters and journals, translated by Mrs. R. W. Felkin.
SPEKE, J. H.: *Journal of the Discovery of the Source of the Nile*, London, 1863.
—— *What led to the Discovery of the Source of the Nile*, Edinburgh, 1864.
STANLEY, H. M.: *How I found Livingstone*, 2nd edn., London, 1872.
—— *Through the Dark Continent*, 2 vols., London, 1878.
—— *The Congo and the Founding of its Free State*, 2 vols., London, 1885.
STUHLMANN, F. (ed.): *Die Tagebücher von Dr. Emin Pasha*, vols. 1–4, 6, Hamburg, 1919–27.
THOMSON, J.: *To the Central African Lakes and Back*, 2 vols., London, 1881.
—— *Through Masai Land*, 2nd edn., London, 1885.
TIPPU TIP: *Maisha ya Hamed bin Muhammed el Murjebi yaani Tippu Tip* (trans. W. H. Whiteley). Supplement to the *East African Swahili Committee Journals*, 28, ii, 1958 and 29, i, 1959.
WAKEFIELD, T.: 'Routes of Native Caravans from the Coast to the Interior of Eastern Africa', *Journal of the Royal Geographical Society*, 40, 1870, pp. 303–38.
WILLOUGHBY, J. C.: *East Africa and its Big Game: the narrative of a sporting trip from Zanzibar to the borders of Masai*, London, 1889.
WILSON, C. T., and FELKIN, R. W.: *Uganda and the Egyptian Soudan*, 2 vols., London, 1882.
WISSMANN, H. VON: *My Second Journey through Equatorial Africa from the Congo to the Zambesi in 1886 and 1887* (trans. M. J. A. Bergmann), London, 1891.

(b) *Local Studies*

(i) Nile Valley and Northern Uganda:

BEATON, A. C.: 'A Chapter in Bari History', *Sudan Notes and Records*, 17, ii, 1934, pp. 169–200.
BERE, R. M.: 'An Outline of Acholi History', *Uganda Journal*, 11, i, 1947, pp. 1–8.
BUTT, A. J.: *The Nilotes of the Anglo-Egyptian Sudan and Uganda*, Ethnographic Survey of Africa, London, 1952.
CRAZZOLARA, J. P.: 'The Lwoo People', *Uganda Journal*, 5, i, 1937, pp. 1–21.
—— *The Lwoo*, 3 vols., Missioni Africane, Verona, 1950–4.
DRIBERG, J. H.: *The Lango. A Nilotic tribe of Uganda*, London, 1923.

GESSI, R.: *Seven Years in the Soudan* (ed. F. Gessi, trans. L. W. Wolffsohn and B. Woodward), London, 1892.
GRAY, J. M.: 'Acholi History, 1860–1901', Part I, *Uganda Journal*, 15, ii, 1951, pp. 121–43.
LAWRANCE, J. C. D.: *The Iteso. Fifty years of change in a Nilo-Hamitic tribe of Uganda*, London, 1957.
SELIGMAN, C. G., and SELIGMAN, B. Z.: *Pagan Tribes of the Nilotic Sudan*, London, 1932.
SOUTHALL, A. W.: 'Alur Tradition and its Historical Significance', *Uganda Journal*, 18, ii, 1954, pp. 137–65.
—— *Alur Society: a study in processes and types of domination*, Cambridge, 1956.
TARANTINO, A.: 'Lango Wars', *Uganda Journal*, 13, ii, 1949, pp. 230–4.
—— 'Notes on the Lango', *Uganda Journal*, 13, ii, 1949, pp. 145–53.

(ii) Interlacustrine Region—Southern Uganda and North-West Tanganyika:

ASHE, R. P.: *Two Kings of Uganda*, London, 1889.
CÉSARD, E.: 'Le Muhaya', *Anthropos*, 32, i/ii, 1937, pp. 32–57.
CORY, H.: *History of Bukoba District*, Dar es Salaam, 1958.
—— and HARTNOLL, M. M.: *Customary Law of the Haya Tribe*, London, 1945.
DUNDAS, Hon. K. R.: 'The Wawanga and other Tribes of the Elgon District, British East Africa', *Journal of the Royal Anthropological Institute*, 43, 1913, pp. 19–75.
FALLERS, L. A.: *Bantu Bureaucracy*, Cambridge, 1956. A sociological account of the Soga.
FISHER, R.: *Twilight Tales of the Black Baganda*, London, 1912.
FORD, J., and HALL, R. de Z.: 'The History of Karagwe', *Tanganyika Notes and Records*, 24, 1947, pp. 3–27.
GORJU, J.: *Entre le Victoria, l'Albert et l'Édouard*, Rennes, 1920. A collation of traditional accounts.
GRAY, J. M.: 'Mutesa of Buganda', *Uganda Journal*, 1, i, 1934, pp. 2–49.
—— 'Ahmed bin Ibrahim—the First Arab to reach Buganda', *Uganda Journal*, 11, ii, 1947, pp. 80–97.
—— 'Sir John Kirk and Mutesa', *Uganda Journal*, 15, i, 1951, pp. 1–16.
KAGWA, A.: *Ebika bya Buganda*, Kampala, 1906.
—— *Mpisa za Baganda*, Kampala, 1908.
—— *Basekabaka be Buganda, Bunyoro, Toro, Ankole ne Koki*, 3rd edn., Kampala, 1927.
—— *Ekitabo kye Bika bya Baganda*, 2nd edn., Kampala, 1949.
—— *Ekitabo kya Basekabaka be Buganda*, 4th edn., London, 1953.
KATATE, A. G., and KAMUGUNGUNU, L.: *Abagabe b'Ankole*, 2 vols., Kampala, 1955.
LOW, D. A.: 'Religion and Society in Buganda, 1875–1900', *East African Studies* No. 8, Kampala, 1957.
LUBOGO, Y. K.: *A History of Busoga*, Jinja, 1960.
MORRIS, H. F.: 'The Making of Ankole', *Uganda Journal*, 21, i, 1957, pp. 1–15.
MUKASA, H.: 'Ebifa ku Mulembe gwa Kabaka Mutesa', *Uganda Journal*, Part I, 1, ii, 1934, pp. 116–23; Part II, 2, i, 1934, pp. 60–65.
NSIMBI, K. B.: *Amannya Amaganda n'Ennono Zaago*, Kampala, 1956.

NYAKATURA, J. W.: *Abakama ba Bunyoro-Kitara*, St. Justin, Canada, 1947.
OBERG, K.: 'The Kingdom of Ankole in Uganda', in Fortes and Evans-Pritchard, 1940, below.
OLIVER, R.: 'The Traditional Histories of Buganda, Bunyoro and Ankole', *Journal of the Royal Anthropological Institute*, 85, 1955, pp. 111–17.
—— 'Ancient Capital Sites of Ankole', *Uganda Journal*, 23, i, 1959, pp. 51–63.
—— 'The Royal Tombs of Buganda', *Uganda Journal*, 23, ii, 1959, pp. 124–33.
PAGÈS, A.: *Un Royaume hamite au centre de l'Afrique*, Brussels, 1933. Rwanda.
REHSE, H.: *Kiziba, Land und Leute*, Stuttgart, 1910.
ROBERTSON, D. W.: *Historical Considerations Contributing to the Soga System of Land Tenure*, Entebbe, 1940.
ROSCOE, J.: *The Baganda*, London, 1911.
—— *The Banyankole*, Cambridge, 1923.
—— *The Bakitara*, Cambridge, 1923.
SOUTHALL, A. W.: 'The Alur Legend of Sir Samuel Baker and the Mukama Kabarega', *Uganda Journal*, 15, ii, 1951, pp. 187–90.
THIEL, P. H. VAN: 'Businza unter der Dynastie der Bahinda', *Anthropos*, 6, 1911, pp. 497–520.
WRIGLEY, C. C.: 'Buganda: an outline economic history', *Economic History Review*, 2nd ser., 10, 1957, pp. 60–80.
K. W.: 'Abakama ba Bunyoro-Kitara', *Uganda Journal*, 3, ii, 1935, pp. 155–60; 4, i, 1936, pp. 75–83; 5, ii, 1937, pp. 53–68.

(iii) Central and Western Tanganyika:

BLEEK, D. F.: 'The Hadzapi or Watindega of Tanganyika Territory', *Africa*, 4, 1931, pp. 273–86.
BLOHM, W.: *Die Nyamwezi, Land und Wirtschaft*, 3 parts, Hamburg, 1931–3.
BÖSCH, F.: *Les Banyamwezi*, Münster, 1930.
CORY, H.: *The Ntemi. Traditional rites of a Sukuma chief in Tanganyika*, London, 1951.
SCHAEGELEN, T.: 'La Tribu des Wagogo', *Anthropos*, 33, 1938, i/ii, pp. 195–217, iii/iv, pp. 515–67.
TREVOR, J. C.: 'The Physical Characters of the Sandawe', *Journal of the Royal Anthropological Institute*, 77, i, 1947, pp. 61–78.
WILSON, M.: *The Peoples of the Nyasa–Tanganyika Corridor*, Communications from the School of African Studies, Cape Town, 1958.

(iv) Hamites, Nilo-Hamites, and Northern-Eastern Bantu in Kenya and North-Eastern Tanganyika:

ALLEN, R. (trans.).: 'The Story of Mbega, by Abdallah bin Hemedi bin Ali Linjjemi [Habari za Wakilindi]', *Tanganyika Notes and Records*, Part I, 1, 1936, pp. 38–51; Part II, 2, 1936, pp. 80–98.
BAKER. E. C.: 'Notes on the History of the Wasegeju', *Tanganyika Notes and Records*, 27, 1949, pp. 16–41.
CAGNOLO, C.: *The Akikuyu. Their customs, traditions and folklore*, Nyeri, 1933.
DUNDAS, C. C. F.: 'History of Kitui', *Journal of the Royal Anthropological Institute*, 43, 1913, pp. 480–549.

DUNDAS, C. C. F.: *Kilimanjaro and its People*, London, 1924.
DUNDAS, Hon. K. R.: 'The Wawanga and other Tribes of the Elgon District, British East Africa', *Journal of the Royal Anthropological Institute*, 43, 1913, pp. 20 sqq.
FOSBROOKE, H. A.: 'An Administrative Survey of the Masai Social System', *Tanganyika Notes and Records*, 26, 1948, pp. 1–50.
GULLIVER, PAMELA, and GULLIVER, P. H.: *The Central Nilo-Hamites*, Ethnographic Survey of Africa, London, 1953.
HOBLEY, C. W.: 'Notes on the Dorobo People and other Tribes', *Man*, 1906, 78.
HONEA, K. H.: *A Contribution to the History of the Hamitic Peoples of Africa*, Vienna, 1958.
HUNTINGFORD, G. W. B.: *Nandi Work and Culture*, H.M.S.O., London, 1950.
—— 'The Social Institutions of the Dorobo', *Anthropos*, 46, i/ii, 1951, pp. 1–48.
—— *The Nandi of Kenya*, London, 1953.
—— *The Northern Nilo-Hamites*, Ethnographic Survey of Africa, London, 1953.
—— *The Southern Nilo-Hamites*, Ethnographic Survey of Africa, London, 1953.
—— 'Bantu Peoples of Eastern Kenya and North-eastern Tanganyika', in Hamilton, 1955, below.
—— *The Galla of Ethiopia*, Ethnographic Survey of Africa, London, 1955.
LAMBERT, H. E.: *The Systems of Land Tenure in the Kikuyu Land Unit. Pt. 1. History of the tribal occupation of the land*, Cape Town, 1950.
—— *Kikuyu Social and Political Institutions*, London, 1956.
LEWIS, I. M.: *Peoples of the Horn of Africa*, Ethnographic Survey of Africa, London, 1955.
LINDBLOM, G.: *The Akamba in British East Africa*, 2nd edn., Uppsala, 1920.
MIDDLETON, J.: *The Kikuyu and Kamba of Kenya*, Ethnographic Survey of Africa, London, 1953.
PRINS, A. H. J.: *The Coastal Tribes of the North-Eastern Bantu*, Ethnographic Survey of Africa, London, 1952.
RECHE, O.: *Zur Ethnographie des abflusslosen Gebietes Deutsch-Ostafrikas*, Hamburg, 1914.
WILSON, G. McL.: 'The Tatoga of Tanganyika', *Tanganyika Notes and Records*, 33, 1952, pp. 34–47.

(v) Southern Tanganyika and Lake Nyasa:

ABDALLAH, Y. B.: *The Yaos* (arranged, edited, and translated M. Sanderson), Zomba, 1919.
CULWICK, A. T., and CULWICK, G. M.: *Ubena of the Rivers*, London, 1935.
FÜLLEBORN, F.: *Das deutsche Njassa- und Ruvumagebiet*, Berlin, 1906.
GULLIVER, P. H.: 'History of the Songea Ngoni', *Tanganyika Notes and Records*, 41, 1955, pp. 16–30.
HATCHELL, G. W.: 'The Angoni of Tanganyika Territory', *Tanganyika Notes and Records*, 25, 1948, pp. 69–71.
JOHNSON, W. P.: 'Seven Years' Travels in the Region East of Lake Nyasa', *Proceedings of the Royal Geographical Society*, New Series, 6, 1884, pp. 512–33.

SELECT BIBLIOGRAPHY

JOHNSON, W. P.: *Nyasa the Great Water: a description of the lake and the life of the people*, London, 1922.
LANCASTER, D. G.: 'A Tentative Chronology of the Ngoni', *Journal of the Royal Anthropological Institute*, 67, 1937, pp. 77–90.
LECHAPTOIS, A.: *Aux rives du Tanganyika*, Algiers, 1913.
NIGMANN, E.: *Die Wahehe*, Berlin, 1908.
READ, M.: 'Tradition and Prestige among the Ngoni', *Africa*, 9, iv, 1936, pp. 453–84.
TEW, M.: *The Peoples of the Lake Nyasa Region*, Ethnographic Survey of Africa, London, 1950.
WILSON, M.: *The Peoples of the Nyasa–Tanganyika Corridor*, Communications from the School of African Studies, Cape Town, 1958.

2. THE FOLLOWING ADDITIONAL SOURCES ARE PARTICULARLY RELEVANT TO CHAPTER III

ANKERMANN, B.: *Das Eingeborenen-recht: Ost-Afrika*, Stuttgart, 1929.
BAUMANN, H.: 'Zur Morphologie des afrikanischen Ackergerätes', *Koloniale Völkerkunde*, 1944, 1, pp. 192, 322.
BRYAN, M. A.: 'The T/K Languages: a new substratum', *Africa*, 29, 1959, pp. 1–21.
—— and TUCKER, A. N.: *Distribution of the Nilotic and Nilo-Hamitic Languages of Africa*, London, 1948.
CLINE, W. B.: *Mining and Metallurgy in Negro Africa*, Menasha, Wis., 1937.
FORTES, M., and EVANS-PRITCHARD, E. E. (eds.): *African Political Systems*, London, 1940.
GILLMAN, C.: 'An Annotated List of Ancient and Modern Stone Indigenous Structures in Eastern Africa', *Tanganyika Notes and Records*, 17, 1944, pp. 44–55.
GREENBERG, J. H.: *Studies in African Linguistic Classification* (reprinted from *South-Western Journal of Anthropology*), New Haven, 1955.
GROTTANELLI, V. L.: 'A Lost African Metropolis', *Afrikanistische Studien*, Berlin, 1955, pp. 231–42.
GUTHRIE, M.: *The Classification of the Bantu Languages*, London, 1948.
HERSKOVITS, M. J.: 'Culture Areas of Africa', *Africa*, 3, 1930, pp. 59–77.
HUNTINGFORD, G. W. B.: 'Local Archaeology in Kenya Colony', *Journal of the East Africa and Uganda Natural History Society*, 24, 1926, pp. 3–25.
—— 'The Azanian Civilization of Kenya', *Antiquity*, 7, 1933, pp. 153–65.
—— 'The "Nilo-Hamitic" Languages', *South-Western Journal of Anthropology*, 12, 1956, pp. 200–22.
—— and BELL, C. R. V.: *East African Background*, London, 1950.
JOHNSTON, H. H.: 'A Survey of the Ethnography of Africa: and the former racial and tribal migrations in that continent', *Journal of the Royal Anthropological Institute*, 43, 1913, pp. 375–421.
LEAKEY, L. S. B.: 'Preliminary Report on an Examination of the Engaruka Ruins', *Tanganyika Notes and Records*, 1, 1936, pp. 57–60.
—— *Stone Age Africa*, London, 1936.

LEAKEY, M. D.: 'Report on the Excavations at Hyrax Hill, Nakuru, Kenya Colony, 1937–1938', *Transactions of the Royal Society of South Africa*, 30, Part 4, 1945, pp. 274–409.
—— and LEAKEY, L. S. B.: *Excavations at the Njoro River Cave*, Oxford, 1950.
LEHMANN, H.: 'Distribution of the Sickle Cell Gene', *Eugenics Review*, 46, 1954, pp. 101–21.
MASSAJA, G.: *I miei trentacinque anni di missione nell'Alta Etiopia*, vol. 8, Milan, 1885.
MOURANT, A. E.: *The Distribution of the Human Blood Groups*, Oxford, 1954.
—— KOPEĆ, A. C., and DOMANIEWSKA-SOBCZAK, K.: *The ABO Blood Groups*, Oxford, 1958.
SELIGMAN, C. G.: *Races of Africa*, 3rd edn., London, 1957.
SHUKLA, R. A., and SOLANKI, B. R.: 'Sickle-Cell Trait in Central India', *Lancet*, 8 Feb. 1958, pp. 297–8.
TOBIAS, P. V.: 'Bushmen of the Kalahari', *Man*, 1957, 36.
TORDAY, E.: *Descriptive Sociology: African races*, London, 1930.
TROWELL, M., and WACHSMANN, K. P.: *Tribal Crafts of Uganda*, London, 1953.
TUCKER, A. N., and BRYAN, M. A.: *The Non-Bantu Languages of North-Eastern Africa*, London, 1956.
WILSON, G. E. H.: 'The Ancient Civilization of the Rift Valley', *Man*, 1932, 298.

3. THE FOLLOWING ADDITIONAL SOURCES ARE PARTICULARLY RELEVANT TO CHAPTER VI

(a) Unpublished

In the George V Memorial Museum, Dar es Salaam, Tanganyika: *Tanganyika District Books*. These are summaries of vital statistics and other reference material of a political, fiscal, and ethnographic nature which were instituted in all Districts of Tanganyika in 1923 and have since been added to from time to time. The references in this chapter are all to the sections called 'Tribal History and Records', somewhat imperfect copies of which are available for consultation in the library of the School of Oriental and African Studies in London. The compilation of traditional information was a task frequently entrusted to cadets and junior officers as a part of their training. The material in these sections is consequently of very varied quality and needs to be used with discrimination.

(b) Published

HAMILTON, R. A. (ed.): *History and Archaeology in Africa: report of a conference held in July 1953 at the School of Oriental and African Studies*, London, 1955.
OLIVER, R.: 'Oral Tradition: East Africa', in Hamilton, 1955, above, pp. 15–19, 41–42.
WESTERMANN, D.: *Geschichte Afrikas: Staatenbildungen südlich der Sahara*, Cologne, 1952.

SELECT BIBLIOGRAPHY 475

4. THE FOLLOWING ADDITIONAL SOURCES ARE PARTICULARLY RELEVANT TO CHAPTERS VIII AND IX

(a) Unpublished

In the Public Record Office, London: *Foreign Office Records. Series F.O. 84 (Slave Trade)*; this contains important correspondence relating to East Africa in the 'Zanzibar' and 'Domestic Various' volumes. *Series F.O. 27 (Egypt)* and *Series F.O. 403 (Confidential Print—Africa)*. The latter is a printed selection of Foreign Office correspondence, fairly full in respect of certain periods and episodes.

In the British Museum, London: Additional Manuscripts, *Gordon MSS*.

At the Headquarters of the London Missionary Society, Livingstone Hall, London: *Correspondence of the London Missionary Society, Central African Archives*.

At C.M.S. House, Salisbury Square, London: *Church Missionary Society Archives*. Series C.A. 5. East Equatorial Mission, 1848–80; Series C.A. 6. Nyanza Mission, 1875–80; Series G 3/A. 5. East Equatorial Africa Mission, 1881–97; Series G 3/A. 6. Nyanza Mission, 1881–6; Series M.C. Minutes of Committees.

In Rhodes House Library, Oxford: *Projet d'un établissement à la côte orientale de l'Afrique, questions de M. Cossigny, réponses de M. Morice*. MS. Afr. r. 6.

—— *Waller Papers*, Vol. II; correspondence with Gordon.

—— *Thornton Papers*. Diaries of journeys inland from Mombasa.

—— D. A. Low: *The British and Uganda, 1862–1900*. Oxford D.Phil. thesis, 1957.

In Cairo: *Abdin Palace Archives*, especially cartons 37–47 and Régistres, Arrivé Abdine 23–43, Départ Abdine 17–31.

At Rubaga, Uganda: *Diaire du Poste de Rubaga, 1879–1882*.

In the George V Memorial Museum, Dar es Salaam, Tanganyika: *Tanganyika District Books* (as for Chapter VI; see above).

In Zanzibar: *Secretariat Archives*. These duplicate for the most part the East African correspondence in *Series F.O. 84*, but they contain in addition some important letters from missionaries.

(b) Published

Primary Sources:

Egypt, Provinces of the Equator: Summary of Letters and Reports of his Excellency the Governor-General, Part I, Year 1874, Egyptian General Staff, Cairo, 1877. Some Gordon correspondence.

Parliamentary Papers (Slave Trade, Classes B and C). *Correspondence with British Representatives and Agents Abroad, and Reports from Naval Officers relative to the Slave Trade*, 1869–89. Reproduces much of the official British correspondence between Zanzibar and London, and contains occasional information of importance.

Secondary Sources:

ALLEN, B. M.: *Gordon and the Sudan*, London, 1931.
CEULEMANS, P.: *La Question arabe et le Congo (1883–1892)*, Brussels, 1959.

DOUIN, G.: *Histoire du règne du Khédive Ismail*, Tome III, *L'Empire africain*, Cairo, 1936, 1939, 1941.
GRAY, J. R.: *A History of the Southern Sudan, 1839–1889*, London, 1961.
HARRISON, J. W.: *A. M. Mackay, Pioneer Missionary of the Church Missionary Society to Uganda*, London, 1890.
HILL, G. B.: *Colonel Gordon in Central Africa, 1874–1879*, London, 1881.
NICQ, A.: *Vie du Père Siméon Lourdel*, Algiers, 1902.
OLIVER, R.: *The Missionary Factor in East Africa*, London, 1952.
RICHARDS, A. I. (ed.): *East African Chiefs: a study of political development in some Uganda and Tanganyika tribes*, London, 1960.

CHAPTERS X, XI, AND XII

The First Impact of Europe on East Africa

1. UNPUBLISHED MATERIALS

In the Public Record Office, London: *Foreign Office Records. Series F.O. 84 (Slave Trade)*. Under 'Domestic Various' (these volumes contain the correspondence relating to East Africa up to the year 1892), 'Zanzibar', 'France' and 'Prussia'. *Series F.O. 2 (Africa)*. This series contains the East African correspondence after 1892, and includes several volumes of correspondence between the Foreign Office and the I.B.E.A. Company for 1893–5. *Series F.O. 83 (General)*. This series contains miscellaneous correspondence relating to East Africa for the years after 1892.
In the Senate House Library, University of London: de Kiewiet, M. J.: *History of the Imperial British East Africa Company, 1876–1895*, London Ph.D. thesis, 1955.
In the School of Oriental and African Studies, University of London: *Mackinnon Papers*. A large collection of Mackinnon's letters and papers, of the first importance for the history of East Africa in this period.
At C.M.S. House, Salisbury Square, London: *Church Missionary Society Archives*.
In Rhodes House Library, Oxford: *Anti-Slavery Society Papers*.
—— *Lugard Papers*.
—— *Portal Papers*. A small collection of private papers.
In Zanzibar: *Secretariat Archives*.
—— *Residency Archives*.

2. PUBLISHED MATERIALS

Primary Sources

ASHE, R. P.: *Chronicles of Uganda*, London, 1894.
BAUMANN, O.: *In Deutsch-Ostafrika während des Aufstandes*, Vienna, 1890.
—— *Usambara und seine Nachbargebiete*, Berlin, 1891.
BOVILL, M., and ASKWITH, G. R.: *Roddy Owen, a Memoir*, London, 1897.

SELECT BIBLIOGRAPHY 477

CASATI, G.: *Ten Years in Equatoria and the Return with Emin Pasha* (trans. Mrs. J. R. Clay, assisted by I. W. S. Landor). 2 vols., London, 1891.
COLVILE, H. E.: *The Land of the Nile Springs*, London, 1895.
DUGDALE, E. T. S. (ed.): *German Diplomatic Documents, 1871–1914*, 4 vols., London, 1928–31.
ELIOT, C.: *The East Africa Protectorate*, London, 1905.
FOSBROOKE, H. A.: 'The Life of Justin: an African Autobiography', with an introduction by Sir Claud Hollis, *Tanganyika Notes and Records*, Part I, 41, 1955, pp. 30–57; Part II, 42, 1956, pp. 19–30.
GÖTZEN, A. GRAF VON: *Durch Afrika von Ost nach West*, Berlin, 1895.
—— *Deutsch-Ostafrika im Aufstand, 1905–6*, Berlin, 1909.
HANSARD: *Parliamentary Debates*.
HEMEDI BIN ABDALLAH BIN SAID AL-BUHRIY: *Utenzi wa Vita vya Wadachi kutamalaki Mrima* (ed. and trans. J. W. T. Allen), Supplement to the *Journal of the East African Swahili Committee*, 25, 1955; rev. edn. Dar es Salaam, 1960.
HERTSLET, E.: *The Map of Africa by Treaty*, 3rd edn., 3 vols., London, 1909.
HÖHNEL, L. VON: *The Discovery of Lakes Rudolf and Stephanie* (trans. N. Bell), 2 vols., London, 1894.
JACKSON, F. J.: *Early Days in East Africa*, London, 1930.
JOHNSTON, H. H.: *The Uganda Protectorate*, 2 vols., London, 1902.
JUNKER, W.: *Travels in Africa during the Years 1875–1886*, 3 vols., London, 1890–92.
KREUZLER, E.: *Ein Jahr in Ostafrika*, Ulm, 1888.
LANGHELD, W.: *Zwanzig Jahre in deutschen Kolonien, Teil 1. Deutsch-Ostafrika 1889–1900*, Berlin, 1909.
LEPSIUS, J., MENDELSSOHN-BARTHOLDY, A., and THIMME, F. (eds.): *Die grosse Politik der europäischen Kabinette, 1871–1914*, 40 vols., Berlin, 1922–7.
LIEBERT, E. VON: *Neunzig Tag im Zelt*, Berlin, 1898.
LUGARD, F. D.: *The Rise of our East African Empire*, 2 vols., Edinburgh, 1893.
MACDONALD, J. R. L.: *Soldiering and Surveying in British East Africa, 1891–1894*, London, 1897.
PARLIAMENTARY PAPERS: 1890, LI [C. 5906], *Correspondence re Stanley's Expedition to relieve Emin Pasha.*
—— 1890, LI [C. 6146], *Correspondence re Anglo-German Agreement on Africa and Heligoland.*
—— 1890, LXXXI [C. 6193], *Anglo-French Declaration re Madagascar, Zanzibar and North Africa.*
—— 1890–91, XCVI [C. 6316], *Anglo-Italian Protocols re Spheres of Influence in East Africa, 1891.*
—— 1893–94, CIX [C. 7203], *Anglo-German East African Boundaries Agreement.*
—— 1894, LVII [C. 7303], *Reports relating to Uganda by Sir G. Portal.*
—— 1894 XCVI [C. 7358] [C. 7390] [C. 7549]. *Agreement between Great Britain and Congo Free State on Spheres of Influence in East Africa.*
—— 1894, XCVI [C. 7388] *Anglo-Italian Protocol on Spheres of Influence in Eastern Africa, 1894.*
—— 1895 [C. 7646], *Correspondence with Imperial British East Africa Company re Revocation of its Charter.*

PARLIAMENTARY PAPERS: 1905, LVI [Cd. 2312], *Report of Survey of New Frontier between British East Africa and Ethiopia.*
PETERS, C.: *Deutsch national: kolonialpolitische Aufsätze.* Berlin, 1887.
―― *Die Deutsch-Ostafrikanische Kolonie in ihrer Entstehungsgeschichte und wirtschaftlichen Eigenart,* Berlin, 1888.
―― *New Light on Dark Africa; being a narrative of the German Emin Pasha expedition,* London, 1891.
―― *Das Deutsch-Ostafrikanische Schutzgebiet,* Munich, 1895.
―― *King Solomon's Golden Ophir,* London, 1899.
―― *The Eldorado of the Ancients,* London, 1902.
―― *Die Gründung von Deutsch-Ostafrika,* Berlin, 1906.
―― *Lebenserinnerungen,* Hamburg, 1918.
PFEIL, J. GRAF VON: 'Die Erforschung des Ulanga-Gebietes', *Petermanns Mitteilungen,* 32, pp. 353–62, Gotha, 1886.
―― *Zur Erwerbung von Deutsch-Ostafrika,* Berlin, 1907.
PORTAL, G. H.: *The British Mission to Uganda in 1893* (ed. Rennell Rodd) London, 1894.
POSCHINGER, H. VON (ed.): *Fürst Bismarck als Volkswirth,* Vol. I, Berlin, 1889.
PRINCE, M.: *Eine deutsche Frau im Innern Deutsch-Ostafrikas,* Berlin, 1903.
PRINCE, T. VON: *Gegen Araber und Wahehe, 1890–1895,* Berlin, 1914.
SCHMIDT, R.: *Geschichte des Araberaufstandes in Ostafrika,* Frankfurt a. M., 1892.
STANLEY, H. M.: *In Darkest Africa,* 2 vols., London, 1890.
―― *The Autobiography of Sir H. M. Stanley* (ed. D. Stanley), London, 1909.
STUHLMANN, F.: *Mit Emin Pascha ins Herz von Afrika,* Berlin, 1894.
TROTHE, L. VON: *Meine Bereisung von Deutsch-Ostafrika,* Berlin, 1897.
TUCKER, A. R.: *Eighteen Years in Uganda and East Africa,* 2nd edn., 2 vols., London, 1911.
VELTEN, C.: *Prosa und Poesie der Suaheli,* Berlin, 1907.
WAGNER, J.: *Deutsch-Ostafrika, Geschichte der Gesellschaft für Deutsche Kolonisation der Deutsch-Ostafrikanischen Pflanzungsgesellschaft nach den amtlichen Quellen,* Berlin, 1886.
WISSMANN, H.: *My Second Journey through Equatorial Africa from the Congo to the Zambesi in 1886 and 1887* (trans. M. J. A. Bergmann), London, 1891.
ZIMMERMANN, A.: *Mit Dernburg nach Ostafrika,* Berlin, 1908.

Secondary Sources

AYDELOTTE, W. O.: *Bismarck and British Colonial Policy. The problem of South West Africa, 1883–1885,* Philadelphia, 1937.
BERE, R. M.: 'Awich—a Biographical Note and a Chapter of Acholi History', *Uganda Journal,* 10, ii, 1946, pp. 76–78.
BRODE, H.: *Tippu Tip,* Berlin, 1905.
BRUNSCHWIG, H.: *L'Expansion allemande outre-mer du XVIème siècle à nos jours,* Paris, 1957.
Cambridge History of the British Empire, vol. iii, Cambridge, 1959.
CECIL, G.: *Life of Robert, Marquis of Salisbury,* 4 vols., London, 1921–32.
COUPLAND, R.: *The Exploitation of East Africa, 1856–1890,* London, 1939.
CREWE, Lord: *Lord Rosebery,* 2 vols., London. 1931.
CROWE, S. E.: *The Berlin West African Conference, 1884–1885,* London, 1942.

DAWSON, E. C.: *James Hannington, . . . a History of his Life and Work, 1847–1885*, London, 1887.
EYCK, E.: *Bismarck and the German Empire*, London, 1950.
FALLERS, L. A.: *Bantu Bureaucracy*, Cambridge, 1956.
FLETCHER, T. B.: 'Mwanga—the Man and his Times', *Uganda Journal*, 4, ii, 1936, pp. 162–7.
GARVIN, J. L.: *Life of Joseph Chamberlain*, vols. 1–3, London, 1932–4.
GOLDSMITH, F. H.: *John Ainsworth, Pioneer Kenya Administrator, 1864–1946*, London, 1955.
GRAY, J. M.: 'Early Treaties in Uganda', *Uganda Journal*, 12, i, 1948, pp. 25–42.
—— 'Rwot Ochama of Payera', *Uganda Journal*, 12, ii, 1948, pp. 121–8.
—— 'The Year of the Three Kings of Buganda, Mwanga—Kiwewa—Kalema, 1888–1889', *Uganda Journal*, 14, i, 1950, pp. 15–52.
—— 'Sir John Kirk and Mutesa', *Uganda Journal*, 15, i, 1951, pp. 1–16.
—— 'Acholi History, 1860–1901', *Uganda Journal*, Part I, 15, ii, 1951, pp. 121–43; Part II, 16, i, 1952, pp. 32–50; Part III, 16, ii, 1952, pp. 132–44.
—— 'History up to the End of the 1914–1918 War', in *Handbook of Tanganyika*. See Moffett, below.
GREGORY, J. W.: *The Foundation of British East Africa*, London, 1901.
GULLIVER, PAMELA, and GULLIVER, P. H.: *The Central Nilo-Hamites*, Ethnographic Survey of Africa, London, 1953.
GWYNN, S., and TUCKWELL, G. M.: *The Life of Sir Charles Dilke*, London, 1917.
HAGEN, M. VON: *Bismarcks Kolonialpolitik*, Stuttgart, 1924.
HILL, M. F.: *Permanent Way*. Vol. I. *The Story of the Kenya and Uganda Railway*, Nairobi, 1950. Vol. II. *The Story of the Tanganyika Railways*, Nairobi, 1959.
HINDORF, R.: *Der Sisalbau in Deutsch-Ostafrika*, Berlin, 1925.
HOBLEY, C. W.: *Kenya from Chartered Company to Crown Colony*, London, 1929.
HOLLINGSWORTH, L. W.: *Zanzibar under the Foreign Office, 1890–1913*, London, 1953.
HOLLIS, A. C.: *The Nandi, their Language and Folklore*, Oxford, 1909.
HUNTINGFORD, G. W. B.: *The Southern Nilo-Hamites*, Ethnographic Survey of Africa, London, 1953.
INGHAM, K.: *The Making of Modern Uganda*, London, 1958.
KARSTEDT, F. O.: *Hermann von Wissmann*, Berlin, 1933.
KENNEDY, A. L.: *Salisbury, 1830–1903, Portrait of a Statesman*, London, 1953.
KOLLMANN, P.: *The Victoria Nyanza* (trans. H. Nesbit), London, 1899.
KOSCHITSKY, M. VON: *Deutsche Colonialgeschichte*, 2 vols., Leipzig, 1888.
LANGER, W. L.: *The Diplomacy of Imperialism*, 2nd edn., New York, 1951.
LOW, D. A.: 'British Public Opinion and the Uganda Question, October to December 1892', *Uganda Journal*, 18, ii, 1954, pp. 81–100.
LUKYN-WILLIAMS, F.: 'Nuwa Mbaguta, Nganzi of Ankole', *Uganda Journal*, 10, ii, 1946, pp. 124–35.
McDERMOTT, P. L.: *British East Africa or I.B.E.A.*, 2nd edn., London, 1895.
MAGNUS, P.: *Kitchener, Portrait of an Imperialist*, London, 1958.
MIDDLETON, J.: *The Kikuyu and Kamba of Kenya*, Ethnographic Survey of Africa, London, 1953.

MOFFETT, J. P. (ed.): *Handbook of Tanganyika*, 2nd edn.; Government of Tanganyika, Dar es Salaam, 1958.
MOON, P. T.: *Imperialism and World Politics*, New York, 1927.
MÜLLER, F. F.: *Deutschland—Zanzibar—Ostafrika: Geschichte einer deutscher Kolonialeroberung, 1884–1890*, Berlin, 1959.
OLIVER, R.: 'Some Factors in the British Occupation of East Africa, 1884–1894', *Uganda Journal*, 15, i, 1951, pp. 49–64.
—— *The Missionary Factor in East Africa*, London, 1952.
—— *Sir Harry Johnston and the Scramble for Africa*, London, 1957.
PERHAM, M. F.: *Lugard*: Vol. I. *The Years of Adventure, 1858–1898*, London, 1956.
PRAGER, M.: *Die Wissmann-Expedition*, Leipzig, 1896.
PRINCE, T. VON: 'Geschichte der Magwangwarwa', *Mitteilungen aus den deutschen Schutzgebieten*, (Supplement to *Das Deutsche Kolonialblatt*), Berlin, 1894.
PRINS, A. H. J.: *The Coastal Tribes of the North-Eastern Bantu*, Ethnographic Survey of Africa, London, 1952.
SCHMIEDEL, H.: 'Bwana Sakkarani (T. von Prince)' [trans. A. Wahmhoff], *Tanganyika Notes and Records*, 52, 1959, pp. 35–52.
SCHNEE, H.: *Deutsches Koloniallexicon*, 3 vols., Leipzig, 1920.
SCHWEITZER, G.: *Emin Pasha, his Life and Work*, 2 vols., London, 1898.
TAYLOR, A. J. P.: *Germany's First Bid for Colonies, 1884–1885*, London, 1938.
—— 'Prelude to Fashoda, the Question of the Upper Nile, 1894–1895', *English Historical Review*, 65, 1950, pp. 255–80.
THOMAS, H. B.: 'More Early Treaties in Uganda, 1891–96', *Uganda Journal*, 13, ii, 1949, pp. 171–6.
—— 'The Baganda Martyrs, 1885–1887', *Uganda Journal*, 15, i, 1951, pp. 84–91.
THOMAS, H. B., and SCOTT, R.: *Uganda*, London, 1935.
TOWNSEND, M. E.: *Origins of Modern German Colonialism, 1871–1885*, New York, 1921.
—— *The Rise and Fall of Germany's Colonial Empire, 1884–1918*, New York, 1930.
WOOLF, L.: *Empire and Commerce in Africa*, London, 1920.
WRIGLEY, C. C.: 'Buganda: an outline economic history', *Economic History Review*, 2nd ser., 10, 1957, pp. 60–80.
K. W.: 'The Kings of Bunyoro-Kitara', Part III, *Uganda Journal*, 5, ii, 1937.
ZIMMERMANN, A.: *Geschichte der deutschen Kolonialpolitik*, Berlin, 1914.

MAIN SERIAL PUBLICATIONS

Das Deutsche Kolonialblatt, Berlin, 1890–1914.
Die Deutsche Kolonialzeitung, Berlin, 1884–1920.
Journal of the East African Swahili Committee, Kampala.
Kolonial-politische Korrespondenz, Organ of the Deutsch-Ostafrikanische Gesellschaft.
Petermanns Mitteilungen aus Justus Perthes Geographische Anstalt, Gotha.
Tanganyika Notes and Records, Dar es Salaam, 1936–.
Uganda Journal. Kampala, 1934–.

INDEX

Abbasid dynasty, 106–7, 126.
Abd al-Malik ibn Marwan, 102.
Abdallah ibn Ahmad al-Mazrui, 159, 213.
Abdullah ibn Khamis, 246.
Abdullah ibn Salim, 270.
Abdullah, the trader, 420.
Aberdare, Lord, 372, 378 n.
Aberdare range, 5, 14–15, 17, 37, 317.
Abi Amran, al-Sayyid al-Hasan ibn Muhammad, 107.
Abu Saud, 326, 339.
Abul Mawahib, 118, 123.
Abushiri ibn Salim al-Harthi, 386, 439–42, 446–7, 450, 452.
Acheulean culture, see Chelles–Acheul.
Acholi, 18, 75, 173, 175–8, 200, 326, 339, 343, 403–6, 429–31.
Adal towns, 118–19.
Aden, 111, 137, 149, 166, 242.
Adowa, 385 n.
Adulis, see Eritrea.
Afar, 66–67, 71, 83 n., 87; see also Hamites.
Afghanistan, 370, 373.
Africa, South, 45–46, 56.
African Ghats, 8.
Africanthropus njarensis, 39.
Agatharchides of Cnidus, 60–61, 70–71.
Agaw, 65, 68–69, 71.
Ahmad ibn Ibrahim, 272.
Ahmad ibn Majid, 106 n.
Ahmad ibn Said al-Busaidi (1744–83), 157–8.
Ahmad ibn Said al-Mazrui, 159.
Ahmad, Simba, 235–6, 248, 365, 369, 371.
Ainsworth, John, 414–15.
Akokoroi river, 326.
Al Hasa, 102.
Ala-ud-din Bahmani, 111.
Alagwa, 201.
Albert lake, 7, 9, 15, 40, 80, 172–4, 199, 337–8, 343, 358–9, 403, 405–6, 426, 429.
Albert Nile, 9, 15, 18, 173–5.
Aleppo, 120.
Alexandria, 96–97, 99, 363.
Algeria, 27, 29, 40, 48, 56, 362.

Ali (of Tumbatu), 215.
Ali Bey, amir, 137, 138, 140.
Ali, chief of Wadelai, 429.
Ali, Husain, 326.
Ali ibn Daud, 118.
Ali ibn Said, see Zanzibar.
Ali (ibn Saif), 157.
Ali ibn Uthman al-Mazrui, 158.
Alibhoy, Adamjee, 415.
Allen, J. W. T., 162.
Almeida, Francisco d', 124, 134.
—Pedro, 141.
Alolo, see Makua.
Alur, 174–6, 178–9, 204–5.
Amadi (Mwenyi Mkuu), 223.
Amala river, 419.
Amir Musa, 102.
Amuria, 326.
Anaya, Pedro d', 134, 143.
Anglo-French Declaration (1862), 357 369, 374, 437.
Angola, 37, 45, 365.
Angoxa, 150–1.
Angra Pequena, 365 & n. 1, 370.
animism, 127, 167.
Ankole, 7, 14, 18, 73 n, 3, 88, 172 n. 4, 181–2 & n. 1, 185–9, 262, 328, 402–3; people, 87, 89, 92; kingdom of, 191, 297, 331–2, 335–6; trade, 405, 422; and Britain, 428–31.
Annam, 121.
Anti-Slavery Act (1807), 156.
Anti-Slavery Society, 379.
Anuak, 172, 175.
Apis Rock, 45–46.
Arabia, 40, 65, 95, 107, 111, 129, 137, 141, 155, 162, 216, 223–4.
Arabi Pasha, 363.
Ardashir I, 97.
Arendt, consul-general, 447.
Arkell, Dr., 82.
Aromata, see Olok.
Arorr valley, 6.
Arusha, 54, 300, 307.
Asia (João de Barros), 102, 104, 112 n., 118–19.
Asselar, 48, 64.
Asu, 130, 202, 318.

INDEX

Atbara river, 385.
Aterian culture, 45.
Athi river, 17, 90.
Atiak clan, 174; *see also* Lwo.
Atlanthropus, 27, 29, 35, 56.
Australopithecus, 24.
Australopithecines, 24-25, 35, 55-56.
Austria, 357, 380.
Avalites, 94.
Awich of Payera, 406, 430.
Axum (Aksum), 68, 99-100; monument, 71.
Ayyubid sultans, 111.
Azanian culture, 73-74, 93, 100.
Azd, 106.
Azzan, 233.

Ba-Amiri clan, 145.
Badari, 48.
Badger, Rev. G. P., 103, 360-1 & n.
Bagamoyo, 131, 136, 147, 167, 244, 301, 442 & n., 454; trade, 248-9, 318, 398; and Germany, 439-40, 450-1; Arab Muslim uprising, 401.
Bagamoyo, The History of Former Times in, 147.
Baghdad, 111, 120, 146.
Bahmanid kings, 111, 121.
Bahr al Ghazal, 84, 325.
Bahram V, 99.
Bahrein, 99.
Bahrey, the priest, 76 n.
Baisengobi, princely houses, 185.
Bajun Islands, 248.
Bajuni, 129.
Baker, Sir Samuel, 338-41, 343, 350, 358, 426.
bananas, 21, 110 & n., 186, 258.
Bandar Abbas, 161.
Bandar Hais, 94.
Bandawe, 289.
Bantu, 55, 58, 60, 81-82, 84, 86-87, 89-93, 106, 110 n. 1, 145, 147, 178-80, 183-4, 195, 197, 199, 202, 206, 254, 260, 262, 264, 302, 308, 328; language, 58, 60-61, 70, 80-81, 101, 117, 168, 171-3, 176 n., 191, 199, 202-3, 254; Interlacustrine, 191, 197, 327; Kavirondo, 66, 86, 89-90, 180, 203, 308-10 & n.
Barabaig, 201.
Barawi, 147.
Barghash ibn Said, *see* Zanzibar.

Bari, 74-75, 175, 177 & n. 1, 200, 339.
Baring, Sir Evelyn, 364, 385.
Baringo lake, 5-6, 300, 302, 305, 307, 315, 317-18, 321, 413.
Barquq, Sultan, 119.
Barreto, 136.
Barya, 65.
Bateko, priestly class, 192.
Baumann, H., 92 & n.
Becker, Jerome, 292.
Beit al-Amani Museum, 97.
Beit Bridge, 47.
Beja, 65-66, 68-69.
Belgium, 249, 270, 283, 291-3, 362, 367, 426, 429, 433; *see also* Leopold, King.
bellows, 81-82.
Bemba, 195, 278 n. 2, 282, 285, 288.
Bena, 195-6, 210, 258, 284-5.
Benadir coast, 225-6, 401.
Bende, 193.
Benecki, 295.
Benguela, 273 n.
Benue, 80.
Berbera, 94, 149, 152.
Berlin, 236, 369, 372, 436, 451.
Berlin Act (1885), 374.
Berlin Conference (1884), 293, 366-7, 369.
Bigo, 181, 182 n.
Bir Gao, Birikau, 89 n. 3, 130; *see also* Shungwaya.
Bisa, 265-7, 282, 286.
Bismarck, Chancellor: colonial policies of, 373-4, 441; fall of, 382, 450; *Schutzbrief* (1885), 293, 369 & n., 391, 436; and German Companies, 364-5, 369, 381, 435; and Abushiri's revolt, 386, 440-1; and Zanzibar, 359, 364, 434; and Britain, 363-4; and the Angra Pequena affair, 365-7.
Bismarckburg, 442.
Bito, 73 n. 3, 87-88, 172-3, 175-6 & n. 2, 178, 181-2, 184-7, 189-90, 195; *see also* Hima; Hinda.
Blantyre, 289.
Bleek, Dr., 80.
Blue Nile, 38, 385.
Bocarro, Gaspar, 153, 207.
Bondei, 130, 436.
Boni, 101.
Bokono, 185.
Bombay, 232, 237, 270; Government, 157, 159-60, 219, 229-30, 233, 238.

INDEX

Boran, 297, 301, 322; *see also* Galla.
Boteler, Lieutenant, 159.
Brassey, Sir Thomas, 378 n., 380.
Brava, 102, 106, 117; trade, 225, 300; and China, 121; and Portugal, 132, 134, 137–8, 142, 144, 152; and Omani rulers, 157, 159, 220, 374; and Egypt, 359.
Brazza, S. de, 362.
Brenner, Richard, 236.
British East Africa Association, 372, 378, 386.
British India Steam Navigation Company, 242, 355.
British North Borneo Company, 372.
Broken Hill, 38–39, 56. [377–8.
Bruce, Alexander L., 379.
Brussels, International Conference (1876), 362.
Bryan, Miss M. A., 77, 79.
Buddu, 184, 190, 421, 423–4.
Bugabula, 185.
Buganda, 8–9, 13–14, 18, 172 & n. 4, 181–2, 184, 187–91, 195, 284, 323, 328, 330, 332–7, 340 ff., 358, 382, 392–4, 399 ff., 420 ff., 456; administration of, 190–1, 332–4, 339, 399, 404, 428; political parties: Protestants, 345, 350, 382, 392–3, 398–402, 407, 420–1, 423–8; Roman Catholics, 290, 300, 382, 400–2, 421, 423–4, 426–30; Muslims, 347–50, 401–2, 420–5; Wa Ingleza, 402–3, 407, 420–2; Wa Franza, 402–3, 407, 420–3; religious wars and land settlements, 420, 422–7; trade, 334–5, 429; from — to coast, 153, 190, 205, 226, 272–3, 334, 410; monopoly, 190, 324; roads, 190–1, 279, 317, 323–4, 343; traders: Arabs, 287, 290, 400–1; Khartoumers, 299; Zanzibaris, 294–9, 317, 323–4, 334, 342–3, 345–7, 350, 420; and Bunyoro, 297, 337, 340–3, 345, 402–8, 421–2, 424–7, 431; and Egypt, 340–3, 346; and Germany, 382–3, 400, 402; and Britain, 297, 383, 393, 402–4, 406–8, 420–31.
Buganza, 187.
Bugerere, 184.
Bugessera, 187.
Bugusu, 309.
Bugwere, 185.
Bugweri, 185.

Buhaya, 328–9, 335.
Buhimba, 329.
Bukara, 330, 344.
Bukedea, 327.
Bukinga, 196.
Bukoba, 8, 13–14, 18, 184–6 n. 2, 443, 448, 452.
Bukoli, 185.
Bukumbi, 443–4.
Bukusu, 310.
Bulamogi, 185.
Bulingugwe island, 401, 423.
Bülow, von, 445, 447.
Buluhya, 310.
Bumura, 147.
Buna, 322.
Bungungu, 184.
Bunyoro, 14, 88, 172–3, 175–6, 178, 181–2, 184–5, 187–91, 194–5, 203 n., 297, 328, 339 ff., 350, 393, 403 ff.; expansion of, 189–91, 323, 344–5, 403–7, 421–6, 431; trade, 324, 327, 337, 343, 345, 430, (in iron), 327, 335, (firearms) 405, 421–2, 425, 430; and Arabs, 405, 407, 425; and Khartoumers, 299, 326–7, 337, 339–40, 430; and Zanzibar, 323, 327, 337, 344; and Egypt, 339–44, 358, 403–4, 406–7, 422, 431; and Britain, 408, 422–7, 430–1; missions in, 347, 428.
Burdett-Coutts, 379.
Buret, 308.
Burji Mamluk sultans, 111, 119–20.
Burma, 121.
Burton, Richard, 153, 170, 193, 226–7, 254, 260–3, 265–7, 271–3, 275, 277, 285, 338, 446.
Buruli, 184.
Burundi, 186, 191–2, 193 n. 1.
Burungi, *see* Alagwa.
Busaidi, 157–9, 212, 235, 439.
Bushmen, 28, 38, 46–47, 55–56, 60–63, 65–66, 77, 83–84, 92, 130.
Busoga, 181, 185, 189–91, 272, 310, 328, 331, 335–6, 400, 426; trade, 429; and Britain, 408, 428–31; *see also* Buganda.
Busongora, 403, 425–6.
Busubi, 186; *see also* Rusubi.
Busuju, 423.
Butambala, 189, 423, 425.
Büttner, C. G., 451.
Butundwe, 344.

Buvuma Islands, 335, 341.
Buwheju kingdom, 187.
Buxton, Sir Thomas Fowell, 248–9, 379.
Buzimba kingdom, 187.
Buzinza, 186, 297, 323–4 & n. 3, 328, 330, 335–6, 344.
Bweni, 214.
Bweni Kuu, 451.
Bwera, 184.
Bweyorere, 182 n.
Byzantium, 96–97, 126, 379.

Cabral, Pedro Alvares, 134.
Cabreira, Francisco Seixas de, 141.
Caffa, 120.
Cairo, 119, 149, 358, 411; see also Cape to Cairo route.
Calabar, 80.
Cambay, 121, 124, 149–50, 166.
Cambier, Captain, 291.
Cambo, 136.
Cameron, Sir Donald, 194.
Cameroons, 80–81, 366.
Cananor, 149.
cannibalism, 53, 129, 138, 296; see also Zimba.
Canning, Lord, 233.
Canton, 108.
Cape to Cairo route, 381 & n., 384.
Caprivi, 450.
Capsian culture (Middle Stone Age), 40.
Casati, Gaetano, 404, 405.
Caucasoids, 42, 48, 54, 56.
Central African Railway, 410–11, 446.
Ceylon, 108.
Chad, Lake, 81–82.
Chaga, 90, 130, 197, 202–4, 266, 274, 307; trade, 225–7, 313–14, 319; and Germany, 436, 445.
Chaillé-Long, C., 341.
Chamberlain, Joseph, 367, 368.
Changamwe, 145.
Channel, The (Mombasa), 95.
Charibael, King, 95.
Cheke, rock shelter, 54–55.
Chelles–Acheul culture, 26–29, 33, 35, 37, 56.
Chewa, 256; see also Maravi.
Chifubwa Stream shelter, 46.
Chikuse, 209.
Chimbii, 313.
China, 107–8, 116 & n. 1, 120–1, 149–50, 155; see also porcelain.

Chingulia, Dom Hieronimo: see Yusuf (of Mombasa).
Chipeta, 209–10.
Chiponda, chief, 207.
Chittick, H. N., 125–7.
Chogwe, 247.
Chola kingdom, 110.
Chope, 173, 174, 178.
Chuka, 90.
Chumbe island, 231.
Churchill, Henry Adrian, 234, 237–8.
Chwa, 181–2, 184; see also Buganda.
Chwezi, 86–87, 181–2, 184–5, 187.
Chyulu range, 5.
circumcision, 64, 70–71, 76, 87, 90, 167, 199.
Clarke, Herbert, 243.
Cline, W., 82.
Coghlan, General, 234.
coins, 97–99, 104, 107, 110 n., 111, 115, 123–4, 127, 131–2, 138.
Colloe, 99.
Colvile, Colonel H. E., 425–7, 431.
Comoro Islands, 108 n. 6, 115, 374.
Congo, 37, 44, 193 & n. 1, 197, 209, 211, 282, 295–6, 406, 420, 425, 433; trade, 264–5, 276, 288–9, 449; King Leopold's claim, 293, 362, 367, 369; Leopold's International Association, 291–3 & n. 1, 362; and Omani, 251, 292–3; and Britain, 362, 366, 369; and Germany, 443–4; see also copper.
Congo Free State, 294 n. 1, 295, 367, 383 n. 1, 426, 429, 433; see also Tippu Tip.
Constantius, Emperor, 99.
Cooke, H. B. S., 16.
copper, 117 & n., 121, 151, 153, 207, 264–5, 267, 273–4, 278, 280, 286.
Cory, H., 205.
Cosmas Indicopleustes, 98–99.
cotton, 225, 257, 260, 294, 415.
Coupland, R., 152.
Covilha, Pedro da, 133.
Crazzolara, Fr. J. P., 175–6, 178.
cremation, 53.
Crispi, Signor, 385.
Cromer, Lord, 411.
Crusades, 112, 133.
Ctesiphon, 97.
Cuama, 150–1.
Curtis, Dr. G. H., 25.
Cushites, see Hamitic peoples; Hamites.

INDEX

Dabida, 114 n.
Dallons, Captain, 148, 155–7, 212.
Damot kingdom, 71.
Dar es Salaam, 12, 97, 144, 147, 222, 236, 247–9, 267, 441, 449–50; and Germany, 371, 374, 376, 437–9, 445, 447, 450–2.
Dar Fung, 63–64, 66.
Dar Fur, 78 n. 2, 82, 337.
Dar Nuba, 64, 66, 78 n.
Darod Somali groups, 321–2.
Dasikera, 318.
Daud ibn Sulaiman, 119, 125.
Daulatabad, 111.
de Barros, João, 102–4, 112 & n. 2, 118, 143.
Debono (Maltese trader), 326.
Deccan, 111, 118, 121.
Decken, Baron von der, 322.
Delgado, Cape, 96, 213, 223.
Denhardt, Clemens, 365, 369, 371, 374.
Derby, Lord, 368.
Didinga, 75–76, 176.
Digo, 90, 114 n. 2, 130, 145.
Dilke, Sir Charles, 367–8, 388.
Din Muhammad, 230.
Dinka, 74–75.
Diocletian, Emperor, 97.
Diogo de Alcaçova, 132.
diseases, 21, 134–5, 142, 227, 308, 353, 417–19, 426, 448.
Disraeli, Benjamin, 358, 363.
Doe, 131.
Dorobo, 62–63 & n. 4, 73 n. 3 & 4, 83 n. 3, 202, 311, 318–19, 419.
drums, royal, 185 n., 186, 328.
Duarte Barbosa, 149, 154, 163–4, 166.
Dufile, 404, 406.
Dunga, 214.
Duruma, 90, 114 n.
Duta, Henry Wright, 428.
Dyson-Hudson, Neville, 321 n.

East Africa Royal Commission (1953–5), 10, 19.
East India Company, 229, 360, 372.
Eastern Telegraph Company, 242.
Eburru mountain, 50, 182.
Edward lake, 7, 331, 383–4, 426.
Egypt, 48–49, 51, 60, 64, 73 n. 3, 74, 88–89, 125, 440; Ptolemaic, 97–99; Islamic, 111; Burji Mamluks, 119; trade with coast, 95, 119–20, 134, 153, 225; monopoly of Upper Nile, 320, 337; and E. Africa Lakes Region, 245, 300, 337–45, 350, 403–7, 422, 431; and Britain, 358–9, 362–6, 370, 374, 376–7, 379–80, 385, 387–8.
El Banes, 106.
Elandsfontein, 38.
Elgeyo escarpment, 5–6.
Elgon, Mount, 5, 15, 17, 53, 86 & n., 89 & n., 179, 181, 201, 301, 309, 317, 419.
Elmenteitan culture, 40, 44, 47, 49–50, 53, 57; industry, 40.
Embu, 90.
Emery, Midshipman, 160.
'Emozaidij', *see* Said ibn Ali.
Emin Pasha, 295, 343, 351, 376–8 & n. 6, 380, 382–3, 403–6, 430, 442–4; relief expedition, 376–8, 382, 406, 442 & n. 2, 444–5.
'Emporia', 94 & n., 96–97, 125, 218.
Enarya kingdom, 71.
Engaruka ruins, 74, 202.
Entebbe, 425.
Equatoria (Equatorial Province), 295, 340, 343, 358, 376–7, 382, 403, 406, 442 n.
Eritrea, 98–100, 385.
Erythrean Sea, *see Periplus of the Erythrean Sea*.
Essina, 96.
Ethiopia, 37, 62–66, 68, 71, 73, 76, 84, 86–87, 93, 99–101, 156, 242, 297, 320; trade, 322; and Italy, 385 & n.
Euan-Smith, consul-general, 399, 411.
Eustace, naval officer, 147.
Evernden, Dr. J. F., 25.
Eyasi, 6, 38–39, 55–56, 61, 74.

Fabri, Frederic, 361.
Fadl al-Mula, 406.
Fajelu, 176.
Famau family, 221; *see also* Siu.
famine, 314, 417–18, 426.
Faqih Muflah al-Malindi, 104.
Fauresmith culture, 36–39.
Fava, Abbé, 243, 244.
Fayum, 48.
Faza, 137, 138, 141, 142, 152, 220.
Fernandes, Antonio, 88, 112 n.
Ferry, Jules, 362.
Fiji, 364.
Fiomi, *see* Goroa.

Fipa, 193, 208, 260.
firearms, 276–7, 280–2, 285–6, 292–4, 316, 325, 330–1, 334–5, 337, 344, 360, 405, 420, 425; trade in, 287, 319, 322–3, 327, 345, 407, 411, 421–2, 430, 436, 440.
Fischer, G. A., 416.
Fitzmaurice, Lord Edmund, 367.
flood plain, 4, 19.
Florisbad (South Africa), 39.
forest, 17–18, 302, 311–12, 318, 415.
Fort George, 425.
Fort Hall, 90.
Fort Jesus, 140–1, 144, 156, 158, 160–1, 165.
Fort Portal, 14, 393, 425.
Fort Smith, 408, 413, 415–17.
France, 224, 290, 293, 352–3, 357, 361, 423, 439; trade with E. Africa, 155–6, 219, 235, 268 n. 2, 269, 276; missions, 401–3, 439; and Omani, 155–6, 161, 228–30, 233; and Britain, 362–6 & n. 2, 370, 373–4, 380, 387, 437.
Freeman-Grenville, Dr. G. S. P., 97, 126.
Frere, Sir Bartle, 239, 242, 249.
Freretown, 245.
Fumambogo sub-dynasty, 184; see also Buddu.
Fumoluti ibn Shaikh, 159, 221, 235.
Fundikira, chief, 262, 268, 271, 277.

Gaboon, 362.
Gabu dynasty, 181; see also Bunyoro.
Gabunga, 335–6.
Galla, 65–69, 71, 73 n. 3, 74, 76–77, 86–87, 89–90, 105 n. 1, 114, 130, 144, 200, 202, 206, 242, 274, 303, 321–2, 413–14, 417, 419.
Gamble's Cave, 40–42, 47, 56.
Ganda, 91, 189, 206, 272, 297, 323–4, 330, 399; see also Buganda.
'Garaganza' traders, 278 & n., 288.
Gare, 451.
Gaseni, 313.
Gazi, 246–7.
Gedge, 413.
Gedi, 132, 137, 139, 144, 163, 165–6.
George, Lake, 7, 14.
German African Society of Berlin, 434.
German East Africa Company, 374, 381, 386, 436–7, 441, 447, 451.
German Witu Company, 378, 381.

Gesellschaft für Deutsche Kolonisation (German Colonization Society), 368–9, 386, 434–6.
Giesecke (German trader), 442–3.
Giraud, V., 293.
Giryama, 114 n.
Gissaka, 186, 329, 331.
Gisu, 53, 89, 203.
Gladstone, W. E., 361–3, 368, 370–3, 377, 388–9 & n. 2.
Glaser, Dr., 99.
Goa, 136, 138, 140–1, 156.
Gogo, 202, 254, 258, 263–4, 274, 296–7, 303, 305, 444–6.
Golbanti, 417.
Gold, 110, 117, 131, 134–6; *see also* Sofala.
Goldie, Sir George (Taubman), 366 n. 2, 409.
Gomani, 209; see also Ngoni.
Gomba, 189, 423.
Gondokoro, 297, 300, 324, 338–9, 341.
Gordon, Charles, 340–3, 345–6, 350, 358–9, 364, 370, 376–7, 388, 403.
Goroa, 201–2.
Grant, J. A., 338.
Granville, Lord, 368, 371–2.
Gravenreuth, Freiherr von, 440, 451.
Gray, Sir John, 207.
Greece, 94–95, 97–99, 101.
Greenberg, J. H., 69, 81.
Guardafui, Cape, 73 n., 94, 98.
Guha, 284.
Guillain, M., 226.
Gujerat, 121, 124.
Gumba, 50–52, 73 n.
Gusii (Kisii), 89, 308, 315.
Guthrie, Professor M., 81.
Gweno, 90.

Ha, 192, 260–1.
'Habari za Wakilindi', 197–9, 205.
Hadhramaut, 162.
Hadimu, 214–15; *see also* Zanzibar.
Hadramauti, 123; *see also* Kilwa.
Hadya kingdom, 71; *see also* Sidama.
Hadzapi, 55, 61–63.
Hamburg, 234, 355, 442.
Hamerton, Atkins, 161, 219–20, 229–31, 234, 238.
Hamid ibn Ahmad, 159.
Hamites, Hamitic peoples, 54, 60, 63–71, 73–74, 77–79, 83–84, 86–88, 91–93, 95, **194**, 199, 202–3, 206, 308, 320.

INDEX

al-Hamziya, 163.
Handeni district, 130.
Hanga (Wanga), 89, 203 n. 1, 310, 319.
Hannington, Bishop, 336 n., 392, 400, 416.
Hanseatic Republics, 234.
Hapsburgs, 357.
Harcourt, Sir William, 388 & n.
Harthi clan, 212-14, 232, 234, 237; see also Zanzibar.
Harun al-Rashid, Khalifa, 102.
Hasan bin Omari, 446.
al-Hasan ibn Abu Bakr, 124 n.
al-Hasan ibn Ahmad, 140.
Hasan ibn Ahmad Alawi, 214-15.
al-Hasan ibn Ali, 102-3.
al-Hasan ibn Ibrahim, 157.
al-Hasan ibn Sulaiman (1005-42), 125.
al-Hasan ibn Sulaiman (1191-1215), 125.
Hasan ibn Sulaiman (c. 1310-33), 119.
al-Hasan ibn Sulaiman (1479-90), 123-4.
al-Hasan ibn Talut, 118-19.
Hassan, Sultan, 223.
Hatimii, 144.
Hatzfeldt, Count, 383-4.
Haya States, 184, 186-7, 336, 443.
Hehe, 195-7, 210, 258, 297, 303; trade, 284-5; and German administration, 433-5, 439, 446.
Heligoland, 383-4.
Hemedi bin Abdullah bin Said al-Buhriy, 438, 440, 452.
Henderson, Sir John, 166 & n.
Henry the Navigator, 133.
Herti, 321; see also Darod Somali.
Hima, 66, 73 n., 86-89, 91, 181-2, 184-7, 189, 194-6, 274, 282, 328; see also Bito; Tutsi.
Hinda, 172 n., 182, 185-7, 194-5 & n., 205, 262; see Ankole.
Hippalus, 98.
Hirth, Monsignor, 423-4.
Hoffmann, Sergeant-Major, 444.
Hohenlohe-Langenburg, Prince, 368.
Holford-Walker, Frank, 304 n., 305 n.
Hollis, Sir Claud, 145.
Holmwood, Frederick, 220, 372, 377, 379.
Holoholo, 260.
hominid, 23, 25-27, 29, 55.
Homo rhodesiensis, 38-39, 56.

Homo sapiens, 29, 35, 39-40, 56.
Hondogo, 187.
Honea, 65.
Hope Fountain industry, 31.
Hopefield (South Africa), 38.
Hormuz, 141.
horn, royal, 145-6, 198.
Horn of Africa, 66, 98, 321.
Horner, Antoine, 244.
Horombo, chief, 313.
Hottentots, 55, 61-62
Houghton, missionary, 417.
Hübbe-Schleiden, W., 361.
Hungary, 357, 416.
Husuni, *see* Kilwa.
Hutton, James, 372-3, 378 n.
Hutu, 86; *see also* Hima.
Hyrax Hill, 50-52, 72, 74 n.

Ibn Ahmed, 212 n.
Ibn Battuta, 113-14, 116-17, 124, 145-7.
Ibn Hawqal, 105, 108.
Ibn Said, 118.
Ibrahim, Sultan, 123.
Idrisi, 106, 108 & n., 115.
Ihangiro, 186, 191, 329.
Il Damat, 306.
Il Khans, 111.
Île de France, 155-6.
Ilogolala, 305; *see also* Masai.
Ilongo, 196.
Imam Sultan ibn Saif, 141.
Imatong Mountains, 176.
Imperial British East Africa Company, 245, 377 n., 378 n., 380-1, 383 n., 387-90 n., 393-6, 398-99, 407-10, 412-14, 416-17, 420, 422, 424, 431.
India, 64, 94, 96, 98-100, 107-8, 110-11, 117, 121, 123, 133-4, 140, 147, 153, 155-6, 162, 166, 212 n. 3, 218 & n. 2, 223, 226, 237, 272, 358, 395; Muslim, 111, 115, 118, 121, 127, 218; British, 235-6, 254, 427; trade with E. Africa, 120-1, 124, 131, 150-1, 155, 163, 219, 223, 225; Indian traders, 115, 158, 219, 222, 224, 227, 269-70, 275, 278, 315, 415, 439, 441, 450.
Indonesia, 95-96, 107-8, 110 & n., 455.
al-Inkishafi, 163, 165-6.
Iramba, 203.
Iran, 111.
Iraq, 111.

Iraqw, 52, 74, 201–2.
Iringa, 7–8, 195–6, 439, 446, 448, 450.
iron, 3, 48–49, 51, 55–56, 63, 74, 81–82, 95, 106–7, 121, 136, 151, 182, 205, 255, 257, 268, 319, 327, 455.
Iron Age, 48, 55, 72, 74, 93, 100, 253, 257.
Iru, 86; see also Hima.
Isanzu, 203.
Iseera, 327; see also Teso.
Ishangian Mesolithic culture, 40.
Isimila, 31.
Isingiro hills, 186.
Isingoma Mpuga Rukidi, 182; see also Bito.
Islam, 102, 110–12, 115, 120, 122, 133, 152, 167, 208, 287, 347–9, 397, 448, 452, 456.
Ismail, Khedive, 338–40, 358–9, 363.
Italy, 357, 362, 365, 380, 401, 440; Ucciali Treaty, 384–5; and British E. Africa Association, 378, 381, 384.
Itawa, 273, 284, 288.
Itigi, 19.
ivory trade, 19, 95–98, 100, 106, 108, 112–13, 115, 120–1, 149–55, 158, 190–1, 205, 225–7, 236, 257, 265, 268–9, 275–6, 285–6, 288–9, 292, 294, 296, 315, 319, 322–3, 325, 339, 410, 433, 449.

Jackson, Frederick, 382, 396, 402, 413, 416, 429.
Jacobs, Dr. Alan, 304 n., 305 n.
Jalairid sultans, 120.
Janjero kingdom, 71, 87–88.
Jaunpur kingdom, 121.
Java, 24, 27, 29, 107, 110, 166.
Jebel Moya, 76.
Jiddah, 141.
Jie, 176–7.
Jiji, 192, 260–1.
Johannes, Captain, 445.
Johnson, Archdeacon, 255.
Johnston, Sir Harry H., 80, 367, 371, 373, 381 n.
Jomvu, 145.
Jopadhola, 179, 200.
Jopaluo, 327, 404.
Juani Island, 163.
Juaya, 403.
Juba, 121, 177 n.; river, 89, 213, 321–2, 410, 412.
Jühlke, Dr. Karl, 371, 435–6.

Juma bin Rajab, 267–8.
Junker, W., 404, 442.

Kabale, 14.
Kabarega, 337, 339–45, 350, 402–8, 421–2, 425–6, 431; see also Bunyoro.
Kaberamaido, 178, 326.
Kabete, 14.
Kaboyo (of Toro), 344.
Kachope Kamurasi, 404.
Kadam, 5.
Kafa, 67–68, 71, 87, 100.
Kafu river, 25, 184, 403, 405, 426.
Kafuan culture, 26.
Kafuro, 324, 330.
Kagei, 287.
Kagera, river, 25, 31, 38, 186, 190.
Kagulu (of Buganda), 190 n.
Kaguru, 197.
Kagwa, Sir Apolo, 153, 402, 424–5.
Kahama district, 194, 210.
Kahe, 313.
Kaiso, 7.
Kajengwa, 110 n.
Kakamega, 310.
Kakwa, 176.
Kalambo Falls, 31, 35.
Kalasa (of Katanga), 277.
Kalema (of Buganda), 401–2.
Kalenga, 446.
Kamasia horst, 6, 419.
Kamba, 90, 130, 153, 202–3, 302, 311–12, 418; trade, 225–6, 267, 314–17, 320, 419; and Britain (administration), 393, 408, 413–15, 417.
Kambatta kingdom, 71.
Kampala, 184, 393, 408, 421, 423–5, 428, 454.
Kamurasi, 337.
Kanam, 27–28, 46.
Kanbalu, 106 & n., 107.
Kanga clan, 193; see also Nyamwezi; Sukuma.
Kangeju, 61.
Kanjera, 28, 35, 56.
Kano plains, 8.
Kapoeta district, 84, 86.
Kaputie plains, 306–7.
Karagwe, 2, 6, 8, 14, 153, 185 n., 186, 191, 279, 329–30, 344, 347, 443; trade, 324, 422.
Karamoja district, 9, 13, 14, 18, 45, 177, 454.

INDEX

Karamojong, 77, 176–7, 298 & n. 1, 309, 320, 326, 419, 429.
Karema, 283.
Kariandusi, 31.
Karuri (Kikuyu), 312.
Kasagama, Daudi, 408, 422, 425–6, 428; *see* Toro.
Kashaija (of Buzinza), 330.
Kassala, 385.
Katanga, 117 & n. 4, 264–6, 273 & n. 2, 277–8, 286, 288.
Kategora, 405.
Katerega, 189.
Kato Kimera, 182.
Katonga river, 181, 328.
Katwe, 403, 421.
Kau, 235.
Kauma, 90.
Kavalli, 405–6, 421, 429.
Kavirondo, 8, 27, 46, 178–9, 297, 307–9, 317–19, 327, 335, 413–15, 417–18, 428.
Kayango, 185 n.
Kazembes, *see* Lunda.
Kazinga channel, 7.
Kazuna (of Kimwani), 344.
Kedah, 108, 110 n.
Keekonyukie, 305–6; *see also* Masai.
Keni, 313.
Kenya, Mount, 5, 14–15, 17, 37, 51, 90, 153, 202, 242, 267, 297, 301, 310–11, 314–15, 317–18, 322, 350, 419.
Kenya Capsian culture, 39–41, 44, 47–50, 56.
Kericho Highlands, 14.
Kerimba Islands, 119.
Kerio, river, 321.
Keyo, 201, 302, 309, 418; *see also* Nandi.
Khalid ibn Said, 230–2.
Khalid ibn Sulaiman, 118.
Khalifa ibn Said, *see* Zanzibar.
Khandesh kingdom, 121.
Khartoum, 299, 324, 337–8, 340, 350, 370, 377, 403; mesolithic culture, 48–9, 56, 64; trade, 325–7, 337–40, 430.
Kiambu district, 90, 310, 318, 415–16.
'Kibinge', *see* Kippingo.
Kibiro, 403, 405–6, 426.
Kibosho, 313.
Kibwezi, 410, 415.
Kiel Canal, 384.
Kiga, 331, 429.
Kigeri IV (Rwanda), 331–2.
Kikulwe (of Buganda), 190 n.
Kikuyu, 52, 73 n., 90, 114 n., 130, 202–3, 302, 306–7, 310, 312–13, 350; trade, 311, 314, 316–18, 320, 334; and Britain (administration), 394, 407–8, 410, 413, 415–17, 431.
Kilema, 204, 313.
Kilifi, 114, 137.
Kilimanjaro, Mount, 6, 14–15, 89, 97, 153, 197, 242, 247, 254, 267, 297, 300–1, 304–5, 307, 313–14, 316–17, 322–3, 367–8, 371, 373–4, 436–7, 445, 448, 450–2.
Kilimatinde, 448.
Kilindi dynasty, 197, 204–5.
Kilombero, 4, 7, 19.
Kilosa district, 197, 448, 450.
Kilwa, Chronicle of, 102–5, 107, 112, 116, 119, 122–6, 134.
Kilwa (Kisiwani), 102, 107, 112–13, 119, 125–7, 131–2, 136, 138–9, 143–4, 146, 148, 150, 155, 163–5, 167, 206, 222, 249, 283; rulers, 103–4, 112, 118–19, 122–4, 135, 148, 155–6, 222; trade, 117–19, 124, 135, 149, 151, 224, 241; slave trade in, 155, 226, 237, 240–1, 265, 269, 274, 286, 290; routes, 112, 117, 133, 255, 265–6, 268 & n., 273–4, 284–5; and Portugal, 134, 156–7; and Oman, 156–8, 222, 224; and Germany, 438–9, 446.
Kilwa Kisiwani, Ancient History of, 148 222 n.
Kilwa Kivinje, 14, 131, 147, 153–4, 162, 164–5, 222–3, 440, 448.
Kimameta, Jumbe, 415, 419.
Kimberley, Lord, 368.
Kimera, 181–2.
Kimnyole, 418.
Kimwani, 344.
Kimweri, 204, 223 n., 237, 247.
Kindiga, 61.
Kinga, 195–6, 209, 254, 257.
Kintu, culture hero, 181.
Kionga, 149.
Kipalapala, 442.
Kipengere range, 8, 257.
Kipini, 153, 374, 378.
Kipsigis, 201, 307–9, 418–19; *see also* Nandi.
Kipumbwi Mkuu, 146.
Kirk, Sir John, 234, 237–41, 249–51, 291–2 n. 3, 357, 359–61, 367–8, 371–2, 379, 399.

INDEX

Kirua, 313.
Kisamajeng, 201.
Kisese, rock shelter, 46, 54–55.
Kisimani Mafia, 110 n. 2, 113, 124, 126, 133.
Kisiwani, *see* Kilwa.
Kismayu, 137, 220, 359, 374, 378, 384, 396, 409.
Kissongo, 305, 306; *see also* Masai.
Kitab al-Zanuj, 102–3, 130, 162.
Kitakwenda, 184, 187.
Kitale, 200, 317.
Kitara kingdom, 181, 184, 329.
Kitchener, Col. H. H., 373, 376, 409, 437.
Kitgum, 430.
Kitosh, 417, 419–20.
Kitui district, 312, 415, 419.
Kivu, lake, 2, 7, 331.
Kiwewa (Buganda), 401.
Kiziba, 184, 187, 189, 191, 329; *see also* Haya.
Kizimkazi, 107, 127.
Knorr, Admiral, 447.
Koikoiti (Masai), 304.
Koitatel (Nandi), 418.
Koki, 184, 185 n. 2, 187, 190, 191, 336.
Kolo rock shelter, 55.
Kolonialverein, 368.
Kolonialzeitung, 368.
Kondoa, 54, 55, 201.
Kony, 201, 301, 419; *see also* Nandi.
Kota Kota, 287.
Krapf, Johann, 204, 223 n., 242, 268 n., 273 n., 322, 338.
Kua, 124, 163, 165, 167.
Kublai Khan, 111.
Kumam, 178, 299, 326–7.
Kunama, 65.
Kunduchi, 147.
Kurasini, 452.
Kuria, 301.
'Kwavi', *see* Masai.
Kyabagu (Buganda), 153, 190, 205.
Kyagwe, 429.
Kyaka, 184.
Kyamtwara, 186, 191, 329.
Kyoga lake, 9, 18, 172, 175, 177–9, 299, 324, 326–7, 335, 337, 429.

La Bourdonnais, 155.
Labongo, 404.
Lacah, Seven Brothers of, 102–4.

Lacerda, Dr., 265–6.
Lado, 343, 358.
Laetolil, 24.
Laikipia, 5, 17, 300, 304–7, 311, 319, 417.
Laitokitok, 305; *see also* Masai.
Lambert, H. E., 90.
Lamu, 12, 95, 102–3, 119, 129, 132–3, 137, 144–6, 152–3, 157, 159, 220, 223, 226, 232, 236, 359, 374, 381, 409.
Lancaster, Sir James, 151.
Langenburg, 442.
Langheld, Wilhelm, 443–4.
Lango, 175–8, 200, 299, 326–7, 337, 429.
Lavigerie, Cardinal, 290, 355.
Lawrance, J. C. D., 177.
Laws, Dr., 289.
Leakey, Dr. L. S. B., 15, 24, 29, 40, 74, 202.
Leakey, Mrs. M. D., 72.
al-Lemki, Sulaiman ibn Nasr, 438.
Lenana, 306 n., 417.
Lendu, 176.
Leo XIII, 355.
Leopold of Belgium, 291–2, 362, 367, 369 n., 383–4 & n., 388 n.; International African Association, 248–9, 283, 291–2, 433–4; *see also* Congo Free State.
Leue, Hauptmann, 438.
Levalloisian culture, 38.
Lewis, Dr. I. M., 322 n. 3.
Licinius, 97.
Liebert, Eduard von, 447.
Linant de Bellefonds, 341–2.
Lindi, 148–9, 154, 223, 243, 249, 440, 446, 448, 450–1, 453.
Liongo Fumo, 105 n.
Lisa clan, 187; *see also* Bunyoro.
List, G. F., 362.
Lister, Sir Villiers, 361.
Livingstone, David, 153, 209, 255, 266, 268, 272–3 & n., 276–8 & n., 288–90, 338, 354–5, 357, 379, 395, 420; Mountains, 8, 195–6.
Loangwa river, 266.
Loita, 306; hills, 300.
Lomami, 293; river, 288.
London Conference on Egyptian Affairs, 366.
Losegallai, 306–7; war, 305; *see also* Masai.
Lotuko, 75.

INDEX

Lualaba river, 282.
Luapula valley, 265, 282, 288.
Luba empire, 264, 336 n., 400.
Lüderitz, F. A. E., 365.
Lugard, Frederick, 383, 387–8, 396, 402–3, 413, 416, 420–5, 427, 431.
Lugbara, 176.
Luguru, 197.
Luhya, see Bantu Kavirondo.
Lulua, 282.
Lunda, 195, 264–6, 268, 273 & n., 278, 282, 288.
Lungu, 260.
Luo, 74–75, 79, 200 & n., 308, 310, 327, 419; see also Lwo.
Lupemban culture, 37, 44–45, 47, 56.
Lwo, 171–85, 187, 189, 196, 200 & n., 208.

[Mbarak, see Rashid.]
Maasinda (Masai), 200.
Macdonald, Captain J. R. L., 413, 416, 424–7.
Machakos, 17, 312, 393, 408, 413–16, 419, 431.
Machame, 204, 313.
Machemba, 446.
Mackay, Alexander, 348, 351, 399, 401.
Mackenzie, George, 396, 398–9.
Mackinnon, Sir William, 242, 245, 248–9, 355, 360–1, 372–4, 377 & n., 378–82, 384 & n. 5, 395, 409; Road, 248, 355, 360.
Maclear, Cape, 289.
MacMahon, President, 362.
Madagascar, 96, 108, 112, 155, 161, 222.
Madi, 177–8.
Mafia Island, 4, 12, 110–11, 113, 115, 119, 124, 126, 131, 133, 136, 150–2, 156, 163, 222, 239 n., 374.
Mafiti, 440.
Magadi–Natron, 13, 17.
Magila, 243, 439.
Magoro, 326.
Magosian culture, 44–46, 49, 55–56.
Magungo, 343.
Magwangwara, see Ngoni.
al-Mahdali, 123 & n.
Mahdists, 343, 376–7, 385, 403.
Mahenge district, 4, 207, 210, 442, 450.
Maji-Maji rebellion, 435, 446, 452.
Majid ibn Said, see Zanzibar.
Makalia river, 47, 50.

Makilo, 204.
Makonde, 131, 148, 154, 255–6; see also Lindi.
Makua, 195, 207, 243, 255–6, 273, 286.
Malabar, 149, 155.
Malacca, 120, 122–4, 149.
Malagarasi river, 260.
Malao, see Berbera.
Malaya, 96, 108, 100–1, 120, 122.
Maldive Islands, 110.
Mali Empire, 117.
Maliknajlu, 105.
Malindi, 104, 119, 123, 129, 132, 134, 136–7, 140, 144, 151, 163, 412–13; trade, 107, 124, 133, 135, 149–50, 153; and Portugal, 103, 116, 134, 138–9, 156; and Britain, 409–10.
Malta, 326.
Mamboya, 249, 436.
Mambwe, 193, 260.
Mamelukes, 101.
Mamkinga, 313.
Manda, 119, 140–1, 144, 381.
Mandara (of the Chaga), 436, 445; see also Rindi of Moshi.
Manga, 210.
Mankorongo, 330, 336, 344.
al-Mansur, Khalifa, 102.
Manua Sera, 279 & n., 281, 288.
Manyema, 276, 282 & n., 284, 296, 425.
Manyoni district, 201, 283.
Mapharitis, 95.
Mapunga, 210.
Mapungubwe, 55.
Marangu, 204, 448.
Maravi, 256–7.
Marehan (Darod Somali), 321–2.
Martin, James, 413.
Masai, 42, 63, 69, 74–77, 86, 90, 200–2, 207, 246, 253, 258, 263, 297, 301–4, 307–13, 322, 414–15, 417–19, 431; Masai-land, 8, 16, 274, 298–300, 303, 307–8, 316, 318–20, 322–3, 350–1, 400, 415, 419; 'Masai proper', 304–9; 'Kwavi', 304–7, 318, 350; trade, 226–7, 268 n., 274, 314, 318–20, 350.
Masaka, 8, 14, 18.
Masaki (of Kilema), 313; see also Kilema.
Masasi, 207, 243, 250.
Mascarene Islands, 158.
al-Masudi, 105–6, 150.
Maswa district, 195 & n., 201.
Matabele, 208.

INDEX

Mataka ibn Mbaraka, 221, 235.
Matambwe, 207.
Matapatu, 305.
Matengo, 207.
Mathew, Dr. Gervase, 89.
Mathews, Lieutenant William Lloyd, 240, 245–7, 360, 371, 436–7, 439.
'Matimula' the Swahili, 283.
Matson, A. T., 414 n., 417 n.
Mau escarpment, 5, 201, 300, 306, 315, 413.
Maupoint, Monseigneur de, 243–4.
Mauritius, 452.
Maviti, 243, 249, 440; *see also* Ngoni.
Mawindi, 207.
Mawokota, 189.
Mazrui, rulers of Mombasa, 145, 157–61, 213 & n., 220–1, 225, 246–8.
Mbaguta, Nuwa, 429.
Mbale district, 185.
Mbatian, 304, 306 & n., 314, 417.
Mbega (of Kilindi), 197.
Mbere, 90.
Mbeya highlands, 195–6.
Mboamaji, 144, 147, 164, 222, 267.
Mbokom, 313.
Mbonani, 209.
Mbulu, 6, 74, 201.
Mchinga, 147.
Mecca, 140, 164, 168.
Meli, 445.
Menelik, Emperor, 385.
Menengai mount, 306.
Mengo, 400, 420, 423–4.
Menouthias island, *see* Zanzibar.
Merca, 95, 115, 157, 220, 225, 374, 378.
Merere, 210, 281 & n.
Merkl, Sergeant, 446.
Meroe, 81–82, 88, 100.
Meru, 114 n., 130, 202, 301–2; mount, 6, 300, 316, 318.
Mesopotamia, 99, 101, 125.
Mexico, 357.
Mgangi, 414.
Mgera, 450.
Mhonda, 244.
Mianzini, 318.
Midgan, 62.
Mikindani, 149, 151, 153–4, 223, 440.
Mirambo, 210, 250, 262, 268, 279–81, 284, 287, 289–91, 295, 433–4.
missions, 136–7, 145, 242–5, 249–50, 255, 260, 266 n. 1, 270, 281, 289–90, 296, 300, 322–3, 345–9, 353–6, 370, 382, 387, 390 n. 1, 391, 393–4, 397–402, 407, 420–1, 423–8, 430, 434, 438–9, 442, 444, 450–2.
Mkwaja, 146–7, 164.
Mkwawa, 210, 434–5, 446, 452.
Mlozi, Arab slave-trader, 433.
Mo Kondi, Shaikh, 135.
Mo Lin, *see* Malindi.
Mocha, 95.
Mogadishu, 101–2, 116–17, 119, 121, 129, 132–4, 137–8, 145–6, 163, 165, 199, 220, 374, 378; trade, 107, 112–13, 125, 152, 157, 225.
Mombasa, 89 n., 95, 114, 122, 124–5, 132, 136, 143–5, 157, 162–3, 167, 235, 242, 245, 297, 301, 314–15, 322–3, 398; rulers of, 132, 140; *see also* Mazrui; trade, 113, 124, 131,133,135, 149–52, 225–6, 320, 398; routes, 153, 266, 315–16, 318, 410, 412–15; and Portugal, 134, 136–42, 157–8, 205, 223; and Muscat, 141–2, 157–8; and Omani, 159–61, 213, 221, 241, 246; and Egypt, 300; and Britain, 159–60, 372, 376, 396, 398–9, 407, 409–11, 454.
Mombasa, History of, 103, 143, 145, 162.
Mombera, 209.
Monclaro, Fr., S. J., 136, 140, 143, 151.
Mongoloba, 422.
Mongols, 111, 120.
Monsoon, 9, 98, 115, 124, 157, 216.
Moorhead, F. J., 108.
Moresby Treaty, 160, 161.
Morice, 148, 155–7, 268–9.
Morley, John, 388.
Morogoro district, 130, 197, 244, 443.
Moroto, 5.
Moru-Madi, 175.
Moshi, 204, 313, 436, 448, 451.
Mosyllon, 94.
Mozambique, 131, 150–1, 155, 207–8, 440; and Portugal, 135–6, 139, 141, 156–7, 223.
Mpezeni, 209.
Mpimbwe, 193.
Mpina, 404.
Mpororo, 187.
Mputa, 208–9.
Mpwapwa, 290, 443–4, 448, 450.
Mrima coast, 225, 227, 268, 271, 274, 287–8, 290.

INDEX

Mruli, 342–3, 403.
Msellem, 215.
Msene, 271.
Msiri of Katanga, 277–8 & n. 1 & 2, 283, 288.
Mtambalika, 210.
Mtambarara, 210.
Mtang'ata, see Tongoni.
Mtitimira, 163.
Mubende district, 9, 184.
Mugwanya, Stanislas, 424.
Muhammad ibn Ahmad Alawi, 214, 215.
Muhammad Kiwab, 124.
Muhammad bin Salim, the trader, 268.
Muhammad ibn Shaikh, 235.
Muhammad ibn Sulaiman, 122–3.
Muhammad ibn Sultan, 222–3.
Muhammad ibn Sultan al-Husain, 104.
Muhammad ibn Uthman al-Mazrui, 157–8.
Muhammad Zubeir, 321.
Mukabirere, 344.
Mumia, 310, 413–15, 420, 428.
Mundos, see Bandar Hais.
Mundu Wazeli, 312.
Murchison Falls, 430.
Musa Mzuri, the trader, 275.
Muscat, 141, 156, 158–61, 214, 219, 222, 229–30, 232–3, 268, 270, 288.
Musoma district, 195 & n., 201.
Mutambukwa, 336.
Mutara II (of Rwanda), 331.
Mutesa I (of Buganda), 333, 335, 340–2, 344–8, 350, 358, 399, 403, 426, 456.
Muyugumba, 210.
Muza, see Mocha.
Muzaffarid sultans, 120.
Mvuleni, 165.
Mwahawangu, 210.
Mwana wa Mwana, 215.
Mwanda, 414.
Mwanga of Buganda, 336 n. 3, 382, 399–403, 420, 422–4, 426–7, 429–30.
Mwanza, 194, 448, 450.
Mwata Yamvo, 264.
Mwele, 246.
Mwenge, 344.
Mwoko, 202.
Mzee ibn Saif Stambuli, 220.

Naberera, 55.
Nabhani dynasty, 104, 116, 119, 144.

Nachikufan culture, 44–46, 49, 54–56.
Nachikufu caves, 46.
Nairobi, 5, 14, 29, 46, 200.
Naivasha, 40, 42.
Naivasha lake, 6, 300–1, 303–5, 413, 419.
Nakuru, 50–52, 72.
Nakuru lake, 6, 40, 47, 52, 300, 303–4, 306.
Namungongo, 400.
Namuyonjo, 345.
Namwanga, 195–6, 260.
Nandi, 62–63, 69 n., 72–78, 86, 89, 201–2, 302, 307–9, 317, 320, 334, 350, 413, 418–19, 431.
Nanyuki, 51.
Nassa, 186.
Natal, 208.
National African Company, see Nigeria.
Natron lake, 6, 202, 307.
Ndagara of Karagwe, 329–30.
Ndagoni manuscript, 143.
Ndamba, 210.
Ndendeule, 207.
Ndonde, 207.
Ndorwa, 187, 189.
Neanderthal man, 28, 39, 56.
Negroes, 63–68, 73, 79, 83–84, 92–93, 101, 106, 116, 148; sickle-cell gene, 58, 60, 64.
Negroids, 42, 48, 53–54, 56, 61, 64–66, 87, 93, 95, 101, 105, 115.
New, Charles, 243, 323.
New Guinea, 365.
Newala, 243, 249.
Ngarwe, 147.
Ngelengwa, see Msiri, Katanga.
Ngidong', 200; see also Masai.
Ngindo, 207, 273.
Ngonde, 195–6, 257.
Ngong hills, 200, 318, 415–16.
Ngoni, 83, 208–11, 237 n. 1, 243, 249, 253, 256–60, 278 n. 2, 280, 284–7, 330, 440, 444, 446.
Ngoratok, 327.
Ngorongoro, 51.
Ngulu, 202–4, 227.
Nguni, 138.
Nguru mountains, 8.
Nguruni, 127.
Niger, 338, 354, 356, 362, 366 & n., 369–70, 372 & n., 386.

INDEX

Nigeria, 80–82, 393; Royal Niger Company, 372, 377–8, 410.
Nikon, 95–96, 100.
Nilo-Hamites, 60, 62–63, 69 & n., 70, 74–79, 82, 84–86, 92–93, 171, 175–6, 199–204, 206, 302, 308, 320, 326, 328, 408.
Nilotes, Nilotic peoples, 58, 60, 69, 74–75, 77–79, 82, 84, 87–88, 92–93, 171, 175, 199 & n. 1, 206, 308, 325, 328; Nile valley, 60, 67–68, 70, 84, 86, 91.
Nimuli, 9.
Nindi, 207.
Njemps, 307, 318, 420.
Njombe, 209.
Njoro river, caves, 50, 52–53.
Njunaki, 185 & n.
Nok, 82.
Nsenga, 256; *see also* Maravi.
Nsongezi, 31, 37–38, 45.
Ntare, 336, 421, 428–9; *see also* Ankole.
Nuba of Kordofan, 63–64.
Nuer, 74–75.
Nur Aga, 342–3, 346.
Nyaika of Toro, 344.
Nyakyusa, 195–7, 257, 259.
Nyambeni range, 14.
Nyamwezi, 153, 170, 193, 195–6, 205, 210, 250, 262–3, 276, 279, 288, 330, 433, 444, 446, 448; trade, 249, 264, 266–7, 270–1, 274, 277–8, 282–4, 286–9, 296; Sukuma, 191 n., 193–4, 196–7, 204.
Nyando valley, 307, 419.
Nyanja, 256, 273; *see also* Maravi.
Nyankole, 331.
Nyanza, 5, 8, 13, 15, 18, 38, 414, 418.
Nyasa (Nyasaland), 147, 154, 207–9, 243, 379, 442.
Nyasa lake, 8, 117, 147, 153, 192, 195–7, 209, 227, 248, 254–6, 265–6, 268, 273–4, 276–7, 284–90, 355, 433, 442, 450–1.
Nyaturu, 203, 280.
Nyiginya, 187.
Nyika, 16, 76–77, 90–91, 140, 195–6, 203, 226, 302–3, 413–14.
Nyikang, 87.
Nyilila, 207.
Nyoro, 89, 172 n., 188–9, 327, 340, 404–7, 421–2, 425–6.
Nyungu, 283.

Nzega district, 193–4.
Nzoia, 300, 413.

obsidian, 2, 41, 42, 44, 53.
ochre, red, 27, 41, 43, 49, 51, 53.
Oja, 134.
Okebo, 176.
Oldowan pebble culture, 25–27, 55–56.
Olduvai, 24–27, 29, 32, 35, 42, 55.
Olok, 94, 101, 125.
Olorgesailie, 29–32, 35.
Oman, 102–3, 108, 116, 119, 129, 154–5, 157–9, 162, 212–13, 216, 222, 230, 232–4, 238, 271.
Omani, 141–2, 152, 156–7, 212, 269.
Omani Busaidi dynasty, 157.
Omar (of Siu), 235.
Omar ibn al-Khattab, Khalifa, 102.
Omar ibn Muhammad 'Fumo Mari', 119.
Opone, *see* Ras Hafun.
O'Swald & Company, 235.
Otto, August, 435.
Owen, Archdeacon, 47.
Owen, Captain, 160, 162, 213.
Owen, Major Roderic, 424–7.
Ozi river, 221, 235.

Pabo, 406.
Padibe, 404, 430.
paintings, rock, 44, 46–47, 54, 56.
Pakwach, 430.
Palembang, 108.
Palestine, 40.
Pallisa, 327.
Pangani, 97, 132, 146, 164, 222, 247–8, 266, 419; trade, 224, 226–7, 237, 247, 318–20; and Germany, 374, 437, 439–40, 448, 450–1: *see also* Rhapta.
Pangani river, 6, 131, 153, 221, 247–8.
Pangwa, 195, 196, 209.
Pano, 94.
Paramerswara, 122.
Paranthropus, 24–25.
Pare mountains, 6, 8, 199, 307, 316.
Pate, 102, 117, 121, 132, 142, 144, 146–7, 156, 159, 162–3, 165; trade, 119–20, 133, 151–3, 155, 157, 163, 225–6; rulers, 104, 116, 119, 144; and Portugal, 120, 134, 136–7, 140–1, 144, 156–7; and Oman, 159, 213, 220–1, 235; and British administration, 381.
Pate, Chronicle of, 102, 104, 116–17, 119.

INDEX

Patiko, 173 & n., 326, 404, 406.
Pawir, *see* Chope.
Payera, 404, 406, 430.
Pelly, Lewis, 234.
Pemba, 4, 12, 16, 51, 106 n., 112, 132–3, 137–8, 140, 144, 150, 152, 158, 163, 213, 217–18, 220, 225–6, 239–40, 374, 383.
Penjdeh, 370.
Penhalonga, 117.
Peramiho, 451.
Periplus of the Erythrean Sea, 94 & n., 96, 125, 149.
Perlak Kingdom, 111.
Perondo, 448.
Persia, 99–100, 103, 107–8, 114–16, 122–3, 143, 147, 167, 455.
Peters, Carl, 368–9 & n., 374, 382–3, 386, 391, 396, 402, 435–6, 441–2, 444–5, 447, 452.
Petherick, J., 338.
Pfeil, Count Otto, 435, 442, 447.
Phillips, Midshipman, 160.
pit dwellings (*tembe*), 51–52, 74.
Pithecanthropus, 27, 29, 35, 56.
Playfair, Colonel, 221.
Pleistocene periods, 25–29, 32–33, 35, 42–43, 64.
pluvial periods, 15, 33, 35–36, 39, 40, 45, 52.
Pogoro, 207, 210.
Po Pa Li, *see* Somali.
Pok, 201; *see also* Nandi.
Pokomo, 90, 114 n. 2, 413–14.
Pokot, *see* Suk.
population, 83, 114, 129, 139, 143, 171, 180, 192, 194, 217, 224–6, 237, 249, 253–6, 260–2, 275, 294–5, 302, 311–12, 327, 334, 352, 363, 408, 413.
porcelain, 110 n., 113, 121–2, 127, 131, 165–6, 455.
Port Durnford, 95, 97, 100, 114, 130.
Portal, Gerald, 388–9 & n., 396, 411, 417, 424–5, 427, 431.
Portugal, 111, 116, 120–5, 127, 129, 131, 133 ff., 162, 165–9, 205, 212, 223, 265, 269–70, 336–7, 396, 433, 440; trade in East Africa, 135, 150–5; and Turks, 137–9; and Oman, 141–2, 156–7.
potassium–argon method of dating, 25.
pottery, 42–45, 47–51, 53, 64, 126, 165, 203.

Prason, *see* Delgado.
Prince, Magdalene von, 446.
Prince, Tom von, 443, 446, 447.
Prins, A. H. J., 148.
Prior, J., 148, 156.
Proconsul africanus, 23–24.
Ptolemaus, Claudius of Alexandria, 96.
Ptolemy, *Geography* of, 96 & n., 101.
Ptolemy III, 98.
Ptolemy VIII, 97.
Pugu, 438.
Pujini, 144, 163.
Purana, 100 n.
Purchas, Samuel, 138.
Purko, 304–6, 417.
Pyralean islands, 95.

Qutb-ud-din Aibak, 111.

Rabai, 242, 314.
radio-carbon dating, 28, 33–34, 40, 45–46, 49, 52–53.
rainfall, 9 ff., 15, 19–20, 32, 144, 254, 258, 261, 266, 327, 332.
Ranzi dynasty, 181.
Ras Hafun, 94–95.
ibn Rashid, Mbarak, 246–8.
Rashid ibn Hamis, 246.
Rashid ibn Salim, 161, 246.
Ravasco, Ruy Lourenço, 134.
Rebmann, Johann, 242, 322, 338.
Red Sea, 40.
Reitz, Lieutenant, 160.
Rendile, 301, 419.
Rengwa (Chaga), 313.
Réunion, 228, 243–4, 276.
Rhapta, 95–97, 101, 125, 132.
Rhodes, Cecil, 381–2, 386, 338 n., 409.
Rhodesias, 31, 35, 37–38, 44–45, 56, 100, 110, 112, 117, 197, 208.
Rhodesioids, 39–40, 56.
Rift valley, 2, 5 ff., 13–19, 35, 47, 49–50, 56, 73 n., 74, 76, 200, 202, 300, 306, 308–9, 318, 321, 418; culture, 29, 31, 33, 39–41, 44, 47.
Rigby, C. P., 219, 231, 234, 237.
Rindi (of Moshi), 313–14.
Rionga (of Bunyoro), 339–40, 342.
rock engraving, 41, 47, 55.
Rohlfs, Gerhard, 367.
Rome, 94–95, 97–99, 101.
Rosebery, Lord, 373, 380, 387–9 & n.
Royal Geographical Society, 290, 367.

INDEX

Ruaha valley, 7, 8, 267.
Rubaga, 342, 425.
Rudini, Marchese di, 385.
Rudolf, lake, 4–5, 16, 76, 84, 86, 175, 200, 317, 321, 419.
Ruemba, 273.
Rufiji river, 4, 12, 97, 113, 126, 131–2, 153, 206, 221, 249, 442.
ruga ruga, 209, 280.
Ruhinda, 185 & n.; *see also* Hinda.
Ruhinda II, 324, 330.
Rukidi (of the Bito), 182.
Rukwa lake, 7, 91, 195, 259, 451.
Rukwa rift, 8, 19.
Rumanika of Karagwe, 329–30, 344, 347.
Rungu, 193.
Rusinga island, 23.
Russia, 357, 370, 380, 404, 442.
Rusubi, 324, 329–31, 336, 344; *see also* Busubi.
Ruvuma river, 4, 129, 131, 142, 206–9, 223, 249, 256, 286, 374, 437–8, 441–2, 450.
Ruwenzori Mountains, 6–7, 14, 80.
Rwanda-Urundi, 6, 8, 18, 172 n., 180–2, 187–8, 191–2, 254, 260, 262, 324, 328–32, 337, 350.
Rwebogora of Rusubi, 330.
Rwesarura of Rusubi, 330.
Rwoma of Bukara, 330, 344.
Rwot Ochama, 404, 406.
Rwot Ogwok, 404.

Saadani, 222, 227, 267, 318, 435, 450.
Sabaki river, 153, 413.
sacrifice, human, 127, 192, 263.
Safwa, 195–6, 209, 254.
Sagali clan, 193.
Sagara, 197, 203, 446.
Said ibn Ali, 102.
Said ibn Habib, the trader, 273.
Saif ibn Ahmad, 157.
Saif ibn Sultan, *see* Zanzibar.
Sakalava, 222.
Sakwa, 310.
Saldanha bay, 38, 56.
Salih ibn Haramil Abri, the planter, 217.
Salil ibn Raziq, 103, 156.
Salim ibn Thuwain, *see* Zanzibar.
Salisbury, Lord, 361, 373, 376, 378–85, 387 & n., 409.
salt, 3, 7, 264, 267, 406.

Salt Lake, 421, 425.
Samburu, 300, 304, 307–8, 315, 321, 417, 419.
Sandawe, 55, 61–62.
Sangoan culture, 36–39, 44–45, 47, 56.
Sangu, 195, 210, 258, 281, 284–5.
Sanje Magoma, 163.
Sanje ya Kati, 110 n.
Santos, Fr. João dos, 138–9, 148.
Sanye, 62.
Sapei, 201, 301, 419–20.
Sassanian kings, 99, 101.
Sayyid Ali, 134.
Sayyid Said ibn Saif, *see* Zanzibar.
Scanlen, Prime Minister of Cape Colony, 365.
Schele, von, 446.
Schmidt, Lieutenant Rochus, 442.
Schweinfurth, G., 338.
Sebwato, Nikodemo, 428.
Segeju, 114 n., 129–30, 136, 139, 145, 153.
Seligman, C. G., 62, 89.
Selim Bey, 406, 421, 425.
Semakokiro of Buganda, 153, 190.
Semitic peoples, 68–69, 71, 77, 87.
Semliki, river, 7, 403, 425.
Sena, 138.
Sendeu, 306 n., 417.
Senegal, 362.
Sengchi, 107–8.
Sennar, 74.
Serapion, 95.
Serengeti plains, 19.
Sese Islanders, 189.
Seward, G. E., 234.
Seychelles, 230.
'Shagh', *see* Songo Mnara.
Shah ibn Mishham, 140.
Shaheinab, 48–49.
Shaka (of Natal), 208.
Shambaa, 90, 130, 191 n., 202–3, 312.
Shambo, 187.
Shangaans, 208.
Shangalla, 63–64.
Shapur II, 99.
Shela, battle of, 159.
Shi'ite, 102.
Shilluk, 75, 82, 84, 87–88, 172.
Shinduru, 177 n.
Shinyanga district, 194.
Shiraz, 100, 102, 104, 105 n., 120, 122–3, 132, 140, 143, 146–7.
Shire river, 208, 243, 276, 286, 289–90.

INDEX

Shiundu (of Hanga), 310.
Shungwaya (Bir Gao, Birikau), 89 & n., 100, 103, 114, 130.
sickle-cell gene, *see* Negroes.
Sidama, 65–66 & n., 68–69, 71, 78–79, 86–88.
Sidamo kingdom, 100.
Siki (of Nyamwezi), 442–3, 452.
Simba, *see* Ahmad.
Sind, 118, 123.
Singida district, 195 & n. 1, 201, 203.
Singo, 184, 189.
Sirhan ibn Said ibn Sirhan, 103.
Siria, 306.
Sita clan, 185 n., 187.
Siu, 132, 140–1, 144, 220–1, 235.
slave trade, 115, 148–52, 191, 267, 270, 274, 276, 294–6, 418–20, 449, 452, 456; origins of, 95, 101, 154–8; steady demand for, 100, 106–7, 120–1; opposition to, 160, 231, 239–41, 338, 353–4, 356–8, 362, 286, 440; and Arabs, 218, 237, 268, 277, 286, 395, 398–9, 433, 439, 443–4; in Kenya, 121 n.; in Kilwa, 227, 241, 269, 273, 290; and Mombasa, 113; and Tanganyika, 121 n.; and the Yao, 208, 265; and Zanzibar, 157, 161, 216–18, 224, 227–8, 234, 238–41, 269, 323, 364.
Smee, Captain, 157.
Smith, Lieutenant Charles, 249.
Smith, Captain Eric, 413.
Smithfield culture, 44, 47.
Smythies, Bishop, 439.
Soares, factor of Sofala, 113 n., 135, 150–1.
Society for the Exploration of Equatorial Africa, 434.
Soden, von, 442, 447.
Sofala, 101–2, 112–13, 117, 125–6, 131, 133–5, 143, 149–51, 154.
Soga, 334.
Somali, 62, 65–69, 71, 87, 94, 96, 101, 103, 107, 110–11, 14–15, 117–19, 130, 144, 149, 206, 221, 236, 248, 250, 297, 300, 317, 320–2, 337, 350, 419–20, 440, 448; and Italy, 401; and Germany, 412.
Somerset Nile, 172–3, 175, 182.
Songea district, 207, 209–10, 450.
Songo Mnara, 107 & n., 119, 148, 163, 193.

Songoro (Arab adventurer), 283.
Sonjo, 202.
Sotik, 308, 413, 419.
Sousa, Thomé de, 139.
Southall, A. W., 179, 205.
Spain, 137, 144.
Speke, 170, 261, 266–7, 271–2, 277, 279, 338, 348, 446.
Sri Vijaya, 108, 110, 112.
Stanley, H. M., 170, 272, 279 n., 282 n., 288, 290, 292–3, 295, 341–2, 345–6, 377–8 & n., 382, 395, 406, 420, 426, 442, 444; Falls, 293–4 & n.
Steere, Bishop Edward, 243.
Stigand, Captain C. H., 104.
Stillbay culture, 36, 38–40, 43–45, 56.
Stokes, Charles, 401.
Stone Bowl culture, 44–45, 49–50, 57.
Strabo, 71, 88 n.
Strandes, J., 135.
Strandloopers, 60.
Stuhlmann, Franz, 443–4, 447.
Subet (of Masai), 306.
Subi, 329.
Sudan, 48–49, 76, 85, 100, 171, 175–6, 200, 294, 297, 321, 324–5, 337, 343, 358, 364, 370, 376, 380, 389, 403–4, 406, 420–2, 424–5, 430, 440, 442 n. 2, 448.
Sudi, 145, 149, 154, 223, 446.
Sudi, 'Fat', 420.
Suez Canal, 241–2, 358.
Suk (Pokot), 78, 201, 297, 302, 317, 321, 417–19.
Sukuma, 193, 195–6, 261–2, 444, 448.
Sulaiman and Said, brothers, 102–3.
Sulaiman al-Hasan ibn Daud, 112 & n.
Sulaiman al-Matun, 119.
Sulaiman ibn Hamed, 235.
Sulaiman ibn Hamid Busaidi, 222.
Sulaiman ibn Sulaiman, 143.
Sulaiman bin Zeher, 401.
Sultan bin Ali, 270.
Sumatra, 108, 111–12.
Summers, R., 117.
Suna, 272, 333–5, 456.
Swahili, 100, 138, 162–3, 167–8, 244–5, 267–9, 272, 287, 439, 448, 450, 452; language, 104–5, 116–17, 129, 132, 143, 145–6, 156, 162–3, 165, 167–8, 245, 447, 451–2; trade, 222–3, 273 n., 274, 283, 315–16, 397, 414–16, 418–20; European administration, 408, 448.

498 INDEX

Swann, Alfred, 260.
Swanscombe skull, 28–29.
Symbiotic Hunters, 60, 62–63, 65–66.
Syria, 102.

Tabae, 94.
Tabora, 13, 193–4, 227, 249–50, 266, 278, 289, 292, 316–17, 420, 436–7, 446; trade, 287, 319, 323; German administration, 442–4, 448, 450–1.
Tabriz, 111, 120.
Taita, 89–91, 130, 202–3, 300, 302, 312, 314, 318, 321, 413–14, 418.
Takaungu, 246, 409.
Tana river, 17, 77, 89–90, 111, 114, 119, 153, 250, 300, 303, 315, 321, 374, 410, 412, 419.
Tanga, 124, 140, 146, 247; trade, 153, 226, 318, 320, 449; German administration, 438–40, 450.
Tanganyika, Lake, 7, 80, 82, 91, 193 & n., 208, 210, 259–60, 264–5, 268, 272, 274–5, 278, 280, 282–5, 288–90, 292–3, 338, 355, 381, 383–4 n.
Tarantino, A., 178.
Taru desert, 413.
Tatoga, 78, 201–3.
Taveta, 307, 318, 413, 419.
Teleki, Count Samuel, 416.
Tel-el-Kebir, 363.
Telemuggeh, 321.
Tepeth, 178.
Teso, 69 n. 1, 176–9, 200, 299, 326, 419, 429.
Tete, 117, 138, 153, 169, 207.
Tetera, 282, 288.
Tewfik, 363.
Tharaka, 301.
Thiney ibn Amir, 273.
Thomson, Joseph, 258–60, 262, 272, 278, 290, 305 n., 307 n., 323, 351, 392, 400, 413, 416.
Thuwain ibn Said, *see* Zanzibar.
Tibesti, 49.
Timbuktu, 48, 64, 338.
Timur, 120.
Tindiga (Tindega), 55, 61.
Tippu Tip, 267, 271, 274–5, 278 n., 288–95, 433.
Tomé Pires, 149.
Tongoni (Mtang'ata), 124 & n., 132, 136, 140, 146.
Tongwe, 260.

Torday, E., 92 & n.
Toro, 6, 181, 184–5, 297, 328, 330, 335, 337, 344–5, 403, 407–8, 421–6, 428, 430–1.
Tozer, Bishop, 243.
Transvaal, 24, 47, 55.
Tsavo, 410, 413–15.
tsetse fly, 21, 48, 254, 257, 259, 262, 450.
Tshitolian culture, 37, 44–45, 56.
Tucker, Professor A. N., 77.
Tucker, Bishop Alfred, 416, 424, 445.
Tuken, 201, 302, 309, 419; *see also* Nandi.
Tukuyu, 451.
Tumbatu, 214–15.
Tumbuka, 256.
Tunduru district, 207.
Tungi Bay, 223, 250, 374.
Turan Shah, 111.
Turi (of Kilindi), 197–9, 205.
Turkana plain, 6, 297, 302, 317, 320, 322, 419, 454.
Turki (of Oman), 238.
Turks, 136–9, 141, 294, 324, 363.
Turnbull, Sir Richard, 322 n.
Tutsi, 86, 181, 192, 194–6, 260, 262, 328, 331; *see also* Hima.

Uasin-Gishu, 5, 72–73 n., 200, 300, 305–10, 318, 418.
Ubena, 197, 210, 266, 442.
Uchibu clan, 174.
Ufipa, 7–8, 209–10, 260, 273, 284.
Uganda, 25, 31, 37–38, 44–46, 55, 66, 75, 77, 80, 83, 86 ff., 153, 171 ff., 297, 324–5, 358, 376, 380 ff., 402, 406, 408, 411, 413, 424–6, 429–32, 443–4.
Ugogo, 195, 197, 204, 261, 263–4, 267, 284, 443.
Ugowe, 268, 279.
Uhehe, 197, 202, 204, 210, 262, 273, 275, 442.
Ujiji, 209–10, 227, 265, 272, 278–9, 282–3, 287, 398, 401, 437, 448, 450.
Ukamba (Ukambani), 312, 415.
Ukami, 369, 435, 442.
Ukerewe island, 186, 283, 335, 344.
Ukwere, 442.
Uliankuru, 279.
Ulungu, 260.
Umayyad dynasty, 106–7.
Umba river, 374, 378, 437–8, 441.
Ungulu, 369, 435.

INDEX

United States, 161, 219, 223 & n., 225–6, 228, 235–6, 266 n., 270, 341, 352–3, 357, 365.
Unyamwezi, 193, 202–3, 209–10, 250, 261–3, 265–7, 274, 278, 280, 284, 290–1.
Unyanyembe, 262, 266–7, 276–9, 282–3, 287–92; and Muslim traders, 270–3, 275, 277–9, 287–9, 291–2.
Upimbwe, 284.
Urambo, 279.
Urori, 281.
Urua, 265, 275, 293.
Usagara, 8, 249, 290, 369, 371, 435–6, 439, 442.
Usagozi district, 193.
Usambara, 4, 6, 8, 153, 197, 204–5, 223 n., 225, 237, 243, 247, 316, 436, 439, 448, 451.
Usambiro, 443.
Usseri, 313.
Usuku, 177.
Utengule, 210.
Utondwe, 136, 140.
Uzigua, 369, 435, 442.
Uzimia, 146.

Vanga, 246–7, 318, 396, 409.
Vasco da Gama, 124, 132–4, 146.
Velten, C., 143, 149, 152, 453.
Venice, 119–20, 166, 219.
Victoria, Lake 8–9, 13, 18, 23, 27, 38, 46, 74, 80, 82, 86, 89, 91, 153, 180, 186, 190–2, 194–5, 197, 254, 266–7, 283, 297, 299–301, 307, 316–17, 323–4, 328, 335, 338, 341, 344, 346–7, 355, 359, 374, 376, 389, 397, 403, 426, 430, 432, 437, 443, 456.
Victoria Nile, 9, 326–7, 339, 403.
Vidal, Captain, 159.
Vinza, 192, 260–1, 282.
volcanics, 2, 5–8, 12, 23, 35, 47, 52, 254.
Vuga kingdom, 198, 204–5, 247.
Vugusu, *see* Bugusu.
Vumba, 132–3, 145.

Wadelai, 343, 382, 404–6, 429–30.
Wakalaganza, 193.
Wakefield, Thomas, 242–3.
Wakoli, chief, 429.
Waller, Gerald, 360.
Wamara (of Chwezi), 185.

Wami river, 19, 435.
Wanga, *see* Hanga.
Wangombe (of the Kikuyu), 312; *see also* Kikuyu.
Wanyamwezi, *see* Nyamwezi.
Warori, *see* Sangu.
Warsheikh, 225, 250, 374, 378.
water-supplies, 20, 31–32, 37, 48, 64, 144, 256, 360.
Watt, James, 415.
Watta (Wayto) of Ethiopia, 60, 62.
Weber, Ernst von, 361.
Webi Shebelle, 133, 152.
Werne, 338.
West Nile, 18, 176, 204.
White Nile, 171, 338.
Willey's kopje, 50–51.
Williams, Captain W. H., 420–1, 423–4, 429.
Wilson, George, 416.
Wilson, Professor Monica, 195.
Wilton culture, 44–46, 49, 56.
Winton, Francis de, 379.
Wissmann, Hermann von, 295, 382–3, 387, 440–2, 445, 447, 453.
Witt, John, 235.
Witu, 104, 235–6, 365, 369, 371, 374, 378, 381, 383–4, 394, 409, 412, 414, 419; *see also* German Witu Company.
Wolamo kingdom, 71.
Wolff, Sir Henri, 379–80.
Wyaki, chief, 416.

Xavier, St. Francis, 136, 151.

Yakut ibn Ambar, 156–8, 212 n.
Yao, 195, 208–9, 227, 255–7, 265, 267, 273, 282, 286–7, 293, 433, 440, 446.
Yemen, 68, 71, 137.
Yorubi dynasty, 157; *see also* Oman.
Yung-Lo, emperor, 121.
Yusuf ibn Hasan, 147.
Yusuf ibn Hasan, 222.
Yusuf (of Mombasa), 140–1.
Yusuf (of Sofala), 134.

Zafar Khan, 121.
Zambezi, 117, 153, 208–9, 357, 442.
Zanj (Zinj or Zanji), 24, 101, 105–6, 108, 113–14.

INDEX

Zanzibar, 4, 12, 16, 51, 95, 97, 106, 110 n., 124–5, 127, 131, 133, 143, 162, 165–6, 271, 297, 420, 435, 444, 448, 454; trade, 111, 151–2, 155–8, 218–19, 223, 265, 269, 274–6, 281, 289–91, 294, 299, 323, 327, 334–5, 337, 344–6, 350, 355–6, 410, 433–4, 447; trade routes, 266–7, 285, 315–18, 320, 324, 338, 441; traders, 265, 274, 285, 294, 398, 401, 410–12, 420.
Rulers, 107, 124 n., 129, 132, 166:
 Sayyid Said ibn Saif, 129, 145, 148, 157–61, 212 ff., 235, 238, 242, 246–8, 269–70, 315; economic policy of, 217–18, 223–4; foreign policy of, 227–9; plantations, 216–17, 238–9, 241, 268–9, 276, 294, 363, 398, 441.
 Ali ibn Said, 339, 411, 441.
 Barghash ibn Said, 216, 220, 223, 230–2, 234, 237–41, 243, 245–51, 289, 291–3, 357, 359–61, 367, 371 n., 374, 377–8, 396, 399, 401, 434, 436–7.
 Khalifa ibn Said, 399, 411, 438, 441.
 Majid ibn Said, 214–16, 228, 230–8, 242–3, 246–7, 275, 291.
 Thuwain ibn Said, 222, 232–3.
 Salim ibn Thuwain, 233.
 and Portugal, 134, 136–7, 141–2, 156; and Oman, 157, 161, 269; and Britain, 161, 216, 218–19, 228–34, 236–41, 245, 248, 290–1, 293, 356–61, 365, 367–74, 377–83, 389, 391, 396, 411–12, 437; and Germany, 229, 234–6, 249, 293, 367, 369–71, 374, 376–7.
Zeila, bay of, 94, 149, 152.
Zelewski, 439, 446–7.
Zigua, 130, 202, 247.
Zimba, 129, 138–9, 142–3, 207.
Zimbabwe, 112–13.
Zingion, 98, 101.
Zinjanthropus boisei, 24–25, 27–28, 35.
Ziwa, 117.
Zongendaba, 209–10.
Zulu, 80, 208, 301, 440, 448.
Zuru (of the Ngoni), 209; *see also* Ngoni.